Excited States in Quantum Chemistry

NATO ADVANCED STUDY INSTITUTES SERIES

Proceedings of the Advanced Study Institute Programme, which aims at the dissemination of advanced knowledge and the formation of contacts among scientists from different countries

The series is published by an international board of publishers in conjunction with NATO Scientific Affairs Division

A Life Sciences **B Physics**	**Plenum Publishing Corporation** **London and New York**
C Mathematical and Physical Sciences	**D. Reidel Publishing Company** Dordrecht, Boston and London
D Behavioral and Social Sciences	**Sijthoff International Publishing Company** **Leiden**
E Applied Sciences	**Noordhoff International Publishing** **Leiden**

Series C – Mathematical and Physical Sciences

Volume 46 – Excited States in Quantum Chemistry

Excited States in Quantum Chemistry

Theoretical and Experimental Aspects of the Electronic Structure and Properties of the Excited States in Atoms, Molecules and Solids

Proceedings of the NATO Advanced Study Institute held at Kos, Greece, June 4–18, 1978

edited by

CLEANTHES A. NICOLAIDES
Theoretical Chemistry Institute, National Hellenic Research Foundation, Athens, Greece

and

DONALD R. BECK
Theoretical Chemistry Institute, National Hellenic Research Foundation, Athens, Greece

D. Reidel Publishing Company

Dordrecht : Holland / Boston : U.S.A. / London : England

Published in cooperation with NATO Scientific Affairs Division

Library of Congress Cataloging in Publication Data

CIP

NATO Advanced Study Institute, Kos Island, Greece, 1978.
Excited States in Quantum Chemistry
(NATO advanced study institutes series : Series C, Mathematical and physical sciences ; v. 46)
Bibliography: p.
Includes index.
1. Excited state chemistry–Congresses.
I. Nicolaides, Cleanthes A. II. Beck, Donald R. III. Title. IV. Series.
QD461.5.N36 1978 541'.28 78-24278
ISBN-13: 978-94-009-9904-6 e-ISBN-13: 978-94-009-9902-2
DOI: 10.1007/978-94-009-9902-2

Published by D. Reidel Publishing Company
P.O. Box 17, Dordrecht, Holland

Sold and distributed in the U.S.A., Canada, and Mexico
by D. Reidel Publishing Company, Inc.
Lincoln Building, 160 Old Derby Street, Hingham, Mass. 02043, U.S.A.

TABLE OF CONTENTS

PREFACE

It is undoubtedly true that much of the progress in the quantum theory of matter is due to the remarkable success of the independent particle model (IPM)--especially in describing ground states. However, the accurate experimental results of the last 10 years or so, on a variety of spectroscopic phenomena and chemical processes which involve the Excited State, and the related failure of the IPM to reproduce accurately--in many cases, even qualitatively--the observed data, have sent to theorists a clear message: There is need to create and/or apply general and useful approaches to the many-electron problem of the excited state which go beyond the IPM, treat electron correlation and relativity and explain or predict all relevant physical or chemical information with consistent accuracy.

This book contains articles devoted mainly to some of the most important new developments in Quantum Chemistry concerning the theoretical foundations and the computational implementation of many-body approaches to the quantitative and detailed understanding of the electronic excited states of atoms, molecules and solids. Furthermore, it contains experimental and phenomenological articles on Photoelectron and Auger spectroscopy, Lifetime measurements and Organic Photochemistry.

In combination or individually, these articles constitute a good description of some current theoretical and experimental work on the electronic structure and spectroscopy of atoms, molecules, polymers, surfaces, metal oxides and amorphous solids. The theoretical models which are reviewed and employed are based on: Configuration Interaction (CI), Green's function and Polarization Propagator techniques, the Relativistic and Non-Relativistic Restricted and Unrestricted Hartree-Fock methods, the Coherent Potential and Random Phase Approximations, Many-Body Perturbation Theory, Cluster expansions of the wave-function, CI in the continuum and the Complex Coordinate Rotation Method.

When it comes to applications of these advanced theories, it becomes clear from the reading of the articles that much progress is underway in the study of small systems. For example, it is indeed impressive to see the accuracy and efficiency with which several theoretical and experimental approaches produce numbers on valence electron excitation and ionization energies and transition probabilities, core electron binding energies or lifetimes of excited states in a variety of atoms and small molecules. Such accuracy was impossible until very recently. Large and extended systems are of course more difficult to treat quantitatively. There, with few exceptions, the nature and role of the excited state is still "terra incognita", although theory has made considerable advances, especially in the qualitative formulation of the problems and in the formal foundations.

The articles are based on the lectures which were given during the NATO Advanced Study Institute (ASI) on "The Electronic structure and properties of the excited states of atoms, molecules and solids" held on the island of Kos, Greece, June 4-18, 1978. These lectures were addressed to European and American graduate students and active researchers in the fields of Electronic Spectroscopy, Organic and Inorganic Chemistry, Physical and Quantum Chemistry and Solid State Physics. The particular blend of up-to-date reviews and original contributions present in these articles should prove educationally and scientifically valuable to a similar but broader audience.

We close by sincerely thanking the lecturers for their contributions which made the ASI and this book realities.

Special thanks also go to the NATO Scientific Affairs Division, Brussels, for the financial assistance and to Professors Ladik, Linderberg, von Niessen, Peyerimhoff, and Öhrn for their encouragement and advice throughout the period of preparation and organization of the ASI.

Athens, August 1978

C.A. Nicolaides
D.R. Beck

Editors

Ἔστω δὴ κατὰ τὸν ὀρθὸν λόγον καὶ κατὰ τὸν εἰκότα τὸ μὲν τῆς πυραμίδος στερεὸν γεγονὸς εἶδος πυρὸς στοιχεῖον καὶ σπέρμα. τὸ δὲ δεύτερον κατὰ γένεσιν εἴπωμεν ἀέρος, τὸ δὲ τρίτον ὕδατος πάντα οὖν δὴ ταῦτα δεῖ διανοεῖσθαι σμικρὰ οὕτως, ὡς καθ' ἓν ἕκαστον μὲν τοῦ γένους ἑκάστου διὰ σμικρότητα οὐδὲν ὁρώμενον ὑφ' ἡμῶν, ξυναθροισθέντων δὲ πολλῶν τοὺς ὄγκους αὐτῶν ὁρᾶσθαι. καὶ δὴ καὶ τὸ τῶν ἀναλογιῶν περί τε τὰ πλήθη καὶ τὰς κινήσεις καὶ τὰς ἄλλας δυνάμεις, πανταχῇ τὸν θεόν, ὅπῃπερ ἡ τῆς ἀνάγκης ἑκοῦσα πεισθεῖσά τε φύσις ὑπεῖκε, ταύτῃ πάντῃ δι' ἀκριβείας ἀποτελεσθεισῶν ὑπ' αὐτοῦ ξυνηρμόσθαι ταῦτα ἀνὰ λόγον.

ΤΙΜΑΙΟΣ

Thus, in accordance with the right account and the probable, that solid which has taken the form of a pyramid shall be the element and seed of fire; the second in order of generation we shall affirm to be air, and the third water. Now one must conceive all these to be so small that none of them, when taken singly each in its several kind, is seen by us, but when many are collected together their masses are seen. And, moreover, as regards the numerical proportions which govern their masses and motions and their other qualities, we must conceive that God realized these everywhere with exactness, in so far as the nature of Necessity submitted voluntarily or under persuasion, and thus ordered all in harmonious proportion.

TIMAEUS
PLATO

EXPERIMENTAL STUDIES OF ATOMIC AND MOLECULAR LIFETIMES

Indrek Martinson

Department of Physics, University of Lund,
S-223 62 Lund, Sweden.

ABSTRACT

A review is given of measurements of atomic and molecular lifetimes and transition probabilities. Emphasis is placed on comparatively new methods such as laser spectroscopy, high-resolution electron excitation and beam-foil spectroscopy. The applications of the results to problems in astrophysics and plasma physics are discussed.

1. INTRODUCTION

Excited levels in atoms and molecules decay spontaneously to lower states, usually by emitting electric dipole (E 1) radiation. The decay follows the exponential law

$$N_a(t) = N_a(0)\exp(-t/\tau_a) \tag{1}$$

where $N_a(0)$ is the initial population of the level a and $N_a(t)$ that at the time t . The lifetime τ_a (or mean life) of the level is the time after which the population has decreased to $1/e = 37\%$ of its initial value. Excited states in neutral atoms have typical lifetimes of 10^{-8}s while much shorter values can be found in highly ionized atoms. In neutral lithium, Li I, the levels 2p ^{2}P and 3p ^{2}P have lifetimes of 27.3 and 216 ns, respectively. For lithium-like silicon, Si XII, the corresponding values are 1.04 and 0.0022 ns (Lindgård and Nielsen, 1977). Lifetimes in the 10^{-8}s range are not unusual for simple molecules but here also much longer values, several μs, can be found.

Atomic and molecular lifetimes (and transition probabilities) are fundamental quantities for which justification of measurement is

Cleanthes A. Nicolaides and Donald R. Beck (eds.), Excited States in Quantum Chemistry, 1–34.

scarcely necessary (Crossley, 1969). Experimental results provide useful tests of various quantum-mechanical calculations of atomic and molecular structure. There are also several applications. In laser physics the conditions for population inversion depend critically on lifetimes for excited levels. Astrophysical studies of the abundances of chemical elements in the sun, stars and interstellar medium require knowledge about the lifetimes for the observed transitions. Research in plasma physics to achieve thermonuclear fusion is presently hampered by plasma impurities, such as metal ions. Here, also, knowledge about lifetimes is necessary for determining the metal concentrations.

Several methods have been developed in recent years for accurate measurement of atomic and molecular transition probabilities. Not all these techniques and results can be covered here, more detailed discussions will be found in several review articles, e.g. by Foster (1964), Wiese (1968), Corney (1969), Erman (1975, 1977), Lehmann (1975) and Imhof and Read (1977). We also recommend a recent monograph by Corney (1977).

2. RELATIONSHIPS

We assume that the level a decays spontaneously to a lower level b, by emitting a photon of energy $\hbar\omega = \hbar c/\lambda$ and angular momentum L (Fig. 1).

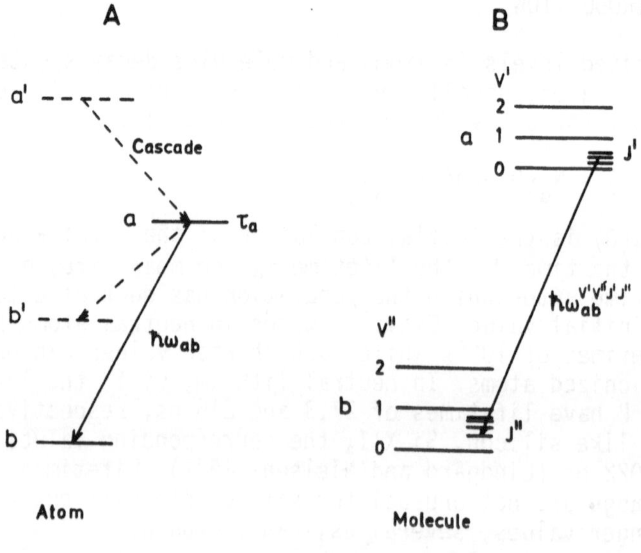

Figure 1. Examples of atomic (A) and molecular (B) decays.

According to the quantum theory of radiation the transition probability for this process, A_{ab}, is proportional to ω^{2L+1} $|\langle b|Q|a\rangle|^2$ where Q is the appropriate (electric or magnetic multipole) operator. In the electric dipole (E 1) approximation the transition probability A_{ab} (s^{-1}) can be expressed as

$$A_{ab} = \frac{4}{3} \frac{e^2\omega^3}{\hbar c^3} |\langle b|\Sigma r_j|a\rangle|^2 \qquad (2)$$

Frequently also the absorption oscillator strength or f-value (dimensionless quantity) is used. The numerical relation is

$$f = 1.499\lambda^2 \cdot A_{ab} \cdot g_a/g_b \qquad (3)$$

Here λ is the photon wavelength (in cm) and g_a and g_b are the statistical weights of the levels involved.

If the excited level a only decays to one final level b the lifetime τ_a is the inverse of the transition probability. If several final states are possible (Fig. 1A) the relation

$$\tau_a^{-1} = \sum_b A_{ab} \qquad (4)$$

must be used. A lifetime measurement now yields the sum of transition probabilities and additional information (relative intensities or branching ratios) is frequently needed.

Similar relations hold in the molecular case (Fig 1B). In the expression for the lifetime for an electronic transition a summation must be made over possible vibrational and rotational final states. The probability $A_{v'v''}$ for a transition between levels with vibrational quantum numbers v' and v" is usually expressed as

$$A_{v'v''} = \frac{4}{3} \frac{e^2\omega^3}{\hbar c^3} \cdot q_{v'v''} R_e^2 (\bar{r}_{v'v''}) \qquad (5)$$

Here q is the Franck Condon factor and R_e the electronic transition moment, similar to the matrix element in the atomic case. In the formula for the lifetime of an electronic state a and vibrational and rotation quantum numbers v' and J' (Fig. 1B) the summation must include all possible final states, i.e.

$$(\tau_a^{v'J'})^{-1} = \sum_{v''} \sum_{J''} A_{ab}^{v'v''J'J''} \qquad (6)$$

3. CLASSICAL METHODS

About 60 years ago Wien (1919) measured intensity decays for spectral lines, using canal rays. His value for the 6p $^3P_1$ level in Hg I, $\tau = 98$ ns, agrees with data obtained much later. In the 1920's and 1930's a number experiments were carried out, based on spectral line intensity studies or optical excitation of atoms with resonance radiation. A review of the early work is given by Mitchell and Zemansky (1934).

Several of the classical methods are still in use although much refined and improved. To this category belong three methods for direct determination of f-values, measurements of emission, absorption and anomalous dispersion, respectively. All these methods are thoroughly described by Foster (1964), Wiese (1968) and Huber (1977).

The emission method is based on determination of spectral line intensities I_{ab} from a plasma light source. The transition probabilities A_{ab} are obtained from the relation

$$I_{ab} = \frac{\omega}{4\pi} \cdot A_{ab} \cdot N_a \tag{7}$$

where N_a is the number of particles in the excited state a and ω the solid angle. If the temperature of the plasma is known and there exists local thermodynamic equilibrium (LTE), N_a can be calculated from Boltzmann formulae. It is difficult to calculate N_a and therefore the emission method is often limited to determination of relative f-values. Absolute values can still be obtained if the data are normalized to an accurately known value, for example from a careful lifetime measurement. In the early work rather simple arcs and photographic detection were used. Modern emission measurements utilize efficient wall-stabilized arcs or shock tubes as well as photoelectric detection. Good examples of such work are found in papers by Garz and Kock (1969), Richter and Wulff (1970), Bridges and Wiese (1970) and Wolnik et al. (1970), all of which deal with iron-group elements. The f-values have typical uncertainties of ±20%.

In absorption measurements light from a continuous source, such as a tungsten lamp (or sometimes a shock tube) is sent through a thin layer of gas The absorption lines are studied spectroscopically and transition probabilities are deduced from the equivalent widths. Using a shock tube Huber and Tobey (1968) measured f-values for many Fe I lines. Accurate results for strong lines in Cr I were recently obtained by Bieniewski (1976). Also in absorption experiments the number density N_a must be known and this requires very reliable vapor-pressure data. An interesting development consists of crossing an atomic beam (in which the density of atoms can be measured with a microbalance) with the continuum light (Bell and Tubbs 1970, Bell and Lyzenga, 1976). The f-values so obtained in Sc I, Ti I and Fe I have uncertainties below 10%.

The method of anomalous refraction (hook method) is based on determination of the refractive index n of a gas in the vicinity of a spectral line of wavelength λ_0. The following relation is valid

$$n - 1 = C \cdot N_b \cdot \frac{\lambda_0^{\,3}}{\lambda - \lambda_0} f \qquad (8)$$

where C is a constant, N_b the number of gas atoms in the ground state and λ the variable wavelength. A two-beam interferometer is used in such experiments. Light from a continuous source passes through the gas under study while a reference beam goes through a compensating tube. Both beams are focussed on a spectrograph slit. The spectrum shows interference fringes, interrupted by characteristic hooks, the separations of the latter giving f-values (Penkin, 1964). In recent years Huber and collaborators have made accurate measurements using this method (Huber, 1977).

In contrast to most lifetime methods, these classical methods give directly transition probabilities, or f-values. As the result of recent developments the data are now quite accurate. It is also important that weak spectral lines can be studied with these methods. In astrophysical applications f-values for weak lines are frequently needed, as well as f-values for very many lines in a given spectrum. A large part of f-value studies using these classical methods have indeed been motivated by needs in astrophysics.

4. OPTICAL EXCITATION EXPERIMENTS

A direct way of populating excited states in atoms or molecules is to irradiate a gas with resonance radiation. For lifetime measurements the excitation should be pulsed or sinusoidally modulated. In the former case the decay of the fluorescence light is measured whereas the phase shift between the excitation and the fluorescence is determined in modulation studies. In the work of Kibble et al. (1967) light from a sodium lamp went via a Kerr cell into Na vapour. The time between excitation and decay of the 3p ^2P level was measured with delayed coincidence methods, routinely used in nuclear physics experiments. An accurate value of $\tau = 16.4 \pm 0.4$ ns was obtained.

The development of lasers has enabled a very fast progress in optical excitation of atoms and molecules. In most experiments a pulsed nitrogen laser and a dye cell are used. Typical pulse lengths are 2 - 10 ns and using various dye cells the spectral region 3 600 Å - 9 000 Å can be covered (Haroche, 1976). With frequency doubling it is possible to extend the region to close to 2 000 Å. The repetition frequencies are of the order of several hundred Hz. Using this method Figger et al. (1974), Siomos et al. (1975), Heldt et al. (1975) have obtained accurate lifetimes in Fe I and Ni I. For the y $^3F^o_4$ in Ni I a lifetime of 16.2±0.4 ns was thus reported. Optical excitation is relatively easily applied to resonance levels which combine with the ground state. However, it is also possible to reach other levels using stepwise excitation. Siomos et al. (1975) thus excited a relatively high-lying level in Fe I with two accurately tuned dye lasers. Even three-step excitation has been successfully applied e.g. by Cooke et al. (1978).

The merits of laser methods include high intensity, tunability and selectivity. No level besides that under study is excited, there is thus no cascading into the level (cf. Fig. 1A where a cascade level is indicated) and the decay follows the simple exponential relation (Eq.1). The selectivity and tunability are particularly valuable when complex atoms or molecules with many close-lying states are studied. The accuracy can be limited by non-linearities in the detection system, and - because of the relatively slow repetition rates - data-taking times can be quite long. To overcome this difficulty Gustavsson et al. (1977) have recently designed a new method for laser excitation. A continuous-wave dye laser is used in their work (Ar^+ laser together with a dye cell) and the pulses are obtained with an acousto-optical modulator or a dentist's air turbine drill. A typical decay curve, for a level in Yb I, is shown in Fig. 2.

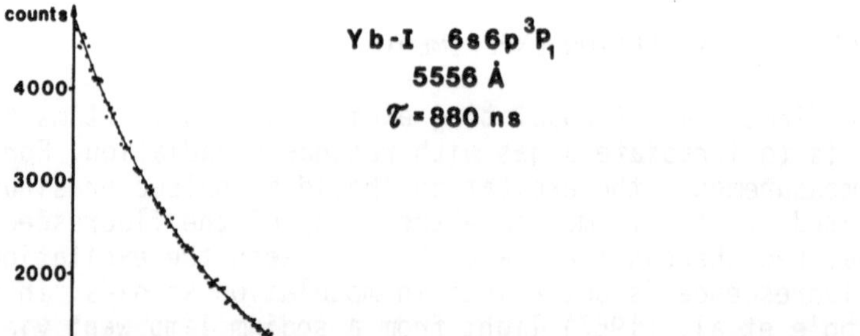

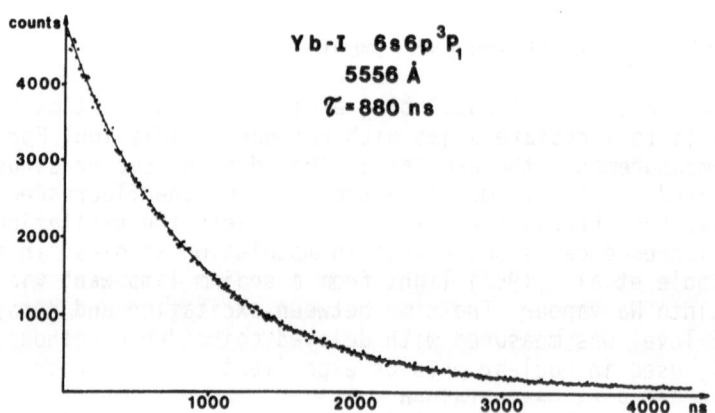

Figure. 2. Decay curves for the 6s6p $^3P_1$ level in Yb I, obtained with the method of Gustavsson et al. (1977); (Svanberg, 1978).

Pulsed dye lasers play an important role also in molecular studies. Sakurai et al. (1971) measured lifetimes for a large number of vibronic states in Br_2 and I_2. Molecular lifetimes may give quantitative information about non-radiative decay modes. The lifetime τ_a of a molecular level can be expressed in the following way

$$\tau_a = (A_r + A_{nr})^{-1} \tag{9}$$

were A_r is the probability for radiative decay and A_{nr} is due to possible non-radiative processes, e.g. predissociation. Such effects can be observed from line broadenings or intensity variations in molecular wavelength spectra - but lifetime measurements are much more sensitive indicators. Lehmann and collaborators (Vigue et al., 1975, Broyer et al., 1975, Lehmann 1977) made detailed studies of predissociation processes in several molecules, e.g. Se_2, Br_2 and I_2. A nitrogen-pumped dye laser which gave 3 ns pulses was used in their work. In I_2 the B $^3\pi$ state was systematically studied and lifetimes (with 3-5% uncertainties) were determined for a large number of vibrational (v') and rotational (J') levels.

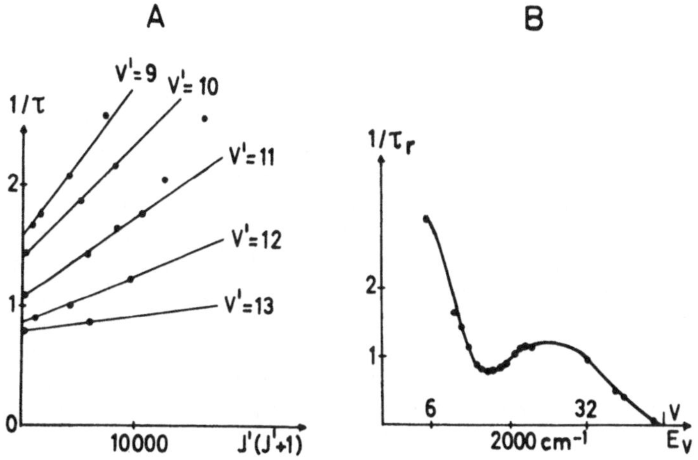

Figure 3. Lifetimes for vibrational levels of the B $^3\pi$ state in I_2 (A) and oscillations, due to hyperfine effects, for the radiative decay probability (B). From Lehmann (1977).

Results are shown in Fig. 3A where the inverse of the lifetime for the vibrational states v' = 9 - 13 is plotted versus J'(J' + 1). In most cases linear relations are found, the so-called gyroscopic predissociation rate being proportional to J'(J' + 1). By extrapolating the lifetime data to J' = 0 the authors obtained values for the radiative decay rate A_r. This factor was found to vary strongly with v' (Fig. 3B) indicating that there are additional predissociation effects. Direct measurements further showed that the ortho states (with total nuclear spin I = 1,3,5) and para states (I = 0,2,4) have markedly different lifetimes, due to hyperfine predissociation. In this case the two predissociation rates are of the order of $10^6\ s^{-1}$, comparable to the radiative decay rate. Even lower predissociation rates can be determined from circular polarization of fluorescence light (Lehmann, 1977).

With modern lasers very short pulses, in the ps range, can be obtained. These have been used in measurements of radiative and non-radiative decay times for a number of complex molecules (Greenhow and Schmidt, 1974).

Atomic and molecular studies using lasers form a dynamic, rapidly developing field. For details the papers quoted above and the review by Walther (1976) should be consulted.

It is difficult to perform time-resolved studies with laser excitation below 3 000 Å. Here, however, a powerful new method is found in synchrotron radiation (Codling, 1973). If an electron storage ring is used, a pulsed light source is available which can reach very short wavelengths. With 4 GeV electron energy the continuous photon wavelength distribution peaks close to 1 Å (12 keV). Despite the continuous photon spectrum it is rather easy to obtain selective excitation, e.g. by using a grating monochromator as a predisperser before exciting the atoms under study. Using the Stanford Synchrotron facility Matthias et al. (1977) made the first atomic and molecular lifetime measurements with synchrotron radiation. The short beam pulses (0.4 ns) and the repetition rate (780 ns) are very favorable for lifetime studies. In the first work, decay times for excited levels in Kr I and Xe I with only 2 - 3% uncertainties were obtained. Also excimers such as Xe_2^* were studied. (The ground states of noble-gas molecules are repulsive while bound excited states exist). A result is shown in Fig. 4 where the radiative decay of a high vibrational level belonging to the 0_n^+ state of Xe_2^* is shown (Matthias, 1978), Note the excellent counting statistics and clean single-exponential decay.

A couple of methods using modulated optical excitation have been used, e.g. the phase-shift technique. Here the excitation has a time-dependence of $\exp(i\omega t)$ (where ω is the modulation frequency) and therefore also the fluorescence radiation varies sinusoidally with the same frequency but a phase shift ϕ due to the finite lifetime τ of the excited level. The following relation gives the lifetime

$$\phi = \arctan(\omega\tau) \qquad (10)$$

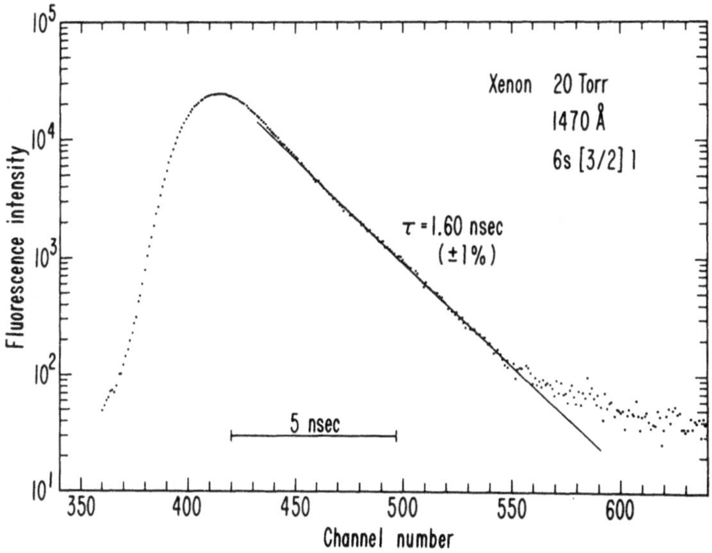

Figure 4. Lifetime measurement using synchrotron radiation (Matthias, 1978). The 0_r^+ state of Xe_2^* was selectively excited and the decay curve, a single exponential, gives the lifetime with 1% uncertainty.

The modulation can be achieved in several ways, e.g. with a rotating diffraction grating, Kerr cell or an acousto-optical modulator (Imhof and Read, 1977). Examples of such lifetime measurements can be found in the papers Link (1966) and Cunningham and Link (1967). In the latter work a substantial number of elements were studied and lifetimes (with about 5% uncertainties) were given for excited levels in e.g. Ga, In, Pb and Bi.

5. EXCITATION WITH ELECTRONS

The electronic excitation methods are largely similar to the optical ones. With electrons the selection rules are much less rigid and many more atomic and molecular levels can be populated. As in photon experiments both pulsed and sinusoidally modulated excitation can be used.

The first modern measurements of this kind was carried out by Heron et al. (1956) who studied lifetimes in neutral helium. A typical experimental arrangement is shown in Fig. 5. The gas atoms are excited with a pulsed beam of electrons, typically of 20-50 eV energy. The electron pulse starts the time-to-pulse height converter. The photons from the excited gas are

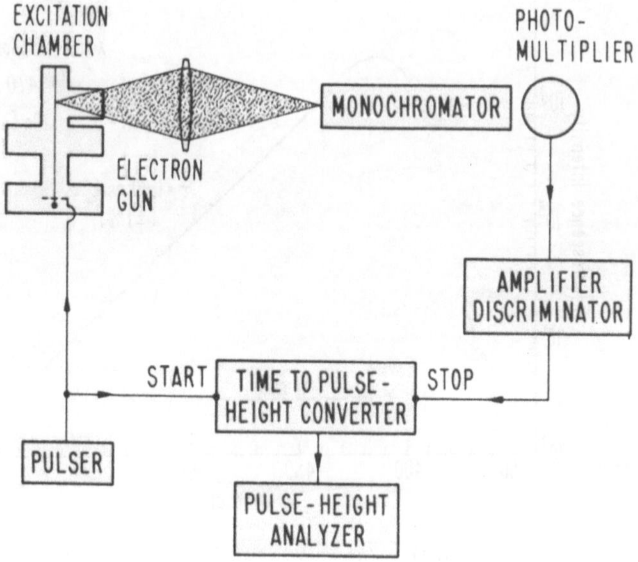

Figure 5. Experimental setup for lifetime measurements using pulsed electron excitation (Klose, 1967).

spectroscopically analyzed and counted with a photomultiplier. The amplified pulse then stops the TPC and the results are stored in a multichannel analyzer (MCA). The time-distribution of pulses is exponential. However, electron excitation is usually not as selective as photon excitation, instead several levels can be populated simultaneously. The decay curves therefore often show a more complex character, due to cascading. The decaying level is now fed from higher levels and the decay curve must be carefully decomposed into several exponentials. Good counting statistics are necessary while the lifetimes for the levels involved should not be too close to each other. The cascading can be eliminated with threshold excitation. In a study of the $2p^5 3p$ levels in Ne I -- which lie 18.3 - 18.9 eV above the $2p^6$ ground state - Bennett and Kindlmann (1966) selected electron energies very close to these values. Single - exponential decay curves (no cascading) were obtained but the method is very difficult. The counting rates are very low, because of low excitation cross sections. Most investigators therefore use higher energies and correct for cascades in data analyses. Important results of this kind can be found in papers by Klose (1967, 1968, 1971) who measured lifetimes in e.g. Ar and Fe to a few per cent accuracy. More recently also a lifetime in U I was determined in this way. For the 27 887 cm^{-1} level Klose (1975) obtained the value $\tau = 7.3 \pm 1.1$ ns. Lifetimes in U I are relevant for studies of isotope separation with lasers. The pulsed electron technique has also applied to ions (Jiménez et al., 1974) as well as molecules and molecular ions. In the latter work, by Möhlmann and de Heer (1977) lifetimes for vibrionic states in e.g. H_2O^+, H_2S^+, HCl^+ and CO^+ and other species, usually in the μs range, were determined with 10% uncertainties.

In measuring so long lifetimes in molecular ions there arise interesting effects due to repulsion of the excited ions. These can leave the viewing volume before the decay takes place and the result is a very misleading distorsion of the experimental decay curves. For example, Möhlmann and de Heer (1977) found that decay times for certain vibrionic level in the important H_2O^+ radical can be off by a factor of 10 from the correct values if this electrostatic repulsion is not corrected for. A thorough survey of such problems is also given by Curtis and Erman (1977).

An important development in electron excitation work consists of using 5 - 10 keV electrons (Erman, 1975) instead of the previously common 30 - 50 eV ones. With high-energy electrons much more favorable beam currents, tens of mA, can be routinely used and the spectral resolution - particularly important in molecular work - can be improved from typically several Å to below 0.1 Å. Another improvement of lifetime measurements is that the electron beam is swept with an oscillator of a very well determined frequency. This gives an excellent duty cycle and greatly increases the counting rates in lifetime experiments. The setup is shown in Fig. 6. By using keV electrons one can also produce highly ionized atoms (Erman and Berry, 1971) but the method has been mainly applied to molecules as well as neutral and singly ionized atoms.

Thanks to the high resolution not only vibrational but also rotational lifetimes in molecules can be measured individually. Examples of molecular decay curves are shown in Fig. 7.

The predissociation effects in molecules, discussed above, can also be efficiently studied with this electron-excitation method. As an example we show the data for the B $^2\Sigma$ state in CH (Fig. 8). A large number of rotational lifetimes for the vibrational states $v' = 0$ and $v' = 1$ are measured and drastic changes in τ are found when the rotational number exceeds a certain value. This is due to predissociation caused by penetration through the potential barrier and it leads to very accurate determinations of dissociation energies. For example, in the CH lifetime work (Brzozowski et al. 1976) a value of 3.465±0.012 eV was obtained, about 30 times as accurate as that found in optical emission spectroscopy.

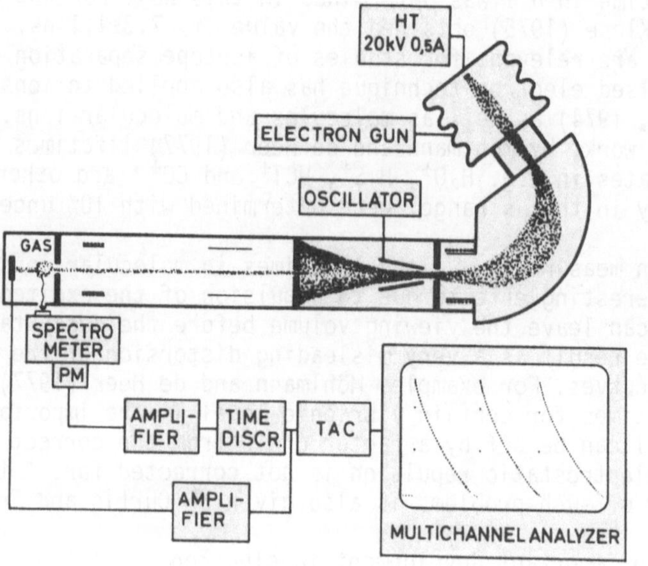

Figure 6. Setup for lifetime measurements using high-energy electrons (Erman, 1975). The electrons excite a gas and the light is observed spectroscopically. The electron beam is further swept with an oscillator and the lifetimes are obtained with delayed coincidence technique.

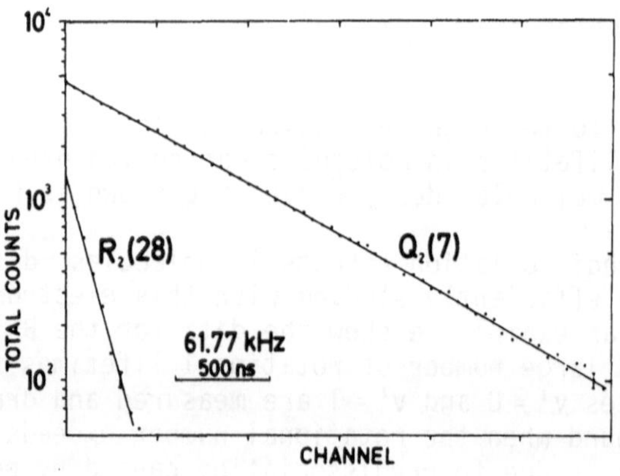

Figure 7. Examples of decay curves, for the OH molecule with the high-energy electron excitation method (Erman, 1978).

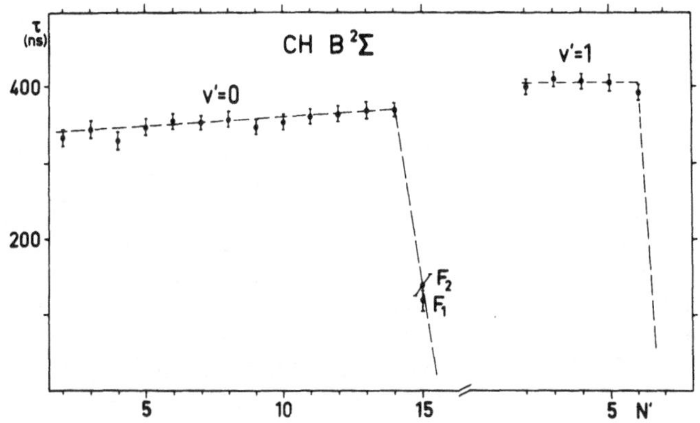

Figure 8. Predissociation in the CH molecule. (Brzozowski et al.,1976). Lifetimes for the v'=0 and v'=1 sequences of the B $^2\Sigma$ state show a slight increase with rotational quantum number before the predissociation sets in, such studies give very accurate dissociation energies.

Such predissociation effects have been observed in a large number of molecules (Erman, 1977). The lifetimes also give important information about molecular configuration as well as isotope effects.

Cascading can generally not be avoided in this work with keV electrons. In most molecules the effects are not serious, however, because higher-lying states often decay be predissociation. Molecular decay curves are in most cases single exponentials.

The photon and electron excitation methods may both be affected by trapping of resonance radiation (Heron et al., 1956). This radiation, from excited atoms, is absorbed by atoms in their ground state and the observable effect is a systematic, pressure-dependent lengthening of the measured lifetime. In practice such studies are carried out at various target gas pressures p and the correct lifetime is obtained by extrapolation to $p = 0$.

The phase-shift method works quite well with electronic excitation. Compared to the optical counterpart the number of available levels is much higher but cascading is the price which must be paid for this improvement.

The phase-shift results must be calibrated using a lifetime which is very accurately known. To obtain satisfactory accuracy the measurements should also be carried out at several frequencies ω. One of the important advantages of the phase-shift method with electrons is that very short lifetimes (below 1 ns) can be measured. It is quite difficult to reach such lifetimes with pulsed excitation.

Both atomic and molecular lifetimes have been studied with the electronic phase-shift technique. For example, interesting data for atoms and ions such as O I (Lawrence, 1970), Si II - IV (Curtis and Smith, 1974) and iron-group elements (Assousa and Smith, 1972, Marek, 1974) can be found in the literature. In the Si study lifetimes ranging from 0.11 to 220 ns were measured. Several astrophysically important molecules, e.g. CH^+ (Brooks and Smith, 1974, 1975) and Si O (Elander and Smith, 1973) have also been investigated in this way. The quoted uncertainties are usually below 10%.

6. LEVEL-CROSSING AND RESONANCE EXPERIMENTS

The methods dealt with in this section also use electron or photon excitation but there are marked differences from the experiments described earlier.

The zero-field level crossing or Hanle-effect technique (Hanle, 1924) belongs to the most accurate methods for measuring atomic and molecular lifetimes. The principle is illustrated in Fig. 9.

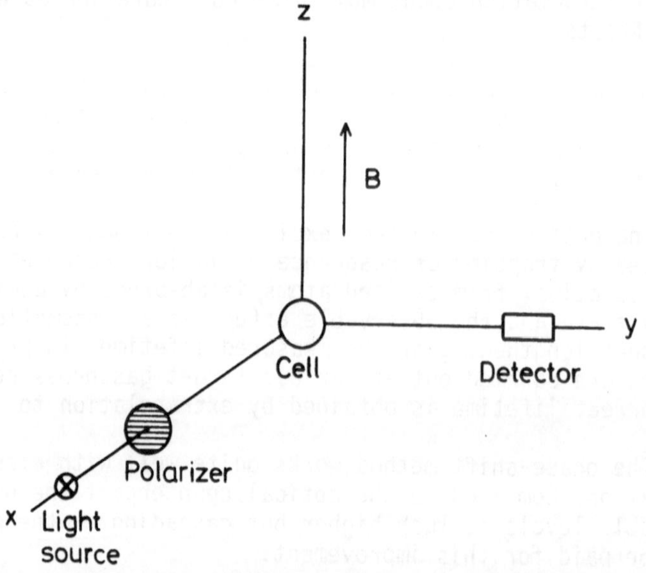

Figure 9. Schematic figure of zero-field level-crossing measurements.

Atoms in a resonance cell are excited with polarized resonance rediation from a lamp. The fluorescence follows the usual dipole pattern damped by the exponential decay. In a constant external magnetic field B the dipoles process with the Larmor frequency ω_L and the time dependance of the fluorescence radiation is

$$I(t) = \exp(-t/\tau)\ \sin^2\omega_L t \tag{11}$$

where $\omega_L = g_J\mu_B \cdot B/h$. If a continuous excitation is maintained while B varies the intensity pattern is an inverted Lorentzian, with a half-width of

$$\Delta B = \hbar/\mu_B\ g_J\ .\ \tau\ . \tag{12}$$

For details about the method see the article by de Zafra and Kirk (1967). Example of Hanle curves are given in Fig.10.

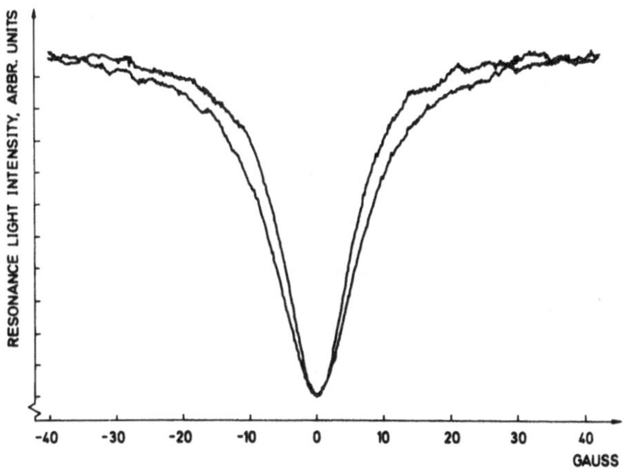

Figure 10. Hanle curves for levels in Pb I (Garpman et al.,1971). The narrower curve is for the 2663 Å line (g_J=-1.494, τ=5.85 ns) and the broader one for the 2802 Å line (g_J=-1.126, τ=6.08 ns).

These are from a study of lifetimes in Pb I (Garpman et al., 1971), determined with uncertainties as low as a few per cent. The Hanle method can be applied to atoms with complicated electron structure where theoretical f-values may be very uncertain. Hilborn and de Zafra (1973) obtained accurate f-values for strong transitions in Fe I and also the complex spectrum of Tm I has been studied in this way (Handrich et al., 1969).

For an example of Hanle-effect studies of molecules we refer to the paper by Broyer et al. (1975 a) in which level lifetimes in I_2 were investigated. The accuracy seems to be somewhat lower than that obtained by the same authors with laser excitation.

To avoid errors in Hanle measurements, due to coherence narrowing, the gas pressure in the cell must be so low that scattering processes can be neglected. Several authors therefore combine Hanle-effect measurements with an atomic beam device.

Optical double resonance (Brossel and Bitter, 1952), mostly used for determinations of atomic fine- and hyperfine-structure, can also give atomic f-values. The atoms are excited with optical resonance radiation while transitions between magnetic sublevels of the excited state are induced with resonant radiofrequency (rf) fields. The advantage of this method - compared to the Hanle technique - is that lifetimes can be determined without knowledge of Landé g_J factors. A representative example of the method is the study of Wagner and Otten (1969), who measured the natural width of a Fe I spectral line. A beam of Fe atoms was excited with light from a Fe hollow cathode and rf-induced transitions between Zeeman sublevels were observed in a weak external magnetic field. This measurement yielded a natural linewidth of 5.35±0.14 MHz for the observed line at 3719.9 Å from which a lifetime of 59.5±1.6 ns was deduced for the z $^5F^o_5$ level of Fe I. (The relation between natural width $\Delta\nu$ and lifetime τ, $\Delta\nu \cdot \tau = 1/\pi$ was used).

More information about optical resonance methods can be found in the reviews by Bucka (1969) and zu Putlitz (1969).

7. ACCELERATOR-BASED METHODS

Beginning with the work of Kay (1963) and Bashkin (1964) heavy ion accelerators play an important role in atomic physics. With the most frequently used method, beam-foil spectroscopy (BFS), lifetimes in neutral as well as highly ionized atoms can be determined.

The experimental arrangement for BFS is shown in Fig.11.

Positive ions from a particle accelerator are directed through a thin foil, usually of carbon. When emerging from the foil the fast ions are often in excited states which decay in vacuum. The light so emitted is analyzed spectroscopically. The decay rates

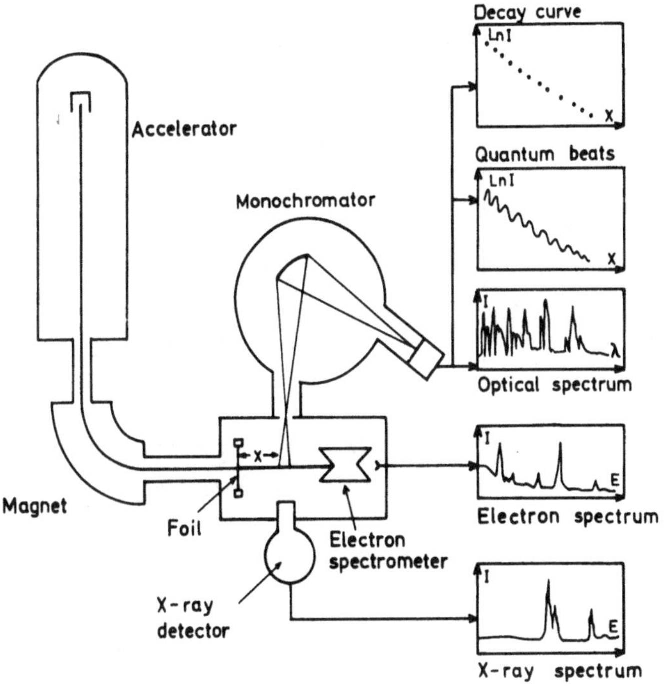

Figure 11. A survey of BFS experiments. Ions from the accelerator are sent through a foil and the radiation is analysed with an optical spectrometer, X-ray-detector or electron spectrometer. Lifetimes are measured by studying spectral line intensities for different values of X_1 *the distance from the foil.*

of the observed spectral lines are measured as the function of the distance x from the foil. Using a simple modification of Eq.(1) the lifetimes are obtained according to

$$I(x) = I(0)\exp(-x/v\tau) \tag{13}$$

where I(x) and I(0) are the counting rates and v is the velocity of ions after the foil. The ions have high velocities, for example ^{12}C ions of 6 MeV have a velocity v = 9.8 mm/ns (v/c = 3.3%), and 0.1 ns therefore corresponds to a distance which can be measured easily.

The lifetime range in BFS experiments is approximately 10^{-12}-10^{-6}s. An example of a very short lifetime is shown in Fig. 12.

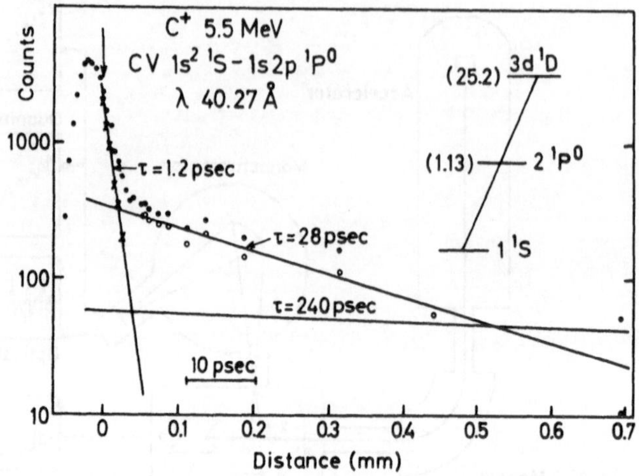

Figure 12. Measurement of a very short lifetime in C V (Knystautas and Drouin, 1976). The prompt decay gives $\tau = 1.2$ *ps which represents the 1s2p* ^1P *lifetime. A longer value, 28 ps, is ascribed to the 1s3d* ^1D *level. Theoretical lifetimes are shown in parentheses.*

Here we observe a multi-exponential behavior, showing the presence of cascading. Such decay curves are very typical in BFS. The cascading can be taken into account by fitting the curves to several exponentials, there are also more sophisticated methods, such as the so-called ANDC analysis. (Curtis et al., 1971). A thorough review of analytical techniques, applicable to many kinds of lifetime measurements is given by Curtis (1976).

The BFS method for lifetime measurements has several unique properties. Practically any element can be accelerated with modern ion accelerators, e.g. Van de Graaff generators, linear accelerators, cyclotrons, isotope separators etc. A large number of ionization degrees is available, there are already data for systems such as Ar XVIII, Fe XXIV, and Kr XXXV. Photons in the range 1 - 10 000 Å can be studied. The non-selective excitation makes a large number of excited states available. Examples of decay curves for highly ionized Fe are shown in Fig. 13.

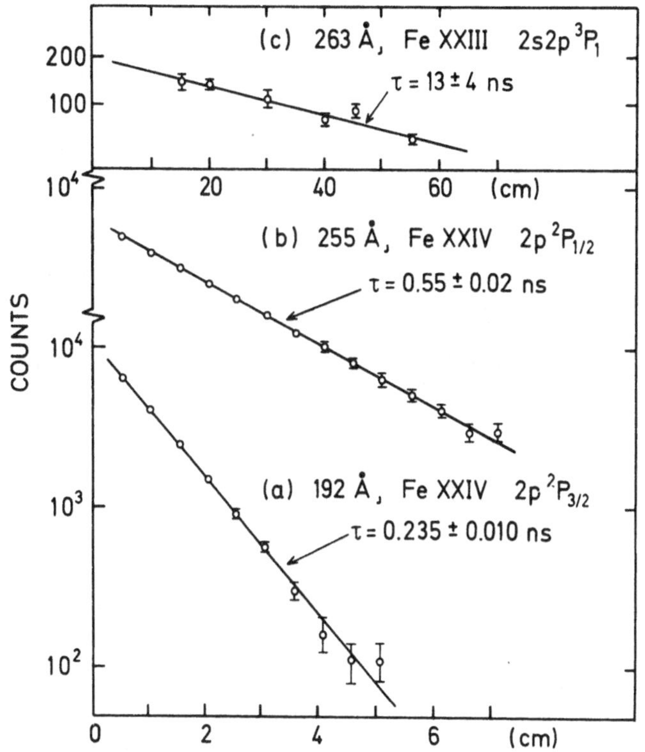

Figure 13. Decay curves for levels in highly ionized Fe (Dietrich et al.,1976). The curves (a) and (b) are for the first excited level in Li-like Fe XXIV. The $^2P_{3/2}$ decay is shortened because of relativistic effects. The curve (c) is for the intercombination line $2s^2\ ^1S_0$ - $2s2p\ ^3P_1$ in Fe XXIII.

Atomic f-values show systematic trends and regularities, e.g. in isoelectronic sequences (Wiese and Weiss, 1968). If the f-values for a given transition are displayed versus 1/Z (Z being the nuclear charge) smooth curves are frequently found. Beam-foil studies have provided much material for such analyses. Examples of the data are shown in Fig. 14 and 15 . The f-values for the 2s ^2S - 2p ^2P doublet in the Li I sequence have been determined from Li I to Fe XXIV. The experimental BFS data for the high Z end (from measurements in Oak Ridge and Berkeley) are shown together with theoretical values (Martin and Wiese, 1976). Note that the relativistic effects are substantial for the $S_{1/2}$ - $P_{3/2}$ branch in Fe XXIV. In the B I sequence the $2s2p^2\ ^2D$ term interacts with the $2s^2nd\ ^2D$ series (n = 3,4,5...), particularly for low

Z. The f-values for the $2s^22p\ ^2P - 2s2p^2\ ^2D$ transition (Fig. 15) are very sensitive indicators of this perturbation. It is worth noting that single-configurational Hartree-Fock calculations for B I give a $2s^22p\ ^2P - 2s2p^2\ ^2D$ f-value that is off by a factor of 10. However, there is very good agreement between BFS data and theoretical calculations (Weiss, 1969, Nicolaides, 1973, Dankwort and Trefftz, 1978) which take into account electron correlation effects.

Many additional isoelectronic sequences have been studied in this way, see e.g. Wiese (1976).

Typical uncertainties in BFS lifetime data are 5 - 10%. If heavy ions are used the errors increase, mainly because the velocity after the foil is less well defined. Much higher precision can be achieved in light elements, in He I the $3p\ ^1P$ lifetime has thus been determined to 0.26% (Astner et al., 1976).

The spectral resolution in BFS experiments is typically 2 - 3 Å in the visible range, inferior to that of laser or high-energy electron work discussed above. In complex spectra line blending causes problems in beam-foil measurements.

Several methods have been developed to increase the accuracy in lifetime experiments with fast ions. These include a) excitation of the beam by laser light b) alignment transfer studies c) ion-solid interaction.

The principle of the beam-laser method, introduced by Andrä et al. (1973) is shown in Fig. 16. The ions from the accelerator are selectively excited with monochromatic laser light. In the first experiments an Ar^+ laser was used to excite Ba^+ ion from an ion accelerator. Careful measurements (Andrä 1976) later yielded $\tau = 6.312 \pm 0.016$ ns for the $6p\ ^2P_{3/2}$ level in Ba II, one of the most accurately determined atomic lifetimes. Also tunable dye lasers by which many levels can be reached have been used (Harde, 1976). An interesting example is the La II study of Arnesen et al. (1977). In this experiment La^+ ions from an isotope separator were excited from their ground state a ^3F into y $^3F^0$ the decay of which into the metastable level a ^3D was measured. The excitation and detection thus occurred at different wavelengths and this fact eliminated stray-light problems which can hamper beam-laser experiments.

The beam-foil interaction often results in a non-statistical population of magnetic sublevels (alignment). It can be shown (Dufay, 1973) that the alignment is not transferred by cascading and this circumstance can be exploited for obtaining simpler decay curves (Berry et al., 1972).

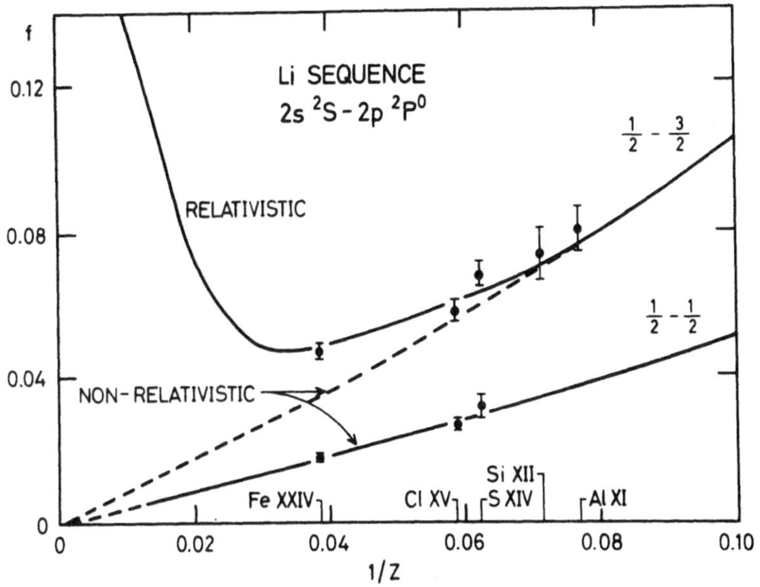

Figure 14. Oscillator strengths for the 2s-2p transitions in the Li I sequence. The experimental data are from Dietrich et al. (1978) and Pegg (1978).

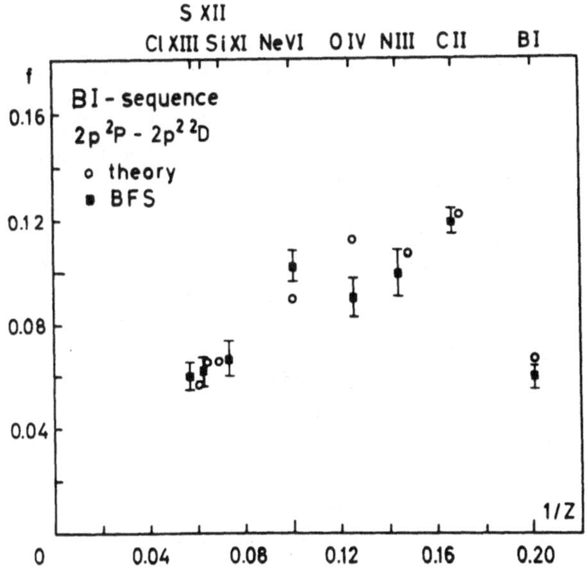

Figure 15. Oscillator strengths for the 2p $^2P - 2p^2\ ^2D$ transitions in the B I sequence. The experimental data are from Martinson and Gaupp (1974) and Pegg (1978).

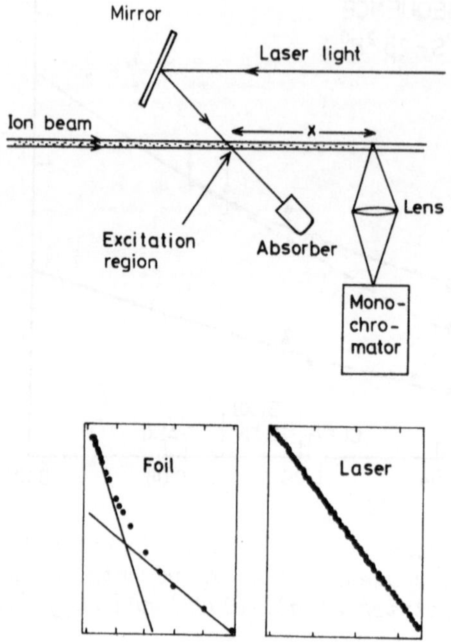

Figure 16. Excitation of fast ions with laser light. The ions are selectively excited from their ground state. A comparison is shown of the decay of the 3p ^2P level in Na I, measured with beam-foil and beam-laser (Harde, 1976).

Another, related method combines Hanle-effect measurements with fast beam techniques. Poulsen et al. (1975) have used this technique to determine lifetimes in Be I and Be II. Compared to conventional Hanle methods this "beam-Hanle" approach is easily extended to levels in ions.

Lifetimes can also be measured if the foil is replaced by a thick solid target. As a result of sputtering processes the target atoms frequently leave the solid in excited states (McCracken 1975). By pulsing the incoming beam and performing a coincidence experiment it is possible to measure lifetimes for excited states in sputtered atoms (Ramanujam, 1977). The cascades are reduced because low-lying excited states are preferentially populated in sputtered, excited atoms.

Molecular lifetimes cannot be measured with the beam-foil method because molecules are dissociated in the foil. However, if the foil is replaced by a gas cell molecular ions can be excited and lifetimes can be determined by time-of flight techniques (Poulizac and Druetta, 1969, Head, 1971). It is also possible to pulse the beam from the accelerator and measure molecular lifetimes with the delayed-coincidence method. In this way Dotchin et al. (1973) obtained lifetimes for levels in N_2^+, N_2, CO^+ and

CO with quoted accuracies as low as 1%.

8. COINCIDENCE TECHNIQUES

We have already mentioned coincidence experiments, in connection with electron or photon excitation. In this section coincidences between two atomic processes are discussed. These experiments can be divided into photon-photon and electron-photon coincidence measurements. The photon-photon method is based on ideas from nuclear physics where $\gamma - \gamma$ delayed coincidences are used for half-life determinations (Fossan and Warburton, 1974). In the atomic case a target gas is excited with electrons. The collision region is viewed with two interference filters which transmit wavelengths λ_1 and λ_2 in a cascade transition (Fig. 17). For example, in a lifetime measurement for the 2p ^2P level in Li I, one of the filters should transmit the 2s ^2S - 2p ^2P transition (6708 Å) while the other accepts the 2p ^2P - 3d ^2D line (6104 Å). The respective signals go to photo-multipliers and - after amplification and discrimination to a time-to-pulse-height converter and a multichannel analyzer. The start pulse is given by a photon from the higher-lying transition and the time-distribution of the lower transitions is measured. The results are free from cascade effects, the decay curve thus being a single exponential. However, like in all coincidence measurements the counting rates are quite

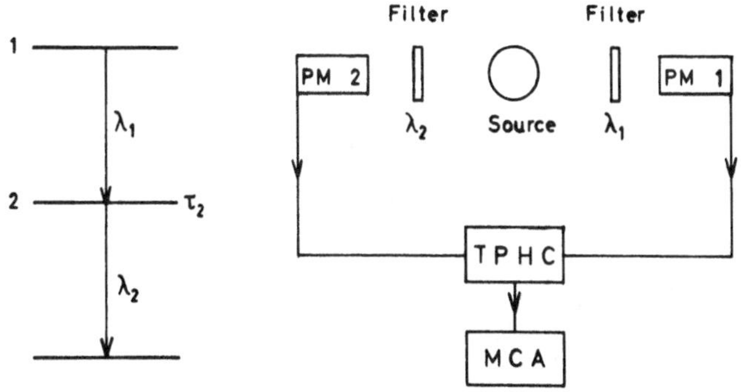

Figure 17. Simplified setup for photon-photon coincidence measurements. Two photons in a cascade are detected with photomultipliers. After amplification and discrimination (not shown) the pulses are fed into a time-to-pulse-height converter and a multichannel analyzer. A single-exponential decay curve for the level 2 is obtained.

low. Good examples of such work is found in Camhy-Val et al. (1975) and King et al. (1976). In the latter experiment a lifetime in Ar II measured to within 1.5%. A variation of this method consists in measuring electron-phonton coincidences. Here the atoms are excited with electrons of accurately known energy. Coincidences are measured between inelastically scattered electrons of a predetermined energy loss (this defines the atomic level that was excited) and photons from the same level. This method, introduced by Imhof and Read (1969) has been applied to both atoms and molecules (Smith et al., 1975).

9. FORBIDDEN TRANSITIONS

When the selection rules for electric dipole (E 1) transitions are not fulfilled other types of decays usually occure. Examples of such forbidden transitions in hydrogen- and helium-like systems are given in Fig. 18. The 2s $^2S_{1/2}$ level in hydrogen-like atoms decays by two-photon emission. These photons have a continuous energy distribution, their sum being equal to the 1s - 2s energy separation. This decay has a probability of $8.23 \cdot Z^6$ (s^{-1}). Using the BFS method Marrus and Schmieder (1972) measured the 2s $^2S_{1/2}$ lifetime in Ar XVII, obtaining the value

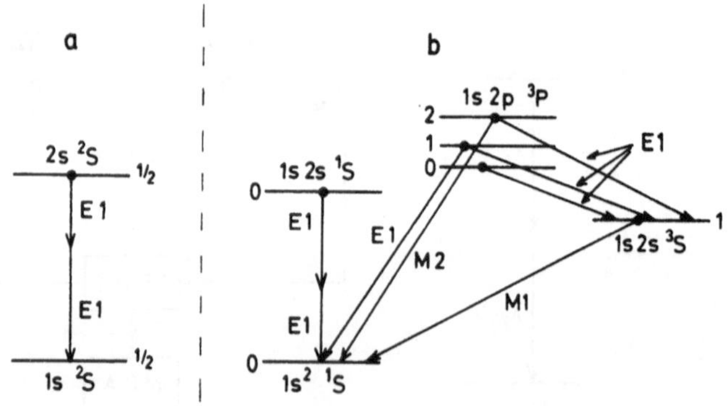

Figure 18. Forbidden decays in hydrogen-like (a) and helium-like (b) atoms and ions. The 2s ^2S and 2s ^1S levels decay by two-photon processes, and 2s ^3S decays with magnetic dipole rationtion. The 2p ^3P levels combine with 2s ^3S allowed transitions while 2p $^3P_1$ and 2p $^3P_2$ also decay to the ground state, with spin-forbidden E 1 and M 2 transitions, respectively.

3.54±0.25 ns, in excellent agreement with the theoretical prediction 3.46 ns (Klarsfeld, 1969). To reach this very high charge state the Ar ions were accelerated to 400 MeV in the Berkeley heavy ion accelerator, HILAC. For hydrogen-like O and F the 2s $^2S_{1/2}$ lifetime was later measured by Cocke et al. (1974) who used the BFS method at somewhat lower energies.

In helium-like systems the 2s $^1S_0$ level also decays by two-photon emission. Using a time-of-flight technique Van Dyck et al. (1971) measured this lifetime in He I, obtaining 19.7±1.0 ms, in excellent agreement with the theoretical value 19.5 ms (Drake et al., 1969). The experimental method consisted in detecting the number of He-atoms in the metastable 2s $^1S_0$ level as a function of the distance from the excitation region. The 2p ^3P term decays to 2s ^3S with allowed transitions. For the 2p $^3P_2$ level there is another decay mode, with magnetic quadrupole (M 2) radiation to the ground state (Fig. 18). This latter process is strongly Z-dependent and competes with the allowed decay in Ar XVII. The 2s $^3S_1$ level de-excites by magnetic dipole (M 1) transitions, these have been observed in the solar corona (Gabriel and Jordan, 1969). This radiation has also been detected in laboratory experiments, (Marrus and Schmieder, 1972, Gould et al., 1974, Cocke et al., 1973). In these BFS experiments, the 2s $^3S_1$ lifetime was measured in He-like S, Cl, Ar, Ti, V and Fe. More recently Gould and Marrus (1976) also observed this decay in He-like Kr XXXV, using 700 MeV ions from the Berkeley Super-HILAC. They obtained a value of τ = 0.11±0.02 ns, which agrees with theoretical predictions.

10. APPLICATIONS OF ATOMIC AND MOLECULAR LIFETIME DATA

Optical spectroscopy, measurements of spectral-line wavelengths is one of the simplest methods for identifying chemical elements. When the concentrations or abundances of elements in a light source (e.g. the sun) are determined the wavelength information must be complemented with knowledge about the corresponding f-values.

The abundances of the chemical elements in the sun, stars and interstellar medium are of considerable interest because they contain information about the creation of the solar system and the stars, various pathways of nucleosynthesis and physical processes such an convection, diffusion or mass loss etc. (Biémont and Grevesse, 1977). In the solar and stellar spectra a large number of spectral lines are observed. From the measured equivalent widths the abundances (relative to hydrogen) are derived if the f-values are known. About 70 chemical elements have been identified in the sun, as can be seen from Fig. 19 ., from Engvold (1977). The abundances are normalized by defining log N_H = 12.00 where

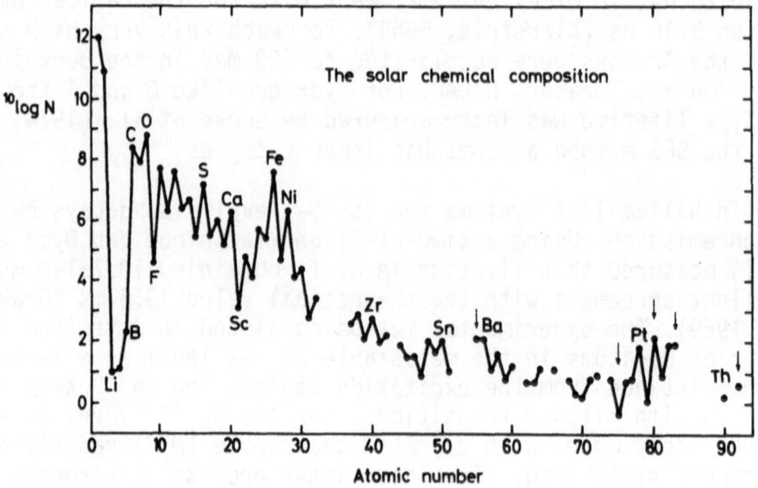

Figure 19. The chemical composition of the sun. About 70 elements have been identified in the solar spectra but the abundances are often very uncertain. (Engvold, 1977).

N_H is the number of hydrogen atoms. In several cases the abundances are highly uncertain because the f-values are only approximately known.

Note that Li, Be, and B have very low abundances, these elements are destroyed by proton reactions. The iron-group elements have presented very interesting problems. About 10 years ago it was found that the Fe abundance (relative to H) seemed to be about 10 times higher in the solar corona and chromosphere (log N_{Fe} = 7.66) than in the solar photosphere (log N_{Fe} =6.51). No good explanation was given for this discrepancy but the problem was solved when Garz and Kock (1969), Huber and Tobey (1968) and Whaling et al. (1969), using different techniques, found that the previously used f-values for Fe I had been erroneous. With correct data the photospheric abundance was increased to log N_{Fe} = 7.5, a value that also agrees with abundance in meteorites. Improved f-values for other 3d-elements have had similar, although slightly less drastic consequences (Smith, 1973, Biémont and Grevesse, 1977, Huber, 1977) for the solar abundances.

Fig. 19 shows that several rare-earth elements have been observed in the sun. The abundances are often very uncertain, because the f-values are poorly known. The observed lines usually belong to singly-ionized atoms. Several beam-foil studies have appeared in later years, e.g. Curtis et al. (1973), Andresen and Sørensen (1974), Lage and Whaling (1976) which give f-values for the lines observed in the solar spectrum. It is also worth noting that certain stars, of the Am, Ap and Ba type show very high abundences of rare earth elements.

Molecular transitions can also be used for abundance determination. Curtis et al., (1976) used the method of high-energy, high-frequency electron excitation to determine f-values in the spectrum of C_2, from which a solar carbon abundance could be derived.

One of the most serious problems in fusion research with magnetically confined high-temperature plasmas (e.g. Tokamaks) is caused by plasma impurities, e.g. highly ionized metal atoms. These are introduced into the plasma when H-ions interact with the surrounding wall. The electromagnetic readiation from the impurity atoms may significantly contribute to the energy loss and cooling of the plasma (Drawin, 1978). It is therefore vital to determine and reduce the impurity concentrations. For this purpose atomic f-values are needed for many strong transitions in highly ionized metal atoms, e.g. Fe, Cr, Ni, Mo, and W. Of particular interest are systems with one or two valence electrons, the Li, Be, Na, Mg, Cu, and Zn isoelectronic sequences which emit strong characteristic lines. Many lifetime studies with BFS have recently been motivated by such demands in plasma physics.

Lifetimes for levels in noble gas ions, Ne II, Ar II, Ar III, Kr II, Kr III and Xe II are finally of importance in laser physics. There are many cases in these spectra where a higher level has a much longer lifetime than the lower level with which it combines. This fact increases the laser gain and is vital for achieving population inversion. Detailed information of atomic transitions of laser interest can be found in the review article by Davis and King (1975).

Although we have discussed many aspects of atomic and molecular lifetime studies, several important areas have been omitted, e.g. inner-shell excited states (Sellin, 1976), highly-excited (Rydberg) levels (Kleppner, 1977) and lifetimes of vacancy states (Nordgren et al., 1976) and a few other topics.

ACKNOWLEDGEMENTS

I am grateful to S. Bashkin, P. Erman, E.J. Knystautas, J.-C. Lehamnn, E. Matthias, D.J. Pegg and S. Svanberg for sending me their recent data. This work was supported by the Swedish Natural Science Research Council (NFR) and the Swedish Agency for Energy Research (NE).

REFERENCES

Andersen, T. and Sørensen, G., 1974 Solar Phys. 38, 343.

Andrä, H.J., Gaupp, A. and Wittmann, W., 1973 Phys. Rev. Letters 31, 501.

Andrä, H.J., 1976 "Beam-Foil Spectroscopy" (I.A. Sellin and D.J. Pegg, eds., Plenum, New York) p.835.

Arnesen, A., Bengtsson, A., Hallin, R., Lindskog, J., Nordling, C. and Noreland, T., 1977 Physica Scripta 16, 31.

Assousa, G.E. and Smith, W.H., 1972 Astrophys. J. 176, 259.

Astner, G., Curtis, L.J., Liljeby, L., Mannervik, S. and Martinson, I., 1976 Z. Physik A279, 1.

Bashkin, S., 1964 Nucl. Instr. Methods 28, 88.

Bell, G.D. and Tubbs, E.F., 1970 Astrophys. J., 159, 1093.

Bell, G.D. and Lyzenga, G.A., 1976 Proc. Roy. Soc. A351, 581.

Bennett, W.R. and Kindlmann, P.J., 1966 Phys. Rev. 149, 38.

Berry, H.G., Curtis, L.J. and Subtil, J.L., 1972 J. Opt. Soc. Am. 62, 771.

Biémont, E. and Grevesse, N., 1977 Physica Scripta 16, 39.

Bieniewski, T.M., 1976 Astrophys. J. 208, 228.

Bridges, J.M. and Wiese, W.L., 1970 Astrophys. J. 161, L71.

Brooks, N. and Smith, W.H., 1974 Astrophys. J. 194, 513.

Brooks, N. and Smith, W.H., 1975 Astrophys. J. 196, 307.

Brossel, J. and Bitter, F., 1952 Phys. Rev. 86, 308.

Broyer, M., Vigue, J. and Lehmann, J.-C., 1975 J. Chem. Phys. 63, 5428.

Broyer, M., Lehmann, J.-C. and Vigue, J., 1975a J. Physique 36, 235

Brzozowski, J., Bunker, P., Elander, N. and Erman, P., 1976 Astrophys. J. 207, 414.

Bucka, H., 1969 J. Physique Suppl. 30, C 1 - 3.

Camhy-Val., C., Dumont, A.M., Dreux, M., Perret, L. and Vanderriest, C., 1975, J. Quant. Spectrocs, Radiat, Transfer. 15, 527.

Cocke, C.L., Curnutte, B. and Randall, R., 1973, Phys. Rev. Letters 31, 507.

Cocke, C.L., Curnutte, B., Macdonald, J.R., Bednar, J.A. and Marrus, R., 1974, Phys. Rev A9, 2242.

Codling, K., 1973, Rep. Prog. Phys. 36, 541.

Cooke, W.E., Gallagher, T.F., Edelstein, S.A. and Hill, R.M., 1978, Phys. Rev. Letters 40, 178.

Corney, A., 1969, Adv. Electron. Electron. Phys. 29, 115.

Corney, A., 1977, "Atomic and Laser Spectroscopy" (Clarendon, Oxford).

Crossley, R.J.S., 1969, Adv. Atom. Molec. Phys. 5, 237.

Cunningham, P.T. and Link, J.K., 1967, J. Opt. Soc. Am. 57, 1000.

Curtis, L.J., Berry, H.G. and Bromander, J., 1971, Phys. Letters 34A, 169.

Curtis, L.J., Martinson, I. and Buchta, R., 1973, Nucl. Instr. Methods 110, 391.

Curtis, L.J. and Smith, W.H., 1974, Phys. Rev. A9, 1537.

Curtis, L.J. in "Beam-Foil Spectroscopy" 1976, (S. Bashkin, ed., Springer, Berlin) p.63.

Curtis, L.J., Engman, B. and Erman, P., 1976, Physica Scripta 13, 270.

Curtis, L.J. and Erman, P., 1977, J. Opt. Soc. Am. 67, 1218.

Dankwort, W. and Trefftz, E., 1978, Astron. Astrophys. 65, 93.

Davis, C.C. and King, T.A., 1975, Adv. Quant. Electron. 3, 169.

Dietrich, D.D., Leavitt, J.A., Bashkin, S., Conway, J.G., Gould, H., MacDonald, D., Marrus, R., Johnson, B.M. and Pegg, D.J., 1978, to be published.

Dotchin, L.W., Chupp, E.L. and Pegg, D.J., 1973, J. Chem. Phys. 59, 3960.

Drake, G.W.F., Victor, G.A. and Dalgarno, A., 1969, Phys. Rev. 180, 25.

Drawin, H.W., 1978, Phys. Reports 37, 125.

Dufay, M., 1973, Nucl. Instr. Methods 110, 79.

Elander, N. and Smith, W.H., 1973, Astrophys. J. 184, 311.

Engvold, O., 1977, Physica Scripta 16, 48.

Erman, P. and Berry, H.G., 1971, Phys. Letters 34A, 1.

Erman, P., 1975, Physica Scripta 11, 65.

Erman, P., 1977, Nukleonika 22, 607.

Erman, P., 1978, private communication.

Figger, H., Siomos, K. and Walther, H., 1974, Z. Physik 270, 371.

Fossan, D.B. and Warburton, E.K., 1974, in "Nuclear Spectroscopy and Reactions C" (J. Cerny, ed., Academic, New York) p.307.

Foster, E.W., 1964, Rep. Prog. Phys. 27, 469.

Gabriel, A.H. and Jordan, C., 1969, Nature 221, 947.

Garpman, S., Lidö, G., Rydberg, S. and Svanberg, S., 1971, Z. Physik 241, 217.

Garz, T. and Kock, M., 1969, Astron. Astrophys. 2, 274.

Gould, H., Marrus, R. and Mohr, P., 1974, Phys. Rev. Letters 33, 676.

Gould, H. and Marrus, R., 1976, in "Beam-Foil Spectroscopy" (I.A. Sellin and D.J. Pegg, eds., Plenum, New York) p.305.

Greenhow, R.C. and Schmidt, A.J., 1974, Adv. Quant. Electron. 2, 157.

Gustavsson, M., Lundberg, H. and Svanberg, S., 1977, Phys. Letters 64A, 289.

Handrich, E., Stendel, A., Wallenstein, R. and Walther, H., 1969, J. Physique Suppl. 30, C 1 - 18.

Hanle, W., 1924, Z. Physik 30, 93.

Harde, H., 1976, in "Beam-Foil Spectroscopy" (I.A. Sellin and D.J. Pegg, eds., Plenum, New York) p.859.

Haroche, S., 1976, in "High-Resolution Laser Spectroscopy" (K. Shimoda, ed., Springer, Berlin) p.256.

Head, C.E., 1971, Phys. Letters 34A, 92.

Heldt, J., Figger, H., Siomos, K. and Walther, H., 1975, Astron. Astrophys. 39, 371.

Heron, S., McWhirter, R.W.P. and Rhoderick, E.H., 1956, Proc. R. Soc. A234, 565.

Hilborn, R.C. and de Zafra, R., 1973, Astrophys. J. 183, 347.

Huber, M. and Tobey, F.L., 1968, Astrophys. J. 152, 609.

Huber, M.C.E., 1977, Physica Scripta 16, 16.

Imhof, R.E. and Read, F.H., 1969, Chem. Phys. Letters 3, 652.

Imhof, R.E. and Read, F.H., 1977, Rep. Prog. Phys. 40, 1.

Jiménez, E., Campos, J. and Sãnchez del Rio, C., 1974, J. Opt. Soc. Am. 64, 1009.

Kay, L., 1963, Phys. Letters 5, 36.

Kibble, B.P., Copley, G. and Krause, L., 1967, Phys. Rev. 153, 9.

King, G.C., Mohamed, K.A., Read, F.H. and Imhof, R.E., 1976, J. Phys. B. 9, 1247.

Klarsfeld, S., 1969, Phys. Letters 30A, 382.

Kleppner, D., 1977, in "Atomic Physics 5" (R. Marrus, M. Prior, and H. Shugart, eds, Plenum, New York) p.269.

Klose, J.Z., 1967, J. Opt. Soc. Am. 57, 1242.

Klose, J.Z., 1968, J. Opt, Soc. Am. 58, 1509.

Klose, J.Z., 1971, Astrophys. J. 165, 637.

Klose, J.Z., 1975, Phys. Rev. A11, 1840.

Knystautas, E.J. and Drouin, R., 1976, in "Beam-Foil Spectroscopy"

(I.A. Sellin and D.J. Pegg, eds., Plenum, New York) p.377.

Lage, C.S. and Whaling, W., 1976, J. Quant. Sepctrosc. Radiat. Transfer 16, 537.

Lawrence, G.M., 1970, Phys. Rev. A2, 397.

Lehmann, J.-C., 1975, in "Frontiers in Laser Spectroscopy, Les Houches 1975" (R. Balian, S. Haroche and S. Liberman, eds., North-Holland, Amsterdam, 1977) p.473.

Lehmann, J.-C., 1977, in "Atomic Physics 5" (R. Marrus, M. Prior and H. Shugart, eds., Plenum, New York) p.167.

Lindgård, A. and Nielsen, S.E., 1977, At. Data Nucl. Data Tabl. 19, 534.

Link, J.K., 1966, J. Opt. Soc. Am. 56, 1195.

Marek, J., 1972, Astron. Astrophys. 17, 83.

Marrus, R. and Schmieder, R.W., 1972, Phys. Rev. A5, 1160.

Martin, G.A. and Wiese, W.L., 1976, J. Phys. Chem. Ref. Data 5, 537.

Martinson, I. and Gaupp, A., 1974, Phys. Reports 15C, 113.

Matthias, E., Rosenberg, R.A., Poliakoff, E.D., White, M.G., Lee, S.-T. and Shirley, D.A., 1977, Chem. Phys. Letters 52, 239.

Matthias, E., 1978, private communication.

McCracken, G.M., 1975, Rep. Prog. Phys. 38, 241.

Mitchell, A.C.G. and Zemansky, M.W., 1934, "Resonance Radiation and Excited Atoms" (Cambridge University Press, Cambridge)

Möhlmann, G.R. and de Heer, F.J., 1977, Physica Scripta 16, 51 and references therein.

Nicolaides, C.A., 1973, Chem. Phys. Letters 21, 242.

Nordgren, J., Ågren, H., Selander, L., Nordling, C. and Siegbahn, K., 1977, Physica Scripta 16, 70.

Pegg, D.J., 1978, private communication.

Penkin, N.P., 1964, J. Quant. Spectrosc. Radiat. Transfer 4, 41.

Poulizac, M.C. and Druetta, M., 1969, Compt. REnd. Acad. Sc. 269B, 114.

Poulsen, O., Andersen, T. and Skouboe, N.J., 1975, J. Phys. B. 8, 1393.

zu Putlitz, G., 1969, in "Atomic Physics" (V.W. Hughes, B. Bederson, V.W. Cohen and F.M.J. Pichanik, eds, Plenum, New York) p.227.

Ramanujam, P.S., 1977, Phys. Rev. Letters 39, 1192.

Richter, J. and Wulff, P., 1970, Astron. Astrophys. 9, 37.

Sakurai, K., Capelle, J.A. and Broida, H.P., 1971, J. Chem. Phys. 54, 1220.

Sellin, I.A., 1976, Adv. Atom. Molec. Phys. 12, 215.

Siomos, K., Figger, H. and Walther, H., 1975, Z. Physik A272, 355.

Smith, A.J., Read, F.H. and Imhof, R.E., 1975, J. Phys. B 8, 2869.

Smith, P.L., 1973, Nucl. Instr. Methods 110, 395.

Svanberg, S., 1978, private communication.

Van Dyck, R.S., Johnson, C.E. and Shugart, H.A., Phys. Rev. A4, 1327.

Vigue, J., Broyer, M. and Lehamnn, J.-C., 1975, J. Chem, Phys. 62, 4941.

Wagner, R. and Otten, E.W., 1969, Z. Physik 220, 349.

Walther, H., 1976, "Laser Spectroscopy of Atoms and Molecules" (Springer, Berlin)

Weiss, A.W., 1969, Phys. Rev. 188, 119.

Whaling, W., King, R.B. and Martinez-Garcia, M., 1969, Astrophys. J. 158, 389.

Wien, W., 1919, Ann. d. Phyik 60, 597.

Wiese, W.L.,1968, in "Methods of Experimental Physics 7A" (B. Bederson and W.L. Fite, eds, Academic, New York) p.117.

Wiese, W.L. and Weiss, A.W., 1968, Phys. Rev. 175, 50.

Wiese, W.L., 1976, in "Beam-Foil Spectroscopy" (S. Bahshkin, ed., Springer, Berlin) p.147.

Wolnik, S.J., Berthel, R.O. and Wares, G.W., Astrophys. J. 162, 1037.

de Zafra, R.L. and Kirk, W., 1967, Am. J. Phys. 35, 573.

EXCITATION ENERGIES AND TRANSITION MOMENTS FROM THE POLARIZATION PROPAGATOR

Jan Linderberg

Department of Chemistry, Aarhus University, Aarhus, Denmark.

DELIMITING THE SCOPE OF THE PRESENTATION.

These lectures will be concerned with the principles behind an algebraic, equation of motion approach to the determination of the polarization propagator. This term is here taken in a general sense to encompass a set of two-time Green's functions, such as used by Yngve Öhrn and the present author in our textbook [1]. There will not be time to present the arguments leading to the appearance of the polarization propagator as the natural quantity for the examination of transition moments and excitation energies, and the concerned reader may find a detailed approach in terms of photon scattering in Chapter 12 of Reference [1].

The formalism in these notes follows in certain respects a previous analysis [2,3]. Only certain limited approximations will be treated but the presentation should hopefully provide a guide to the recent literature on higher order approximations [4].

Results from calculations by the Aarhus group will be used as illustrations of the theory.

PROPAGATORS, SPECTRAL REPRESENTATIONS, AND MOMENTS EXPANSIONS.

Polarization is, in classical dielectric theory, a measure of the dipole moment density. Its divergence is identified with the negative of the density of so-called apparent charges:

$$q(\underline{r}) = - \operatorname{div} \underline{P}(\underline{r}) . \qquad (1)$$

Cleanthes A. Nicolaides and Donald R. Beck (eds.), Excited States in Quantum Chemistry, 35–43.

A microscopic theory is developed directly in terms of the charge density operator $\hat{q}(\underline{r})$ and no explicit reference is made to the polarization, except that it often is convenient to consider the fluctuation operator $\hat{q}(\underline{r})-\langle\hat{q}(\underline{r})\rangle$ as the primary dynamic entity to be studied. The polarization propagator describes how a disturbance, which couples with the charge density, influences the system under consideration and gives a measure of the induced charge density.

Linear response theory [5] yields the result that the induced charge density is determined from the two-time retarded Green's function

$$\langle\langle q(\underline{r}t) \, ; \, q(\underline{r}'t')\rangle\rangle = (i/h)\Theta(t-t')\langle[q(\underline{r}t),q(\underline{r}'t')]\rangle \quad (2)$$

with the Fourier integral representation

$$\langle\langle q(\underline{r}) \, ; \, q(\underline{r}')\rangle\rangle_E = \int d(t-t')\langle\langle q(\underline{r}t);q(\underline{r}'t')\rangle\rangle \, e^{iE(t-t')/h} \quad (3)$$

This function is analytical in the upper half of the complex E-plane, and is, according to Kramers and Kronig, uniquely defined in terms of its spectral density function

$$J(\underline{r},\underline{r}';E) = \mathrm{Im} \, \langle\langle q(\underline{r}) \, ; \, q(\underline{r}')\rangle\rangle_E/\pi. \quad (4)$$

We will only consider cases where the average value formation in Eq. (2) refers to a molecular electronic ground state in the Born-Oppenheimer picture and involves the electronic charge density operator. A rather elementary calculation leads then to the explicit form

$$J(\underline{r},\underline{r}';E) = \Sigma_n[\langle 0|q(\underline{r})|n\rangle\langle n|q(\underline{r}')|0\rangle \, \delta(E-E_n+E_0) - \langle 0|q(\underline{r}')|n\rangle\langle n|q(\underline{r})|0\rangle \, \delta(E-E_0+E_n)] \quad (5)$$

in terms of eigenstates $|n\rangle$ and eigenvalues E_n of the hamiltonian. We see that if we can determine $J(\underline{r},\underline{r}';E)$ it contains the information needed for the calculation of transition moments and energies.

The charge density operator is a one electron operator and is thus, in second quantization, a linear combination of the elementary operators $a_r^\dagger a_s$ defined for an orthonormal spin orbital basis $\{u_s\}$. It is convenient to use hermitian operators and we define

$$b_{rs} = \begin{cases} a_r^\dagger a_s + a_s^\dagger a_r \ , & r \leq s, \\ i a_r^\dagger a_s - i a_s^\dagger a_r \ , & r > s. \end{cases} \tag{6}$$

The determination of the spectral density function (4) is then referred to the calculation of a set of similar quantities,

$$J(rs,r's';E) = \mathrm{Im} \langle\langle b_{rs} \ ; \ b_{r's'} \rangle\rangle_E / \pi. \tag{7}$$

It is at this point we will start to make some assumptions that will allow a direct calculation of an approximate spectral density function.

We will consider the zeroth and first moments of the functions (7). They are obtained from the sum rules

$$\int J(rs,r's';E)dE = \langle [b_{rs}, b_{r's'}] \rangle \equiv S_{rs,r's'} \ , \tag{8}$$

$$\int J(rs,r's';E)EdE = \langle [[b_{rs},H], b_{r's'}] \rangle \equiv A_{rs,r's'} \ , \tag{9}$$

which follow from Eq. (5). The matrices **S** and **A** can be calculated from a knowledge of the ground state. Both are hermitian and **A** is positive semi-definite and they may be brought to a simultaneously diagonal form by means of a linear transformation. Since **A** is real and **S** purely imaginary, one finds that a set of operators may be defined in conjugate pairs,

$$Q_m = \Sigma_{rs} \ Q_{m,rs} b_{rs} \ , \tag{10}$$

$$Q_m^\dagger = \Sigma_{rs} \ Q^*_{m,rs} b_{rs} \ , \tag{11}$$

such that

$$\langle [Q_m, Q_{m'}^\dagger] \rangle = \delta_{mm'} \ , \tag{12}$$

$$\langle [Q_m, Q_{m'}] \rangle = \langle [Q_m^\dagger \ , \ Q_{m'}^\dagger] \rangle = 0, \tag{13}$$

$$\langle [[Q_m,H], Q_{m'}^\dagger] \rangle = W_m \delta_{mm'}, W_m > 0, \tag{14}$$

and

$$\langle [[Q_m,H], Q_{m'}] \rangle = \langle [[Q_m^\dagger,H], Q_{m'}^\dagger] \rangle = 0. \tag{15}$$

The transformation expressed by Eqs. (10) and (11) may be singular in that there may exist linear combinations of the operators b_{rs} which commute with the hamiltonian, that is constants of the motion, or which have the ground state as eigenstate. Such operators will correspond to zero eigenvalues of **A** and **S** and may be eliminated without loss of generality [2].

The canonical form of the moments of the spectral density functions becomes interesting if we assume that the function,

$$J_m(E) = \mathrm{Im} \langle\langle Q_m;Q_m^\dagger\rangle\rangle_E/\pi \quad , \tag{16}$$

may be represented by the "one state approximation"

$$J_m(E) = \delta(E-W_m), \tag{17}$$

which exhibits the same first two moments. This implies that the annihilator condition holds,

$$Q_m|0\rangle = 0, \tag{18}$$

and that

$$|m\rangle = Q_m^\dagger|0\rangle, \tag{19}$$

is an approximate eigenstate of the hamiltonian with energy eigenvalue E_o+W_m. That is,the manifold of states, $\{|0\rangle, |m\rangle|\ m = 1,2,\ldots\}$, diagonalize the hamiltonian within their subspace of the total state space.

The procedure outlined above can then be summarized as follows: In order to achieve a partial diagonalization of the hamiltonian H for a given system, we introduce a set of operators, $\{b_{rs}\}$, and construct the matrices $\mathbf{S}$ and $\mathbf{A}$ from some assumption about ground state expectation values (generally in the form of density matrices). Diagonalization of the matrices gives excitation operators $Q_m^\dagger$ and deexcitation operators Q_m. The ground state should have the property that it is annihilated by the Q_m's. When this is the case we have a set of excitation energies, W_m, and transition moments may be calculated from the form

$$\langle 0|\underline{P}|m\rangle = \langle[\underline{P},Q_m^\dagger]\rangle \tag{20}$$

Calculations according to this program have been carried out in various degrees of sophistication and the molecular applications by the Caltech group with McKoy and by Jørgensen, Oddershede, Elander and Beebe at Aarhus have shown that accurate results may be obtained at lower cost than by methods based on the method of superposition of configurations in many cases [4].

The iterative ground state determination from the annihilator condition (18) is a critical point in the procedure and no accurate self-consistent results have been obtained so far. Most applications have been satisfied with the one step procedure to use the Hartree-Fock state to generate the matrices and to disregard the conditions (18), but recently [2,3] we have shown that a self-consistent solution implies a particular form for the ground state when the operator set consists of the one electron operators of Eq. (6). The more general case where the operator set also includes so-called two-particle, two-hole operators still awaits a definitive solution. The schemes of this nature derive from the random phase approximation and carry names derived thereof or the linearized time-dependent Hartree-Fock method. The equation of motion method is also a term used in this context, but since all dynamical features of a system must derive from an equation of motion this name is less appropriate.

THE SECULAR PROBLEM.

Two features of the diagonalization problem will be considered in this section. Firstly we examine the form of the separation theorem which applies and secondly we draw some conclusions concerning symmetry blocking of the matrices.

The real symmetric matrix $\mathbf{A}$ may be brought to diagonal form by means of an orthogonal matrix $\mathbf{O}$ and the simultaneous transformation of the purely imaginary hermitian matrix $\mathbf{S}$ by $\mathbf{O}$ leaves it purely imaginary and hermitian:

$$\mathbf{A} \to \mathbf{A}' = \mathbf{O}^t\mathbf{A}\mathbf{O} = \text{diagonal}, \; \mathbf{O}^t\mathbf{S}\mathbf{O} = \mathbf{S}'. \tag{21}$$

Vanishing diagonal elements in $\mathbf{A}'$ correspond to vanishing rows and columns in $\mathbf{S}'$, and we assume from now on that the operator giving rise to these zeroes are eliminated from the treatment since they can have no matrix elements connecting the ground state to other states. It is thus admissible to form the dynamical matrix

$$\mathbf{\Lambda} = (\mathbf{A}')^{-\frac{1}{2}}\mathbf{S}'(\mathbf{A}')^{-\frac{1}{2}} \tag{22}$$

which also is hermitian, purely imaginary and exhibits eigenvalues in pairs of opposite sign and with conjugate eigenvectors. This is the result quoted above.

Most applications of random phase approximation like schemes employ operator sets $\{b_{rs}\}$ where the two indices refer to one occupied and one unoccupied spin orbital in a reference Hartree-Fock state. The dimension of the set is limited by the capacity of the computer, the investigator, or the sponsor. We may conclude that an increase in the dimension of the set will lead to an increase in the width of the spectrum of the matrix $\mathbf{\Lambda}$[2] and that consequently the smallest excitation energy W_1 will be monotoneously decreasing as a function of the operator basis dimension. If the matrices are calculated from approximate representations for the ground state there is no guarantee that W_1 is an upper bound to the first excitation energy. A ground state approximation which satisfies the annihilator condition (18) and has the energy expectation value $\langle H \rangle$ ensures, however, that $\langle H \rangle + W_1$ is an upper bound for the first excited state.

A partial diagonalization of the matrices **A** and **S** into blocks along the diagonal may be accomplished by means of irreducible tensor operators as a basis rather than the simple b_{rs}'s. An effective procedure requires that the ground state transforms as a one dimensional irreducible representation under the operations of the appropriate group, <u>or</u> the use of an ensemble average over the degenerate states. Tensor operators are defined so that they transform according to the formula,

$$R\, b^j_m R^{-1} = \Sigma_{m'} b^j_{m'} D^j_{m'm}(R), \tag{23}$$

where R is an element of the group and $D^j_{m'm}(R)$ is an element of the representative of R in the j'th irreducible representation. We can then perform the following calculation:

$$\begin{aligned} \langle[[b^j_m,H],b^k_n]\rangle &= M\langle[Rb^j_m R^{-1},H,Rb^k_n R^{-1}]\rangle \\ &= \Sigma_{m'n'}\langle[[b^j_{m'},H],b^k_{n'}]\rangle\, MD^j_{m'm}(R)D^k_{n'n}(R), \end{aligned} \tag{24}$$

where M denotes the operation of taking the invariant mean over the group. The right hand side equals zero unless k is the contragradient representation to j [6]. Accordingly we have that the excitation operators can be classified in terms of the irreducible representations of the group and the symmetry labels of the excited states are then simply deduced from the one dimensional representation of the ground state. The case of a degenerate ground state requires a more extended analysis [7].

EXAMPLES

An area where the propagator calculations have shown their utility is in the determination of radiative lifetimes for excited states. Oddershede and his collaborators [4,8] have presented accurate results for some excited states of the SiO and CH^+ systems and a summary of their results is presented in Table 1.

Knowledge of excitation energies for a range of nuclear displacement offers a possibility for the determination of potential energy surfaces for a set of states in relation to one known surface. These calculations show, as presented by Oddershede [4], errors of 2-16% in the spectroscopic constants. The least accurate values are those referring to the anharmonic behaviour.

The spectral density function (4) determines the frequency dependent polarizability $\alpha(E/h)$ for an electronic system. Goscinski has shown [9] that accurate estimates of the long range London dispersion interaction between molecules may be obtained from a

Table 1: Comparison between calculated and measured radiative lifetimes, (nsec).

system	state	τ(nsec)	
		calc.	experiment
He	$2p(^1P)$	0.591	0.557±0.015[a]
Be	$2p(^1P)$	1.83	1.85±0.07[b]
CH^+	$A^1\Pi(v'=0)$[c]	598[d]	630±50[e]
	$A^1\Pi(v'=2)$[c]	845[d]	850±70[e]
SiO	$A^1\Pi(v'=0)$[c]	31.6[f]	9.6±1.0[g]
	$A^1\Pi(v'=3)$[c]	32.7[f]	9.6±1.0[g]

[a]Burger, J.M., and Lurio, A. (1971). Phys.Rev.A 3, 64.

[b]Martinson, I., Gaupp, A., and Curtis, L.J. (1974). J.Phys.B: Atom.Molec.Phys. 7, L463.

[c]v' is the vibrational quantum number of the $A^1\Pi$ state.

[d]Elander, N., Oddershede, J., and Beebe, N.H.F. (1977). J.Chem. Phys. 66, 2344.

[e]Erman, P. (1977). Astrophys.J. 213, L89.

[f]Oddershede, J., and Elander, N. (1976). J.Chem.Phys. 65, 3495.

[g]Smith, W.H., and Liszt, H.S. (1972). J.Quant.Spectrosc.Radiat. Transfer 12, 505.

Reprinted with permission from J. Oddershede: Polarization Propagator Calculations. Aarhus 1968.

few of the moments of its spectral density function. The approximate moments obtained from random phase type calculations yield expressions which may give useful estimates of the London term [10]. The magnitude of the Faraday effect is dependent on the Verdet constant, which is proportional to $Ed\alpha/dE$. Jørgensen and collaborators have found the proportionality to be excellent for their calculated polarizability and the observed Verdet constant [11] in a calculation for the carbon monoxide molecule.

A general feature of the calculations within the random phase approximation framework is the apparent difficulty of obtaining accurate results for spin dependent features. It often happens that the regular random phase approximation with exchange or, equivalently, the linearized time-dependent Hartree-Fock method exhibits instabilities [12] and these are generally most prominent in the triplet operator matrix problem. Improved versions of the procedure are then required, particularly with regard to the annihilator condition. The instability is present whenever an estimate of the matrix A has negative eigenvalues. There exists then a hermitian operator B, such that

$$\langle[[B,H],B]\rangle<0, \tag{25}$$

and thus we may conclude that the normalized state $|x\rangle = e^{ixB}|0\rangle$, x real, has a lower energy expectation value than $|0\rangle$ for some x different from zero. The reference state is then an insufficient approximation to the ground state. The need for a higher approximation is evident here, but is also present in many other cases which give unreasonably small excitation energies to triplet levels. One manifestation of the difficulties with triplets is the error encountered in the determination of nuclear spin-spin coupling constants [13].

The cost of these propagator calculations compare favorably, as pointed above, with those invoking alternative methods. Timing data for three calculations are given in Table 2.

Table 2: Timing data for propagator calculations at RECAU, Aarhus on a CDC 6400 installation. Compiled from J. Oddershede: Propagator Calculations. Aarhus 1978 with permission.

	Hartree-Fock part		Integral transformation		Propagator	
	integrals	total	"old"	present	matrix	secular
CH^+	1897s	2349s	10591s	1780s	2457s	276s
45 STO	28%	34%		26%	36%	4%
He	277s	520s	9867s	2067s	233s	40s
48 STO	10%	18%		72%	8%	1.5%
SiO	2055s	2762s	7968s	1780s	4358s	502s
45 STO	22%	29%		19%	47%	5%

We see that even at a rather high level of approximation, including two-particle, two-hole corrections, one spends less than half the cost on the propagator part of the calculation. The fraction appears even less when the "old" integral transformation routine based on the Yoshimine algorithm [14] is employed. The presently used method, which is based on a Cholesky decomposition [15], brings the integral transformation into the same range of effort as the other parts.

EPILOGUE.

I have allowed myself to concentrate this exposition on work done at Aarhus, but it is clear that the fruitful collaboration with the groups at Gainesville, Florida, Salt Lake City, Utah, Evanston, Illinois, Waterloo, Ontario, and Uppsala, Sweden has stimulated the development greatly. The experience gained with propagators has supported the contention that they are a resource for theoretical development in chemistry [1] and that they are particularly adequate for the discussion of excitation phenomena. Lectures at this institute by Öhrn and von Niessen will further examplify the theoretical unification and powerful computational tools provided by propagators.

It is a great pleasure to thank professor C.A. Nicolaides for the invitation to lecture at the Advanced Study Institute.

REFERENCES:

1. J. Linderberg and Y. Öhrn, Propagators in Quantum Chemistry (Academic Press, London 1973).
2. J. Linderberg and Y. Öhrn, Int.J.Quant.Chem. 12, 161 (1977).
3. Y. Öhrn and J. Linderberg, Int.J.Quant.Chem. (to be published).
4. P. Jørgensen, Ann. Rev. Phys.Chem. 26, 359 (1975), J. Oddershede, Adv.Quant.Chem. (to be published).
5. D.N. Zubarev, Nonequilibrium Statistical Thermodynamics (Consultants Bureau/Plenum Press, New York 1974) p. 154.
6. P.O. Löwdin, Rev.Mod.Phys. 39, 259 (1967).
7. E. Dalgaard, J.Phys.B:Atom.Molec.Phys. 8, 695 (1975).
8. N. Elander, J. Oddershede, and N.H.F. Beebe, Astrophys.J. 216, 165 (1977).
9. O. Goscinski, Int.J.Quant.Chem. 2, 761 (1968).
10. J. Oddershede, P. Jørgensen, and N.H.F. Beebe, J.Chem.Phys. 63,2996 (1975).
11. P. Jørgensen, J. Oddershede, P. Albertsen, and N.H.F. Beebe, J.Chem.Phys. (to be published 1978).
12. D.J. Thouless, Nucl.Phys. 21, 225 (1960), J. Paldus and J. Cisek, Phys.Rev. A2, 2268 (1970).
13. J. Oddershede, P. Jørgensen, and N.H.F. Beebe, Chem.Phys. 25, 451 (1977).
14. M. Yoshimine, IBM Res.Rep. RJ555 (1969).
15. N.H.F. Beebe and J. Linderberg, Int.J.Quant.Chem. 12, 683 (1977).

REMARKS.

I have allowed myself to concentrate this exposition on work done at Aarhus, but it is clear that the fruitful collaboration with the groups at Gainesville, Florida; Salt Lake City, Utah; Evanston, Illinois; Waterloo, Ontario; and Uppsala, Sweden has stimulated the development greatly. The experience gained with propagators has supported the contention that they are a resource for theoretical development in chemistry [1] and that they are particularly adequate for the discussion of excitation phenomena. Lectures at this Institute by Öhrn and von Niessen will further exemplify the theoretical unification and powerful computational tools provided by propagators.

It is a great pleasure to thank professor C.A. Nicolaides for the invitation to lecture at the Advanced Study Institute.

REFERENCES:

1. J. Linderberg and Y. Öhrn, Propagators in Quantum Chemistry (Academic Press, London 1973).
2. J. Linderberg and Y. Öhrn, Int.J.Quant.Chem. 12, 161 (1977).
3. Y. Öhrn and J. Linderberg, Int.J.Quant.Chem. (to be published).
4. P. Jørgensen, Ann.Rev.Phys.Chem. 26, 359 (1975); J. Oddershede, Adv.Quant.Chem. (to be published).
5. D.N. Zubarev, Nonequilibrium Statistical Thermodynamics (Consultants Bureau/Plenum Press, New York 1974), p. 154.
6. P.O. Löwdin, Rev.Mod.Phys. 39, 259 (1967).
7. E. Dalgaard, J.Phys.B:Atom.Molec.Phys. 8, 695 (1975).
8. R. Elander, J. Oddershede, and N.H.F. Beebe, Astrophys.J. 216, 165 (1977).
9. O. Goscinski, Int.J.Quant.Chem. 2, 761 (1968).
10. J. Oddershede, P. Jørgensen, and N.H.F. Beebe, J.Chem.Phys. 63, 2996 (1975).
11. P. Jørgensen, J. Oddershede, P. Albertsen, and N.H.F. Beebe, J.Chem.Phys. (to be published 1978).
12. D.J. Thouless, Nucl.Phys. 21, 225 (1960); J. Paldus and J. Čížek, Phys.Rev. A2, 2268 (1970).
13. J. Oddershede, P. Jørgensen, and N.H.F. Beebe, Chem.Phys. 25, 451 (1977).
14. M. Yoshimine, IBM Res.Rep. RJ555 (1969).
15. N.H.F. Beebe and J. Linderberg, Int.J.Quant.Chem. 12, 683 (1977).

ACCURACY, TIMING AND GENERAL APPLICABILITY OF THE MRD-CI METHOD

Robert J. Buenker
Lehrstuhl für Theoretische Chemie
Gesamthochschule Wuppertal, 56 Wuppertal 1,
W. Germany

Sigrid D. Peyerimhoff
Lehrstuhl für Theoretische Chemie
Universität Bonn, 53 Bonn
W. Germany

I. INTRODUCTION

The configuration interaction method is generally acknowledged to be a quite useful means of obtaining correlated wavefunctions and energies for the electronic states of atoms and molecules. The theory behind this approach is very straightforward but in applying it to practical problems of chemical interest certain computational problems arise, particularly if it is desired to extend the calculations to the limit of a full CI in a large AO basis. In essence all the CI method involves is the formation of a matrix representation of the non-relativistic electronic Hamiltonian (within the framework of the Born-Oppenheimer Approximation [1]), followed by solution of the associated secular equation (diagonalization).

If a basis consisting of a complete (orthonormalized) set of the eigenfunctions of the electronic Hamiltonian (H_{el}) is employed for this purpose it is clear that such a matrix representation would be diagonal by virtue of the Schrödinger equation. Use of any other basis for the same linear space leads to a generally non-diagonal H_{el} matrix but upon applying the diagonalization procedure in this case it follows that the same eigenvalues and characteristic vector spaces must result as before (both for discrete and continuous solutions). For

Cleanthes A. Nicolaides and Donald R. Beck (eds.), Excited States in Quantum Chemistry, 45–61.

practical computations one simply limits the number of n-electron basis functions and hopes that such a simplification will not produce errors of undesirable magnitude relative to the corresponding exact results.

The manner in which the configuration space is restricted (truncated) is thus clearly of paramount importance in designing CI calculations and must be done with three goals in mind: a) to keep the level of accuracy of the computations as high as possible, b) to put a minimal number of restrictions on their range of applicability and c) to nonetheless allow for the design of a theoretical treatment which is economically feasible for molecular systems of reasonable size. These objectives are obviously closely interrelated and hence under the circumstances it is difficult to come up with a truly optimal design for a general CI treatment, but in what follows a method will be described [2-4] which has been tested over a period of the last four years and which appears to satisfy the three criteria of accuracy, general applicability, and computational simplicity to a good degree. The present work will be concerned almost entirely with examining the design of this CI procedure from a theoretical point of view, but in three subsequent papers by the authors at this Institute a more general survey of the actual results of such calculations will be undertaken.

II. DETAILS OF THE MRD-CI PROCEDURE

The basic idea in CI calculations is to start out with a fixed AO basis and to form from such one-electron functions (or some transformed set thereof) Slater determinants of spin orbitals (products of spatial orbitals with α or β spin), i.e. the n-electron basis functions needed for the H_{el} matrix representation. Because of the antisymmetry of the electronic wavefunctions no spin orbital may appear more than once in a given product function, so it is easy to show that for m (spatial) AO's and n electrons a total of $\binom{2m}{n}$ distinct Slater determinants can be formed, corresponding to the full CI space in this AO basis. It is also not difficult to see that the eigenvalues and characteristic vector spaces for such a full CI are invariant to a transformation of the one-electron basis, e.g. from AO's to MO's or natural orbitals.

Since only a subset of the full CI space can be handled in practice, in which case the choice of the one-electron basis becomes a factor in determining the results of the calculations, the first of a series of decisions connected with the design of CI methods must be dealt with. Upon noting that a single covalent VB configuration (structure) gives a better description of the H_2 ground state than does a single determinant using (orthonormal) MO's, one might conclude that it is simply best to use the AO's directly in forming the n electron basis. This procedure has the advantage that electron repulsion integrals over AO's can be used directly in the computation of H_{el} matrix elements but it also has a very critical disadvantage, namely that the resulting determinantal species are themselves not mutually orthonormal; as a result the calculation of the H_{el} matrix in a VB formulation is still very complicated. By choosing an orthonormal one-electron basis (MO's or NO's) the latter problem is greatly alleviated as is well-known [5], but only at the expence of carrying out a four-index integral transformation and of having to deal with generally larger sets of determinantal functions than in a non-orthogonal approach. Experience indicates that the speed with which the H_{el} matrix can be constructed is such an important factor in this type of treatment that the advantage of the orthonormal one-electron basis functions in this regard are decisive, and it is for this reason that the great majority of CI calculations being done at the present time employ such basis sets. Such considerations naturally do not rule out the possibility that a competitive VB-CI method using unrestrictively non-orthogonal orbitals may yet be developed, but they do give an accurate indication of the relative importance of the difficulties encountered in the two approaches in practical applications.

Having opted for the choice of an orthonormal basis it is very important to be able to deal with a large number of configurations in a rapid manner. Since the direct solution of secular equations of order $\binom{2m}{n}$ is out of the question for even moderately large systems when reasonably flexible AO bases are employed, it is useful to note that for most types of electronic states a single-configuration (SCF or Hartree-Fock) wavefunction is already capable of giving a reliable description of the energy and properties of a given system. The problem with using this type of wavefunction is that it does not provide for a balanced representation for all states of interest and for a suitably broad range of nuclear geometries, but at least it offers a good starting point for carrying out CI calculations. Since the determinantal functions are orthonormal it also follows that H_{el} matrix elements vanish between species which differ by more than a double spin orbital substitution, and as a result it is natural to expect that the most important secondary configurations in a CI expansion for a given state should differ from the corresponding SCF species by at most a double excitation. Taken together these observations suggest that use of a CI space which contains the SCF configuration and all species related to it by a single or double orbital substitution is an effective means of introducing correlation into electronic wavefunctions.

Since the number of doubly excited configurations is roughly proportional to $n^2.(2m-n)^2$ it is possible to carry out such double-excitation CI treatments for fairly large basis sets and numbers of electrons but it is unfortunately not difficult to find situations in which such a theoretical treatment constitutes a rather poor approximation. The problem is that indirect interactions can take place because a more highly excited configuration has a non-zero H_{el} matrix element with one of the singly or doubly substituted species which in turn undergoes a <u>strong direct interaction</u> with the SCF or leading term in the CI expansion. To account for such effects it is necessary to bring more highly excited configurations into the CI space, but it is very impractical to deal with entire classes of such excitations. Thus while the order of the double-excitation space varies as roughly the fourth power of $\sqrt{mn}$, the number of triple and quadruple excitations varies as the sixth and eighth power of this quantity respectively.

This difficulty forces what may be called the second major decision in designing such calculations, namely how to safely restrict the excitation classes included in the CI space. Since the problem gets quickly out of hand by adding whole excitation classes another solution must be found, but once the reasons for the importance of configurations with such a high degree of excitation are recognized there is reason to believe that all that is really needed is to include species of this type which undergo a strong indirect interaction with the leading term because of the presence of key secondary configurations in the CI state of interest. This analysis suggests a more controlled approach for including more highly excited species, namely to add all singly and doubly substituted configurations relative to a series of the most important terms appearing in the CI expansions of each of the desired eigenfunctions. For p such reference configurations the dimension of the corresponding CI space would vary as $p \cdot n^2(2m - n)^2$, i.e. fourth order in $\sqrt{mn}$.

In practice one hopes to keep p as small as possible but this eventuality is generally ensured by constructing the determinantal functions from the SCF MO's of some representative state. If secondary configurations are quite important an NO transformation may prove useful in reducing the size of the reference or generating set but further discussion of this point will be reserved until later when more detailed computational procedures are outlined. For now the main point is that by systematically increasing the size of the reference set one can build toward the full CI in a controlled manner and furthermore by simultaneously improving the AO basis the true solutions of the Schrödinger equation can be approached as closely as desired thereby. Whether the latter point is of practical significance needs to be discussed in terms of actual calculations. In any event in view of the definition of this type of theoretical treatment it is appropriate to refer to it as the multi-reference double-excitation CI (MRD-CI) method*.

---

*Other alternatives that should be mentioned involve the use of a large reference set and only the singly excited configurations related thereto [9], or else a single reference species, all singly and doubly excited configurations and a restricted number of triple and quadruple substitutions (determined by narrowing the list of orbitals which can be occupied in the more highly excited configurations) [10].

The difficulty remains, however, that even after truncating the full CI space in this way the number of configurations remaining is still quite large, typically in the 10000-50000 range for relatively simple problems and easily up to 500000 if AO basis sets containing 70 - 100 functions are employed with several reference configurations and a moderately large number of correlated electrons (15 - 25). Since present-day methods [6-8] for the solutions of secular equations of general constitution are conveniently applied for orders up to 5000 - 10000 it is obvious that additional simplifications have to be made. One means of doing this is to artificially restrict the magnitudes of p, m and n in designing such a CI treatment, but before relying on this eventuality entirely it is well to consider whether the entire MRD-CI space needs to be treated explicitly via a large secular equation to achieve the main objectives sought in the calculations.

The possibility of considering only a selected subset of some generated space thus represents the third major decision which must be made in the organization of CI methods. For the present purposes it is best to side-step this issue by noting that it is simply more general to employ a configuration selection, since obviously a null selection can be achieved with such methods as well by simply setting some threshold value equal to zero or by circumventing the routine entirely.

Once the idea of making a configuration selection is accepted, there is then the obvious question of how to go about ordering the configurations in an effective way. A number of variations on this theme are available in the literature but to simplify matters it is helpful to categorize them according to whether they involve comparison of secular equation results with and without test configurations [2,11] or some form of perturbation theory [12] on the one hand, and on the other whether an individualized [2,12] or a group selection technique [11] is employed thereby. Since the evaluation of the relative importance of the configurations depends on how the selection is carried out it is necessary to consider what criteria are important in judging the various possibilities. Before turning to this point, however, an additional aspect of the calculations should be considered, namely whether one should be satisfied with results for <u>selected</u> MRD-CI spaces even when such procedures are employed.

Since a selection method generally assigns an estimated energy-lowering capability ΔE_i to each test configuration it is natural to ask to what extent the sum over such quantities for all unselected species at a given threshold value T approximates the true energy lowering which would result from abandoning the selection entirely. In practice it is found [2,3] that the family of curves defined by $E(T) + \lambda\Sigma_i \Delta E_i(T)$, where λ is a simple linear parameter, has useful properties, for example. Since by definition the sum term in such expressions vanishes for all finite values of λ, it follows that all such curves converge to the same value at T = 0. As a result an energy extrapolation scheme can be employed whenever a selection is carried out, whereby the accuracy thereof depends almost exclusively on the magnitude of the smallest threshold value considered; in other words, the less drastic the selection the more accurate the prediction of the eigenvalue at the limit for the entire MRD-CI space; a recently completed example for the ethylene $^1(\pi,\pi^*)$ state demonstrates this point quite well (Table 1). Furthermore it is found that property results for the individualized type of selection scheme used in this work are quite stable with respect to lowering of the threshold value [4]. Basically when an energy extrapolation is carried out one finds that the choice of the associated selection scheme is far less critical than might otherwise have been thought; for the most part such considerations have little more effect than to produce a different ideal value of the parameter which gives the most nearly constant curve in the family of $E(T) + \lambda\Sigma_i \Delta E_i$ species. From this point of view it can be concluded that the selection should simply be done in a systematic fashion and as rapidly as possible.

In conclusion in answer to the five questions raised in the foregoing discussion an MRD-CI method with an individualized configuration selection scheme and an associated energy extrapolation procedure has been devised. In the last analysis it is probably true that it can never really be proven that one choice of CI techniques is truly optimal but at the same time it should not be concluded therefrom that a method cannot be found which has clear advantages over any other procedure with the same function. In trying to explore the various possibilities as well as any other type of quantum mechanical treatment it is well to keep in mind all three

Table 1 Comparison of extrapolated (T = 0) energy and property results of several $^1B_{1u}$ π— π^* states of ethylene obtained for various selection threshold values T_{min} (using a four reference-two root MRD-CI treatment employing natural orbitals; all values in atomic units).

	$T_{min}=5x10^{-5}$	$T_{min}=2x10^{-5}$	$T_{min}=1x10^{-5}$	$T_{min}= 2x10^{-6}$	T = 0
Order of secular equation[a)]	1334 (11.2%) (1204,1058,951 904)	2285 (19.2%) (1821,1527, 1334,1204)	3216 (27.0%) (2285,1821, 1527,1334)	5446 (45.8%) (2942,2176, 1773,1485)	11896 (100%)
Number of (m_s=0) determinants in largest secular equation	4450	7898	11302	19454	43180
E(1^1B_{1u}) extrapolated to T = 0	-77.9240	-77.9236	-77.9246	-77.9248	-77.925211
$<\psi_1\|\Sigma_i x_i\|\psi_1>$[b)]	29.855	29.859	29.699	29.632	29.476
E(2^1B_{1u}) extrapolated to T = 0	-77.8685	-77.8672	-77.8680	-77.8683	-77.86915
$<\psi_2\|\Sigma_i x_i\|\psi_2>$[b)]	30.116	30.165	30.347	30.496	30.687
Total CPU time for CI[c)] (IBM 370/168)	1 min 57 s	3 min 43 s	5 min 35 s	11 min 58 s	35 min 27 s
CPU time for property calculation	16 s	30 s	47 s	1 min 50 s	7 min 6 s

a) Numbers in parentheses correspond to orders for additional secular equations for the extrapolation procedure.

b) The C_2H_4 molecule is located in the yz plane; property results correspond to the CI wavefunction obtained from T_{min}

c) Includes configuration generation and selection, construction of Hamiltonian matrix, solution of secular equations for two roots and corresponding energy extrapolation.

of the key objectives in such calculations of economy (timing), accuracy and generality. In what follows these points will be discussed in this order, with primary emphasis on the question of timing since this factor ultimately determines to what extent the CI method can be applied, i.e. how large an AO basis can be conveniently employed, as well as how many reference configurations and how small a threshold value can effectively be used in a given situation. Since a full CI in a nearly complete AO basis must closely approach the results obtainable at the Schrödinger equation limit, such timing considerations ultimately play a decisive role in determing how satisfactorily the other two goals cited above can be achieved in practical computations.

III. TIMING OF THE MRD-CI CALCULATIONS

To carry out the type of CI treatment outlined above the following computation steps are necessary:

1) Calculation of one- and two-electron integrals over AO's;

2) Transformation of the AO integrals over symmetry orbitals* (done for each AO integral case);

---

*In the present work only Abelian symmetry orbitals are used with coefficients of $\pm$ 1 but there is also the possibility of carrying out a subsequent transformation from these simple species to any desired set of full point group symmetry orbitals.

---

3) Calculation of SCF MO's for both closed- and open-shell configurations (most conveniently carried out using the results of step 2, i.e. with symmetry;

4) Transformation of the symmetry orbital integrals to a basis of SCF MO's (or a set of natural orbitals; see step 8 below);

5) Generation of the MRD-CI space and selection of configurations (calculation of ΔE_i for randomly ordered sets of test species);

6) Construction of the H_{el} matrix for a selected portion of the MRD-CI space (for some minimal threshold value T_{min});
7) Diagonalization of the above matrix for the desired roots at a series of values of $T \geq T_{min}$ and energy extrapolation using these results and the $\Sigma_i \Delta E_i(T)$ counterparts from step 5;
8) Construction of the CI first-order density matrix and diagonalization to obtain natural orbitals for use in step 4.

In addition if properties associated with the resulting MRD-CI wavefunctions are desired two other steps are required:

9) Calculation of state properties using the density matrices of step 8;

10) Calculation of transition density matrices for various pairs of states for use in computing oscillator strength values.

The AO integrals are generated in our CI package by using the IBMOL Vb program of Popkie and Clementi, while all other routines have been developed at the University of Bonn by the authors and coworkers*. Many details concerning the timing of

----

*The special diagonalization routines have been written by W. Butscher and E. Kammer [8] and the overall testing of the program system has been done in close collaboration with S. Shih.

----

these routines may be found in a recent paper [4] and only a short summary of these results will be given in the present work.

The CI part of the procedure begins with step 5 and there it has been found that symbolic generation of the MRD-CI space proceeds at the rate of 2000 - 10000 SAF's (symmetry-adapted functions) per CPU sec with an IBM 370/168 single processor system; the large variation is due to the fact that more checking for duplicate configurations needs to be made if there are many reference species. The corresponding selection procedure requires the calculation of MN H_{el} matrix elements, where M is the number of reference SAF's and N is the number of generated SAF's. In practice roughly 3000 such quantities can be calculated in a CPU sec (including the time for secular equation solutions as overhead), or about 200000 per CPU min (the rate is 500000/min on a CDC 7600 machine). As a result it is possible to handle MRD-CI spaces of dimension 500000 with several reference species in 5-10 CPU min on the IBM 370/168; furthermore very little IO processing is required for this part of the calculations. The manner in which the reference configurations are choosen for generating the MRD-CI space is clearly of importance in this discussion but this choice is most conveniently made on the basis of the results for larger secular equations.

The next step then involves calculating the H_{el} matrix for a selected portion of the generated space; this procedure generally requires more IO processing but this is still by no means a dominant feature of the calculations. The CPU times for this step run typically as: 1 min for order 1000, 5-6 min for order 4000 and 24 min for the largest case yet attempted (order 11896; see Table 1). Diagonalization times for the latter case were 4 CPU min/root and in general are less than what is needed for the matrix element generation. Overall the total time for the CI procedure is found to increase as about $\eta^{4/3}$, where η is the number of selected SAF's, i.e. considerably less sharply than η^2. The total CI times are also observed to vary as roughly $m^{5/2}$ for an increase of basis set size of from 40 to 72 AO's.

To choose the reference configuration set the simplest procedure is to run a single-reference CI for a high selection threshold value (50 - 100 μhartree) with parent SCF MO's (or at least some closely related set). Secondary configurations which show up with large expansion coefficient (typically ≤ 0.05) are then added to the original reference set. If the new configurations are dominant, i.e. the SCF description is poor, it is advisable to go a step further and generate NO's for the state in question through diagonalization of its associated first-order density matrix (step 8 above [14,15]). Even for the order 11896 case discussed earlier such a procedure takes only 3.5 CPU min per density matrix and for order 1000 species the time reduces

Table 2 Timing results for sample calculations for planar C_2H_4 in a basis containing 84 contracted functions. Times refer to an IBM 370/168 single processor system.

Computational Procedure	Remarks	Timing
AO Integral Generation	IBMOL Vb	28 min 45 s
Symmetry Transformation	Needed only once	20 min 20 s
SCF Calculation	Closed-shell Ground State (20 iterations)	2 min 38 s
SCF calculation	Open-shell $^1(\pi,\pi^*)$ State (23 iterations)	3 min 2 s
MO Transformation	Needed for each set of MO's or NO's	4 min 34 s
Small CI Case (Excited States) 7 reference-3 root treatment, T_{min} = 1.0x 10^{-4} hartree, $^1(\pi,\pi^*)$ SCF MO's	Generation of Configurations (183873 SAF's)	2 min 20 s
	Configuration Selection	7 min 15 s
	H_{el} Matrix Construction (1143 SAF's)	47 s
	Diagonalization (Three Roots)	57 s
	First-order Density Matrix Construction (Three Cases) + NO generation	25 s
Small Ground State CI Case	Generation of Configurations (39904 SAF's)	29 s
3 reference-1 root treatment, T_{min} = 1x10^{-3} hartree, $^1(\pi,\pi^*)$ NO's from above	Configuration Selection	52 s
	H_{el} Matrix Construction (672 SAF's)	34 s
	Diagonalization (One Root)	17 s

Large CI Case (Excited States) 7 reference-3 root treatment, $T_{min} = 3x10^{-5}$ hartree, $^1(\pi,\pi*)$ NO's from above	Generation of Configurations (183873 SAF's)	2 min 20 s
	Configuration Selection	7 min 15 s
	H_{el} Matrix construction (3675 SAF's)	4 min 13 s
	Diagonalization (Three Roots)	3 min 55 s
	First-order Density Matrix Construction (Three Cases)	1 min 45 s
	Properties Calculation (11 operators)	18 s
	Transition Moment Calculation (Three Cases)	1 min 50 s
Large CI Case (Excited State) 6 reference-1 root treatment	Generation of Configurations (167168 SAF's)	1 min 20 s
	Configuration Selection	5 min 42 s
	H_{el} Matrix Construction (4968 SAF's)	9 min 43 s
	Diagonalization (One Root)	3 min 9 s
	First-order Density Matrix Construction (One Case)	53 s
	Properties Calculation (11 Operators)	18 s
	Transition Moment Calculation (One Case)	56 s

to about 10 CPU sec (an order 1720 case require 20 CPU sec, for example). The normal procedure is then to run a second calculation with a larger reference set for a suitably small T_{min} value and to use NO's if they are called for. In principle one must allow for the possibility of additional macroiterations of this type but in practice none is usually needed. One-electron property calculations can be carried out very rapidly once the first-order density matrix is at hand and the corresponding transition moment results are obtained with similar expenditures of CPU time (step 10).

The speed with which the pre-CI steps can be done depends to a fair degree on the amount of symmetry present in H_{el}. For N_2 and a basis of 72 functions [16] the AO integral generation requires 12.5 CPU min on the 370/168, while the symmetry transformation takes only half as long (5.7 min); subsequent MO or NO transformations then use only 2 CPU min. For 86 AO's and the same system 10 CPU min are needed for the symmetry transformation, while for planar ethylene and m = 84 the time increases to 20.2 min. By lowering the symmetry one generally notes a decrease in the symmetry transformation times but an increase in the corresponding MO transformation time. As long as there is a moderate amount of symmetry at hand, basis sets of 100 AO's can thus be conveniently handled in the CI treatment with these programs. Finally a summary of timing results for a complete run for the 84 function ethylene basis is given in Table 2 for further consideration.

IV. ACCURACY AND GENERALITY

The timing results discussed above allow a large amount of flexibility in choosing the level of theoretical treatment to be employed. For high accuracy large reference sets are called for as well as high-quality AO basis sets. For small systems such as He_2^+ [17] or HeH_2 [18] it has been possible to to obtain 95 % of the total correlation energy with this method and to generate very accurate potential surfaces for these systems. In the case of BH it has been found that for the same AO basis the MRD-CI method (85 CPU sec) obtains an energy of -25.2273 hartree (T = 0.5 μhartree, five reference species) compared to -25.22153 hartree for PNO-CI [4,19]. The corresponding extrapolated energy is -25.2275 hartree for the MRD-CI compared to a CEPA (non-variational) estimate of -25.2259 hartree [19]. Similar results for N_2 calculations have also been obtained [4,20].

Another indication of the accuracy of the MRD-CI method is the fact that when relatively large reference sets are used it is found that the extrapolated energy results do not vary strongly with the choice of MO's or NO's in the calculations, as one would expect to be the case once the full CI limit is closely approached. In the long run,however, the best test of accuracy is made by comparing the calculated results with experiment but this topic will be taken up in detail in the authors' following contributions in this Institute.

As for the generality of the method one again expects that as the full CI limit is approached for a reasonably flexible basis that no restrictions of this nature should be present. One of the severest tests which can be put forward for this point is the calculation of the dissociation energy of the N_2 ground state, with the SCF or Hartree-Fock value thereof showing an error of nearly 5 eV. Even with a relatively large basis (72 AO's) containing two separate d functions on nitrogen it is found that an error of 0.6 eV remains in this D_o value once MRD-CI calculations are carried out [21], but at least a large percentage of the correlation error is removed thereby. The extrapolation scheme is very important in achieving this result, however, as is seen by the fact that for relatively low T values (10 μhartree) the multiplets for the various asymptotes of the ground state dissociation (at R = 20 a_o) are separated by as much as 2.0 eV, whereas after extrapolation the true physical situation of essentially equal energies for all such species is closely approximated. Similarly clear examples for the need of going beyond the results of truncated CI calculations can be found in the treatment of the vertical spectrum of ozone [22], for which errors in transition energies of more than 1.0 eV are obtained at conventional T values if no extrapolation is effected.

In the remaining lectures by the authors many more concrete examples will be discussed which indicate that the theoretical method outlined above is in fact applicable to all types of electronic states, including inner-shell and shake-up excited species, regardless of multiplicity, net charge or nature of the geometrical conformation. To conclude this paper, however, it is perhaps well to mention that independent evidence for the overall reliability of the MRD-CI procedure can be obtained by observing how stable its results are relative to further expansion of the AO basis and reference configuration sets, as well as upon lowering the magnitude of the selection threshold. Whenever tests of this sort have been carried out, especially by the time that DZP AO basis sets have been employed, it has invariably been found that a high degree of stability is characteristic of this theoretical method. Furthermore the fact that results obtained at this level are quite generally in very good agreement with the corresponding experimental data, as will be demonstrated in the following papers, bodes well for the future of this type of CI calculation.

References

1. M. Born and E. Oppenheimer, Ann. Phys. 84,457 (1927)

2. R.J. Buenker and S.D. Peyerimhoff, Theor. Chim. Acta 35, 33 (1974)

3. R.J. Buenker and S.D. Peyerimhoff, Theor. Chim. Acta 39, 217 (1975)

4. R.J. Buenker, S.D. Peyerimhoff and W. Butscher, Mol. Phys. 35, 771 (1978)

5. M. Tinkham, "Group Theory and Quantum Mechanics", McGraw-Hill Book Co., Inc., New York (1964)

6. I. Shavitt, C.F. Bender, A. Pipano and R.P. Hosteny, J. Comput. Phys. 11, 90 (1973)

7. E.R. Davidson, J. Comput. Phys. 17, 87 (1975)

8. W. Butscher and W.E. Kammer, J. Comput. Phys. 20, 313 (1976)

9. H.F. Schaefer III, J. Chem. Phys. 54, 2207 (1971)

10. L.B. Harding and W.A. Goddard III, J. Chem. Phys. 67, 1777 (1977)

11. Z. Gershgorn, I. Shavitt, Int. J. Quantum Chem. 2, 751 (1968)

12. J.L. Whitten and M. Hackmeyer, J. Chem. Phys. 51, 5584 (1969)

13. R.J. Buenker and S.D. Peyerimhoff, Chem.Phys. Letters 29, 253 (1974)

14. C.F. Bender and E.R. Davidson, J. Phys. Chem. 70, 2675 (1966)

15. K.H. Thunemann, J. Römelt, S.D. Peyerimhoff and R.J. Buenker, Intern. J. Quantum Chem. 11, 743 (1977)

16. S.K. Shih, W. Butscher, R.J. Buenker and S.D. Peyerimhoff, Chem. Phys. 29, 241 (1978)

17. J.G. Maas, N.P.F.B. van Asselt, P.J.C.M. Nowak, J. Los, S.D. Peyerimhoff and R.J. Buenker, Chem. Phys. 17, 215 (1976)

18. R. Römelt, S.D. Peyerimhoff and R.J. Buenker, Chem. Phys. in press

19. W. Meyer and P. Rosmus, J. Chem. Phys. 63, 2356 (1975)

20. R. Ahlrichs, H. Lischka, B. Zurawski and W. Kutzelnigg, J. Chem. Phys. 63, 4685 (1975)

21. W. Butscher, S.K. Shih, R.J. Buenker and S.D. Peyerimhoff, Chem. Phys. 52, 457 (1977)

22. K.H. Thunemann, S.D. Peyerimhoff and R.J. Buenker, J. Mol. Spectry (1978)

CALCULATION OF VIBRATIONAL WAVEFUNCTIONS AND ENERGIES USING MRD-CI TECHNIQUES

Robert J. Buenker

Lehrstuhl für Theoretische Chemie
Gesamthochschule Wuppertal, 56 Wuppertal 1
W. Germany

Sigrid D. Peyerimhoff and Miljenko Perić

Lehrstuhl für Theoretische Chemie
Universität Bonn, 53 Bonn, W. Germany

I. INTRODUCTION

The calculation of electronic wavefunctions is an important objective in quantum chemistry but it is evident that if even this could be accomplished without any approximation it would still not be sufficient to explain many types of molecular structure phenomena observed experimentally. In the spectra of molecular systems, for example, the intensity associated with a given electronic transition is often spread over a wide range of wavelength and in this instance the most that one can hope to obtain from a calculation of the electronic energy for the participating states at a single geometry is the approximate location of an absorption (or emission) maximum found therein. To go beyond the simple calculation of vertical spectra it is clearly necessary to introduce at least vibrational motion into the theoretical treatment and this objective is most easily accomplished using the Born-Oppenheimer Method [1], or, as it is more commonly referred to among spectroscopists, the Franck-Condon Approximation [1]. In this paper a short review of this theoretical procedure will be given and results of some ab initio calculations of this type will be discussed.

Cleanthes A. Nicolaides and Donald R. Beck (eds.), Excited States in Quantum Chemistry, 63–77.

II. CALCULATION OF VIBRATIONAL WAVEFUNCTIONS

The basic formulation of the Born-Oppenheimer is well-known and will only be briefly reviewed in this work. In essence the total molecular (or global) Hamiltonian can be divided into two parts. One is the familiar electronic Hamiltonian H_{el}, containing electronic kinetic energy and all nuclear and electronic attraction and repulsion terms, while the other is simply the nuclear kinetic energy H_{nuc}:

(1) $$H = H_{el} + H_{nuc}.$$

If one neglects the fact that the electronic wavefunctions (eigenfunctions of H_{el}) are dependent on the choice of nuclear conformation it is possible to solve the corresponding Schrödinger equation by a simple separation of variables, with approximate eigenfunctions written as products of electronic wavefunctions and functions of the nuclear coordinates only. It is important to note that this approximation is equivalent to saying that all derivatives of the electronic wavefunctions relative to the nuclear coordinates (i.e. coming from the H_{nuc} term) are of vanishing magnitude. In reality such an assumption is generally quite far from the truth; in fact, if it were the case no vibrational calculations would be possible at all, as will be discussed below. In justifying this procedure they simply noted the obvious, namel that such nuclear derivatives are always multiplied with reciprocal nuclear masses, and since electronic masses are several thousand times smaller than those of protons and neutrons, it is nevertheless a good idea to neglect such terms under normal circumstances.

Having made this approximation it is clear that the problem is separable into two simpler types of Schrödinger equations, one involving exactly the same H_{el} operator as discussed in connection with the CI treatment, and the other having the form, in the simplest case of a diatomic molecule, of:

$$(-E_{el}(R) + \frac{1}{2\mu}\frac{d^2}{dR^2})\psi_v(R) = -E_v(R)$$

In the latter expression R is the bond distance, $E_{el}(R)$ is the eigenvalue of H_{el} obtained by pointwise solution of the electronic Schrödinger equation, μ is the reduced mass and ψ_v and E_v are the vibrational eigenfunction and eigenvalue (usually referenced to the electronic energy potential minimum)*.

---

*Analogous rotational and translational equations can also be derived in this manner, with the simplifying feature that the electronic energy is constant with respect to these nuclear variables.

The method of solution used in this work for the vibrational equations is very similar to that described earlier for the purely electronic motion. A fixed basis of vibrational functions is assumed (usually Hermite polynomials or Fourier series) and a matrix representation of the vibrational potential and kinetic energy operator is formed; for this purpose a polynomial fit of $E_{el}(R)$ is usually obtained. The desired approximate eigenvalues and eigenvectors are then obtained by diagonalization, exactly as in the CI method. More details of this procedure may be found in an earlier paper by the authors [3]. When more than one vibrational coordinate is present the simplest approach is to assume that there is effectively no coupling between such modes; this assumption reduces the vibrational problem to a series of one-dimensional Schrödinger equations, whereby the total vibronic wavefunction is then taken to be a product of a given electronic eigenfunction with appropriate vibrational functions of each coordinate:

$$\Psi(r;R_1...R_q) = \Psi_{el}(r;R_1,R_2...R_q)\times\psi_{v_1}(R_1)\cdot\psi_{v_2}(R_2)\cdots\psi_{v_q}(R_q) ,$$

where v_1 ..... v_q are the corresponding vibrational quantum numbers. If cross terms in the potential and kinetic energy expressions are not neglected the procedure must be somewhat more complicated but as long as the number of vibrational coordinates is not very large practical methods are available for improving the theoretical treatment in this manner [4].

To calculate properties for such vibronic functions it is only necessary to evaluate matrix elements involving appropriate operators; in the case of transition probabilities two different vibronic species are needed for each vibrational transition in the band system, using the expression:

$$\langle\vec{r}\rangle_{e'v_1'...v_q'e''v_1''...v_q''} = \int \Psi_{el}'(r;R_1...R_q)\times\psi_{v_1}'(R_1)...\psi_{v_q}'(R_q)\times \\ \times\vec{r}\Psi_{el}''(r;R_1...R_q)\times\psi_{v_1}''(R_1)...\psi_{v_q}''(R_q)\, d\tau_{el}d\tau_{v_1}...d\tau_{v_q}$$

To evaluate these integrals it is convenient to first integrate over electronic coordinates and obtain what is conventionally referred to as the electronic transition moment $\vec{R}_{e'e''}$ as a function of all desired vibrational coordinates:

$$\vec{R}_{e'e''}(R_1 \ldots R_q) = \int \Psi_{el}'(r;R_1 \ldots R_q)\vec{r}\Psi_{el}''(r;R_1 \ldots R_q)d\tau_{el} \,.$$

Subsequent integration of the product of this function (which is usually expressed as a polynomial fit in terms of the vibrational coordinates) with appropriate combinations of ground and excited vibrational wavefunctions then gives the desired vibronic transition moment results. The corresponding intensity or oscillator strength distribution for the entire electronic band system is then obtained using the relationship:

$$f_{e'v'e''v''} = \frac{2}{3} |\langle\vec{r}\rangle_{e'v'e''v''}|^2 \cdot \Delta E$$

with all quantities expressed in atomic units (ΔE is the corresponding vibronic transition energy). As long as the underlying assumptions of the Franck-Condon Approximation remain valid and suitably accurate electronic, vibrational functions are employed in the integral evaluations, such a procedure should be capable of delivering a very detailed prediction of the appearance of band systems*, observed in molecular

*At least those which are not spin-forbidden or are only quadrupole-allowed.

spectra.

III. EXAMPLES OF CALCULTATIONS FOR INTENSITY DISTRIBUTIONS

Beginning with a study of the ethylene spectrum in 1972 [3] a number of calculations of the type outlined above have been carried out by the authors and coworkers and the results of these investigations will now be discussed in some detail. A more thorough discussion of such findings can be found in the original literature in each instance.

A. Oxygen

The first example to be considered is the O_2 Schumann-Runge band system and neighboring transitions [5]. Potential curves have been generated for the ground and various excited states of this system using a moderately large AO basis and the CI method discussed in the first lecture in this series; and from these data it is apparent that several strongly avoided crossings occur in the 9-10 eV region of this spectrum as a result of the mixing of various valence states, including the $B^3\Sigma_u^- \pi_u^3\pi_g^3$ species known to be the upper state in the Schumann-Runge transition, and certain Rydberg states of like symmetry (such as the $^3\Sigma_u^-\pi_u^4\pi_g3p\pi_u$ component). When such potential crossings occur it is necessary to scrutinize the assumptions underlying the vibrational treatment discussed above, but since the calculated minimal splitting of the two adiabatic $^3\Sigma_u^-$ curves is quite large (about 1 eV) there is good reason to expect that this aspect of the calculations is quite satisfactory. After taking account of the avoided crossing (Fig. 1 of Ref. [5]) and calculating the associated vibrational energies and eigenfunctions it is found that the vibrational frequencies of the two $^3\Sigma_u^-$ states are quite different (758 cm^{-1} for the lower and 2970 cm^{-1} for the upper). The first value is in satisfactory agreement with the well-known experimental frequency of the Schumann-Runge bands (709 cm^{-1}) and the location of the corresponding 0-0 transition also agreed quite well with the measured spectrum (6.07 eV calc. vs. 6.12 eV exptl.).

The much higher frequency for the second $^3\Sigma_u^-$ state was not known, however, although the calculations of Yoshimine et al. [7] had recently led to much the same result. Since the ground state frequency of O_2 is only about half as large (1580 cm^{-1} exptl. vs. 1621 cm^{-1} calc.) the high excited state value was at first received with a certain amount of skepticism; in one of the more colorful formulations of this viewpoint it was argued that such a result was extremely unlikely since it would be equivalent to attributing triple bond character to an O_2 excited state. Nonetheless when the intensity distribution was calculated in the manner described above three strong vibrational transitions to the $2^3\Sigma_u^-$ upper state were predicted at 9.93, 10.29 and 10.63 eV*, which coincided almost exactly with measured

---

*Qualitatively similar results were also reported by Yoshimine et al. [7].

(and previously unassigned) absorption peaks at 9.96, 10.28 and 10.57 eV respectively (known in the literature as longest band, second band and third band). Furthermore the calculated intensities for these features agreed acceptably well with the measured oscillator strengths of Huebner et al. [8], particularly in the case of the two strongest species.

The intensity distribution calculated for the vibrational transitions to the $B^3\Sigma_u^-$ state also agreed quite well with experiment,whereby in this case it is shown to be quite important to evaluate the transition moment explicitly as a function of internuclear distance rather than assume constancy for this quantity for all geometrical conformations; this situation is easily understandable in the present case since a strong interaction between a valence and a Rydberg state is involved thereby and the intensities of transitions to these two types of states are quite different. In addition another of other weak vibrational transitions to upper $^3\Pi_u$ were found in the calculations which also appear to fit in quite well with other unassigned features in the O_2 spectrum. Hence altogether it appears that the ab initio calculations give a quite accurate description of the spectrum in question and are able to explain a number of experimental observations thereof which hitherto had not been understood.

B. HSO and SOH

The first triatomic molecule to be discussed in the HSO system, which was first identified as a product of the reaction between O_3 and H_2S [9]. An emission spectrum was obtained in the latter experiments and considerable vibrational structure was observed. The first two electronic states of such a 13-valence-electron system are well known to be the $^2A''$ and $^2A'$ components of a $^2\Pi$ linear species and from the location of the measured spectrum it was assumed that a transition between these states was involved.

To study this question further potential curves for the two stretching and the ∢HSO bending vibrations in both electronic states were carried out, in this case with SCF calculations [10]. Vibrational energies and wavefunctions were calculated under the assumption of no coupling among the various modes and in addition it was simply decided to compute the intensity distribution without taking account of the variation of the electronic transition moment with nuclear geometry. The calculated 0-0 transition energy is 1.56 eV in this treatment compared to the corresponding measured result of 1.778 eV [9]. The emission spectrum measured by Becker et al. indicates that the $v''_3 = 1$ and $v''_3 = 2$ transitions in the SO stretching mode from $v'_3 = 0$ are the most intense of this type while the strongest SH stretching and ∢HSO bending species occur from $v'_i = 0$ to $v''_i = 0$ in both instances. These findings are mirrored quite well in the calculated results (Table 6 of Ref. [10]) and on this basis there seems little doubt that the previous experimental identification of the emitting system as HSO is correct.

In addition analogous calculations were carried out for the SOH isomer [11] to determine whether a similar spectral distribution is to be expected in this case. Instead, however, it was found that the T_0 value for this system is much smaller than for HSO (0.57 eV) and furthermore that a notably less diffuse spectrum is probable for SOH (see Table 4 of Ref. [11]). Nonetheless the CI calculations indicate that SOH is more stable than HSO by some 12 kcal/mole and so the strong indication is that the thermodynamically less preferred system is the first to have been synthesized in the laboratory.

C. Methylene and its negative ion

For some time now one of the most actively pursued objectives in molecular spectroscopy has been to determine the triplet-singlet splitting between the lowest two states of methylene [12]. For a time the question appeared largely settled as photochemical measurements gradually converged to a value of 9-10 kcal/mole for this quantity while ab initio calculations started out at much higher values before 1970 but in more recent times came out rather uniformly with a result in the 11 - 13 kcal/mole range. The discussion took an abrupt turn in 1976, however, when new electron detachment measurements for the CH_2^- ion [13] were interpreted in a convincing manner in terms of a $^3B_1-^1A_1$ CH_2 0-0 splitting of 19.5 kcal/mole, in significant disagreement with both previous experiments and also with the best calculations (judged on a purely technical basis)as yet reported.

These new data suggested very strongly that the T_0 value for the $^1A_1 \leftarrow ^2B_1$ ionization of CH_2^- is 1.05 eV, for which a very strong energy loss peak (G) is observed in the electron detachment spectrum. Since both the $^2B_1$ and $^1A_1$ states are expected to have nearly equal bond angles and distances it is expected that only a single strong vibration transition should be observed for this electronic system and that is exactly what is found. The geometry of the more stable $^3B_1$ CH_2 state is significantly different, however, and thus one expects to see a progression of lines (in the bending frequency) in its spectrum, again as found [13]. The origin of this spectrum was claimed to correspond to a weak peak (A) corresponding to an energy loss of only 0.20 eV, however, which by substraction led to a value of 0.85 eV or 19.5 kcal/mole for the $^3B_1$-$^1A_1$ T_0 value of methylene itself. Because a very detailed intensity distribution for the electron detachment process was now available it was of interest to calculate the corresponding results using the ab initio methods outlined above and this work was carried out independently by Harding et al. [4] and by the authors and coworkers [15].

The resulting theoretical data agreed quite well with one another and in both cases the rather surprising conclusion was drawn that neither of the previous experimental assignments for the 0-0 $^1A_1$-$^3B_1$ transition of CH_2 is correct. Instead a value of 0.50 eV was indicated in the calculations, some 0.35 eV below the electron detachment result and some 0.14 eV above that assumed on the basis of the most recent photochemical data [16]. On the other hand there was general agreement that the $^1A_1$-$^2B_1$ electron affinity of 1.05 eV is quite accurate; in the authors' own work, for example, a direct EA value of 0.81 eV is obtained, which from experience with analogous calculations for C and C^- should be 0.20 eV too low, i.e. after making a straightforward correction for a similar error in the CH_2-CH_2^- calculations one arrives at a value of 1.01 eV, only 0.04 eV below what is observed.

Furthermore clear possibilities for reconciling the disagreement between the two types of experimental results for the $^3B_1$-$^1A_1$ 0-0 splitting were readily available from the new calculations, namely that the true 0-0 peak had not been seen in the photochemical work (the discrepancy here is equal to one quantum of the bending frequency in the upper $^1A_1$ state or roughly 0.15 eV) but instead only the 1 ⟵ 0 counterpart calculated at 0.36 eV, and that the electron detachment measurements had found either two hot band $^1A_1 \leftarrow {}^2B_1$ or cold band $^1A_1 \leftarrow {}^2A_1$ transitions at low kinetic energy loss and instead the true 0-0 peak of interest was the C species of Zittel et al., corresponding to a splitting of 0.50 eV. Under the circumstances it would be premature to say that the calculations have settled this question entirely but at least it is clear that the two experimental interpretations are in conflict with one another in this instance and that the theoretical predictions for the corresponding intensity distribution are very consistent with what is actually measured in the experimental studies.

D. HCN and DCN

The HCN molecule is well-known to be linear in its ($^1\Sigma^+$) ground state and to possess a number of low-lying (π,π^*) states. In 1953 Herzberg and Innes reported a detailed spectrum for this system as well as DCN in the 1700 - 2000 A° region which appears certain to involve transitions between these two types of states. On the basis of this work these authors concluded that two distinct transitions were involved therein, which they referred to as the $\alpha \leftarrow X$ and $\beta \leftarrow X$ band systems. A controversy has also developed over the assignment of these transitions, however, with Herzberg and Innes concluding on the basis of a rotational analysis of the spectrum that the α and β upper states are both bent A" species and theoretical calculations [18,19] indicating that they are instead $^1A''$ and $^1A'$ states respectively. The problem is that only one $^1A''$ (π,π^*) species possesses a bent geometry, namely that correlating with $^1\Sigma^-$ for the linear molecule.

Recent MRD-CI calculations lead to intensity distributions for the $^1A'' - {}^1\Sigma^+$ and $2\,^1A' - {}^1\Sigma^+$ transitions which are in very good agreement with the measured absorption pattern, although a reassignment of the original bending progressions [17] is necessary to obtain consistency between experiment and theory in this case. Furthermore the observed predissociative behavior of these transitions is also explained on the basis of these calculations. Nonetheless consideration of the rotational manifold in both the α-X and β - X systems have led Herzberg and Innes to conclude that no A' state can be involved in either case, even though it is extremely unlikely that more than one bent $^1A''$ HCN state exists in the spectral region in question. The matter is thus open to further investigation, with an explicit treatment of the rotational intensity distribution in the (π,π^*) transitions of HCN and DCN having been undertaken. A recent suggestion based on semiempirical calculations [20] that in fact one of the two upper states is a $^3A''$ species also needs to be considered in this regard, but the fact that both band systems have been seen in UV absorption studies certainly would need to be reconciled quantitatively with an assignment in terms of such a spin-forbidden transition from the $^1\Sigma^+$ ground state.

E. N_2H_2 and N_2D_2

The methods employed above are easilyextendable to systems with more than three atoms, particularly as long as one continues to assume that there is no coupling between the various vibrational modes. The first example of this type to be discussed is the $^1(n,\pi^*)$ transition in N_2H_2 and N_2D_2. An attempt to predict the vertical transition energy for this system with MRD-CI calculations [21] was somewhat unsettling because the value obtained was some 0.6 eV below the location of the most intense transition observed for the corresponding band system by Back, Willis and Ramsay (BWR) [22]. Since the ground and $^1(n,\pi^*)$ state possess very similar characteristics it was expected that an error of at most 0.2 eV in the vertical electronic energy could be expected, in agreement with the experience obtained with this CI method in numerous other examples. As a result Vasudevan et al. speculated that the measured absorption maximum did not correspond to a vertical transition and suggested a different assignment for the vibrational structure than that given by BWR.

In order to test this hypothesis potential curves were calculated for each of the six vibrational coordinates of N_2H_2 in both the ground and $^1(n,\pi^*)$ excited states, and vibrational wavefunctions, energies and corresponding vibronic intensities were determined from these results [23]. It was clear from these investigations that the earlier speculation of Vasudevan et al. for the expected non-verticality was incorrect, but nevertheless a closer analysis of the calculated intensity distribution showed a great similarity to the measured data. The main qualitative point which came out of the calculations was that the formally dipole-forbidden transition in the trans-equilibrium conformation becomes vibronically allowed through the NH antisymmetric stretching vibration ν_5, i.e. it was found that $|\vec{R}_{e'e''}|$ increased fairly quickly from zero once a vibration of this type was executed. Under these circumstances the strongest vibrational transition involving the NH antisymmetric coordinate is $1 \leftarrow 0$ and not $0 \leftarrow 0$, even though the latter corresponds much more closely to a strictly vertical transition. Because the ν_5' excited state frequency is in the order of 0.4 eV it was clear that this observation reduced the apparent discrepancy from the previously assumed value of 0.6 eV to somewhere in the 0.2 eV range, in agreement with the earlier experience with the MRD-CI method.

Nevertheless difficulties remained in trying to match up the calculated and measured intensity distributions since the observed ∢HNN bending progressions were notably longer from maximum to tail than was indicated in the computations. It was noted, however, that by making a change of three quanta in the previous experimental assignment that a very good correlation between calculation and experiment was forthcoming, except that in a few instances this adjustment seemingly forced the invoking of <u>negative</u> values for the bending quantum number. On the other hand a still closer look at the calculated results showed that another explanation for such extraneous transitions was readily available, namely as being induced by a different antisymmetric mode, namely the ν_4 torsion species. Since the ν_4' and ν_5' frequencies are quite different, this meant a one-to-one correspondance should exist between certain vibrational transitions which are separated by the difference $(\nu_4' - \nu_5')$ or some 2900 cm^{-1}. Because of the fact that the quantity $4\nu_2' - \nu_3'$ was also equal to around 2900 cm^{-1} it was clear that it would be very difficult to tell whether the spectrum consists of two parts, induced by ν_4 and ν_5 respectively, or whether there were only one inducing mode, as implied in the original BWR interpretation. In other words the calculations did appear to be consistent with the experimental findings but is was difficult to be certain if this agreement did simply come about by accident.

Because of the near equality of the frequency combinations mentioned above, however, it was clear that certain doubling phenomena were to be expected in the experimental spectra if the two-inducing-mode interpretation were correct, but not if only one mode was responsible. This observation led to a reexamination of the experimental data by BWR to test the various possibilities and the result was that line doubling was found in numerous instances in the spectral region predicted by the calculations [24]. As a result there is now strong evidence that the theoretical vibrational treatment does give a very accurate description of the band system in question without having made any assumptions of a semi-empirical nature.

F. Photoionization of Ethane

As a final example of this type of calculation the photoionization spectrum of ethane will be discussed [25]. Over the years there has been a controversy about the order of the IP's in ethane and similarly over the assignment of its PES. Because there are 18 vibrational coordinates in this instance no attempt was made to calculate potential curves for all such species in the ground and various ionic states; instead emphasis was placed on those vibrations which are expected to be excited upon ionization, i.e. for which the equilibrium values for such quantities are thought to be significantly different before and after ionization. The latter (qualitative) information can be obtained to a satisfactory degree from the Mulliken-Walsh Model for moleculargeometry [26].

The procedures employed are others wholly similar to those discussed above, with the exception that the transition moments (or photoionization cross-sections) are assumed to be independent of changes in nuclear geometry. The subject is fairly complicated but a detailed discussion of the calculated results may be found in the original reference [25]. What was found is that the regular structure with a 0-0 transition observed at 11.56 eV [27] is due to the ^2E_g-^2B_g ionization, with progressions in both the CC stretching (ν_3) and ∢HCH bending (ν_{11}) vibrations coinciding very closely in the spectrum (average exptl. spacing of about 1170 cm^{-1} [27]). In some ways such an assignment is surprising because the calculations show clearly that the lowest energy for $C_2H_6^+$ is obtained for the $^2A_{1g}$ state but for a much larger CC bond length than for the neutral ground state; the latter fact causes the ionization spectrum for this system to be too diffuse to be assigned to a nearly 1200 cm^{-1} progression [28], however, and so the alternative ^2E_g-^2B_g assignment seems quite reasonable, especially since very good agreement is again found between the calculated and measured intensity distributions.

The fact that the other component of the ^2E_g state has the same symmetry in a distorted C_{2h} nuclear conformation as does the $^2A_{1g}$ counterpart complicates the interpretation of this spectrum beyond 12.6 eV, however. The problem is that in the high-symmetry arrangements their two potential curves can cross but for slightly distorted geometries they undergo a sharply avoided crossing. As a result the usual

Franck-Condon Approximation of complete separability of nuclear and electronic motion is no longer acceptable and a more thorough treatment of the non-adiabatic coupling needs to be carried out. Nevertheless the fact that the vibrational structure becomes very irregular in the region where such sharply avoided crossings are predicted to occur is at least a good qualitative indication that the electronic (CI) calculations are accurate to within the usual error limits of a few tenths of an eV in this case as well.

In conclusion there is a great deal of evidence that the results of MRD-CI calculations with reasonably large AO basis sets can be used to give quite accurate predictions of the vibrational structure in general electronic transitions. There are still some unresolved questions about the interpretations of certain of the calculated spectra but it does not appear that simply by enlarging the basis sets or improving the CI treatments that the calculated results will change to any great degree. Certainly one objective in the future is to go beyond the Born-Oppenheimer Approximation in making such predictions and work in this direction is already in progress.

References

1. M. Born and E. Oppenheimer, Ann. Phys. 84, 457 (1927)

2. G. Herzberg, "Spectra of Diatomic Molecules", D. van Nostrand Co., New York (1950)

3. S.D. Peyerimhoff and R.J. Buenker, Theor. Chim. Acta 27, 243 (1972)

4. M. Perić, Mol. Phys. 34, 1675 (1977)

5. R.J. Buenker, S.D. Peyerimhoff and M. Perić, Chem. Phys. Letters 42, 383 (1976)

6. R.J. Buenker and S.D. Peyerimhoff, Chem. Phys. Letters 34, 225 (1975)

7. M. Yoshimine, K. Tanaka, H. Tatawaki, S. Ohara, F. Sasaki and K. Ohno, J. Chem. Phys. 64, 2254 (1976)

8. R.H. Huebner, R.J. Celotta, S.R. Mielczarek and C.E. Kuyatt, J. Chem. Phys. 63, 241 (1975)

9. K.H. Becker, M.A. Inocencio and U. Schurath, Int. J. Chem. Kinetics 51, 205 (1975); K.H. Becker, U. Schurath and M. Weber, J. Chem. Phys. in press; M. Weber, Ph. D. thesis Bonn (1976)

10. A.B. Sannigrahi, K.H. Thunemann, S.D. Peyerimhoff and R.J. Buenker, Chem. Phys. 20, 25 (1977)

11. A.B. Sannigrahi, S.D. Peyerimhoff and R.J. Buenker Chem. Phys. 20, 381 (1977)

12. J.F. Harrison, Accounts Chem. Res. 7, 378 (1974)

13. P.F. Zittel, G.B. Ellison, S.V. O'Neil, E. Herbst, W.C. Lineberger and W.P. Reinhardt, J. Am. Chem. Soc. 98, 3731 (1976)

14. L.B. Harding and W.A. Goddard III, Chem. Phys. Letters 55, 217 (1978); J. Chem. Phys. 67, 1777 (1977)

15. S.K. Shih, S.D. Peyerimhoff, R.J. Buenker and M. Perić, Chem. Phys. Letters 55, 206 (1978)

16. J. Danon, S.V. Filseth, D. Feldmann, H. Zacharias, C.H. Dugan and K.H. Welge, Chem. Phys. 29, 345 (1978)

17. G. Herzberg and K.K. Innes, Can. J. Phys. 35, 842 (1957)

18. G.M. Schwenzer, S.V. O'Neil and H.F. Schaefer III, J. Chem. Phys. 60, 2787 (1974)

19. M. Perić, S.D. Peyerimhoff and R.J. Buenker, Can. J. Chem. 55, 3664 (1977)

20. L. Åsbrink, C. Fridh and E. Lindholm, in press

21. K. Vasudevan, S.D. Peyerimhoff, R.J. Buenker, W.E. Kammer and H. Hsu, Chem. Phys. 7, 187 (1975)

22. R.A. Back, C. Willis and D.A. Ramsay, Can. J. Chem. 52, 1006 (1974)

23. M. Perić, R.J. Buenker and S.D. Peyerimhoff, Can. J. Chem. 55, 1533 (1977)

24. R.A. Back, C. Willis and D.A. Ramsay, preprint communicated prior to publication

25. A. Richartz, R.J. Buenker, P.J. Bruna, S.D. Peyerimhoff, Mol. Phys. 33, 1345 (1977)

26. R.J. Buenker and S.D. Peyerimhoff, Chem. Rev. 74, 127 (1974)

27. O.W. Turner, C. Baker and C.R. Brundle, "Molecular Photoelectron Spectroscopy", Interscience New York (1970)

28. C. Sandorfy, "Chemical Spectroscopy and Photochemistry in the Ultraviolet", Reidel Publishing Co., Dordrecht, Holland (1974)

CI CALCULATIONS OF VERTICAL EXCITATION ENERGIES AND OSCILLATOR STRENGTHS FOR RYDBERG AND VALENCE STATES OF MOLECULES

S.D. Peyerimhoff
Lehrstuhl für Theoretische Chemie
Universität Bonn, 53 Bonn, W. Germany

R.J. Buenker

Lehrstuhl für Theoretische Chemie
Gesamthochschule Wuppertal, 56 Wuppertal 1
W. Germany

INTRODUCTION

Procedures for the calculation of transition energies in molecules should be designed in such a manner that they can treat states of any multiplicity and characteristics, i.e. valence-shell or Rydberg states (or a mixture thereof) or those corresponding to inner-shell excitations, for example. Furthermore such methods should be applicable equally well to the study of valence-shell or inner-shell ionization as well as to the calculation of shake-up states. The MRD-CI method is in principle capable of performing according to such requirements.

The treatment of Rydberg states requires inclusion of the respective long-range functions in the AO basis set employed in order to represent the expanded charge density of the upper orbital in such states. These orbitals can thereby be placed at the nuclear centers or in the "midpoint" of the molecule [1,2], whereby the latter location is conceptually more appealing (united-atom character of the orbital) and has sometimes computational advantages (reducing the number of AO's); orbital exponents (for gaussian functions) for the lowest members of the Rydberg series have been optimized in various instances [1,2,3] and are found to be relatively insensitive to the specific molecular environment. (The same exponents are adequate, to

Cleanthes A. Nicolaides and Donald R. Beck (eds.), Excited States in Quantum Chemistry, 79–103.

describe equivalent Rydberg series in C_2H_2 [4], C_2H_4 [2] and C_4H_6 [5], for example).

The success of the calculation of transition energies depends on the extent to which differences in correlation energy between the two states involved in the transition are accounted for in the treatment employed. A major factor thereby is again the AO basis set (since the MRD-CI can be designed to closely approach the limit of a full CI) and present-day routine calculations make generally use of a double zeta (DZ) basis augmented by various polarization functions in the form of nuclear-centered d functions and/or bond functions. A broader discussion of this AO basis set question can be found in a recent review [6]. At the same time it is also obvious that it is not always required to correlate all the electrons if energy differences between states are to be calculated; in particular it is common usage to keep the 1s-shell (and in second-row atoms the K and L shell) doubly occupied in such investigations. This is of course not possible if inner-shell excitations are studied; such calculations require correlation of all electrons and thus lead to somewhat larger CI spaces than necessary for the investigation of outer-shell phenomena.

Finally in the calculation of ionization energies by CI methods it is more difficult to account for the proper balance of the correlation energy difference between the two states (neutral molecule and the ion) than in excitation processes, simply because in one system only n electrons are to be correlated whereas there are (n+1) in the other; in addition the contraction in the MO's upon ionization (especially in inner-shell ionization) plays a role. This observation, however, does not exclude the application of CI methods to the calculation of ionization potentials (indeed they seem to yield for all practical purposes the same results as the corresponding Green's function treatments once the same AO basis is employed [7]), it simply suggests that larger AO basis sets including more polarization functions must be employed to obtain equivalent accuracy as in the calculation of transition energies.

In the following sections a variety of examples will be presented in order to show in line with the above discussion the applicability of the MRD-CI method to the calculation of electronic transition and ionization energies in polyatomic molecules. In addition to the electronic energy difference ΔE_e between various states intensities will also be calculated in numerous instances. They are obtained in form of the oscillator strength f, evaluated in the so-called dipole and velocity form of the operator:

$$(1) \quad f(\vec{r}) = 2/3 \left| \langle\psi'|\vec{r}|\psi''\rangle \right|^2 \cdot \Delta E_e \quad \text{and}$$
$$(2) \quad f(\vec{\nabla}) = 2/3 \left| \langle\psi'|\vec{\nabla}|\psi''\rangle \right|^2 / \Delta E_e .$$

II. SPECTRA OF SATURATED SYSTEMS

The first excited states in simple saturated molecules such as water, hydrogen sulfide, ethane and propane have essentially Rydberg character and can be described quite often by a single dominant term in the CI expansion.

A. Water

The vertical transition energies ΔE_e (determined as difference between the electronic energy of ground and respective excited state at the ground state equilibrium geometry) to various states of H_2O obtained by MRD-CI calculations are compared with the corresponding experimental quantities in Table 1; the AO basis set employed consists of 24 contracted gaussians [8].

It is obvious from the table that discrepancies between calculated ΔE_e values and experimental transition energies are in every instance smaller than 0.2 eV, which is a typical error limit experienced in many other MRD-CI studies of molecular transition energies employing an AO basis of DZ quality including some polarization functions. It should also be mentioned that at the time the results of calculations were published the experimental value given for $^3B_1$ was 7.2 eV; only the more recent electron impact measurement [9] brings the experimental and the prior predicted theoretical value of 6.9 eV in very good accord. The characterization of the upper states in the transitions is in terms of (the dominant) Rydberg MO's although the wavefunction expansion show (especially for the 3s) also some hydrogen admixture. It is furthermore obvious from the calculations that the 4.5 eV feature observed in a number of experimental studies [10] does not correspond to a vertical electronic H_2O transition. Finally it is seen that the calculated IP's underestimate the corresponding measured values by a relative large margin in this treatment. Both

Table 1 Comparison of calculated and experimental transition energies (in eV) in H_2O[a]).

State		present MRD-CI	Experimental Ref. [9]
$^1A_1$	ground state	0.0	0.0
$^3B_1$	$b_1 \longrightarrow 3s$	6.90	7.0 (7.2)
$^1B_1$		7.30	$\tilde{A}$ 7.4
$^3A_2$	$b_1 \longrightarrow 3py$	9.04	8.9
$^1A_2$		9.20	9.1
$^3A_1$	$3a_1 \longrightarrow 3s$	9.01	9.3
$^1A_1$		9.80	$\tilde{B}$ 9.7
$^3A_1$	$b_1 \longrightarrow 3px$	9.65	9.81
$^1A_1$		10.32	$\tilde{D}$ 10.16 (000)
$^3B_1$	$b_1 \longrightarrow 3pz$	9.84	9.98
$^1B_1$		9.90	$\tilde{C}$ 10.01 (000)
$^3B_2$	$3a_1 \longrightarrow 3py$	10.99	11.1
$^1B_2$		11.21	(11.46)
$^3B_1$	$3a_1 \longrightarrow 3px$	11.68	
$^1B_1$		11.72	(11.77)
$^3A_1$	$3a_1 \longrightarrow 3pz$	11.53	
$^2B_1$	$b_1 \longrightarrow \infty$	12.12	(12.61)
$^2A_1$	$3a_1 \longrightarrow \infty$	14.06	(14.7)

a) Notation: pz possesses a_1, py possesses b_2 and px transforms like b_1 symmetry.

the magnitude as well as the sign of the deviation (the ion with n electrons is represented somewhat "better" compared to the X^1A_1 ground state molecule with n+1 electrons) is expected from the discussion given in the Introduction. Furthermore there is always some uncertainty whether the two values (vertical electronic energy difference in the theoretical treatment and experimental IP) can strictly be compared with one another.

Oscillator strengths for the various transitions in H_2O are contained in Table 2. There is relatively good agreement between corresponding $f(\vec{r})$ and $f(\vec{\nabla})$ values, and it is also seen that the calculated data represent the measured quantities quite satisfactorily. It is furthermore interesting that there are relatively little deviations when the wavefunction corresponding to the large threshold of T = 100 μhartree is employed for evaluation of the f value instead of the more appropriate expansion at T = 20 μhartree.

In summary then it can be said that the MRD-CI calculations at the present level (as well as CI calculations of various other groups on this molecule [11]) give a quite satisfactory description for the gross features of the H_2O spectrum.

B. Hydrogen Sulfide

Similarly reliable results as for H_2O are obtained from an equivalent treatment of hydrogen sulfide [12]. In this case transitions into valence-like sulfur 3d MO's are also found (at quite low energy) in addition to the various Rydberg series of s, p and d character (lowest member 4s, 4p and 4d) originating from the $5a_1$ and $2b_1$ valence MO's; the latter series are wholly analogous to those in H_2O. In fact, the first broad feature between 40000 and 60000 cm^{-1} in the H_2S spectrum [13] which has been explained under the assumption of a predissociation mechanism [14] involving two or more states [13] is predicted by the MRD-CI work to originate from the Rydberg $^1B_1(2b_1, 4s)$ and valence-like $^{3,1}A_2(2b_1,3dxz)$ states which both possess the same symmetry upon execution of an asymmetric stretching vibration and are hence capable of interaction. The maximum in absorption for this first broad band is found to coincide almost exactly with the calculated vertical transition energy of the $^1B_1(2b_1,4s)$ state, and the calculated oscillator strength of 0.06 is also in good agreement with the

Table 2 Oscillator strengths for various electronic transitions in water obtained from MRD-CI treatments truncated at a selection threshold of T = 20 μh and T = 100 μh respectively

State	T = 100 μh $f(\vec{r})$	$f(\vec{\nabla})$	T = 20 μh $f(\vec{r})$	$f(\vec{\nabla})$	Experimental
$\tilde{A}\ ^1B_1$	0.0604	0.0765	0.0592	0.0779	0.044[a)] 0.060 ± 0.006[b)] 0.052 [c)]
$\tilde{B}\ 2^1A_1$	0.0742[e)]	0.0722[e)]	0.0689	0.0680	0.05[a)]
$\tilde{C}\ 2^1B_1$	0.0133	0.0079	0.0120	0.0083	d)
$\tilde{D}\ 3^1A_1$	0.0141	0.0165	0.0130	0.0150	d)
3^1B_1	0.0001	0.0002	0.0002	0.0002	-
$^1B_2$	0.0035	0.0057	0.0036	0.0071	-

a) K. Watanabe and M. Zelikoff, J. Opt. Soc. Am. 43(1953)753.

b) E.N. Lassettre and A. Skerbele, J. Chem. Phys. 60(1974)2464.

c) A.J. Harrison, B.J. Cederholm and M.A. Terwilliger, J. Chem. Phys. 30(1959)30.

d) Observed ratio $\tilde{D}$ / $\tilde{C}$ estimated to be 1.2 according to b), the value of the present work is 1.084 (using $f(\vec{r})$).

e) If all contributions with $|c_i c_j| < 10^{-4}$ are neglected, $f(\vec{r})$ = 0.0714, $f(\vec{\nabla})$ = 0.0708.

value of 0.04 estimated from experiment for this quantity.

Furthermore the effect of adhering to a core of diverse closed shells in the entire MRD-CI expansion can be seen quite clearly from Table 3. The total ground state energy is lowered by as much as 0.0908 hartree (2.47 eV) if only one orbital is kept doubly occupied in constructing the configurations instead of both the sulfur K and L shells; nevertheless energy differences between $X\,^1A_1$ and the first four $^1B_1$ states differ only by as little as 0.02 eV upon the same change in the theoretical treatment. Since the number of configurations to be processed has increased thereby from 1477 (core = 1) to 5737 (core = 5) in the ground state and from 9503 to a total of 44841 in the $^1B_1$ states such a procedure which also correlates the 2s and 2p electrons of sulfur is certainly not necessary if emphasis is placed only on attainment of transition energies. Finally it should be pointed out that a similar CI study has been performed independently by another research group [15] and the results agree remarkably well with those discussed here, emphasizing the fact that present day theoretical CI calculations have reached the point at which they can be employed to give quite reliable information complementary to experiments.

C. Ethane and Propane

The third example in this section deals with ethane and propane; details can be found in the original references [16,17].

In ethane the two highest-lying occupied MO's $3a_{1g}$ and $1e_g$ are al- most isoenergetic and hence the various Rydberg series with the same upper MO originating from these two valence-shell lower species also lie very close in energy. Indeed it is seen from Table 4 that a total of five allowed transitions falls in the small energy region between 9.86 and 10.0 eV. The calculated total f value is 0.288 ($f(\vec{r})$ is evaluated in the ground state MO basis but if the excited state MO basis is used throughout $f(\vec{r})$ = 0.280) and compares quite well with the intensity for the first broad ethane band estimated to possess an oscillator strength of approximately 0.3.

It is also seen from Table 4 that the often-heard statement (based on arguments that the corre-

Table 3 Comparison of calculated vertical energy differences ΔE_e between the H_2S ground state and various excited states employing a core of one and 5 MO's respectively.

State	core = 1 ≙ $(1s)^2$ ΔE_e (eV)	core = 5 ≙ $(1s)^2(2s)^2(2p)^6$ ΔE_e (eV)
Ground State	0.0 (-398.8894 hartree)	0.0 (-398.7986 hartree)
1^1B_1	6.21	6.22
2^1B_1	7.76	7.78
3^1B_1	10.98	10.96
4^1B_1	12.23	12.22

Table 4 Calculated vertical excitation energies (in eV) and oscillator strengths f for C_2H_6

State	Excitation	Polarization	$f(\vec{r})$	SCF	MRD-CI
$^1A_{1g}$	Ground State		-	0.0	0.0
^3E_g	$1e_g \rightarrow 3s$		-	9.29	9.01
$^3A_{1g}$	$3a_{1g} \rightarrow 3s$		-	8.69	9.05
^1E_g	$1e_g \rightarrow 3s$		0.0	9.38	9.16
2^1A_{1g}	$3a_{1g} \rightarrow 3s$		0.0	-	9.21
^3E_u	$1e_g \rightarrow 3p\sigma$		-	10.01	9.85
$^3A_{2u}$	$3a_{1g} \rightarrow 3p\sigma$		-	9.34	9.73
^1E_u	$1e_g \rightarrow 3p\sigma$	(x,y)	0.056	10.08	9.91
$^1A_{2u}$	$3a_{1g} \rightarrow 3p\sigma$	z	0.144	9.44	9.86
2^3E_u	$1e_g \rightarrow 3p\pi$		-	-	9.88
2^3A_{2u}	"		-	-	9.77
$^3A_{1u}$	"		-	-	9.97
2^1E_u	"	(x,y)	0.002	-	9.99
2^1A_{2u}	"	z	0.020	-	9.99
$^1A_{1u}$	"		0.0	-	10.04
3^3E_u	$3a_{1g} \rightarrow 3p\pi$		-	9.53	9.97
3^1E_u	"	(x,y)	0.058	9.50	10.00
$^2A_{1g}$	$3a_{1g} \rightarrow \infty$		-	11.72	12.22
^2E_g	$1e_g \rightarrow \infty$		-	12.34	12.25

lation energy is always larger for closed-shell ground than for open-shell excited states) that SCF transition energies always underestimate the true ΔE values is incorrect; indeed the opposite situation holds for all the ethane states of Table 4 depopulating the $1e_g$ MO. This constitutes one (of many) examples in which the correlation energy of the excited state is larger than that of the closed-shell ground state.

The assignment of the ethane IP's has long been a matter of controversy: Koopmans' theorem yields essentially the same energy for both ^2E_g and $^2A_{1g}$ ions, the SCF procedure finds $^2A_{1g}$ to be more stable while the MRD-CI calculations [16] give essentially equal stability to both species (at the ethane ground state equilibrium geometry). Since the experimental PES shows a distinct structure with various peaks it is not sufficient to only investigate the vertical IP; instead a treatment which takes into account the various geometrical perturbations associated with ionization is called for in this case in order to obtain a reliable relationship between the theoretical $C_2H_6^+$ electronic structure and the measured PES, and such a treatment will be dealt with in the subsequent lecture.

Finally, in one respect the situation in propane is quite analogous to that calculated in ethane: in this molecule there are three (highest occupied) orbitals which are of very nearly equal stability namely $6a_1$, $4b_2$ and $2b_1$, and hence the low-energy spectrum of propane is calculated to consist of a variety of closely overlapping Rydberg transitions originating from these orbitals [17]. There is one difference however: while in ethane the first Rydberg members to which transitions are allowed by the dipole selection rules are of 3p symmetry, excitations to the lower 3s Rydberg MO in propane is dipole allowed for all three series.

Before concluding this section a remark on Rydberg states of positive ions seems in order. While the term values for Rydberg s, p and d series in neutral molecules are fairly constant from one molecule to another (roughly 19000 cm^{-1} for 3p and 13000 cm^{-1} for 3d, for example), those for positive ions are expected to be considerably larger since the outer electron moves effectively in a field of two positive charges rather than one. According to this simple model R is then replaced by $Z_{eff}^2 . R = 4R$ in the Rydberg formula and a similar change must be made for the effective charge in the radial part of the wavefunction. In fact calculations on $C_2H_6^+$ show that optimized Rydberg functions are considerably more contracted in $C_2H_6^+$ than in ethane itself and predict term values in $C_2H_6^+$ of approximately 68000 cm^{-1} and 54000 cm^{-1} for 3s and 3p respectively, i.e 2.5 to 3 times larger than in neutral molecules.

III. MOLECULES WITH LOW-LYING RYDBERG AND VALENCE STATES

The vertical electronic spectrum of a large number of molecules which possess relatively low-lying unoccupied valence-shell MO's has been studied by the MRD-CI treatment (or similar CI techniques). Examples are formaldehyde, thioformaldehyde, acetone, thioacetone, ethylene, butadiene, benzene, pyrrole, a large number of triatomic species (like O_3, NO_2, CO_2^-, C_3), various molecules of the type ABH or AH_n (NH_3, CH_2, CH_2^-, NH_2, HOCl, HO_2, HN_2, HC_2, HF_2, HS_2, HCO, HCN, HSO, HSiN, HCS, HCO^+, HCS^+ with the isomers HClO, HOC etc.) as well as several diatomic molecules (C_2, O_2, N_2). For the larger species only the vertical spectrum has generally been obtained while sections of the various potential curves have been calculated for the molecules with fewer degrees of vibrational freedom. In the present context only a few examples will be treated.

A. Ozone

Experimentally the ozone spectrum between 2 and 5 eV is well-characterized by the weak Chappuis (2.1 eV), Huggins (3.5 - 4.2 eV) and the strong and broad Hartley bands (max. 4.86 eV). Various peaks at lower energy are observed in the electron impact spectrum [19] and numerous features at higher energies, all of which are unassigned, are also known [20]. The recent MRD-CI calculations [21] separate the ozone transitions in essentially three groups:
a) transitions in the 1 to 5 eV area arise predominantly from excitations out of the energetically neighboring $1a_2$, $4b_2$ and $6a_1$ MO's into the unoccupied $2b_1$ species
b) transitions in the energy range to follow (up to 8 eV) result from double-excitations from the three orbitals just mentioned into the $2b_1$ MO
c) higher energy transitions (> 8 eV) involve Rydberg states (the lowest singlet state $4b_2 \longrightarrow 3s$ is calculated at 9.21 eV) as well as excitations into an antibonding σ^* type orbital.

Again the agreement between the measured absorption maximum and the calculated vertical ΔE_e values is very good: 1.95 eV for the $^1B_1(6a_1,2b_1)$ Chappuis band (2.1 eV exptl.), 3.60 eV for the Huggins band (exptl. 3.7 eV) which is predicted to arise from essentially $4b_2^2 \longrightarrow 2b_1^2$ and $6a_1^2 \longrightarrow 2b_1^2$ configurations, and 4.97 eV for the $^1B_2(1a_2, 2b_1)$ Hartley transition (exptl. maximum 4.86 eV). The energy extrapolation procedure is especially important for the treatment of this latter state, since the energy difference for the $^1B_2-^1A_1$ states at a selection threshold of T = 30 μh, (raw secular equation result) for example, is 6.7 eV, i.e. 1.7 eV higher than the value obtained by the use of the extrapolation procedure. This result points out very clearly the danger in neglecting the large number of weakly interacting configurations in the CI entirely . Hence the main difference between the present work and another CI study [22] which finds the Hartley band considerably higher in energy (between 5.6 and 6.09 eV depending on AO basis) seems to lie in the CI truncation in the latter study. It is also interesting that upon asymmetric distortion both the Huggins and Hartley states will have the same spatial symmetry giving rise to interaction which seems to play an important role in the ozone photolysis.

The broad maximum in the 7.18 eV area is predicted to arise predominantly from $4b_2, 1a_2 \longrightarrow 2b_1^2$ excitations (calculated $f = 0.14 \times 10^{-2}$, $\Delta E_e = 7.26$ eV) although the $4b_2, 6a_1 \longrightarrow 2b_1^2$ (calc. $\Delta E_e = 6.87$ eV and $f = 0.2 \times 10^{-3}$) and a combination of $4b_2^2 \longrightarrow 2b_1^2$ and $6a_1^2 \rightarrow 2b_1^2$ (calc. $\Delta E_e = 7.34$, $f = 0.5 \times 10^{-3}$) might also add to the intensity. The strong intensity feature around 9.32 eV is predicted to arise from $4b_2 \longrightarrow 3s$ ($f = 0.3 \times 10^{-1}$), $6a_1 \rightarrow 3s$ ($f = 0.1 \times 10^{-1}$) and $6a_1 \longrightarrow \sigma^*$ ($f = 0.24 \times 10^{-2}$) transitions.

Among technical details [21] it should be mentioned that in all calculations a core of 3 orbitals is held doubly occupied corresponding to the oxygen inner shells; the AO basis (37 functions) is of DZ quality with s bond polarization and Rydberg functions. The results for the Hartley band, for example, are practically identical (deviation 0.02 eV) if oxygen-centered d polarization functions (α = 0.8) are employed instead of bond polarization functions, although the total energy of both ground and excited $^1B_2$ states is lowered by as much as 4.8 eV upon this change in the theoretical treatment. Similarly small differences in the calculated $^1B_2-^1A_1$ transition energy are found upon increasing the core by 2 or 3 to a total of 5 or 6 shells held always doubly occupied. The number of SAF's is generally in the order of 30000 to 175000 in the present work and the number of reference species between 1 and 8 while selection is carried out for up to three roots; NO's are employed for most valence like states while MO's are used for many of the Rydberg species.

B. Acetone

A comparison between the experimental acetone spectrum [23] and the respective calculated data is best seen from Fig. 1 ; details are again found in the

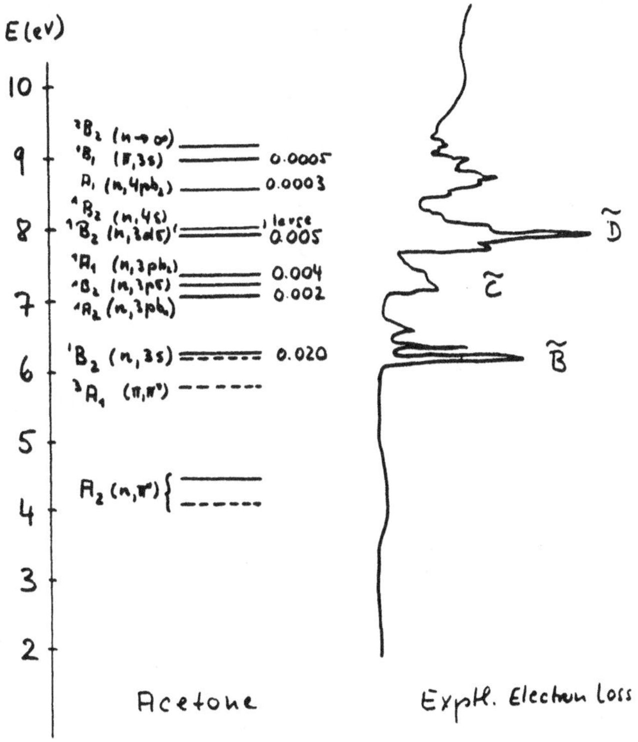

Fig. 1 Comparison of the experimental electron loss spectrum and the calculated energy levels. The numbers in the center of the figure are calculated oscillator strengths.

Table 5 Several technical details of the MRD-CI calculations of acetone. Given are for various states the orders of the full MRD-CI space to which extrapolation is carried out and the orders of the secular equations which have actually been solved.

State		Secular Equation Size generated	solved
$^1A_1$,	Ground State	43004	1569
$^3A_2$,	(n, π^*)	201275	2736
	$(n, 3pb_1)$	171457	2659
	$(\pi, 3pb_2)$	139223	2294
$^3A_1$,	(π,π^*)	209537	2438
	$(n, 3pb_2)$	76049	2722
$^3B_2$,	$(n, 3s)$, $(n, 3p\sigma)$, $(n, 3d\sigma)$	108960	4286
$^1B_2$,	$(n, 3s)$, $(n, 3p\sigma)$, $(n, 3d\sigma)$	64248	4433
$^2B_2$,	(n,∞)	59446	4678

original reference [24]. The electron loss spectrum shows a weak feature in the 4 eV area in which the calculations predict the dipole-forbidden A_2 (n,π*) transition. The $\tilde{B}$ band clearly corresponds to the first Rydberg transition with a calculated f value of 0.02; the calculations assign the $\tilde{C}$ state to two dipole-allowed Rydberg transitions with considerably smaller f values. The calculations would further assign the observed $\tilde{D}$ band to a Rydberg 3d and 4s species whereby some intensity is underlying from the higher-energy (π,π*) state (not shown in the figure). The vibrational structure seen in the electron loss spectrum can of course not be obtained in a theoretical treatment in which only vertical energy differences are sought.

Some technical details of the acetone MRD-CI treatment are found in Table 5 and demonstrate the importance of the selection and extrapolation procedure: while configuration spaces are routinely in the 100000 to 200000 range, the secular equations actually solved for acetone are below 5000 in each case (a core of 7 orbitals is chosen thereby).

C. Thiocompounds

A comparative study of acetone has been undertaken with the sulfur-containing compound [25]. While relatively little experimental information is available on thioacetone, however, measurements have been published on thioformaldehyde quite recently [26] and it is seen from Table 6 that the data obtained by CI calculations a number of years earlier [27] constitute an excellent prediction of the actual H_2CS transition energies. The only discrepancy in this comparison is the (π,π*) state which is at present a matter of further experimental investigation [28]. It is also interesting that apparently the emission spectrum for this band is quite different from that in absorption, a finding which is a strong indication that large geometrical changes occur upon (π,π*) excitation; in turn this fact suggests also that the vertical ΔE_e value does not necessarily coincide with the measured absorption maximum.

D. Butadiene

Butadiene is one of the molecules which has been under intense investigation by experimentalists and theoreticians for many years. There has always been agreement on the location of the first triplet

Table 6 Comparison of theoretically predicted vertical transition energies in H_2CS with the recently observed data (values in eV).

State	Excitation	ΔE_e (vertical) theoretical prediction [27]	observed peak
$^1A_1$	Ground State	0.00	0.00
$^3A_1$	$n \longrightarrow \pi^*$	1.84	1.80[a)]
$^1A_1$	$n \longrightarrow \pi^*$	2.17	2.03[a)]
$^1A_1$	$\pi \longrightarrow \pi^*$	7.92	5.72[b)]
$^1B_2$	$n \longrightarrow 4s$	5.83	5.83[b)]
$^1A_1$	$n \longrightarrow 4py$	6.62	6.59

a) R.H. Judge and G.W. King, Can. J. Phys. 53(1975)1927.

b) Ref. [26].

Table 7 Calculated vertical transition energies (in eV) to various low-lying states of trans-butadiene and comparison with existing experimental results. Corresponding oscillator strengths are also given.

State	Excitation	all-valence electron MRD-CI	$f(\vec{r})$	Experimental[b)]	
^1A_g	ground state	0.0	-	X	0.0
^3B_u	$x_2 \rightarrow x_3$	3.31	-	a	3.22[c)], 3.2[d)], 3.3[e)]
^3A_g	$\{x_1 \rightarrow x_3,\ x_2 \rightarrow x_4$	4.92	-	b	4.8[e)], 4.93[c)], 4.9[d)]
^1B_g	$x_2 \rightarrow 3s$	6.20	-	B	6.22
^1A_u	$x_2 \rightarrow 3p\sigma$	6.53	0.0002		probably not observed
^1B_u	$x_2 \rightarrow 3p\pi$	6.67	0.07	C	6.657, series (3), n = 3
2^1A_u	$x_2 \rightarrow 3p\sigma$	6.72	0.05		
2^1A_g	valence mixture	7.02	-	E or F	7.06, series (2), n = 3
2^1B_g	$x_2 \rightarrow 3d\sigma$	7.29	-		7.328, series (4), n = 3
3^1A_g	$x_2 \rightarrow 3d\pi$	7.53	-		{7.481, series (1), n = 3
4^1A_g	$x_2 \rightarrow 3d\delta$	(7.78)	-		
n^1B_u	$x_2 \rightarrow x_3$	7.67	0.99	A	5.71-6.29[a)]
$2^1B_u{}^*$	$x_2 \rightarrow 4p\pi$	7.96	0.09	G	7.857, series (3), n = 4 or 8.002, series (2). n=4
^2B_g	$x_2 \rightarrow \infty$	8.68	-		9.08

a) The theoretical work suggests that the intensity maximum of this $^1B_{1u}$ (π,π^*) state does not correspond to a vertical transition [33].

b) Values for the Rydberg states are taken from Ref. [31].

c) O.A. Mosher, W.M. Flicker and A. Kuppermann, J. Chem. Phys. 59(1973)6502.

d) J.H. Moore, J. Phys. Chem. 76(1972)1130.

e) H.H. Brongersma, Ph. D. thesis, Leyden (1968).

in this system, whereas the discrepancy on the location of the second such state has only been resolved after the experimental transition energy was reevaluated. A number of Rydberg states have also been predicted by early CI calculations [29,30] and have quite recently been studied in detail by various experimental procedures [31,32].

A summary of the data obtained from the best present all-valence electron CI calculation on this system is given in Table 7 in comparison with experiment. A cursory glance at the data again points out the good performance of the calculations; a detailed discussion is found in ref. [33]. There is one important unresolved discrepancy between measurements and calculations in the butadiene spectrum, namely the strong and broad ^1B_u (π,π^*) transition; since calculations indicate that geometrical changes in the nuclear framework as well as interaction with a close-lying (in the perturbed geometry) Rydberg state play an important role in the entire description of the upper ^1B_u state, it is clear that calculation of the vertical ΔE_e value is not sufficient to represent the major features of this transition.

IV. IONIZATION POTENTIALS

At various points in the present paper ionization potentials have been calculated routinely together with excitation energies of the various molecules discussed. A more systematic MRD-CI study has been undertaken for the first five IP's of ethylene [7] in order to compare with the equivalent results of the Green's function approach which has been very successful in obtaining molecular IP's. In particular the two AO basis sets (DZ plus d polarization functions) employed in the Green's function study by v. Niessen et al. [34] are chosen and the results are collected in Table 8.

It is seen that for all practical purposes both treatments (CI and Green's function approach) perform equally well and furthermore, that the calculated IP's are closer to the corresponding experimental quantities, (especially π ionization) if the AO basis with the less contracted polarization species (α = 0.3 instead of α = 0.8) is employed. This finding has been attributed [34] to a better description of π

Table 8 Comparison of ionization potentials (in eV) for various states of ethylene obtained from the MRD-CI method and the Green's function approach (MBPT).

	Basis III[a)] (α_d=0.8)			Basis IV (α_d=0.3)			Experiment[c)]
Ion	MRD-CI	Full-CI[b)]	MBPT	MRD-CI	Full-CI[b)]	MBPT	
$^2B_{3u}$	10.26	10.36	10.24	10.33	10.46	10.45	10.51
$^2B_{3g}$	12.93	12.98	12.98	12.97	13.01	13.04	12.85
^2A_g	14.48	14.55	14.59	14.58	14.66	14.77	14.66
$^2B_{2u}$	15.79	15.80	16.03	15.85	15.88	16.10	15.87
$^2B_{1u}$	19.12	18.98	19.47	19.17	19.04	19.45	19.23 } 19.10 }

a) The two AO basis sets denoted by III and IV in the original work [34] are employed; they differ only in the exponent of the d polarization function.

b) Estimated full CI limit of the AO basis evaluated from the MRD-CI treatment: as $\Delta E = (1 - \Sigma c_0^2) \times (E_{MRD-CI} - E_{ref})$, whereby the sum runs over all reference species.

c) As quoted in Ref. [34].

correlation effects but the CI study clearly points out [7] that the ionic state is described somewhat less satisfactorily with respect to a total correlation in this basis (Table 9), thereby increasing the energy difference between ionic and ground state to the desired magnitude. Further CI calculations on ethylene show that basis sets with a large number of polarizationtype functions are required for a truly unbiased calculation of vertical ionization potentials [7].

Table 9 Correlation energy accounted for (as the difference between corresponding SCF and CI calculations) in the various C_2H_4 states.[a)]

State	Basis III (α_d=0.8) MRD-CI	Basis III (α_d=0.8) full CI estimate	Basis IV (α_d=0.3) MRD-CI	Basis IV (α_d=0.3) full CI estimate
^1A_g	0.28476	0.30549	0.26167	0.28128
$^2B_{3u}$ (π)	0.23729	0.25435	0.21138	0.22645
$^2B_{3g}$	0.29274	0.31171	0.26510	0.28345
^2A_g	0.27941	0.29777	0.25327	0.26977
$^2B_{2u}$	0.31726	0.33747	0.29139	0.30968
$^2B_{1u}$	0.34472	0.37078	0.31961	0.34408

a) The SCF energy for the C_2H_4 ground state is -78.04326 hartree in basis III and -78.03120 hartree for basis IV.

V. INNER-SHELL PHENOMENA

The MRD-CI package can also be successfully employed for the calculations of inner-shell excitation processes and corresponding shake-up states; examples treated so far are N_2, ethylene and acetylen. The computations are thereby carried out in a delocalized framework so that gerade and ungerade states can be distinguished; configuration spaces up to order 300000 are considered.

In the most flexible AO basis studied (69 AO's) for N_2 the MRD-CI treatment yields a value for the 1s IP of 410.01 eV which is in very good agreement with the corresponding experimental value of 409.9 eV; transformation to NO's is thereby found to be beneficial, and inclusion of quite contracted p and d type functions ($\alpha(p) = 15$, $\alpha(d) = 2.0$) in the AO basis are found to be important. It should be pointed out that the SCF energy for the nitrogen molecule $(1\sigma_g)^{-1}\ {}^2\Sigma_g^+$ ion in the same basis is as large as 419.77 eV.

Transition energies to core-valence excited states are uniformly overestimated in this study [35] but the relative location of such states is predicted very satisfactorily (Table 10). The energy difference

Table 10 Transition energies ΔE_e (in eV) relative to the N_2 ground state for various core-valence excited states obtained from an MRD-CI treatment [35] and comparison with corresponding experimental quantities.

State	ΔE_e	Excitation exptl.	Relative excitation calc.	Relative excitation exptl.
$X\ {}^1\Sigma_g^+$	0.0	0.0	-	-
${}^1\Pi_u(1\sigma_u \rightarrow \pi_g)$	402.23	400.84	0.0	0.0
${}^1\Sigma_u^+(1\sigma_u \rightarrow 3s)$	407.05	405.59	4.82	4.75
${}^1\Pi_u(1\sigma_g \rightarrow 3p\pi_u)$	408.11	406.50 406.72	5.88	5.66 5.88
${}^1\Pi_g(1\sigma_g \rightarrow \pi_g)$	402.01	≈400	-0.22	-

between the $1\sigma_g^{-1}$ ion and its lowest shake-up state $(1\sigma_u^{-1})\pi_u \rightarrow \pi_g\ {}^2\Sigma_g^+$ is calculated to be 9.39 eV versus 9.3 eV experimentally, pointing out again that the CI procedure is a very powerful tool for the study of molecular excitation and ionization phenomena of quite general characteristics.

VI. MIXED VALENCE-RYDBERG STATES

In various instances the calculations have detected interaction between Rydberg and valence-shell states. Examples include various states in N_2 and O_2, the V state of ethylene, states in butadiene and acetone as well as dissociative states in ammonia.

Since the amount of mixed character is very much dependent on the nuclear conformation of the molecule, such mixed states will be treated together with the discussion on vibrational features and on photochemical decomposition.

VII. SUMMARY

The present lecture has attempted to show by a number of examples that the electronic energy difference ΔE_e between various electronic states of molecules as well as the (electronic) oscillator strengths obtained from MRD-CI calculations can generally be quite satisfactorily related to corresponding experimental data, i.e. to transition energies or ionization potentials. It is clear however, that this correspondence does not hold if vibrational effects are of considerable importance. This will be the case if potential surfaces of ground and excited states are quite different in shape, for example, or if the electronic transition moment depends strongly on geometrical parameters so that the Franck-Condon approximation is not a valid description for the process under discussion; the latter situation will often derive from interaction of various (upper) states or from geometrical distortions which change the symmetry of the nuclear framework. In all such cases it is important to calculate a representative portion of the entire potential energy surface for the electronic states involved together with vibrational wavefunctions and energy levels; this procedure is of course always necessary for the prediction of any vibrational progression in the spectrum, and this subject will be taken up in a further lecture.

References

1. S.D. Peyerimhoff and R.J. Buenker, in "Chemical Spectroscopy and Photochemistry in the Vacuum Ultraviolet" (ed. C. Sandorfy, P.J. Ausloos and M.B. Robin), Reidel, Dordrecht (1974), ASI series C8

2. U. Fischbach, R.J. Buenker and S.D. Peyerimhoff, Chem. Phys. 5, 265 (1974)

3. T.H. Dunning, Jr., and P.J. Hay, in "Methods of Electronic Structure Theory" (ed. H.F. Schaefer III), Vol. 3, Plenum Press New York London (1977)

4. W.E. Kammer, Chem. Phys. 5, 408 (1974)

5. R.J. Buenker, S.K. Shih and S.D. Peyerimhoff, Chem. Phys. Letters 44, 385 (1976)

6. S.D. Peyerimhoff and R.J. Buenker in "Advances in Quantum Chemistry" (ed. Per-Olov Löwdin), Vol. 9, Academic Press, New York, (1975)

7. K.H. Thunemann, S.K. Shih, S.D. Peyerimhoff and R.J. Buenker, in press

8. R.J. Buenker and S.D. Peyerimhoff, Chem. Phys. Letters 29, 253 (1974)

9. A. Chutijan, R.I. Hall and S. Trajmar, J.Chem. Phys. 63, 892 (1975)

10. S. Trajmar, W. Williams and A. Kuppermann, J. Chem. Phys. 54, 2274 (1971); 58, 252 (1973); F.W.E. Knoop, H.H. Brongersma and L.J. Osterhoff, Chem. Phys. Letters 13, 20 (1972); A. Skerbele, M.A. Dillon and E.N. Lassettre, J. Chem. Phys. 49, 5042 (1968); E.N. Lassettre, and W.M. Huo, J. Chem. Phys. 61, 1703 (1974)

11. N.W. Winter, W.A. Goddard III and F. Bobrowicz, J. Chem. Phys. 62, 4325 (1975); R.F. Hausman, Jr., and C.F. Bender in "Methods of Electronic Structure Theory" (ed. H.F. Schaefer III), Vol. 3, Plenum Press New York, London (1977)

12. S.K. Shih, S.D. Peyerimhoff and R.J. Buenker, Chem. Phys. 17, 391 (1976)

13. M.B. Robin, "Higher Excited States of Polyatomic Molecules" (Academic Press, 1975), Vol. 1, p. 276

14. R.S. Mulliken, J. Chem. Phys. 3, 506 (1935)

15. M.F. Guest and W.R. Rodwell, Mol. Phys. 32, 1075 (1976)

16. R.J. Buenker and S.D. Peyerimhoff, Chem. Phys. 8, 56 (1975)

17. A. Richartz, R.J. Buenker and S.D. Peyerimhoff, Chem. Phys. 00, 000 (1978)

18. A. Richartz, R.J. Buenker and S.D. Peyerimhoff, Chem. Phys. 28, 305 (1978)

19. R.J. Celotta, N. Swanson and M. Kurepa, presented at the Xth IPEAC Conference, July 1977

20. R.J. Celotta, S.R. Mielczarek and C.E. Kuyatt, Chem. Phys. Letters 24, 428 (1974); Y. Tanaka, E.B.Y. Inn and K. Watanabe, J. Chem. Phys. 21, 1651 (1953)

21. K.H. Thunemann, S.D. Peyerimhoff and R.J. Buenker, J. Mol. Spectry, 00, 000 (1978)

22. P.J. Hay, T.H. Dunning Jr., and W. A. Goddard III, J. Chem. Phys. 62, 3912 (1975);, Chem. Phys. Letters 23, 457 (1973).

23. R.H. Huebner, R.J. Celotta, S.R. Mielczarek and E.C. Kuyatt, J. Chem. Phys. 59, 5434 (1973)

24. B. Hess, P.J. Bruna, R.J. Buenker and S.D. Peyerimhoff, Chem. Phys. 18, 267 (1976)

25. P.J. Bruna, R.J. Buenker and S.D. Peyerimhoff, Chem. Phys. 22, 375 (1977)

26. R.H. Judge, C.R. Drury-Lessard and D.C. Moule, Chem. Phys. Letters 53, 82 (1978)

27. P.J. Bruna, S.D. Peyerimhoff, R.J. Buenker and P. Rosmus, Chem. Phys. 3, 35 (1974)

28. D.C. Moule, private communication

29. S. Shih, R.J. Buenker and S.D. Peyerimhoff, Chem. Phys. Letters 16, 244 (1972)

30. R.P. Hosteney, T.H. Dunning, Jr., R.R. Gilman, A. Pipano and I. Shavitt, J. Chem. Phys. 62, 4764 (1975)

31. R. McDiarmid, J. Chem. Phys. 64, 514 (1976); Chem. Phys. Letters 34, 130 (1975)

32. K.P. Gross and O. Schnepp, J. Chem. Phys. 68, 2647 (1978)

33. R.J. Buenker, S.K. Shih and S.D. Peyerimhoff, Chem. Phys. Letters 44, 385 (1976)

34. W. von Niessen, G.H.F. Diercksen, L.S. Cederbaum and W. Domcke, Chem. Phys. 18, 469 (1976)

35. W. Butscher, R.J. Buenker and S.D. Peyerimhoff, Chem. Phys. Letters 52, 449 (1977)

THEORY OF THE ELECTRONIC STRUCTURE OF EXCITED STATES IN SMALL SYSTEMS WITH NUMERICAL APPLICATIONS TO ATOMIC STATES

DONALD R. BECK and CLEANTHES A. NICOLAIDES

Theoretical Chemistry Institute
National Hellenic Research Foundation
Vas. Constantinou 48, Athens 501/1, Greece

ABSTRACT

In this lecture we discuss an approach to the understanding and efficient calculation of electron correlation of valence or inner hole excited states in atoms and small molecules. We start with the formal subshell cluster expansion of the wave-function and we analyze the various correlation effects which appear in the wave-function whose form is dictated by first order perturbation theory. I.e. only single and pair symmetry adapted correlation functions are considered. This analysis allows: 1) The formal decoupling, to a good approximation, of the subshell correlation vectors into groups and their economic computation from small Variational Configuration-Interaction (VCI) procedures. 2) The systematic optimization of the different for each group virtual one-electron functions--expressed in terms of STO's or GTO's--by minimizing the corresponding to each group energy functional. 3) The consequent determination of compact but accurate total wave-functions from basis sets which contain the Hartree-Fock and only a few more virtual orbitals. 4) The recognition and isolation of important for electronic spectroscopy and chemical bonding correlation effects from those which contribute mainly to total energies. 5) The recognition of the importance of triple and quadruple correlation effects for certain inner hole excited states, even in small systems. We present previously unpublished numerical results on a) the position of the H^- $2p^2$ ^3P metastable state whose study supports our suggestion that variationally optimized (VO) one-electron basis sets are competitive with r_{ij} dependent basis sets in terms of fast convergence. b) The position of the H^{--} $2p^3$ $^4S^o$ state which is found to be unbound. c) The pair correlations of C where reasonably accurate calculations

Cleanthes A. Nicolaides and Donald R. Beck (eds.), Excited States in Quantum Chemistry, 105–142.

indicate that VO-GTO virtuals are reasonably competitive with VO-STO virtuals. This suggests that the size of current molecular calculations using Gaussians can be reduced considerably at no expense of accuracy. d) The effect of electron correlation on the term structure of Ni III, of current importance in solid state physics.

I. INTRODUCTION

In this article we discuss some aspects of the electronic structure of excited states of atoms and molecules and a computational methodology for the calculation of their wave-functions and energies.

In the zeroth order description, excited states are usually open shell systems. Exceptions may be states where electrons are missing completely from an inner shell, e.g. Ne^{++} $1s^2$ $2p^6$ ^1S. In general, open shell excited states have multi-determinantal zeroth order configurational descriptions. Furthermore, as one goes up in energy for a fixed Born-Oppenheimer (BO) geometry, or as one reaches the dissociation region, the density of states per energy interval increases, near degeneracies become more pronounced and often affect the zeroth order description considerably. A manifestation of such degeneracies is the occasional heavy valence-Rydberg mixing.

Our emphasis will be on the economic and consistent generation of compact, well correlated wave functions for small systems. Such functions render the physically interesting features of the electronic structure distinctly transparent, allow the recognition of important, property dependent correlation effects by inspection and are possible to handle easily for the calculation of other properties and phenomena such as polarizabilities and electron-molecule scattering. Furthermore, small atomic basis sets containing information about correlation at the atomic level can prove extremely useful as starting points for rigorous studies of the molecular and solid state.

The approach to be discussed below is based on notions of a cluster expansion of the exact N-electron wave-function, analysis of the important excited state correlation effects and the calculation of one - and two - electron correlation functions via Variational Configuration-Interaction (VCI) procedures which employ Slater or Gaussian type orbitals. We argue that, by systematically and consistently decoupling portions of the total wave-function and by optimizing variationally in small CI the virtual orbitals describing the specific correlation effects of each portion before the total wave-function is put together, the magnitude of computation is reduced considerably without significant loss of accuracy. This approach yields nonorthonormal Hartree-Fock and virtual functions which result in the well-known computational problem of "nonorthonormality" which is more complicated but straight-forward to solve if the wave-functions are small.

The numerical examples are taken from the realm of atoms. Thus we have considered: 1) The H^{--} $2p^3$ $^4S^o$ state, whose energy position is of importance to the formal spectral theory of the Schrödinger operator as well as to recent experimental observations. 2) The H^- $2p^2$ ^3P state, where our results indicate that, in general, energy optimized single particle expansions of pair correlations can compare well, in terms of accuracy, compactness and convergence, with r_{ij} dependent basis sets, usually thought to be much more effective. 3) The Ni III $3d^8$ term structure which is of much interest in transition metal oxide solid state physics. 4) The CI ^3P ground state where the correlation functions are expanded in terms of variationally optimized Gaussians.

II. SIMPLE CLASSIFICATION OF GROUND AND EXCITED STATES IN TERMS OF THEIR SHELL STRUCTURE

Before proceeding with the theory and calculation of excited state wave-functions, we outline certain of their gross features in terms of the shell model.

As is well known, the commonly accepted zeroth order representation of atomic and molecular states is based on a single configurational assignment obtained from the aufbau principle. For ground states, this principle works well in most cases yielding reasonably good representations of the wave-function. Exceptions can be found in near degeneracy cases, in large atoms where the hydrogenic filling model breaks down and in molecules with small singlet-triplet separations or far from the equilibrium distance.

Excited states can be classified roughly as follows:

a) Low lying Unperturbed Rydberg Configurations: For low energy spectroscopic measurements, such states behave essentially like pseudo one-electron systems modified by core polarization. The perturbation due to other Rydberg low lying configurations of the same orbital structure is negligible, due to nearly zero off-diagonal matrix elements. For atoms, the wave-functions of these states are easily computed within the spherically symmetric Independent Particle Model (IPM). For molecules, their recognition is based on spectroscopic considerations and on large values of the matrix element $\langle r^2 \rangle$. Their computation can be carried out satisfactorily at the LCAO-Self Consistent Field

(SCF) level, provided the atomic basis set contains Rydberg-like (i.e. diffuse) orbitals. Of course, if properties other than low energy excitation energies and intensities are desired (e.g. hyperfine structure), an IPM description may turn out to be insufficient.

b) Unperturbed Valence Configurations: These are cases where, due to symmetry, there is no mixing with other low lying excited valence or Rydberg configurations. These are usually states with maximum spin multiplicity (e.g. Carbon $1s^2\ 2s\ 2p^3\ ^5S^0$). Here, the Hartree-Fock SCF function constitutes a good global description of the exact wave-function.

c) Perturbed Rydberg States: These are cases where mixing with a different Rydberg series and/or, more important, with a nearby valence configuration of the same symmetry, results in an overall mixed N-electron wave-function with different characteristics from the original Rydberg functions. Therefore, properties sensitive to the character of the wave-function, such as dipole or transition moments, can be obtained reasonably accurately only after a many-electron treatment of such states.

d) Perturbed Valence States: These are cases which arise from the same type of situations as those described in (c). The single configuration Hartree-Fock (HF) description is not sufficient. Furthermore, if separate SCF calculations are carried out for the valence and Rydberg configurations, the resulting orbitals exhibit serious nonorthonormality (NON) effects which complicate the computational aspects of a many-body treatment.

e) Inner Hole Rydberg or Valence States: These states usually lie in the continuum. They are created during collisions of atoms or molecules with high energy photons, electrons or other atoms. Depending on the mixing with nearby bound or scattering configurations and the degree of hole localization, the zeroth order HF description may or may not be a good representation.

The above shell-structure classification leads to the notion of a species dependent zeroth order wave-function in a physically motivated many-body theory which allows a much faster convergence to the desired answer for a particular property. For example, if one is aware of a near degeracy or otherwise important mixing (e.g. hydrogenic mixing in the alkaline earths or molecular configurations dictated by the correct dissociation), he can easily define and obtain the zeroth order vector accordingly from a trivial CI, rather than allow a standard, single configuration based many-body algorithm to introduce the important correction to the single configuration as a correlation effect.

III. ELECTRON CORRELATION IN EXCITED STATES

Electron correlation (EC) is defined as the difference between the Hartree-Fock (HF) approximation and the exact nonrelativistic solution. With this definition, EC is often more important and, at the same time, more difficult to compute in excited states than in ground states. Thus, accurate information about excited states requires, in most cases, the application of many-body theories. Apart from the straight-forward, big computations which use standard algorithms, it is desirable to have theoretical treatments which address themselves to the task of analyzing the mathematical structure of the wave-function of classes of states and of producing quantitative and interpretive results on physical phenomena with a minimum amount of effort. Such treatments may also decrease the chances of systematic errors. The analysis and computational approaches presented below are in this spirit.

A. The Fermi-Sea (FS) Zeroth-Order Vector

Suppose we have a complete set of one electron functions, $\{\varphi_n\}_{n=1}^{\infty}$, and we examine a nonrelativistic N-electron system. We divide them as follows:

$$\underbrace{\varphi_1, \varphi_2, \ldots \varphi_N; \varphi_{N+1}, \ldots \varphi_k}_{\text{Fermi-Sea spin-orbitals}} \Big/ \underbrace{\varphi_v, \varphi_{v+1}, \ldots}_{\text{virtual orbitals}} \qquad (1)$$

The first N spin-orbitals constitute the functions which represent the single configuration, shell model description of the state of interest. They are assumed calculable in the restricted HF approximation. The remaining $\varphi_{N+1}, \ldots, \varphi_k$ functions complete the Fermi-Sea (FS) of this state. These functions, although unoccupied in the single configurational approximation, are very important in determining the optimum zeroth order description of the wave-function. They form configurations which, in the HF approximation, either have diagonal matrix elements close (in a relative sense) to the energy of the single configuration, Φ_{HF}, or have large off-diagonal matrix elements with Φ_{HF}. The FS configurations have relatively large coefficients and hence their importance for an optimum description of the zeroth order vector. We shall call the set of orbitals out of which the zeroth order vector configurations are constructed, the Fermi Sea [1,2]. These orbitals can be established for each state separately using criteria such as: a) filling of shells as revealed by optical spectra, b) photoabsorption experiments where vacant orbitals near the occupied ones play an important role in the interpretation of the observed spectra, c) approximate HF calculations of of atomic or molecular orbital energies for a few configurations (suspected of) having nearly the same energies, d) in atoms and diatomics, near-degeneracies suggested by the exact solutions of H and H_2^+, which usually represent two-electron replacements but big off-diagonal matrix elements (these are always included), e) trial and error calculations with a priori fixed coefficient size thresholds [3,4], f) formation of configurations which assure the correct molecular dissociation.

In table 1 we give a few samples of atomic and molecular states, their single configuration representation and the corresponding Fermi-Seas. We take three states from Be to show the dependence of the Fermi-Sea on the symmetry of the state, especially in systems of high symmetry such as the atoms.

In conclusion, we abandon the single-configurational description of the zeroth order state and instead we choose the Fermi-Sea vector, Φ_{FS}, whose configurations consist of H-F orbitals, in practice computed from separate H-F calculations on each FS configuration. For atoms this is feasible and inexpensive using numerical functions [7]. For molecules it is more difficult, especially for states with many open-shells [8]. We note that once a list of FS orbitals is created from different configurations, a reasonable choice of radial functions should then be made to minimize non-orthonormality (NON) effects which must be taken into account. We shall return to the choice of the FS orbitals later.

Thinking about atomic and molecular states in terms of Fermi-Seas rather than the standard single configuration, allows not only a different computational procedure but also, and most important, a physical interpretation and prediction of phenomena such as photoabsorption and satellite peaks [9,10], the sharp increase of the dipole polarizability from Ne to Ar and the related chemical reactivity, shape resonances [11], general bonding properties etc.

B. The Subshell Cluster Expansion of the Wave-Function

Having constructed the single configuration HF function, Φ_{HF}, we may start replacing the occupied subshells with unoccupied ones from the Fermi-Sea or from the virtual space. The resulting configurations can be made by direct diagonalization spin-eigenfunctions transforming irreducibly under the molecular group. For nonrelativisitc atoms we diagonalize L^2, S^2 (see section VIII). If we perform all the permutations with all possible replacements of occupied with unoccupied subshells and group the resulting symmetrized configurations according to whether they contain single, pair, triple,..., N-tuple, subshell replacements (excitations), we obtain for an arbitrary open shell state

Table 1

Standard Configuration	Additional FS orbitals	FS zeroth order configurations	Criteria.
Be $1s^2\ 2s^2\ {}^1S$	2p	$1s^2\ 2s^2$, $1s^2\ 2p^2$	off-diagonal matrix element
$1s^2\ 2p^2\ {}^3P$	—	$1s^2\ 2p^2$	
$1s^2\ 2p^2\ {}^1D$	2s, 3d	$1s^2\ 2p^2$, $1s^2\ 2s3d$	near-degeneracy off-diagonal
$B^+\ 1s^2\ 2p^2\ {}^1D$	—	$1s^2\ 2p^2$	
Ar^+ KL $3s\ 3p^6\ {}^2S$	3d, 4s	KL ($3s\ 3p^6$, $3s^2\ 3p^4\ 3d$, $3s^2\ 3p^4\ 4s$, $3s\ 3p^4\ 3d^2$, $3s\ 3p^4\ 4s^2$)	near-degeneracy off-diagonal
F_2 core $3\sigma_g^2\ 1\pi_u^4\ 1\pi_g^4\ {}^1\Sigma_g^+$ (see ref, 5)	$3\sigma_u$	core ($3\sigma_g^2\ 1\pi_u^4\ 1\pi_g^4$, $1\pi_u^4\ 1\pi_g^4\ 3\sigma_u^2$, $3\sigma_g^2\ 1\pi_u^2\ 1\pi_g^4\ 3\sigma_u^2$, $3\sigma_g^2\ 1\pi_u^4\ 1\pi_g^2\ 3\sigma_u^2$)	near-degeneracy off-diagonal
O_2 core $3\sigma_g^2\ 1\pi_u^3\ 1\pi_g^3\ {}^3\Sigma_u^-$ (see ref. 6)	$3\sigma_u$	core ($3\sigma_g^2\ 1\pi_u^3\ 1\pi_g^3$, $3\sigma_g 1\pi_u^4\ 1\pi_g^2\ 3\sigma_u$)	near-degeneracy off-diagonal

the exact form of the nonrelativistic wave-function as a subshell cluster expansion:

$$\Psi = \Phi_{HF} + a_1\Phi_{HF}^{-1}\,\Sigma(r_1) + a_2\Phi_{HF}^{-2}\,\Pi(r_1,r_2) + a_3\Phi_{HF}^{-3}\,T(r_1,r_2,r_3) + \ldots = \Phi_{HF} + \chi \qquad (2)$$

with

$$\langle\Phi_{HF}|\chi\rangle = 0 \ ; \ \langle\Psi\,|\,\Phi_{HF}\rangle = 1 \qquad (3)$$

$\Sigma(r_1)$ is a single electron correlation function which, coupled to the (N-1) configuration Φ^{-1}, yields the overall correct symmetry. $\Pi(r_1,r_2)$ is a symmetry adapted subshell pair correlation function, $T(r_1,r_2,r_3)$ is a symmetry adapted subshell triple correlation function and so on.

Formal expansions such as that represented by expression (2) have appeared in both statistical (partition functions) and quantum (wave-functions) mechanics of many particle systems and have formed the foundations for systematic analyses and approximations to the many-body problem [12-21]. In quantum chemistry, they were developed and applied in terms of spin-orbitals. However, since most methods start with the restricted HF approximation where the radial functions correspond to subshells as assigned by the configuration and not to spin-orbitals, it appears at least as useful to work with a many-body theory expressed in terms of subshells which symmetrized are to form configurations and correlation vectors.

As a simple example consider the "classic" case, the Be $1s^2\,2s^2\,{}^1S$ state. Eq. 2 becomes:

$$\begin{aligned}\text{Be } {}^1S = &|1s^2\,2s^2\rangle_{HF} + [(1s\,2s^2)^2S\ \Sigma_{1s}(r_1)\,{}^2S] + [(1s^2\,2s)^2S\ \Sigma_{2s}(r_1)\,{}^2S] + [(2s^2)^1S\ \Pi_{1s^2}(r_1,r_2)\,{}^1S] + [(1s\,2s)^{3,1}S\ \Pi_{1s2s\,{}^{3,1}S}(r_1,r_2)^{3,1}S] \\ &+ [(1s^2)\,{}^1S\ \Pi_{2s^2}(r_1,r_2)\,{}^1S] + [(1s)^2S\,T_{1s2s^2}(r_1,r_2,r_3,)^2S] + [2s\,{}^2S\ T_{1s^22s}(r_1,r_2,r_3)^2S] + K(r_1,r_2,r_3,r_4)\end{aligned}$$

where we have absorbed the coefficients into the correlation functions.

In the following sections we analyze eq. (2) in the case of

excited states and demonstrate how one can implement the calculation of the one-and two-electron correlation functions using variational procedures and one electron basis sets.

C. The First Order Form of the Wave-Function

If the exact wave-function of eq. (2) were known, the exact nonrelativistic energy E would be obtained in terms of the first three terms only. This follows from the fact that the N-electron Hamiltonian is a one-and two-electron operator and the following manipulations:

$$E = \frac{\langle \Psi | H | \Psi \rangle}{\langle \Psi | \Psi \rangle} \quad , \quad \Psi = \Phi + \chi \tag{4}$$

$$\langle \Psi | \Psi \rangle = 1 + \langle \chi | \chi \rangle \tag{5}$$

$$E = \frac{\langle \Phi | H | \Psi \rangle + E \langle \chi | \chi \rangle}{1 + \langle \chi | \chi \rangle} \tag{6}$$

$$= \langle \Phi | H | \Phi \rangle + \langle \Phi | H | a_1 \Phi^{-1} \Sigma(r_1) + a_2 \Phi^{-2} \Pi(r_1, r_2) \rangle \tag{7}$$

$$= E_{HF} + E_{Corr} \tag{8}$$

We note that the same formula holds for approximate CI calculations (and therefore approximate a_1, a_2, $\Sigma(r_1)$ and $\Pi(r_1,r_2)$) where, for a finite set of configurations Φ_n, n=0,...,S, with

$$\Psi_{CI} = \Sigma_{n=0}^{S} c_n \Phi_n, \quad H_{nm} \equiv \langle \Phi_n | H | \Phi_m \rangle, \quad \underset{\sim}{H} \underset{\sim}{C} = E_{CI} \underset{\sim}{C} \tag{9}$$

after diagonalization:

$$E_{CI} = \langle \Phi_o | H | \Phi_o \rangle + \Sigma_{n=1}^{S} \frac{c_n}{c_o} \langle \Phi_o | H | \Phi_n \rangle \tag{10}$$

Equation (7) says that the exact energy depends on the coefficients a_1 and a_2 and the single and pair correlation functions. Of course, these can be obtained exactly only after a complete solution of the Schrödinger equation. However, first order perturbation theory, which truncates the cluster expansion at the pair correlation level, suggests a form of the first order wave-function (which yields the total energy up to third order) which coincides with that of the portion of the exact function surviving the matrix element $\langle \Phi | H | \Psi \rangle$ of eq. 7. Thus, we could write for the form of a trial function:

$$\Psi_1' = \Phi + a_1' [\Phi^{-1} \Sigma'(r_1)] + a_2' [\Phi^{-2} \Pi'(r_1, r_2)] \tag{11}$$

where a_1', a_2', $\Sigma'(r_1)$ and $\Pi'(r_1,r_2)$ are to be computed within a certain scheme, equal to, equivalent to or approximating total CI within this restricted Hilbert space. Such calculations [e.g. 22-33,2] on ground states suggest that for small systems the form (11) yields very good approximations to the exact function and correlation energy (about 10%-20% error). Therefore, Ψ_1' appears to be decoupled very well from the higher order terms (triple, quadruple clusters) at least in small, ground state systems so that a theoretical treatment of the electronic structure of few electron systems based on the form of Ψ_1' seems justified. On the other hand, from arguments which we give below, it appears that higher order clusters will prove important in certain highly excited states even for small systems and therefore for answers of chemical accuracy (.05 eV-.1 eV) they should be considered explicitly.

D. Comments

1. Although the form of the exact cluster expansion always remains the same, its exact contents depend on the choice of the zeroth order function, Φ, the reference state. In most cases this is a single configuration from which all single and double subshell excitations are taken. However, Φ can be a Fermi-Sea, multiconfigurational reference state built from a common set of basis functions. Such is the case of the CI model developed and employed by Buenker and Peyerimhof [4,35]. In this case, the size of the calculation increases considerably - and so does the accuracy [4,35] - since it includes excitations which would be termed triples or quadruples in the case of the single configurational reference state.

2. The importance, or lack of it, of each of the cluster functions in eq. 2, assumed to be calculable exactly, depends on the choice of the zeroth order, initially occupied subshells. Thus, for a closed shell system, the first order function of the Rayleigh-Schrödinger [25] perturbation theory, which yields the energy to third order, equals the sum of the first order pair functions, each one calculated separately [17,30,31,36]. If Φ is the HF function, the total energy of a small closed shell system is approximated well by the sum of the corresponding pair energies. If, however, Φ is, say, a hydrogenic vector (for an atom) pair correlations are not sufficient and more weight is put on single, triple and higher excitations which correct for the non-optimal IPM choice. This effect has also been found in molecules [36]. Similar observations can be made for HF canonical vs. localized molecular orbitals (LMO) [24] where physical intuition suggests that in many cases the LMO's will help minimize pair-pair interaction corrections to a calculation based on pair energies only.

We now come to a description of the Σ and Π cluster functions for excited states and analyze briefly their physical significance before we go into the computational aspects of the theory.

E. Single Subshell Correlation Functions $\Sigma(r_1)$

Each one electron function Σ can be expanded formally in terms of FS and virtual orbitals. Thus,

$$\Sigma(r_1) = \Sigma\, a_n \varphi_{FS} + \Sigma\, b_n \varphi_v$$
$$= a\Sigma_{FS} + b\Sigma_v \tag{12}$$

with

$$\langle \Sigma \mid \Phi_{HF} \rangle = 0 \tag{13}$$

Subshell single excitations are much more important for properties than for total energies. Even though a Brillouin type theorem for symmetry preserving single subshell excitations in open shells may occasionally work [38] their mixing with certain subshell pair correlations makes them more important than first order theory implies, especially for FS_2 hole filling replacements (e.g. for $H_2O^+(1a_1^2\ 2a_1\ 1b_2^2\ 3a_1^2\ 1b_1^2\ )^2 A_1$, $3a_1 \rightarrow \Sigma_{FS} = 2a_1$).

In the usual language of spin-orbital excitations and determinants, certain types of single subshell correlations are equivalent to spin-orbital pair excitations. For example, take Carbon $1s^2\ 2s^2\ 2p^2\ {}^3P$. Excitation of the HF 2s orbital to a Σ_v of d symmetry, (call it V_d), yields the correlation vector $[1s^2\ 2s\ 2p^2\ {}^{2S+1}L \otimes v_d\ {}^2D]^3P$ which is a multideterminantal L^2, S^2 eigenfunction. The core configuration can have ^4P, ^2P or ^2D symmetries. When we excite in terms of spin-orbitals we have the following situation: For the ground state,

$^3P_{HF} = (K\ 2s\alpha\ 2s\beta\ 2p_0\ \alpha 2p_1\ \alpha)$. For, say, the core configuration of ^2D symmetry,

$^2D_{HF} = (K\ 2s\alpha 2p_1\alpha 2p_1\beta)$. Therefore, the single orbital excitation $2s \rightarrow V_d$ yields a correlation vector corresponding to pair excitation $(2s\beta 2p_0\ \alpha) \rightarrow (2p_1\beta v_{d_{-1}}\ \alpha)$. This type of spin-orbital pair correlation has been called "semi-internal" [19] and has been computed to different degrees of accuracy for several atoms and molecules [39-43]. This equivalence of certain single subshell excitations with spin-orbital pair correlations which are known to be important, underlines the significance of $\Sigma(r_1)$ especially when symmetry allows excitation to a vacant FS orbital (e.g. Al I KL $3s\,3p^2\,{}^2D \rightarrow 3p^2\,3d\,{}^2D$).

F. Pair Subshell Correlation Functions, $\Pi(r_1,r_2)$

The pair function $\Pi(r_1,r_2)$ can also be expanded formally as a sum of symmetrized subshell products:

$$\Pi(r_1,r_2) = \Sigma_{n,m}\, a_{n,m}\, |n\rangle_{r_1} |m\rangle_{r_2} \tag{14}$$

with

$$\langle \Pi | \Phi_{HF} \rangle = 0 \tag{15}$$

The sum can be divided into three meaningful parts: One is the part with no Fermi-Sea subshells, i.e. both electrons occupy virtual subshells, and the others with one or two FS subshells and one or no virtual subshells. The total pair function can then be divided into three orthogonal parts:

$$\Pi = a\Pi_{FS} + b\Pi_{FS-V} + c\Pi_{V} \tag{16}$$

Π_{FS} constitutes the larger part of Fermi-Sea correlations whose spectroscopic and chemical importance has already been mentioned.

Π_{FS-V} represents Rydberg and continuum configurations which may lie above or above and below the Φ_{HF} configuration. This is seen immediately if the virtual V is expanded in terms of spectroscopic orbitals. The significance of Π_{FS-V} is then great in that it gives rise to correlation vectors with real spectroscopic significance. Since they usually lie close to Φ_{HF}, these vectors mix heavily with it causing serious deviations from the IPM description and affecting proportionally the intensities of observed spectra [e.g. 44-47]. Furthermore, if Φ_{HF} is a highly excited state embedded in the continuum of the corresponding ion, Π_{FS-V} for a particular pair of valence electrons represents a well-known physical phenomenon, the Auger effect or autoionization.

The spin-orbital counterpart of Π_{FS-V} is the "semi-internal" correlation of ref. 19. However, there is no equivalence since, due to symmetry, Π_{FS-V} gives rise to correlation vectors corresponding to three (or more) spin-orbital excitations not included in the original work, [19,40], while, as we saw, a part of the spin-orbital "semi-internal" correlation is included in our subshell single correlation. Furthermore, the notion of the Fermi-Sea is more general and complete than that of the "Hartree-

Fock sea" [19,40,41] on which the classification of refs. 19,40 are based.

Π_V is important for energy reasons. It contains the additive contributions of an infinity of high energy wave-packets and only rarely of a doubly excited spectroscopic configuration. Depending on the symmetry of the state, Π_V correlation vectors contribute 50%-90% of the total correlation energy in small systems. However, since their diagonal matrix elements are far from E_{HF} and their off-diagonal with Φ_{HF} are small, their effect on the total wave-function in terms of magnitude of their coefficients is small and often negligible for calculation of various expectation values [1]. (For certain one-body properties (e.g. hfs, valence-Rydberg transition probabilities), bi-virtual correlation can play a significant indirect role--by affecting the symmetry preserving single excitations.) Furthermore, they often seem to play a secondary role in bonding in the sense that they affect the total energy by an approximately constant amount independent of geometry. This is so because, roughly speaking, bonding implies electron reorganization from the atomic to the molecular and solid state. This reorganization requires virtual transitions to low lying states, something that the bi-virtual correlation Π_V does not represent. This property of Π_V might allow their neglect when there is interest only in the shape of a potential energy surface (from which energy differences can be deduced) but not in the absolute minima. However, for diatomic excited state surfaces with avoided crossings caused by valence-Rydberg interaction, the bivirtual correlation is different for each type. Therefore if accurate information about the exact properties of the wave-functions at the crossing is required, Π_V must be considered. In atoms, this additional correlation can be taken into account by a semiempirical effective Hamiltonian approach which shifts the energy of the valence state with respect to the Rydberg series and yields correct mixing coefficients, provided the zeroth order functions are represented by true SCF HF solutions [2,45].

The calculation of pair correlations by traditional CI methods has been known to be slowly convergent requiring an overwhelming number of configurations and fixed virtual orbitals. This is due to the difficulty in calculating accurately Π_V. In fact it has been repeatedly stated that only r_{ij} dependent basis sets can hope to handle Π_V efficiently. Our work of the past few years suggests that even a few virtual orbitals--Gaussians or STO's--can yield good convergence provided they have been optimized by minimizing the total energy. We shall return to this point in section IX.

IV. TRIPLE (T) AND QUADRUPLE (K) CORRELATION FUNCTIONS IN HIGHLY EXCITED STATES: ENERGY CRITERION

By now, it is fairly well established that for ground states of small systems, single and pair correlations computed to all orders (i.e. total CI) suffice to obtain about 80%-90% of the correlation energy. Higher order clusters become important with increasing number of electrons and dominate in extended systems [48-52]. Quadruples could also be relatively more important if their unlinked pairs represent Fermi-Sea correlations. This can be deduced from the work of ref. 28 and implies that in molecules the dissociation region is less well described by a wave-function of the form Ψ_1' based on a single configuration reference function.

The importance of higher order clusters in excited states has not been studied yet. The argument of the number of electrons holds here as well of course. However, it appears that orbital clusters such as triple and quadruple will prove important for certain highly excited states even for small systems. These are states with holes in the HF configuration (section II, category 5). The reason is that triple and quadruple rearrangements are here favored energetically much more than in ground states where the energy denominators in the perturbation expansion become very large when exciting simultaneously three or four electrons. Now, the filling of the lower lying hole or two holes by valence electrons is a "source of energy" for the other electrons which are virtually excited to unfilled higher lying orbitals of the Fermi-Sea or of the virtual space. For example, consider the decoupling of the triple and quadruple cluster functions into their unlinked terms;

$$T \rightarrow \Sigma \times \Pi + \ldots \tag{17a}$$

$$K \rightarrow (\Sigma \times T) + (\Pi \times \Pi) + \ldots \tag{17b}$$

Take (17a). If the orbital excitation of Σ is such that it fills an inner hole, particularly if it is of different symmetry, (e.g. (AOs) $d \rightarrow s$, (MOs) $\pi^* \rightarrow \sigma$), and if Π is relatively important (e.g. near-degeneracy pair correlation), the triple cluster function will be important since the corresponding term in the perturbation expansion will have a large denominator as well as numerator (relatively speaking). Similar arguments hold for K as well as for higher clusters provided the energy gain by filling inner vacancies is large enough to compensate for the virtual excitation of the outer electrons.

As an example, consider the inner hole state $Ar^+\ 1s^2\ 2s\ 2p^6\ 3s^2\ 3p^6\ {}^2S$. The triple cluster $T = [2s \times \Pi(r_1, r_2)]$, representing correlation in the $3p^6$ shell, will give rise to the correlation vector $[(1s^2\ 2s^2\ 2p^6\ 3s^2\ 3p^3)^{2S+1} \otimes \Pi(r_1,r_2)\ ^{2S'+1}L']\ ^2S$ which may have a diagonal energy close to the hole state thus contributing in a non-negligible way.

We note that a separation of T similar to that of Π (eq. 16) is also meaningful and straightforward, with similar observations applying to it.

Although we don't have numerical proof of the importance of higher clusters in excited states of small systems, this can be inferred from the observed physical counterpart of the triple cluster, the double Auger effect [53] where two, instead of one electrons are ejected.

We note that in assessing the importance of higher order clusters, one must also define the observable of interest. Thus, although for total energies these effects may be small, for energy differences, e.g. one electron binding energies, they will be more crucial. An analogous situation occurs with the calculation of electron affinities of first row atoms [28].

In conclusion, the above heuristic considerations suggest that apart from the ordinary criterion of the number of electrons (i.e. proximity to the electron gas situation) there is also an energy criterion which applies to excited states: the higher the energy of the excited, inner hole state, the greater the possibility for multielectron reorganization and hence for multielectron correlations.

V. RECOGNITION OF CERTAIN IMPORTANT CORRELATION EFFECTS BY INSPECTION

The value of knowing, before doing any calculations, when correlation and/or which types of correlation play an important role is considerable. It enables the experimentalist to anticipate the magnitude of the gross deviations from the independent particle model without large scale computations; and it enables the theorist to determine whether an expensive calculation of χ (eq. 2) is really necessary and also, given the frequently occurring limitations due to expense and computer size, to determine for which problem χ can be treated approximately and still yield accurate answers.

The decoupling of the cluster expansion into the form of Ψ_1 (eq. 11) and subsequently of $\Sigma(r_1)$ (eq. 12) and $\Pi(r_1,r_2)$ (eq. 16), allows the recognition of the importance of certain correlation effects which affect the wave-function and therefore properties such as photoabsorption spectra. These effects can be easily seen in terms of orbital replacements. The most common ones are:

1) Fermi-Sea configurations (see discussion in III A)

eg. $O_2\,|{}^3\Sigma_g^-\rangle$: $(3\sigma_g^2\,1\pi_u^4\,1\pi_g^2,\ 3\sigma_g^2\,1\pi_u^2\,1\pi_g^4,\ 3\sigma_u^2\,1\pi_u^4\,1\pi_g^2)$

$Mg\,|{}^1D\rangle$: (KL $3p^2$, KL $3d^2$, KL 3s 3d)

2) Hole-Filling pair correlations (Occur in photoelectron spectroscopy): These are of the Π_{FS-V} type (eq. 16) for HF configurations with an inner hole. The virtual orbital V is a sum of Rydberg and continuum orbitals.

eg. $O_2\,|{}^3\Sigma_u^-\rangle$: $(3\sigma_g^2\,1\pi_u^3\,1\pi_g^3 \leftrightarrow 3\sigma_g^2\,1\pi_u^4\,1\pi_g n\pi_u)$

$B\,|{}^2S\rangle$: $(1s^2\,2s\,2p^2 \leftrightarrow 1s^2\,2s^2\,ns)$

3) Symmetric Exchange of Orbital Symmetry: These are the most important of the Π_{FS} and Π_{FS-V} type pair correlations. They occur both in the discrete and the continuous spectrum provided the quantum numbers or energies are close. (See also our accompanying paper "Theory of One Electron Binding Energies" for additional discussion of these effects.)

eg. $p^2 \leftrightarrow sd$

$d^2 \leftrightarrow pf$

$\pi^*\sigma \leftrightarrow \pi\sigma^*$

Examples:

a. Binding energy of the 3s electron in K. The energy shift due to $3p^2 \rightarrow 3s\,3d$ FS correlation is 6.6 eV [2].

b. Fluorine $1s^2 2s\,2p^6\ {}^2S$ Autoionizing State. The energy shift of the HF energy due to the surrounding continuum is dominated by the interaction $2p^2 \rightarrow 2s\,\varepsilon d$ which is 3.3 eV [54]. Here, εd are continuum HF orbitals calculated in the HF ionic core of $F^+\ 1s^2\,2s^2\,2p^4\ {}^1D$ [54].

c. Double Electron Photoionization.

eg. O $1s^2\,2s^2\,2p^4\ {}^3P \xrightarrow{h\nu} 1s^2\,2s\,2p^3\,(\varepsilon p\varepsilon' p)\,{}^3P^o$

The CI in the continuum will be most important for the channels representing the mixing $\varepsilon p\varepsilon' p \leftrightarrow \varepsilon s\varepsilon' d$.

d. Ethylene $^1(\pi,\pi^*)$ state [55]. O_2 $^3\Sigma_u^-$ state [56]. In both of these excited states, the mixing $\sigma\pi^* \leftrightarrow \sigma^*\pi$ is suggested by the authors [55,56] as being important - in agreement with the symmetry exchange idea of this classification.

e. Rb^{++} photoelectron spectroscopy satellites [57]. The inner hole $4s\ 4p^6\ ^2S$ state of Rb III is found [57] to mix much more heavily with $4s^2\ 4p^4\ 4d\ ^2S$ than with $4s^2\ 4p^4\ 5s$, in agreement with the rule $p^2 \leftrightarrow sd$. This mixing shows up in the emission spectrum as a satellite peak.

More examples from atomic theory can be found in ref. [1]. These were based on our previous computational work on excited states of the second and third row [e.g. ref. 41 p. 545, 42] and an analysis of various HF R^k integrals and of the then available literature.

The importance of some of these effects in atoms has also been noted by other groups working in the field of photoelectron spectroscopy [e.g. 58-60].

VI. THE CALCULATION OF Ψ_1'. APPLICATION TO ATOMS

We now turn to the problem of how to calculate Ψ_1' accurately and efficiently. No doubt, for small systems the surest approach from the point of view of accuracy would be to carry out as large a CI as possible - provided a large computer and computer budget is available. However, we have already argued for an alternative approach which is more economical, equally accurate, physically instructive and completely general and feasible for large N systems. It is outlined below.

A. The Restricted Hartree-Fock Function, Φ_{HF}

Most many-body theories start with a HF zeroth order function. This usually applies to the ground state. The excited states are then obtained through the particular algorithm used. For diagrammatic perturbation theory and Green's functions methods now employed, excited state information relies upon a calculation based on a single basis set having as reference a single state. For straight forward CI, excited state configurations are built from orbitals also corresponding to a single reference state. This practice is not without danger of serious errors. For example, Nitzsche and Davidson [60] have recently pointed out the inadequacy of using ground state or triplet $\pi\pi^*$ orbitals to describe the singlet $\pi\pi^*$ state of amides. In atoms, such deficiencies of a single HF set to describe more than one state are also existent. For example, the alkaline earths singlet and triplet (nsnp) $^{3,1}P^o$ first excited states are represented by quite different HF orbitals. The same often holds for HF orbitals of valence and Rydberg configurations of the same symmetry.

The theory which we present in this lecture is state specific. This means that the zeroth order function is radially optimized for each state and for each geometry separately. This allows consistent and systematic treatment of electron correlation in excited states. (In this way, EC also becomes state specific).

From the computational point of view, there are two distinct HF methods: The numerical one, originated with Hartree and extensively analyzed and computerized by Froese-Fischer [7], and the analytic one first introduced and developed by Roothaan and his school [61]. The numerical approach solves the Hartree-Fock integrodifferential equations directly. The analytic approach transforms these equations into a matrix eigenvalue equation using as basis sets STO's or GTO's.

Both methods have been applied to very many atomic states and a few diatomic molecules [62]. For polyatomic molecules only the analytic method is feasible at present. The numerical HF is more efficient because the integrations are done numerically and has the advantage that it can be made to converge to the correct excited state with relative ease by fixing the nodal structure of the desired root. It is also more accurate in general, especially for large atoms or for diatomic molecules [62].

The analytic HF constitutes, of course, the backbone of molecular LCAO-MO studies. Nevertheless, there are still problems with excited configurations of many-open shells which require special treatment [e.g. 8] as well as with convergence to the correct root in molecular excited states having lower states of the same symmetry. On the other hand, the analytic method can also allow the user more control over the computational process, often making evaluation of certain highly excited (autoionizing) states easier, at least in atoms. Such functions have been obtained for example for the states He^- $2s^2$ 2p $^2P^o$, Li 1s $2p^2$ ^2D etc. [44,63]. We note that analytic functions of autoionizing states at the H-F level are required in a recently developed many-body theory of resonances using the complex coordinate rotation method [11,64,65].

In general, the situation regarding applicability of state specific SCF HF procedures to arbitrary excited states and availability of related computer programs is satisfactory for atoms [60] while significant progress has recently been made for molecules [8,66,67]. We note that if accurate correlation studies are to become "easy" in the future, computer costs of the SCF calculations should not exceed ~ 1/4 of the total cost. Thus, a development such as that of McCullough's on diatomics [62] is indeed encouraging.

B. The Fermi-Sea Wave-Function, Φ_{FS}

The Φ_{FS} function is a linear combination of a few (two to four) zeroth order configurations which mix heavily (see section III A). In the ideal case, this function should be calculated by the Multi-Configurational HF (MCHF) method which optimizes both the radial functions and the mixing coefficients. Again, this method can be numerical [71, 32] or analytic [5,68]. We follow the practice of obtaining separate HF functions for each Fermi-Sea configuration which are then used together with other correlation vectors to be derived below, to diagonalize small CI matrices. This approach is cheaper and more stable. From some test cases on Be and C we have found that there is very little difference between the two procedures. The NON difficulties which arise with our method are accounted for explicitly since the size of the CI within the FS is very small (2x2-4x4). We note that in molecules, the FS configurations should be radially optimized for different geometries of interest so that each configuration represents the best diabatic state for each geometry.

C. The Cluster Functions, Σ and Π

The expansions 12 and 14 together with perturbation-variation theory, are the guidelines for the computational scheme which we follow for the calculation of Σ and Π. There are three basic steps in the theory:

a) Decouple (formally) Σ_{FS}, Σ_V, Π_{FS}, Π_{FS-V} from Π_V. This implies a decoupling of the (roughly) low lying "states" from the roughly high lying "states"(Π_V) in the sense that the radial characteristics of these correlation functions are assumed essentially unchanged by the interaction of (Σ_{FS},Σ_V,Π_{FS},Π_{FS-V}) with Π_V. We compute the two parts separately by variationally optimizing non-linear parameters of one electron functions. (See below.)

b) Decouple the total Π_V into separate symmetry adapted pairs and optimize variationally their radial functions from small iterative 2x2 CI, i.e. $\Phi_{HF} + \Phi_{HF}^{-2}\Pi_V$. The notion of a "pair at a time" approach was first suggested by Sinanoğlu [17,23] after he showed that, to first order, the correlation vector χ for closed shell systems is rigorously decoupled into pairs.

c) Having obtained the radial characteristics of the cluster functions in this way, we form the total CI matrix of the total function Ψ_1' and diagonalize once, without any further optimization, to obtain the correct (within the model) coefficients of the cluster expansion.

D. Radial Optimization of the Virtual Orbitals in Excited States

In the usual CI calculations, the virtual functions are computed in one of the following ways:

1) Use the unoccupied orbitals of LCAO-MO SCF calculation on a ground state to construct excited state configurations and correlation vectors. The common belief that one-electron function basis set CI is slowly convergent has its foundation in this type of calculation. The Improved Virtual Orbital (IVO) method of Hunt and Goddard [69] which used a V^{N-1} potential to create virtuals, is indeed an improvement when only one STO per symmetry is necessary. However, it will probably be much less successful if more than one STO of the same symmetry are required. This is because of the following: In atomic terms, $\langle r \rangle$ is proportional to n^2/Z^* where n is the principal quantum number and Z^* is the effective nuclear charge. Since the potential essentially fixes Z^*, $\langle r \rangle$ increases with n quickly in large intervals. Hence the radials become diffuse rapidly as n increases and don't contribute to electron correlation efficiently. This is a characteristic of any V^{N-m} potential!

2) Use of Natural Orbital techniques [70]. The main computational procedures are those of Bender and Davidson [70]--called the Iterative Natural Orbital (INO) technique--and of Edmiston and Krauss [71] called the Pseudo-Natural Orbital (PNO) (see also ref. 30). Both methods are based on the diagonalization of the first order density matrix of a function built with canonical HF orbitals. In the INO this is done iteratively in order to improve the energy while in the PNO one pair is first obtained from an ordinary CI, and then transformed to a NO pair which is then used for all the other excitations. NO techniques are becoming more or less standard practice because of their significant improvement in convergence upon the calculation of type 1. The number of NO per pair is between four and ten. However, in spite of their relative success, INO procedures may still need improvement regarding convergence, especially when describing correlation in the bond region.

3) In our work, the virtual orbitals are spectroscopic, i.e. state specific, as well as simple STO's and GTO's whose non-linear exponents are optimized variationally by minimizing the relevant energy functionals. Since we have already mentioned the calcu-

lation of Φ_{FS} (i.e. the calculation of Σ_{FS} and Π_{FS}) we proceed with the calculation of the cluster functions Σ_V, Π_{FS-V} and Π_V.

a) The Π_{FS-V} and Σ_V functions: The Π_{FS-V} are parametrized and calculated according to the type of state and its position in the spectrum. To see what this means, let us consider a formal expansion of a particular Π_{FS-V} in terms of symmetrized orbitals:

$$\Pi_{FS-V} = \Sigma\, a_n \,|FS\rangle|n\rangle + \int d\varepsilon\, a_\varepsilon \,|FS\rangle|\varepsilon\rangle \qquad (18)$$

$|FS\rangle$ represents the FS orbital, which is a fixed HF orbital, and $|n\rangle$, $|\varepsilon\rangle$ are Rydberg and continuum orbitals in terms of which the virtual orbital V is expanded. The actual choice of V depends on the energy position of the state under consideration:

1) The state is the lowest of its symmetry. In this case, all the single and pair excitations yield correlation vectors which push the HF energy down from above. The virtual can then be expanded in terms of one, two, or three STO's or GTO's (depending on the compactness of the system), and then optimized variationally by iteratively minimizing the CI energy of eq. 10 of the diagonalized matrix containing the configurations Φ_{HF},$\Phi^{-1}_{HF}\Sigma_{FS}$, $\Phi^{-1}\Sigma_V$, $\Phi^{-2}_{HF}\Pi_{FS}$ and $\Phi^{-2}_{HF}\,\Pi_{FS-V}$ as a function of the nonlinear parameters of the STO's or GTO's. The same goes for the virtual orbitals of Σ_V. Due to first order theory symmetry restrictions, the virtual orbitals of Σ_V and Π_{FS-V} have only a few fixed symmetries and thus this type of iterative variational CI is small (around 5-30 on the average) and relatively inexpensive, especially when experience with a few states is gained.

2) The state is not the lowest of its symmetry. In this case, depending on the type of the excited state, Π_{FS-V} (not Σ_V), together with Σ_{FS} and Π_{FS}, can give rise to configurations and correlation vectors lying below the state of interest. Examples: a) Al^+ KL $3d^2\ {}^1D$; Σ_{FS} = 3s, i.e. $\Phi^{-1}_{HF}\,\sigma_{FS}$= KL 3s 3d ${}^1D$ which is below. b) B^+ $1s^2\ 2p^2\ {}^1S$; Π_{FS} = $(2s^2)$, i.e. $\Phi^{-2}_{HF}\,\Pi_{FS}$= $1s^2 2s^2\ {}^1S$. c) B $1s^2$ 2s $2p^2\ {}^2S$; $\Pi_{FS-V} = \Sigma\, \alpha_n |2s\rangle|ns\rangle$. The first five HF ns represent the configurations $1s^2$ $2s^2$ ns ${}^2S$ which are below the valence state $1s^2$ 2s $2p^2\ {}^2S$. d) He 2s 2p ${}^1P^o$; Π_{FS-V} = $\Sigma\, \alpha_n |1s\rangle|np\rangle + \int d\varepsilon\, \alpha_\varepsilon |1s\rangle|\varepsilon p\rangle$. This state not only has all the Rydberg configurations $|1s\rangle|np\rangle\ {}^1P^o$ below it, but also is embedded in the continuum $|1s\rangle|\varepsilon p\rangle$ into which it decays-- i.e. He 2s 2p ${}^1P^0$ is autoionizing. These cases are handled in the following way: The Rydberg and continuum states surrounding the configuration of interest are computed separately and are included

explicitly in the expansions of Σ_{FS} and Π_{FS-V}. Since, for atoms, the matrices are small, the NON problem which ensues is solved exactly (see section VII). The remaining of the expansion of Π_{FS-V}, which represents states above the state of interest, is parametrized with STO's or GTO's which are optimized, as before, variationally.

Comments: 1) Due to symmetry restrictions, the virtuals in Π_{FS-V} can only have a finite number of orbital symmetries ($\ell_V \leqslant 3 \times \ell_{max}$, where ℓ_{max} is the highest FS symmetry).

2) In optimizing Π_{FS-V} for large systems, one can decouple it from Σ_V, Σ_F and optimize the non-linear exponents in Σ_V and Π_{FS-V} separately. Numerical experience [42] suggests that this decoupling and separate optimization determines fairly accurately at what values of the non-linear parameters, ζ, the minimum total energy occurs, although the true minimum is obtained from a total CI where the ζ, being already accurate, are adjusted only slightly or even kept fixed.

3) The repeated decoupling procedures result in a series of small problems none of which requires simultaneous optimization of more than 3-4 parameters. For example, a 5000 determinant, spin-orbital based wave-function corresponding to Σ_F, Σ_V, Π_{FS}, Π_{FS-V}, was decoupled [42] so that no virtual section had more than 200 dets. in the second row (Na-Ar). Molecular decoupling procedures similar in spirit to comment 2 have been used by Das and Wahl [5] and Guberman [5].

b) The Π_V functions. Symmetry adapted Pair energies. One of the first calculations on a large part of Π_V and their energies in excited states, was done on three electron autoionizing states [63] in the spirit of the Silverstone-Sinanoğlu classification [19]. The calculation was approximate in the sense that the Π_V were composed of the virtual functions from the Π_{FS-V} calculation and just a single diagonalization was carried out. The results [63,44] were reasonably accurate.

For an arbitrary N-electron atom, one must have a consistent and efficient procedure for calculating these subshell pairs. This can be accomplished as follows (similar procedures for molecular ground states are described in [30]).

Taking into account symmetry considerations, it can be shown [eg. 40] that the bivirtual symmetry adapted energy in first order has the form:

$$E_{VV'} = \Sigma_k c_k^2 \, \Sigma_{k_a<k_b} \, \Sigma_{S,L} \, T^2_{k_a k_b;SL} \, \varepsilon(n^a\ell^a n^b\ell^b;SL) \quad +$$

$$\Sigma_{\substack{k,L\\ k\neq L}} \; c_k \, c_L \, (-1)^{m_{kL}} \; \Sigma_{SL} \; T_{k_\alpha k_\beta;SL} \; T_{\ell_\alpha \ell_\beta;SL} \; \varepsilon(n^\alpha \ell^\alpha n^\beta \ell^\beta;SL) \tag{19}$$

where k_a, k_b are the spin-orbitals belonging to the determinant Δ_k, of the subshell $n^a\ell^a$, $n^b\ell^b$ which have been replaced. The second sum arises from two determinants Δ_k and Δ_L which differ in only two spin-orbitals (k_α, k_β of Δ_k and ℓ_α, ℓ_β of Δ_L) which have their N-2 common spin-orbitals aligned producing the phase factor $(-1)^{m_{kL}}$. These may arise from either the same or different configurations (if Φ is multi-configurational), although for simplicity we will now restrict Φ to be single configurational. The ε are the bi-virtual pair energies.

The factor $T_{ij;SL}$ is that which linearly combines two electron determinants (all with a common M_L and M_S) such that two electron symmetry eigenstates (S,L) are formed. In ref. 40 no explicit formula for T was given because the few results on $s^m p^n$ configurations were obtained on a case by case basis.

It turns out that T is well known [72], i.e.

$$T_{k_a k_b;SL} = (-1)^{\ell^a+\ell^b + M_L+M_S} \sqrt{\frac{(2S+1)(2L+1)}{N}} \times$$

$$\times \begin{pmatrix} 1/2 & 1/2 & S \\ m_s^a & m_s^b & -M_S \end{pmatrix} \begin{pmatrix} \ell^a & \ell^b & L \\ m^a & m^b & -M_L \end{pmatrix} \tag{20}$$

where N=2 if the subshells (a and b) are equivalent, and is 1 otherwise. The ( ) are 3j symbols.

We evaluate equation (19) explicitly for two general cases:

(i) Let a be a closed subshell and b be an arbitrary (open or closed) subshell. The second term in (19) vanishes because k_α, k_β, ℓ_α, ℓ_β must all belong to the open subshells. In the first surviving term, the sum over k_a becomes :

$\Sigma_{m_s^a=-1/2}^{+1/2} \; \Sigma_{m^a=-\ell^a}^{+\ell^a}$ and as only T^2 depends on these quantities (because a is closed and common to all Δ), these sums may be performed using standard methods [72], yielding $1/(2(2\ell^b + 1))$. Since nothing remains which is dependent on m^b and m_s^b, the sum over these quantities yields o_b, the occupation number of the b subshell. Finally, as $\Sigma_k c_k^2=1$ we get

$$E_{VV'}(a,b) = \frac{o_b}{2(2\ell^b+1)} \; \Sigma_{S,L} \; (2S+1)(2L+1) \; \varepsilon(n^a\ell^a, n^b\ell^b;SL) \tag{21}$$

where the only restriction is that a is closed (b may be identical to a). The sum may be extended to cover all such a and b subshell pairs, leaving us only with contributions between all subshell pairs involving just open subshells.

(ii) The above analysis can be extended to include the contributions within (b=a) a subshell having one hole. The result is:

$$E_{VV'}\left[(n^a\ell^a)^{4\ell^a+1}\right] = \frac{2\ell^a}{2\ell^a+1} \Sigma_{SL} (2S+1)(2L+1)\ \varepsilon(n^a\ell^a, n^a\ell^a; SL) \tag{22}$$

These remaining cases have to be evaluated numerically. This has been done [73] for all optical configurations of atomic species through N=22.

The value of the above analysis is the fact that it can be used for systems where Π_V appears to be state nonspecific. In this case, Π_V is transferable and can therefore by used to build N-electron wave-functions and energies without much computation. On the other hand, we note that this chemically very desirable feature of parts of the correlation depends on the definition of the Fermi-Sea to which Π_V is orthogonalized and therefore it is not clear to what extent this is satisfied in excited states of arbitrary atoms and small molecules. We point out that a semi-quantitative idea of the extent of transferability of pair energies (not pair functions) for configurations $1s^2\ 2s^m 2p^n$ can be found in the semiempirical work of ref. [40] (for comparison see refs. 28 and 41).

Once the symmetry adapted pairs have been defined, their calculation is carried out individually, by expanding them in two-three STO's or GTO's and minimizing the energy

$$E_V = \frac{\langle \Phi_{HF} + \Phi_{HF}^{-2}\Pi_V | H | \Phi_{HF} + \Phi_{HF}^{-2}\Pi_V \rangle}{1 + \langle \Phi_{HF}^{-2}\Pi_V | \Phi_{HF}^{-2}\Pi_V \rangle}$$

. This procedure converges quickly, even for tight pairs [2], and fixes the radial characteristics of Π_V. These are then assumed constant as we diagonalize the total matrix to obtain the coefficients in the expansion of each Π_V (see equation 24).

Ψ_1' is now a compact, well correlated function. According to the problem of interest, the method of computation allows the flexibility for it to contain spectroscopic (mainly through Π_{FS} and Π_{FS-V}) as well as purely energetic information (mainly through Π_V). We note that the whole calculation involves analytic (virtual orbitals) as well as numerical (FS orbitals) functions and the integrations are carried out numerically.

VII. THE NONORTHONORMALITY (NON) PROBLEM

Usually, CI calculations in the MO picture employ a common set of orthonormal spin-orbitals. However, there are physical and chemical problems whose solution may be optimized in terms of deduced chemical information using nonorthonormal basis sets. These are cases where the system under examination (e.g. valence-Rydberg, atom-molecule reaction) can be divided into groups, each one optimized separately within its own function space, which are then allowed to interact. Thus, valence bond functions or calculations of off-diagonal matrix elements of wave-functions obtained from different CI give rise to NON between the basis functions [e.g. 74,75]. This "problem" is, of course, only computational not conceptual. For these cases we have handled it using the method of King et al [76].

For the theory we have outlined above, the separate variational calculation of the pair functions and valence Rydberg HF orbitals gives rise to NON within a single calculation. The same NON procedure can be used. However, for reasons of economy, the complete NON calculation can be reduced to the calculation of the "direct" and the most important "exchange" overlap only.

VIII. DETAILS CONCERNING CONSTRUCTION OF L^2, S^2 EIGENSTATES

For small and moderate numbers of open subshell electrons ($\leqslant$100 determinants) the symmetry adapted basis functions can be obtained by numerical diagonalization of the operator

$$\hat{A}^2 = \Sigma_{i=1}^{m} (\alpha_i \hat{S}_i^2 + \beta_i \hat{L}_i^2) \tag{23}$$

For i=1, which corresponds to the Schaefer and Harris algorithm [77], we span the full complement of the open subshell electrons. All other i's refer to sub-groups of electrons (with the only restriction being that all equivalent electrons must belong to the same subgroup), chosen to provide unique labels (up to d^3 subgroups) for the $\hat{L}^2$ S^2 vectors. The α_i and β_i are chosen so the eigenvalues of $\hat{A}^2$ are distinct.

While it is not essential to parentalize (any orthonormal set would do), it is useful for the purposes of analysis and sometimes efficiency. For example, by coupling the symmetries of two virtual subshells to a specific symmetry (S_V, L_V), one can ensure that the vector belongs to a specific FS pair--i.e. $(n\ell\, \bar{n}\, \bar{\ell})\, {}^{2S_V+1}L_V$ where $n\ell\, \bar{n}\, \bar{\ell}$ are the subshells being replaced.

Or, one can assign the parents in such a way as to correspond to optical spectroscopy conventions. Finally, when the number of possible couplings (N) for a given configuration exceeds the number (M) of distinct I and R^k radial integrals associated with all the matrix elements $\langle\Phi|H|\chi\rangle$, then we may recouple to remove N-M basis functions, provided such couplings are not important for the calculation of properties other than the energy. Examples of such reductions are available for energies [78] and transition probabilities [9].

For large numbers of open subshell electrons, a new procedure is desirable, due to diagonalization costs. Generally, these cases will involve a group of two electrons (e.g. bi-virtual) and a "core" of N-2 electrons which are distinct (no common subshells). It is possible to separately obtain eigenstates for the "core" and two electron subsystems with ease. These then can be combined using 3j symbols, also with ease (for an example, see ref. 79). This should eliminate this step as a bottleneck.

IX. APPLICATIONS

Below we present a few, previously unpublished results from calculations on small as well as large atoms which are of current physical interest. Each example demonstrates one or more computational features which, we believe, contribute to the field of computational Quantum Chemistry.

A. H^- $2p^2$ ^3P

We discuss this calculation at length to illustrate implementation details.

The HF energy for this state is E_{HF}=-0.115884 a.u., i.e. it is predicted to be unbound. The correlation vectors are: $(2p\Sigma_p)$, Π_{2p^2}. For Σ_p we took $\Sigma_p = v_p$ where v_p is a virtual orbital of p symmetry while for Π_{2p^2} we wrote:

$$\Pi_{2p^2}(r_1,r_2) = a_p(v_{p_1} v_{p_2}) + a_d(v_{d_1} v_{d_2}) + a_f(v_{f_1} v_{f_2}) + \sum_{\ell=4}^{9} a_\ell v_\ell^2 \qquad (24)$$

The virtual p-space was found to need six virtual STO's and the d-space three. Estimates for the first compact STO per symmetry were made according to IX C. The second and third STO's, when necessary, were initialized by adding them and plotting the energy as a function of the non-linear parameters (no minimization). Typically, two deep, apparently well separated minima were found. They corresponded to compact virtuals, although with substantially different parameters than that of the first STO, partly due to the extra orthogonalization. Using these initial estimates we iterated the total CI until minimization. We note that the net energy lowering was not the sum of the increments obtained in separate calculations by the addition of the STO's.

During iteration, the cross terms $v_{p_1} v_{p_2}$, $2pv_{p_2}$ were found to play an important role. After iteration was complete within the pdf space, these were transformed away by diagonalizing the first order density matrix, leaving 12 configurations of the type ℓ^2. (During iteration, one could only use the ℓ^2 form if a true MCHF calculation was performed (all coefficients and radials changing).) Configurations of the type ℓ^2 ($4 \leq \ell \leq 9$) were then added, and their energies were found to vary as $\sim \ell^{-6}$. Our final result is 0.0934 eV for the BE of H^-, in comparison with Drake's 50x50 CI with Hylleraas basis set value [57] of 0.0953 eV. The remaining difference of 0.002 eV is likely due to the incompleteness of the p and d spaces.

We believe that the results of this calculation demonstrate that single particle expansions can be highly accurate and efficient within reasonably short expansion lengths and that they can compete effectively with functions involving r_{ij}, traditionally thought to be the only way to obtain compact functions.

B. H^{--} $2p^3$ $^4S^o$

Recently, the possibility [81] of a long lived ($\sim 10^{-8}$ sec) H^{--} state which results in the ejection of one electron has been the subject of considerable controversy [82].

Shell structure considerations suggest $2p^3$ as the possible most stable structure, with the $^4S^o$ term selected as the only one for which autoionization is forbidden non-relativistically. This configuration is embedded in the two electron continuum $2p$ (ε,ε'). H^{--} is diffuse and presents an interesting theoretical problem.

At first a standard [7] HF calculation was attempted, but even with careful extrapolation, we were unable to reduce Z below 1.4. Eventually ($Z \rightarrow 0$) the potential will get too small to support a single level without modification [83] of it (this difficulty might be equivalent to the production of positive orbital eigenvalues observed [84] in the matrix SCF method-- something not permitted in the numerical version).

To proceed, we chose to replace the standard SCF procedure with an "SCF-CI" method, involving the two configurations $2p^3$ and $2p^2$ v_p for which the energy was minimized subject to the requirement that the $2p^3$ character of the root was preserved. The CI solution was then condensed into an SCF one, creating a new 2p, such that Brillouin's theorem was satisfied. The process was repeated until little change in the energy was noted. This method has been used for other autoionizing states with success [65], and a similar one has been applied to molecules [85]. Here, a very poor virial theorem was noted for the "SCF-CI" results, which however was considerably improved when correlation was added via the configurations $2p\ (v_p^2 + v_d^2 + v_f^2)\ {}^4S^o$. All three virtuals were iterated, and a local minimum was found (as we are in the double continuum, the absolute minimum corresponds to two free electrons). The final result is that the $H^{--}\ 2p^3\ {}^4S^o$ "state" is 1.083 eV above the H 2p threshold. After completing this work, we received a preprint from the Bunges [86] in which an "instability" appeared about 0.1 eV lower than our "state". These results suggest strongly that no H^{--} state exists.

C. Atomic Optimized Virtual Nuclear Centered GTO's for Use in Molecular Studies

Virtual functions are always initially represented by us in analytic form, and have only a few (usually one) non-linear parameters (exponents) to be optimized. This representation allows easy adjustment during the iterating process. Our past atomic work has used STO's and to a lesser extent (and less successfully) screened hydrogenic functions for virtuals. One STO per symmetry (STO/ℓ) is usually capable of picking up 70-90% of the correlation energy in the ℓ part of the pair in question. Usually two STO's /ℓ are sufficient. In fact the agreement with full multi-configurational virtuals obtained using standard [7] methods is quite good at this level. It turns out that rather good estimates for the exponents can be obtained cheaply--in particular for the first STO/ℓ which we discuss below (the estimation process for the additional STO's, if needed, was discussed in Section IX A).

First, the error we make (with the HF function) is due to the presence of the $1/r_{ij}$ terms and should be largest when $r_i \approx r_j$. Secondly, the largest error within this region will occur for $r_i \approx r_j \approx \langle r \rangle_{HF}$, i.e. the average HF radius for the subshell(s) being replaced. So, by setting $\langle r \rangle$ for the virtual function equal to $\langle r \rangle_{HF}$, we emphasize the proper region. Prior to orthogonalization, the virtual function ($R_{n\ell}$ form) is:

$$r^{n-1}\ e^{-ar}\ Y_{\ell m}(\vartheta,\varphi) \begin{pmatrix} \alpha \\ \beta \end{pmatrix} \qquad (25)$$

and our estimate is determined by:

$$\frac{2n + 1}{2a} = \langle r \rangle_{STO} = \langle r \rangle_{HF} \qquad (26)$$

which determines a, given n. We find that different (a,n) sets are about as effective as one another (except for hfs, where we always take n=1 for s STO's, as only such STO's survive at the origin), and we normally use n equal to the principle quantum number of the lowest unfilled subshell. Equation (26), which is always used, is so successful that frequently no further adjustment of a is needed. However, if the virtual must be orthogonalized to the FS, then a few (5-6) iterations may be required. The use of non-integer values of n is not found necessary which is fortunate as this complicates the calculation (by introducing Gamma functions).

We would like to transfer some of these results directly to molecules. For this to be valuable, three different propositions must be essentially viable. The first is that no present method of generating molecular virtuals is fully satisfactory. The second is that a portion of the significant correlation effects must be essentially the same for atoms and molecules. Finally, we need to develop a Gaussian Type Orbital (GTO) virtual representation, as essentially all non-linear polyatomic codes use GTO's and not STO's.

The generation of optimum virtual functions was mentioned in VI D. We believe that the variational procedure combined with systematization regarding initialization and extrapolation (see e.g. ref. 41 for systematics of virtual orbitals along isoelectronic sequences--related to $\langle r \rangle$) holds promise for molecules as well.

Concerning the question of transferability of functions from atom to molecule, things are more uncertain. One might expect correlation associated with core electrons (significant for binding energies, etc.) to be reasonably transferable. In the same spirit, the bi-virtual correlation will probably be the most transferable of the sections as it depends least on degeneracy effects--which are quite sensitive to their environment. Atomic virtuals might also be expected to describe the ionic part of the bond better than they do the covalent part.

The simplest virtual functions which can appeal both to atomic and molecular theorists are nuclear centered single particle GTO's with atomic optimized parameters, viz:

$$r^{n-1} e^{-ar^2} Y_{\ell m}(\vartheta,\varphi)\binom{\alpha}{\beta} \qquad (27)$$

To our knowledge, this is the first time such functions have been explored as variationally optimized single particle virtual functions. Once again we may use $\langle r \rangle$ to estimate the parameter values, i.e.

$$\frac{(1)(2)\ldots(2n)}{(3)(5)\ldots(2n-1)\sqrt{2\pi a}} = \langle r \rangle_{GTO} = \langle r \rangle_{HF} \qquad (28)$$

As we will restrict n to correspond to the lowest unoccupied subshell (this is not a necessity, only a convention. Thus $n-\ell$ can be restricted to have a definite parity, as is preferred in molecular work.), equation (28) determines the estimate for a.

In Table 2, we present results for the bi-virtual correlation of He and the L-shell of C. Both STO's and GTO's are used and compared where possible with accurate values. We find the following:

(1) GTO's are nearly as good as STO's (which are quite good) in picking up the correlation. (Internal evidence suggests the STO of d symmetry for the (2s 2p) pair is incompletely optimized, so this result is not an abberation. On the contrary, it suggests the importance of the optimizing process).

(2) For both GTO's and STO's where no orthogonalization to the FS is required, the optimized values are in excellent agreement with the estimates.

(3) The estimation process predicts the pd exponents of the $1s^2$ C pair to be ~ 13.1 (which has been confirmed for other states). This supports the statement that different pairs (e.g. K vs. L) require different exponents.

(4) It may be noted that the estimation process is uncertain for $n\ell n'\ell'$ pairs when $n' \neq n$ (when $n'=n$, then $\langle r \rangle_{n\ell} \approx \langle r \rangle_{n\ell'}$). This is symptomatic of the need for a modification. In practice, these pairs are found to require at least two STO's/ℓ —one of which is approximately near $\langle r \rangle_{n\ell}$ and the other near $\langle r \rangle_{n'\ell'}$. Development of a valid estimatory process for such pairs--which in their totality play a larger role in molecules than do the intra-shell pairs [24]-is in progress.

(5) Although it is not obvious, we found that in general, the most important correlation configuration corresponds to increasing the orbital angular momentum by one unit (for single excitations, by two units). Thus, for example, ns^2, ns np, nd^2, ns subshells have v_p^2, $v_p v_d$, v_f^2, and v_d as the most important virtual replacements.

Table 2

Comparison of Bi-Virtual Pair Energies Using One STO/ℓ ($\ell \leq 3$), One GTO/ℓ ($\ell \leq 3$), and a Nearly Complete Basis ("Exact")

	GTO	STO		Pair Energy Ratios	
Pair Energy	Exponent (power)	Pair Energy	Exponent (power)	GTO/STO	STO/"Exact"
		He $1s^2$			
-0.033 a.u.	s 0.46(2)	-0.0375 a.u.	s 1.72(2)	88%	89%[a]
	p 1.23(3)		p 2.42(3)		
	d 1.79(3)		d 3.51(3)		
	f 2.77(4)		f 4.61(4)		
		C I ground state; $2s^2$ pair			
-0.009 a.u.	s 0.29(3)	-0.0095 a.u.	s 1.75(3)	94%	75%[b]
	p 0.24(3)		p 3.19(3)		
	d 0.66(3)		d 2.18(3)		
	f 0.91(4)		f 2.81(4)		
		C I ground state; 2s 2p ($^1P^0$ + 3x $^3P^0$) pairs			
-0.017 a.u.	s 0.29(3)	-0.016 a.u.	s 2.9(3)	107%	53%[b]
	p 0.22(3)		p 2.9(3)		
	d 0.59(3)		d 2.18(3)		
	f 0.74(4)		f 2.51(4)		
		C I ground state; $2p^2$ pair			
-0.0074 a.u.	----	-0.0079 a.u.	----	94%	74%[c]
	p 0.22(3)		p 3.86(3)		
	d 0.56(3)		d 1.94(3)		
	f 0.74(4)		f 2.54(4)		

a) C.L. Pekeris, Phys. Rev. 115, 1216 (1959).
b) F. Sasaki and M. Yoshimine, Phys. Rev. A9, 26 (1974), (Table V, set B).
c) A.W. Weiss, Phys. Rev. A3, 126 (1971), (Table I, F^-).

D. Ni III $3d^8$ Term Structure

Many interesting solid state systems (bulk and surface) involve Ni in its various ionization states (I-IV). At present these are treated within the IPM--for example at the UHF level with possibly long-range correlation added [87]--but it is clear that a thorough understanding must involve consideration of atomic like correlation effects. While it will be difficult to justify, on a quantitative basis, the use of purely atomic treatments, the qualitative aspects of the correlation in the free atom or ion can be of considerable value in guiding solid state treatments. In fact recent work [87] based on localized orbitals indicates that at least semi-quantitative results can be obtained through the use of a rather simple external potential which can be directly incorporated into the atomic procedures. On the other hand, few purely atomic studies of species with more than eighteen electrons exist, which is probably due to the seeming computational complexity involved.

In this section, we will identify the principle correlation effects associated with the Ni III ion term structure and discuss efficient methods for obtaining results accurate to $\sim$0.1 eV. This species was selected as it plays an important role in the solid state (NiO), and yet is ionized enough to depress the strong 3d,4s degeneracies present in the neutral atom.

To begin with, only non-relativistic calculations were done to simplify matters (see ref. 88 for examples of how relativistic effects are currently treated). This is probably consistent with the level of approximation present in the remaining part of this treatment (in any case, the full use of relativity, even at the SCF level, is currently prohibited in the solid phase).

Next, we have restricted our attention to the valence portion of the configuration, i.e. the 3s,3p,3d space to which we have added a 4s, to complete the FS. It is known [89] that at least for some neutral and lightly ionized species $3d^q$ of low q, that the $3d\leftrightarrow 4s$ interaction is moderate to strong, with the MCHF 4s being substantially different than the single configuration 4s. What is the situation for $3d^8$ of Ni^{++}? This question was examined here in two ways--(1) by seeing how the 4s in $3d^{8-n}\,4s^n$ varied with n, and (2) by seeing whether a large polarization correction ($4s\rightarrow v_s$) could be found. No large effects were found, so the reference configuration was solely $3d^8$, and the 4s was generated from the configuration average $3d^6\,4s^2$.

Full SCF calculations were made for each of the five terms (^{1}SDG, ^{3}PF) using the code of Froese-Fischer [7]. These were found to give term splittings accurate to 0.5 eV (Table 3) when compared to experiment, except for the ^{1}S which has not been observed (for this term only, the results based on the radials of the configuration average were quite far off ($\sim$1.5 eV) the SCF ones).

In constructing the internal correlation (Σ_{FS} and Π_{FS}, the only excitations allowed to 4s were those involving the 3d electrons. Specifically we included 3d$\rightarrow$4s, 3s$\rightarrow$3d, $3d^2\rightarrow 4s^2$, $3p^2\rightarrow 3d^2$, and $3s^2\rightarrow 3d^2$. Where allowed, (^1DS, ^3P) $3p^2\rightarrow 3d^2$ is the largest, followed (50% smaller) by 3s$\rightarrow$3d where allowed (^1D). As can be seen from Table 3, the effects were particularly large for ^1S, and to a lesser extent ^1D.

Next we directed our attention to the bi-virtual correlation believing it to be the most significant remaining correlation (this is discussed below). To do this, we assume pair transferability within terms. This involves only a change of state (Z and N are fixed), and is probably worst for the well separated terms (e.g. ^3F and ^1S). Under these restrictions, the contribution of the pairs to the excitation energy are found, by numerically evaluating equation 19, to be:

$$E(3d^8\ S,L) - E(3d^8\ S',L') = \varepsilon(3d^2 S,L) - \varepsilon(3d^2 S',L')$$

Note that only the $3d^2$ pairs enter. This is due to the fact that the interaction of $(m\ell')^q$ with a closed subshell $(n\ell)^{4\ell+2}$ depends only on q (see Section VI) and not on the terms. Correlation functions then take the form:

$$(3d^6\ (S_c,L_c)\ v\ v'\ (S_v,L_v))S\ L$$

This correlation vector however, contains a considerable number of determinants (172 for $S_vL_v={}^3F$, $SL={}^1S$) which, for our standards, amounts to a substantial computational effort. As this was an exploratory calculation, we reduced the labor by assuming the (approximate) transferability of the pairs (which we note has not yet been computationally demonstrated for $3d^2$ pairs in transition metals), to reduce the complexity by looking at a smaller species (i.e. $3d^2$). Since Z can be varied, we chose it in such a way as to make the calculation most nearly correspond to the original system. Recalling that most virtuals are essentially determined by $\langle r\rangle_{HF}$, we chose Z for $3d^2$ such that $\langle r\rangle_{3d^2,Z^*} = \langle r\rangle_{3d^8,Z=28.0}$. Z* was found to be $\sim$24.5 by a series of SCF calculations on the average energy.

The pair energies were then obtained using the methods discussed earlier, with 1 STO/ℓ, and $\ell \leqslant 4$. The ℓ cutoff was chosen according to the observation that the maximum contribution comes from an ℓ one unit greater than the ℓ we are replacing, (in this case ℓ =2), and we include one higher ℓ as well to achieve reasonable convergence. All permitted combinations of the virtuals were allowed, e.g. $v_p v_g$, v_g^2, etc. and all exponents were optimized.

The pair energies ($3d^2$, Z=24.5) for the ^{1}DGS, ^{3}FP in eV are: -0.320, -0.217, -0.991, -0.139, -0.320. The main contributors to these pairs was the v_f^2 with the v_d^2 following this. For the ^{3}P and ^{1}D pairs, v_g contributed a non-negligible -.0.006-8 a.u. and for the ^{1}S -0.007 a.u.. Clearly, for the ^{1}S the v_g plays an important role, so much so that v_h should be explored as well. We believe these pair results for $3d^2$ electrons to be the most accurate ones available at this time.

The results are then added to the previous ones, and the term structure is brought into ~0.1 eV agreement with experiment (Table 3). It should be noted that there is a 2 eV correction to the ΔSCF results for the ^{1}S term.

The remaining correlation, presents us (especially Π_{FS-V}) with even greater difficulties (e.g. larger number of dets per LS state) with no corresponding simplifications apparent (e.g. transferability). Based on the bi-virtual results and past experience, we suggest that the excitations $3s \rightarrow v_d$, $3p \rightarrow v_f$, $3p^2 \longrightarrow 3d\ v_d$, $3s\ 3p \rightarrow 3d\ (v_p + v_f)$, $3d \rightarrow v_s + v_d + v_g$ are probably the most significant.

Since this is an exploratory calculation, and several other approximations have been made, we sought to determine through examination of previous results (primarily the ground state $s^m p^n$ configurations for first row atoms) to what extent these could be neglected when comparing term structure arising from the same ground state configuration. The conclusion was that (a) if all terms have internal correlation (here $s^2 \rightarrow p^2$) or all terms have no such correlation (true for only some terms of Ni III) and (b) if symmetry does not exclude an otherwise important virtual polarization or hole-virtual configuration (true here), the contributions of Σ_V and Π_{FS-V} pretty much balance out between terms.

In summary, we have identified the most important correlation effects contributing to the Ni III term structure and developed methods for obtaining them efficiently.

ACKNOWLEDGEMENTS

We wish to thank Mr. George Aspromallis for the computational aid he provided for some of these results.

Table 3

Term Splitting in Ni III $3d^8$ (in eV) Relative to ^3F.

Calculation Type	Upper Term			
	^1D	^3P	^1G	^1S
Av. Energy	2.0358	2.4663	3.1817	6.313
ΔSCF	2.0356	2.4663	3.1841	7.8292
+Internal Fermi-Sea	1.6846	2.2276	3.0598	6.7064
+Bi-Virtual	1.5132	2.0313	2.9808	5.8539
Error	-0.12	+0.07	+0.12	------

REFERENCES

1. D.R. Beck and C.A. Nicolaides, Int. J. Qu. Chem. S8, 17, (1973).
2. D.R. Beck and C.A. Nicolaides, Int. J. Qu. Chem. S10, 119, (1976).
3. R.J. Buenker and S.D. Peyerimhoff, Theor. Chim. Acta 35, 33, (1974).
4. R.J. Buenker and S.D. Peyerimhoff, this volume.
5. G. Das and A.C. Wahl, J. Chem. Phys. 50, 3532, (1972); S.L. Guberman, J. Chem. Phys. 67, 1125 (1977).
6. R.J. Buenker and S.D. Peyerimhoff, Chem. Phys. Letts. 34, 225, (1975).
7. C. Froese-Fischer, Comp. Phys. Comm. 4, 107, (1972).
8. H.J. Silverstone, J. Chem. Phys. 67, 4172, (1977).
9. C.A. Nicolaides and D.R. Beck, Chem. Phys. Letts. 36, 79, (1975).
10. D.R. Beck and C.A. Nicolaides, Phys. Letts. 65A, 293, (1978).
11. C.A. Nicolaides and D.R. Beck, "Time Dependence, Complex Scaling and the Calculation of Resonances in Many-Electron Systems", to be published in the Volume "Complex Scaling in the Spectral Theory of the Hamiltonian", ed. by P.O. Löwdin (Int. J. Qu. Chem. supplement volume 1978).
12. B. Kahn and G.E. Uhlenbeck, Physica 5, 399, (1938); the term "linked cluster"apparently first appeared in this article.
13. V. Fock, M. Vesselov and M. Petrashen, J. Exp. Theor. Phys. (U.S.S.R.) 10, 723, (1940).
14. F. Iwamoto and M. Yamada, Prog. Theor. Phys. 17, 543, (1957).
15. W. Brenig, Nucl. Phys. 4, 363, (1957).
16. R. Brout, Phys. Rev. 111, 1324, (1958).
17. O. Sinanoğlu, J. Chem. Phys. 36, 706, (1962).
18. L. Szasz, Phys. Rev. 126, 169, (1962).

19. H.J. Silverstone and O. Sinanoğlu, J. Chem. Phys. 44, 1898, 3608, (1966).
20. J. Cizek, J. Chem. Phys. 45, 4256, (1966); Adv. Chem. Phys. 14, 35, (1969).
21. R.K. Nesbet, Phys. Rev. 155, 56, (1957).
22. C. Møller and M.S. Plesset, Phys. Rev. 46, 618, (1934).
23. O. Sinanoğlu, Proc. R. Soc. (london) A260, 379, (1961).
24. C.F. Bender and E.R. Davidson, Phys. Rev. 183, 23, (1969).
25. A.N. Weiss, Phys. Rev. A3, 126, (1971).
26. J.W. Viers, F.E. Harris and H.F. Schaefer III, Phys. Rev. A1, 24, (1970).
27. C.F. Bunge, Chem. Phys. Letts. 42, 141, (1976).
28. F. Sasaki and M. Yoshimine, Phys. Rev. A9, 26, (1974).
29. C.M. Moser and R.K. Nesbet, Phys. Rev. A6, 1710, (1972).
30. R. Ahlrichs, H. Lischka, B. Zurawski and W. Kutzelnigg, J. Chem. Phys. 63, 4685, (1975); W. Meyer, J. Chem. Phys. 58, 1017, (1973) and references therein.
31. F.W. Byron and C.J. Joachain, Phys. Rev. 157, 7, (1967).
32. C. Froese-Fischer and K.M.S. Saxena, Phys. Rev. A9, 1498, (1974).
33. H.P. Kelly and A. Ron, Phys. Rev. A4, 11, (1971).
34. J. Paldus, J. Cizek and I. Shavitt, Phys. Rev. A1, 50, (1972).
35. R.J. Buenker, S.D. Peyerimhoff and W. Butscher, Mol. Phys. 35, 771, (1978).
36. R.E. Knight, Phys. Rev. 183, 45, (1969).
37. A. Pipano and I. Shavitt, Int. J. Qu. Chem. II, 741, (1968).
38. C. Froese-Fischer, J. Phys. B6, 1933, (1973).
39. H.P. Kelly, Phys. Rev. 144, 39, (1966).
40. I. Öksüz and O. Sinanoğlu, Phys. Rev. 181, 42 (1969).
41. C.A. Nicolaides and D.R. Beck, J. Phys. B6, 535, (1973).
42. D.R. Beck and O. Sinanoğlu, Phys. Rev. Letts. 28, 945, (1972).
43. H.F. Schaeffer III, J. Chem. Phys. 55, 176, (1971).
44. C.A. Nicolaides and D.R. Beck, J. Chem. Phys. 66, 1982, (1977).
45. D.R. Beck and C.A. Nicolaides, Phys. Letts. 61A, 227, (1977).
46. C.A. Nicolaides and D.R. Beck, J. Phys. B9, 1259, (1976).
47. R.J. Buenker and S.D. Peyerimhoff, Chem. Phys. Letts. 36, 415, (1975).
48. E.R. Cooper and H.P. Kelly, Phys. Rev. A7, 38, (1973).
49. E.R. Davidson, in "The World of Quantum Chemistry", eds. R. Dandel and B. Pullman, Reidel Publ. Co. Dordrecht, (1974).
50. A. Mennier, B. Levy and G. Berthier, Int. J. Qu. Chem. 10, 1061, (1976).
51. F. Sasaki, Int. J. Qu. Chem. S11, 125, (1977).
52. R.J. Bartlett and I. Shavitt, Int. J. Qu. Chem. S11, 165, (1977).
53. T.A. Carlson and M.O. Krause, Phys. Rev. Letts. 17, 1079, (1966).
54. Y. Komninos, D.R. Beck and C.A. Nicolaides, unpublished.

55. L.E. McMurchie and E.R. Davidson, J. Chem. Phys. 66, 2959, (1977).
56. R.J. Buenker and S.D. Peyerimhoff, Chem. Phys. 8, 324, (1975).
57. J. Reader, Phys. Rev. A7, 1431, (1973).
58. D.A. Shirley et. al., "Electron-Correlation Satellites in Electron Spectroscopy", preprint, presented at the 2nd Int. Conf. on Inner Shell Ionization Phenomena, Freiburg, (1976).
59. E.K. Viinikka and Y. Ohrn, Phys. Rev. B11, 4168, (1975).
60. L.E. Nitzsche and E.R. Davidson, J. Chem. Phys. 68, 3103, (1978).
61. C.C.J. Roothaan and P.S. Bagus, Methods of Comp. Phys. 2, B. Alder, S. Fernbach and M. Rotenberg, eds. Acad. Press (1963).
62. E.A. McCullough, Jr., J. Chem. Phys. 62, 3991, (1975).
63. C.A. Nicolaides, Phys. Rev. A6, 2078, (1972); Nucl Inst. Methods 110, 231 (1973).
64. D.R. Beck and C.A. Nicolaides, Phys. Rev. Letts. submitted May 1978.
65. C.A. Nicolaides and D.R. Beck, this volume.
66. E.R. Davidson and L.Z. Stenkamp, Int. J. Qu. Chem. S10, 21 (1976).
67. L.A. Yaffe and W.A. Goddard III, Phys. Rev. A13, 1682, (1976).
68. J. Hinze and C.C.J. Roothaan, Supp. Prog. Theor. Phys. 40, 37, (1967).
69. W.J. Hunt and W.A. Goddard III, Chem. Phys. Letts. 3, 414, (1969).
70. E.R. Davidson, "Reduced Density Matrices in Quantum Chemistry", Acad. Press (N.Y.) (1976); C.F. Bender and E.R. Davidson J. Phys. Chem. 70, 2675 (1966).
71. C. Edmiston and M. Krauss, J. Chem. Phys. 45, 1833 (1966).
72. e.g. D.R. Beck, J. Chem. Phys. 51, 2171 (1969).
73. D.R. Beck and C.A. Nicolaides, unpublished.
74. D.R. Beck, C.A. Nicolaides and J.I. Musher, Phys. Rev. A10, 1522 (1974).
75. C.A. Nicolaides and D.R. Beck, Can. J. Phys. 53, 1224 (1975).
76. H.F. King et. al., J. Chem. Phys. 47, 1936 (1967).
77. H.F. Schaefer and F.E. Harris, J. Comp. Phys. 3, 217 (1968).
78. A. Bunge, J. Chem. Phys. 53, 20, (1970).
79. D.R. Beck and H. Odabasi, Ann. Phys. 67, 274, (1971).
80. G.W.F. Drake, Phys. Rev. Letts. 24, 126, (1970).
81. R. Schnitzer and M. Auber, J. Chem. Phys. 64, 2466 (1976).
82. W. Aberth, J. Chem. Phys. 65, 4329 (1976); M.L. Vestal, ibid p. 4331; J. Durup, ibid. p. 4331; R. Schnitzer and M. Aubar, ibid. p. 4432.
83. J.F. Liebman, D.L. Yeager and J. Simons, (Chem. Phys. Letts. 48, 227, (1977)) have added an attractive external potential to introduce binding, in their study of shape resonances.
84. S. Huzinaga and A. Hart-Davis, Phys. Rev. 8A, 1734 (1973). Positive eigenvalues were also obtained in certain HF calculations of He^- compound states (C.A. Nicolaides, Phys.

Rev. A6, 2078 (1972); C.A. Nicolaides, unpublished).
85. D.R. Yarkony, H.F. Schaefer, III, and C.F. Bender, J. Chem. Phys. 64, 981 (1976).
86. C.F. Bunge and A.V. Bunge, "Calculations of Atomic Electron Affinities", preprint.
87. A.B. Kunz, this volume; Phys. Rev. B7, 5369 (1973).
88. D.R. Beck and C.A. Nicolaides, this volume.
89. C. Froese-Fischer, Can. J. Phys. 49, 1205 (1971).
90. C.E. Moore, "Atomic Energy Levels Vol II", NBS circular 467 (1952).
91. R.L. Kelly and L.J. Palumbo, "Atomic and Ionic Emission Lines below 2000 Å", NRL Reprint 7599 (1973).

MANY-BODY THEORY OF PHOTOABSORPTION IN ATOMS AND MOLECULES

CLEANTHES A. NICOLAIDES and DONALD R. BECK

Theoretical Chemistry Institute
National Hellenic Research Foundation
48 Vas. Constantinou Ave.
Athens 501/1, Greece

ABSTRACT

We present an approach to the calculation of photoexcitation and photoionization transition probabilities which first derives, consistently, the important correlation effects and then computes them efficiently and accurately. The emphasis is on the correct evaluation of the transition matrix element and not of the exact Schrödinger equation for initial and final wave-functions. The theory is implemented through Configuration-Interaction techniques which allows the practical consideration of any type of state. Both initial and final states are treated at the same level of approximation due to the similarity of restrictions imposed upon them. The basic conceptual and computational characteristics of the theory are simple: The zeroth order vectors for initial and final states are the Fermi-Sea (FS) wave-functions. The transition operator is then applied to the FS vectors and selects the additional correlation effects in initial and final states which contribute to the transition amplitude the most. These correlations are expressed in terms of Hartree-Fock and variationally optimized virtual orbitals for each state. The resulting very small wave-functions are then employed for the calculation of the transition probabilities. This First Order Theory of Oscillator Strengths (FOTOS) is applicable to any system with a shell structure and symmetry (including nuclear transitions). Its application to a variety of atomic transitions has yielded accurate results, some of which are presented here. Also presented are brief discussions on the question of which form of the electric dipole operator is the appropriate one to use in computations, nonorthonormality, extraordinary absorption properties of certain systems, polarizability calculations within FOTOS and the

Cleanthes A. Nicolaides and Donald R. Beck (eds.), Excited States in Quantum Chemistry, 143–182.

extension of FOTOS to the relativistic domain.

I. QUALITATIVE INTERPRETATION OF PHOTOABSORPTION AND PHOTOEMISSION SPECTRA OF ATOMS AND MOLECULES

Photoabsorption and photoemission spectra vary considerably according to the system examined and the energy at which they are carried out. The literature on the subject is big and is growing fast. As a result, the same basic phenomena are often interpreted using different types of nomenclature--a fact which may have caused a slight obfuscation in the field. Therefore, we believe it is useful for the nonexpert reader to start this paper by presenting briefly, in a shell model language, a simple interpretation of the most interesting and general transition processes and their significance.

The electronic transition processes in atoms and molecules which are of interest to Quantum Chemistry usually involve the absorption or emission of one photon in the range (roughly) 1 eV-5000 eV. The low energy side involves the spectra of the valence electrons in neutrals and singly ionized species (e.g. low lying valence-or Rydberg- Rydberg transitions, "forbidden" decays within the same configuration (e.g. electric quadrupole)) while the high energy side involves the spectra of highly ionized atoms or deep inner electron transitions, which are very important for the quantitative examination of relativistic and quantum electrodynamic effects, and may serve to identify super-heavy transient nuclear species.

Most low intensity source, one photon processes are of the electric dipole type. When nonrelativistic symmetry forbids this type, the nonrelativistically "forbidden" transitions (spin-orbit allowed electric dipole (SOAED), electric quadrupole (E2), magnetic dipole (M1), etc.) take place whose probabilities are very small in the neutrals but--except for the E2--increase rapidly with Z (e.g. $\sim Z^{10}$ for M1 transition probabilities compared with $\sim Z^4$ for E1). Forbidden transitions have most often been seen in emission. With the application of the high intensity laser sources they can now be seen in absorption as well, although accurate ($\sim$10%) measurement of longer lived species ($>$ 0.1 sec.) still presents a challenge.

In emission, both initial and final states may be in the discrete or in the continuous spectrum. In absorption, the initial state is usually the ground or a low lying excited state. The final state may again be in the discrete or in the continuous spectrum.

In contrast to the spontaneous photoemission processes which occur almost exclusively via one or zero electron jumps, the induced photoabsorption processes may involve multielectron transitions as well. In theories which use a common one electron basis set for initial and final states, (standard CI, Many-Body Perturbation Theory, Green's functions, etc.), multielectron transitions are attributed to electron correlation exclusively, since the dipole operator is monoelectronic. In theories such as the one described in this paper, where initial and final states are assigned their own SCF Hartree-Fock functions, multielectron transitions can be attributed partly to electron correlation and partly to the change of the HF potential which yields nonorthonormal orbitals--provided the orbital symmetries of initial and final configurations allow it.

The samples of transitions given below demonstrate some of the points made above and give us the opportunity to comment, whenever necessary, on the importance--or lack of it--of the most conspicuous electron correlation effects.

1. Ne $1s^2 2s^2 2p^6\ {}^1S \rightarrow Ne^+[1s^2 2s^2 2p^5\ {}^2P^o_{1/2,3/2} + e^-]\ {}^1P^o$

Single electron photoionization from the outer valence shell of a closed shell system. Electron correlation effects are small. In noble gases, only one channel is open. In larger systems, the spin-orbit interaction splits the (1/2), (3/2) levels so that there are two distinct photoionization thresholds.

2. Ar $1s^2 2s^2 2p^6 3s^2 3p^6\ {}^1S \rightarrow Ar^+[KL\ 3s^2 3p^5\ {}^2P^o_{1/2,3/2} + e^-]\ {}^1P^o$

Same as above only that now electron correlation is more important since Ar does not have its Fermi-Sea [1] closed. Thus, the $3p^2 \leftrightarrow 3d^2$ pair correlation is nonnegligible.

3. Li $1s^2 2s\ {}^2S \rightarrow Li\ \overline{2s}^2 \overline{2p}\ {}^2P^o$

Two-electron excitation to an autoionizing state. Correlation is important in the initial (e.g. $1s^2 \leftrightarrow 2sV_s$, $2pV_p$) as well as in the final state (e.g. $\overline{2s}^2 \leftrightarrow \overline{2p}^2$). This transition is allowed in SCF-HF theory as well, since $\langle 1s|\overline{2s}\rangle \neq 0$ [2].

4. Zn KLM $4s^2\ {}^1S \rightarrow$ Zn KLM $4snd\ {}^1D$

Electric quadrupole transition from the ${}^1S$ ground state of the alkaline earths to the nd ${}^1D$ Rydberg series. Seen in photoabsorption [3]. For the initial state, the Fermi-Sea correlations $4s^2 \leftrightarrow 4p^2 \leftrightarrow 4d^2 \leftrightarrow 4f^2$ are important while for the final states the "symmetry exchange" [1] correlation $4snd \leftrightarrow 4p^2$ creates a valence-Rydberg series whose effect in neutrals can be (n = 4) so large [4-6] that single configuration assignments to certain states lose meaning.

5. $O_2\ (3\sigma_g^2\pi_u^4\pi_g^2)\ X\ ^3\Sigma_g^- \longrightarrow O_2\ (3\sigma_g^2\pi_u^3\pi_g^3, 3\sigma_g^2\pi_u^4\pi_g 3p\pi)\ ^3\Sigma_u^-$

Photoexcitation to low lying excited states which exhibit "localized" valence-Rydberg mixing. In molecules this mixing is a function of geometry [7,8], while in atoms, it is a function of the nuclear charge [9,10]. The mixing coefficients and the transition probabilities are sensitive to the relative positions of the diagonal matrix elements and the types of basis sets used [6,10].

6. Ag KL $3s^2 3p^6 3d^{10} 4s^2 4p^6 4d^{10} 5s\ ^2S \rightarrow Ag^+ [KL\ 3s^2 3p^5 3d^{10} 4s^2 4p^6 4d^{10} 5s\ ^{3,1}P^o + e^-]\ ^2P^o$

Inner electron photoionization. Due to the open valence shell (5s electron) there is multiplet structure. Electron correlation is, in general, symmetry dependent i.e. multiplet structure dependent. Thus, its exact effect on photoionization processes in open shell systems varies from term to term. For the above process, apart from the Fermi-Sea correlations for initial and final states, the major correlation effect is the "hole-filling", "symmetry exchange" [1] pair correlation $3d^2 \leftrightarrow 3pnf, \varepsilon f$, where (nf, εf) are discrete and continuum orbitals. This correlation corresponds to configurations $[KL\ 3s^2 3p^6 3d^8 (^1S, ^1D, ^1G, ^3P, ^3F) 4s^2 4p^6 4d^{10} 5snf, \varepsilon f\ (^{3,1}P^o) + e^-]\ ^2P^o$ which represent two electron excitations. Depending on the amount and type of the overall correlation and configuration mixing, experiments on atoms, molecules and solids record a multitude of peaks usually called "satellite", "shake-up", "multiplet" etc. [11-17].

7. $C\ 1s^2 2s2p^3\ ^5S^o \xrightarrow{\text{emission}} C\ 1s^2 2s^2 2p^2\ ^3P$

Spin-orbit allowed electric dipole transitions (SOAED). The quintet state decays be mixing with the $1s^2 2s2p^3\ ^3P^o$ term via the spin-orbit perturbation [18]. For the ground state, the Fermi-Sea correlation $2p^2 \leftrightarrow 2s^2$ and the $2s \leftrightarrow V_d$ are the most important. For the $^5S^o$ term there is no Fermi-Sea or hole-filling correlation because symmetry forbids it. On the other hand hole-filling correlation is allowed for the $^3P^o$ term ($2p^2 \leftrightarrow 2sV_d, V_s$) and affects the $^3P^o$-^5S mixing as well as the $^3P^o \rightarrow {}^3P$ transition matrix element. In the work of ref. [18] the effect of electron correlation on the spin-orbit mixing was neglected. Current work on the fine structure operators [19] takes electron correlation into account. The simpler case of singlet-triplet mixing and the corresponding SOAED transitions are those which are better known in Quantum Chemistry [20].

8. $$\mathrm{S\ KL}\ 3s^2 3p^4\ {}^3P \rightarrow S^{++}\ [3s^2 3p^2 ({}^3P, {}^1D, {}^1S) + 2e^-]\ {}^3S^o, {}^3P^o, {}^3D^o$$
$$\rightarrow S^{++}\ [3s3p^3 ({}^5S^o, {}^3S^o, {}^{1,3}P^o, {}^{1,3}D^o + 2e^-]\ {}^3S^o, {}^3P^o, {}^3D^0$$

<u>Double photoionization</u> of valence shells. This process contributes substantially to the total photoionization of noble gases [21-23,13]. Open shell cases show of course the main features of closed shell photoionization and in addition those specific to their multiplet structure. For double photoionization processes: a) The problem of correctly computing the scattering functions of two free electrons simultaneously is outstanding. b) Fermi-Sea correlations for the bound functions are important: For S ^3P these are: $3s^2 \leftrightarrow 3d^2$, $3p^2 \leftrightarrow 3d^2$, $3s \leftrightarrow 3d$, and for S^{++} we have in addition $3p^2 \leftrightarrow 3s3d$. However, the important "symmetry exchange" correlation $3p^2 \leftrightarrow 3s3d, nd$ does not apply to the $^5S^o, {}^3S^o$ terms. We note that this correlation effect has been given special emphasis by Chang et al [22] who called it "virtual Auger transition" in their study of Ne photoionization. c) The important Fermi-Sea correlations in the <u>continuum</u> should be again of the "symmetry exchange" type especially for high energies. Thus for the first case we have: $\varepsilon p \varepsilon' d \leftrightarrow \varepsilon s \varepsilon' f$ and for the second: $\varepsilon p \varepsilon' p \leftrightarrow \varepsilon s \varepsilon d$. Their interference and mixing with autoionizing states and single photoionization channels can in principle be accounted for by CI in the continuum [24,25].

9. $$\mathrm{Li}\ 1s2p^2\ {}^2P \xrightarrow{\text{emission}} Li^+\ 1s^2 + e^-\ \ {}^2P^o$$

<u>Radiative Autoionization</u> of certain metastable states [26]. An emission process where the final state is in the continuum and the transition energy is distributed between the photon and the free electron.

Although most of the examples were from the realm of atoms, the same phenomena and similar correlation effects are, of course, present in molecules. Furthermore, the general rules [1] about the importance of certain correlation effects characterize <u>all</u>

atoms and many molecules. Specific cases, such as the $3p^2 \leftrightarrow 3s3d$ example of "symmetry exchange" and valence-Rydberg series perturbation example of "hole-filling" correlation, were first discovered in optical spectroscopy in the 1930's. With the recent and current numerous applications of theoretical models and experimental spectroscopic techniques, such effects have been seen in a variety of cases and most probably will continue to be discovered in investigations of new systems.

We now proceed with our theory of transition probabilities and its results.

II. WHAT KIND OF A PHOTOABSORPTION THEORY AND WHY IS IT NECESSARY?

Absorption and emission of radiation have formed the basis for formulating the rules of Quantum Mechanics and detecting the energy spectra of quantized systems. However, until recently, the accurate measurement of absolute intensities of these spectra or of the related quantity, the lifetime of the excited state, had been extremely difficult and rare. Thus, the details of the response of the electronic structure of matter to the perturbation of the electromagnetic field, from the X-Ray to the optical region, remained hidden and so theory was kept comfortably unchallenged.

Fortunately, due mainly to progress achieved recently with lifetime measurements of singly and multiply excited atomic and molecular states (eg. [27-29]and refs. therein) as well as with photoabsorption cross-section measurements (e.g.[11-15,30-33] and refs. therein) which have been producing experimental information of impressive detail and accuracy, the atomic and molecular electronic structure has revealed its qualitative and quantitative nature with considerable generosity. This unprecendented and continuing experimental success has imposed rather stringent criteria on theory, as photon absorption and emission in many-electron systems are processes which involve all the electrons to a greater or lesser degree. Theory must then be able to treat the dynamical effects of electron-electron interactions with such accuracy as to allow not only explanation of the observed but also quantitative prediction of the observable.

In spite of occasional attempts to restore its respectability, the independent particle model (IPM) is in general, (excluding special cases such as the K-shell photoeffect in the KeV region or certain Rydberg transitions in unperturbed Rydberg series) not good enough to treat absorption and emission of radiation quantitatively. Thus, for practical applications in other fields (e.g. application to elemental abundance determination in the stars, laser physics, fusion related research in plasma physics, etc.) where oscillator strengths of 1%-10% accuracy are needed, as well as for purposes of contributing to the understanding of the electronic structure of atoms and molecules, the IPM is of limited use.

Therefore, one has to resort to perturbational or variational many-body methods [4, 6-8,16,22,23,34-57] whose usefulness and necessity lie in the fact that they can produce accurate numbers for properties and processes other than total energies of ground states or low lying excited states which, in most cases, are known to a high accuracy experimentally, especially for atoms [e.g. 58].

A reasonably complete many-body theory of photoabsorption should be able to deal with initial→(final) discrete, autoionizing (predissociating) or purely continuous state transitions, whether the initial state is the ground or an excited state. As such, it should have the following characteristics:

a) It should be based on some well-founded guiding principle and not be a "hit or miss" approach which sometimes works and sometimes does not.

b) It should be able to deal with single- as well as many-determinantal zeroth-order functions with the same ease.

c) It should be able to predict cross-sections for single or multiple excitations in closed as well as in open-shell systems.

d) It should be able to account to the same order and in a consistent manner, electron correlation in both initial and final states which mostly contributes to the <u>amplitude</u> of excitation (i.e. other types of correlation effects may contribute to the total energy but not necessarily--within 10%--to the cross-section). This requirement, which must be the fundamental characteristic of a true many-body theory of oscillator strengths, is closely related to the last characteristic:

e) It should be of physical consequence and computationally manageable. This allows a more direct understanding of the electronic decay or excitation dynamics and the consistent for each system, optimum, and theory-guided calculation of the required transition matrix elements.

If (d) and (e) are satisfied,"overkill" calculations which aim at obtaining nearly exact solutions of the Schrödinger equation for initial and final states and are therefore extremely difficult to carry out for arbitrary transitions, are seen to be unnecessary and, perhaps, of little theoretical-chemical value since there is no emphasis on the transition process itself. Similarly, existing procedures which calculate excitation amplitudes directly, such as the Random Phase Approximation (RPA) and other, similar in spirit, algorithms, apart from the difficulties in satisfying characteristics (b) and (c) treat every system in the same numerical way, without due consideration for its electronic structure.

In this paper we formulate and apply a theory (FOTOS-First Order Theory of Oscillator Strengths) which is based on few fundamental notions characterizing the mathematics and physics of the electronic structure of atoms and molecules and appears to be a very good approximation to the complete theory mentioned above [6,10,54-57].

FOTOS contains both phenomenological and computational aspects. Both are conceptually simple and straightforward to apply. The phenomenology explains and predicts the quantitative features of photoabsorption spectra such as those due to multiplet splittings and multiple excitations (satellite, shake-up peaks, etc.). The computational part involves the prediction and efficient calculation to all orders via Variational Configuration Interaction (VCI) procedures [1] of the correlation effects which affect the amplitude of excitation the most.

FOTOS is applied here quantitatively as well as semi-quantitatively. The quantitative application has yielded a variety of accurate discrete-discrete and discrete-autoionizing state allowed transition probabilities corresponding to single and multiple excitations. The semiquantitative applications deal with the derivation of the important correlation effects for a variety of transitions in atoms and molecules.

Finally we comment briefly on a) the choice of the computationally most useful form of the electric dipole operator, b) extraordinary absorption properties in cases of heavy valence-Rydberg mixing, c) application of FOTOS to polarizabilities, d) proposed relativistic variants of FOTOS.

III. FOTOS (FIRST ORDER THEORY OF OSCILLATOR STRENGTHS)

A. Physical Considerations: The choice of the zeroth order functions according to the Fermi-Sea

The formal development and numerical application of a many-body theory of a property of a quantized system depends crucially on the choice of the zeroth-order approximation Ψ_o. This is because the actual calculations are by necessity approximate and therefore the rate of convergence to the correct answer depends on Ψ_o. Thus, a choice of Ψ_o as hydrogenic, Hartree-Fock Slater, minimal basis MO, or Hartree-Fock function, will yield different results if say a perturbation treatment to the same order is applied. For properties such as photoabsorption cross-sections, different choices of Ψ_o within the RPA may even lead to different physical interpretations of the observed spectra [39]. In general, the zeroth-order choice has been based on computational convenience with the object of adjusting to the necessities of a particular algorithm to be used.

In FOTOS, the zeroth order wave-function Ψ_0 is based on the Fermi-Seas of the states involved in the transition. This choice is derived from considerations of the electronic structure as revealed by the shell model. The Fermi-Sea concept [1], goes beyond the single-configuration zeroth-order model and refers to many essential configurations as describing the fundamental shells where, from a pictorial point of view, electrons spent the overwhelming part of their time.

As we shall show later on, the F-S notion is instrumental in establishing FOTOS as a theoretical approach capable of quantitative as well as easily arrived at semi-quantitative predictions. In addition, it appears that it is also useful for qualitative predictions related to other properties such as polarizabilities, electron-atom scattering, chemical bonding, etc. For example, let us consider the phenomenology of properties of the inert gases which, due to their characteristic shell structure and symmetry, are commonly thought of as quantum billiard balls. A mean field theory notion would then imply that the polarizability of the noble gases should increase more or less linearly, as a function of the number of electrons. This is indeed what is found from calculations which use statistical models [60]. Yet, experiment shows [60] that there is a sharp increase from Ne to Ar and from then on a linear increase. Within the Fermi-Sea concept we have the following qualitative explanation: Since both He and Ne in their ground states have all their F-S orbitals occupied, virtual excitations of the electrons due to the application of the external field are much more difficult than in the other noble gases where F-S orbitals are vacant. Thus the polarizability, a quantity which measures the response of the atom to the external perturbation, is expected to increase markedly from Ne to Ar, the first "pseudo-closed" shell noble gas (see examples 1 and 2 of section I).

The above example of a qualitative application of the Fermi-Sea notion to a physical observation illustrates the physical relevance of this concept.

In the many-body theory of photoabsorption to be developed below, this concept forms the basis for the choice of the zeroth-order functions, in both the semiquantitative as well as the strictly quantitative applications of the theory. This means that the zeroth order functions for the states participating in the transition are taken to be a linear combination of all FS configurations:

$$\Psi_o = \Psi_{FS} = \Sigma_K c_K \Phi_K \tag{1}$$

$$\Phi_K = \Sigma_L d_L \Delta_L^K(\alpha_1^K, \alpha_2^K, \alpha_3^K, \ldots \alpha_N^K) \tag{2}$$

Φ_K corresponds to a configuration, built in terms of SCF Hartree-Fock orbitals, with index K. Δ_L^K are Slater determinants built in terms of the corresponding spin-orbitals, α^K. The coefficients d_L are obtained by direct diagonalization of the symmetry operators. The c_K are obtained by direct diagonalization of the total Hamiltonian [1].

B. The Form of the Transition Matrix Element

We start by reminding the reader that what is observed in photoabsorption (emission) processes is proportional to the square of the amplitude $\langle \Psi_i | e^{i\vec{K}\cdot\vec{r}} \vec{p}\cdot\vec{\varepsilon} | \Psi_f \rangle$, where $\vec{K}$ is the wave-vector and $\vec{\varepsilon}$ the polarization vector of the photon, $\vec{r}$ the position and $\vec{p}$ the momentum of the electron. The expansion of $e^{i\vec{K}\cdot\vec{r}}$ ($= 1 + \vec{K}\cdot\vec{r} + \ldots$) yields the various tensor components (an ad hoc introduction of spin is required for M1, etc.) and the resulting symmetry selection rules for electric and magnetic dipole, quadrupole etc. radiation. These selection rules do not only refer to quantum numbers of the exact wave-functions. They also apply to the electronic coordinates of the individual electrons treated as occupying atomic or molecular orbitals of a particular symmetry.

Thus, it seems desirable to develop a photoabsorption theory which looks at the form of the transition matrix elements (i.e. the form of the wave-functions for initial and final states) using as basis these selection rules and the radial form of the corresponding transition operators rather than emphasizing the accurate solution of the time independent Schrödinger equation for initial and final states which, as it turns out, is unnecessary.

Let us consider the most common type of transition, the electric dipole process. We write for the exact wave-functions of the initial and final states:

$$\Psi_i = \Psi_i^o + \chi_i, \qquad \Psi_f = \Psi_f^o + \chi_f \tag{3}$$

so that the exact transition matrix element is:

$$\langle D \rangle \equiv \langle \Psi_i | D | \Psi_f \rangle = \langle \Psi_i^o | D | \Psi_f^o \rangle + \langle \Psi_i^o | D | \chi_f \rangle + \langle \chi_i | D | \Psi_f^o \rangle + \langle \chi_i | D | \chi_f \rangle \tag{4}$$

The principal contribution to $|\langle D \rangle|^2$ comes from the first term. Since Ψ_i^o and Ψ_f^o have, in general, the largest coefficients in the wave-function expansion, the next two terms dictate the part of the function space of χ_i and χ_f which is connected directly with the zeroth order Fermi-Sea vectors and therefore contributes to $|\langle D \rangle|^2$ the most.

Let χ^{IN} depict the correlation function space of $|i\rangle$, $|f\rangle$ which interacts directly via D with Ψ_f^o, Ψ_i^o respectively. Let χ^{NON} be its noninteracting orthogonal complement. Eq. 4 is then equivalent to:

$$\langle \Psi_i | D | \Psi_f \rangle = \langle \Psi_i^o | D | \Psi_f^o \rangle + \langle \Psi_i^o | D | \chi_f^{IN} \rangle + \langle \chi_i^{IN} | D | \Psi_f^o \rangle + \langle \chi_i^{IN} | D | \chi_f^{IN} \rangle + \langle \chi_i^{NON} | D | \chi_f^{IN} \rangle + \langle \chi_i^{IN} | D | \chi_f^{NON} \rangle + \langle \chi_i^{NON} | D | \chi_f^{NON} \rangle \tag{5}$$

Analysis of a few atomic systems has revealed that χ^{NON} corresponds to correlation vectors with small coefficients which are reasonably well decoupled from Ψ^o and χ^{IN}. This implies: a) their direct contribution to $|\langle D \rangle|^2$ is very small. I.e., the last three terms of eq. 5 are negligible because of the small coefficients and because not all correlation vectors in χ_i and χ_f are connected via D. b) Their neglect does not affect the coefficients of Ψ^o and χ^{IN} significantly.

The above observations together with an analysis of previous results on transition probabilities [35,36,54], has led us to the following theory:

The <u>form</u> of the parts of χ_i and χ_f which contribute the most to $\langle \Psi_i | D | \Psi_f \rangle$ can be obtained from the first order terms, $\langle \Psi_i^o | D | \chi_f \rangle$ and $\langle \chi_i | D | \Psi_f^o \rangle$, by applying the one electron dipole selection rules to the molecular or atomic orbitals of the Fermi-Sea configurations. The derived new χ_i^{IN} and χ_f^{IN} constitute a projected Hilbert space which, as predicted by electronic structure theory [1] and confirmed by computation, contains all vectors of spectroscopic significance together with others which contribute to the character of the wave-function. These have the largest coefficients in the correlation vector space and therefore contribute the most to $\langle \chi_i | D | \chi_f \rangle$ as well. Since off-diagonal properties, i.e. transition probabilities, involve the square of the corresponding matrix element (in contrast to diagonal properties), the remaining vectors with small coefficients have only a small contribution either through the last three terms of eq. 5 or through normalization. Thus eq. 5 is approximated well by:

$$\langle \Psi_i | D | \Psi_f \rangle \approx \langle \Psi_i^o | D | \Psi_f^o \rangle + \langle \Psi_i^o | D | \chi_f^{IN} \rangle + \langle \chi_i^{IN} | D | \Psi_f^o \rangle + \langle \chi_i^{IN} | D | \chi_f^{IN} \rangle \qquad (6)$$

so that the FOTOS wave-functions for initial and final states depend on the transition under examination and have the form

$$\Psi_i^{FOTOS} = \Psi_i^o + \chi_i^{IN}, \quad \Psi_f^{FOTOS} = \Psi_f^o + \chi_f^{IN} \qquad (7)$$

where

$$\langle \Psi_i^o | D | \chi_f^{IN} \rangle \neq 0, \quad \langle \Psi_f^o | D | \chi_i^{IN} \rangle \neq 0 \qquad (8)$$

Conditions (8) depend on symmetry and on the manner by which the radial functions in Ψ^o and in χ^{IN} are chosen (i.e. nonorthonormality cases).

Given this general analysis we proceed with the implementation steps of FOTOS:

C. Implementation of FOTOS

Step 1: Calculate the Fermi-Sea zeroth order vectors, Ψ_i^o and Ψ_f^o. In practice, Ψ_i^o and Ψ_f^o contain 1-4 spectroscopic configurations with orbitals computed at the H-F or the MCHF level [1].

Step 2: The response of the F-S orbitals of the ground state implies a "symmetry quantization" due to the symmetry of the dipole operator D. We write symbolically:

| Fermi-Sea | ⊗ D ⟶ One-electron First Order Symmetries(FOS)
ground state   excited states

The same holds for the excited states:

| Fermi-Sea | ⊗ D ⟶ First Order Symmetries (FOS)
excited state(s)   ground state

We note that this step refers to subshell symmetries.

Step 3: Apply "energy quantization" i.e. radial quantization. This means that we formally associate with each one-electron subshell symmetry a complete set of radial functions. We write e.g.

$$\text{for atoms:} \qquad |\ell\rangle = 1\ell + 2\ell + 3\ell + \ldots + \varepsilon\ell + \varepsilon'\ell$$

$$\text{for diatomics:} \qquad |\lambda\rangle = 1\lambda + 2\lambda + 3\lambda + \ldots + \varepsilon\lambda + \varepsilon'\lambda \tag{9}$$

At this point, these expansions are only formal. The numbers 1, 2,3... signify bound orbitals and ε,ε' scattering orbitals with energy ε,ε'.

Step 4: Depending on the type of transition and the energy region of the excited state, we assign to the orbitals of the formal expansions specific radial functions. Thus, if there is significant valence-Rydberg mixing, the Rydberg configurations are calculated at the H-F level. Other types of FOS are approximated with one or two virtual orbitals (STO's or GTO's) with as yet unoptimized exponents. When strong valence-Rydberg (continuum) mixing is absent, explicit inclusion of H-F orbitals is unnecessary except in cases of transitions to Rydberg states where single subshell excitations in the ground state must have some Rydberg character even though their energy contribution is small.

Step 5: Using the approximate one electron basis set of step 4 we create all possible configurations for initial and final states and keep only those which survive nonorthonormality (NON) [61]. Strong orthogonality between initial and final basis sets would eliminate two electron excitations of virtual type out of the main configuration. Usually, NON is present in neutrals and is negligible in the ions where the charge distribution is compact [61]. Energy considerations from first order perturbation theory [54] also eliminate a few correlation vectors. Thus, the final number of transition participating configurations is very small, on the average 10-20 only. They contain mostly single subshell excitations and one or two pair correlations.

Step 6: For bound functions, the virtual orbitals are optimized by minimizing the appropriate energy functionals of the FOTOS wave-functions (eq. 7) according to the theory of ref. 1. For autoionizing states, when the continuum contribution is represented by square integrable functions, additional constraints are implicitly or explicitly imposed and care to avoid the region nearly degenerate with the autoionizing state is taken [62,63]. Purely continuum states can be approximated either by square-integrable functions whereby a smoothing procedure must be introduced in order to create the continuous absorption spectrum [64] or by explicit scattering functions obtained in a frozen correlated core of the ion [65]. Once the FOTOS wave-functions have been computed, the corresponding transition probability is obtained by evaluating eq. 6.

Step 7: Evaluation of the transition matrix element (eq. 6). Nonorthonormality (NON). The theory of evaluating off-diagonal matrix elements of transition operators between sums of Slater determinants constructed from a common set of orthonormal basis functions is, of course, well known and is used extensively. In FOTOS, the functions of eq. 7 are optimized separately and so their basis sets are nonorthonormal between themselves. This necessitates the evaluation of N-electron instead of one-electron integrals.

A useful NON computational procedure is that of King et al [66] which allows any two sets of spin orbitals to be transformed into equivalent sets such that their overlap matrix is diagonal. It was first adopted to the calculation of E1 transition probabilities by Westhaus [38] and later to E2 by Nicolaides and Westhaus [67]. However, since the brute force application of the NON method requires a very large number of matrix diagonalizations of determinantal overlaps, we have further simplified the computational effort by orders of magnitude, as follows:

a) The inactive deep core is left intact (orthogonality is satisfied there).

b) Radial overlaps below 10^{-5} are set equal to zero.

c) Pseudo-configurations which ignore the radial parts (e.g. $1s^2 2s^2 2p^2 v_p \rightarrow s^4 p^3$) are constructed and the corresponding determinants are considered only if the initial and final state pseudoconfigurations interact.

d) The overlap matrix O (and OO^+, O^+O) is symmetry-blocked and this allows the breaking of N electron matrices into a series of smaller ones.

e) If zeros exist in the non-dipole symmetry blocks computation ceases.

f) For the dipole symmetry sub-block, we solve a system of linear equations rather than diagonalizing, since we already know the eigenvalue (=0).

The net result is that we have to solve one very small set (1-5 typically) of linear equations for less than 10% of the possible det-det' interactions, and evaluate one determinant involving all the active electrons. Further study [61] has shown that NON can very often be neglected without significant loss of accuracy.

In conclusion we emphasize that the idea to select the part

of Hilbert space which contributes to the transition probability the most by simply employing the forementioned symmetry arguments, is not sufficient without the correct computational implementation which must be based on proper considerations of many-body aspects of the electronic structure. By this we mean that if the zeroth order vector were, say, a hydrogenic function, and the virtual space were chosen in a very approximate way, the same selection rules would not yield accurate results.

IV. HEURISTIC EXAMPLES

A. The formal application of the FOTOS procedure gives rise to correlation vectors in the excited state usually described as "shake-up transitions", "satellites", "resonances" etc. Below we treat two simple cases from the popular groups of the alkaline earths and the noble gases.

1. The Be Atom: Photoabsorption from the Ground State.

Fermi-Sea orbitals: 1s,2s,2p

Fermi-Sea configurations: $1s^2 2s^2$, $1s^2 2p^2$ ($2p^4$ is insignificant)

Fermi-Sea symmetry configurations: (ssss), (sspp)

Then,

Symmetry quantization:

$$(ssss) \otimes D \longrightarrow sssp$$

$$(sspp) \otimes D \longrightarrow sssp, sspd, sppp \tag{10}$$

i.e. the excited states reached from the ground state by electric dipole photoabsorption are restricted by the symmetry selection rules to have the above orbital symmetries;

Radial quantization:

$$|s\rangle = \alpha_1 |1s\rangle + \alpha_2 |2s\rangle + \ldots + \alpha_\varepsilon |\varepsilon s\rangle + \alpha'_{\varepsilon'} |\varepsilon' s\rangle + \ldots$$

$$|p\rangle = \beta_1 |2p\rangle + \beta_2 |3p\rangle + \ldots + \beta_\varepsilon |\varepsilon p\rangle + \beta'_{\varepsilon'} |\varepsilon' p\rangle + \ldots$$

$$|d\rangle = c_1 |3d\rangle + c_2 |4d\rangle + \ldots + c_\varepsilon |\varepsilon d\rangle + c'_{\varepsilon'} |\varepsilon' d\rangle + \ldots \tag{11}$$

By substituting expansions (11) into the first-order symmetry configurations we obtain the possible observable excited states in terms of configurations. Thus, when energy allows it, not only the singly excited 2snp, εp series appear but also the doubly excited, autoionizing sequences nsmp and mpnd ($n > 2$, $m \geqslant 2$), continuum components, and excitations from the K-shell. The same observations hold of course for the rest of the alkaline-earths, where, due to their larger ground state Fermi-Seas ($ns^2 + np^2 + nd^2 + \ldots$) a larger variety of valence electron excitations can occur [68] as well as excitations involving the core subshells.

2. The Ne Atom: Photoabsorption from the Ground State. Due to its closed shell, inert structure, Ne is attractive to both theoreticians and experimentalists investigating the photoabsorption process. Recently, its valence shell photoionization cross-section was treated by diagrammatic Many-Body Perturbation Theory (MBPT) [22,23] and RPA [44] techniques. The MBPT approaches did not agree with each other. In their analysis of the double photoionization process, Chang and Poe singled out "three physical effects--core rearrangement, ground-state correlations and a virtual Auger transition--(which) contribute significantly to the double photoionization in Ne". In this paragraph we apply the FOTOS analysis to this problem.

First of all, the FOTOS zeroth order Hartree-Fock matrix element, which includes direct and exchange (N-1) electron overlaps, accounts in a straight forward manner for "core rearrangement".

We now proceed with the derivation of the important correlation effects in ground and excited states.

a) Excited State Correlations:

Ground state F-S configurations: $1s^2 2s^2 2p^6\ {}^1S$

F-S symmetry configurations: $s^4 p^6$

Symmetry quantization

$$(s^4p^6) \otimes D \longrightarrow s^3p^7,\ s^5p^5,\ s^4p^5d \qquad (12)$$

Radial quantization

$$|s\rangle = \alpha_1|1s\rangle + \alpha_2|2s\rangle + \ldots + \alpha_\varepsilon|\varepsilon s\rangle + \ldots \qquad (13)$$

.

.

.

same expressions as in (11)

By substituting (13) into (12) we see that the function space of ${}^1P^o$ symmetry is spanned by configurations of the type: $1s^22s^22p^5$ ns,εs, $1s^22s^22p^4$npms,εpε's,$1s^22s2p^6$np,εp,$1s^22s^22p^5$nd,εd,$1s^22s^22p^4$

npmd,$\varepsilon p\varepsilon'd$,$1s^2 2s2p^5$npmp,$\varepsilon p\varepsilon'p$, etc. corresponding to single, double or inner electron excitations. It is obvious that the amplitudes of all these types of excitations interfere, provided that the energies required are of the same order of magnitude.

Consider for example the Ne^+ $1s^2 2s2p^6$ ^2S configuration. The "symmetry exchange" correlation $2p^2 \leftrightarrow 2sd$ is the most important as can be seen from the mixing coefficients in a small CI expansion of single subshell correlations with optimized virtual STO's (v_ℓ) [1]:

$$|Ne^+\ ^2S\rangle = |2s2p^6\rangle_{HF} + .205|2s^2 2p^4 3d\rangle_{HF} + .140|2s^2 2p^4 v_d\rangle$$
$$- .056|2s^2 2p^4 3s\rangle_{HF} + .0023|2p^6 v_s\rangle + .056|2s(2p^5 v_p)\ ^3S\rangle$$
$$- .0023|2s(2p^5 v_p)^1S\rangle \qquad (14)$$

This correlation effect, whose importance and easy calculation has been emphasized in this and in our electronic structure paper [1], gives rise to the "virtual Auger transition" [22] and the "shell interference" [44] diagrams, i.e. to mixing of a subset of FOTOS configurations derived above: $1s^2 2s2p^6$np,$\varepsilon p \longleftrightarrow$ $1s^2 2s^2 2p^4$npmd,$\varepsilon p\varepsilon'd$. It should be expected that for the larger noble gases it becomes more important [44] since the d symmetry can give rise to a Fermi-Sea orbital, e.g. $3p^2 \leftrightarrow 3s3d$ in Ar^+.

b) Ground State Correlations: The symmetry and orbital structure of the ground state correlation vectors which should enter a full scale FOTOS calculation of double photoionization is derivable from the Fermi-Sea of the continuum excited states:

$K2s^2 2p^4 \varepsilon p\varepsilon'd$, $K2s^2 2p^4 \varepsilon p\varepsilon's$, $K2s^2 2p^4 \varepsilon s\varepsilon'f$, $K2p^6 \varepsilon p\varepsilon'd$,

$K2p^6 \varepsilon p\varepsilon's$ --/Continuum Fermi-Sea for $^1P^o$ doubly ionized states

We note that the mixing coefficients are energy dependent.

Following the FOTOS analysis on the above continuum excited states and disregarding a few very high lying states, we obtain the type of correlated wave-function for the ground state:

$$|^1S\rangle_{FOTOS} = a_1|2s^2 2p^6\rangle + a_2|2s2p^5 \varepsilon p\varepsilon d\rangle + a_3|2s2p^5 \varepsilon p\varepsilon s\rangle$$
$$+ a_4|2s^2 2p^4 \varepsilon d\varepsilon s\rangle + a_5|2s^2 2p^4 (\varepsilon s)^2\rangle + a_6|2s^2 2p^4 (\varepsilon d)^2\rangle$$
$$+ a_7|2s^2 2p^4 \varepsilon d v_s\rangle + a_8|2s^2 2p^4 \varepsilon d v_d\rangle + a_9|2s^2 2p^5 \varepsilon p\rangle$$

$$+ a_{10}|2s^2 2p^4(\varepsilon p)^2\rangle + a_{11}|2s^2 2p^4 \varepsilon p v_p\rangle + a_{12}|2s^2 2p^4 \varepsilon p v_f\rangle$$
$$+ a_{13}|2p^6(\varepsilon p)^2\rangle + a_{14}|2p^6 \varepsilon p v_p\rangle + a_{15}|2p^6 \varepsilon d v_d\rangle$$
$$+ a_{16}|2p^6(\varepsilon d)^2\rangle + a_{17}|2p^6(\varepsilon s)^2\rangle + a_{18}|2p^6 \varepsilon s v_s\rangle \qquad (15)$$

where ε represent continuum or diffuse square-integrable orbitals. Comparison with refs. [22,23] indicates that quite a few ground-state correlations which are predicted here to be important have been neglected. For example, in [22] they considered diagrams corresponding to a_6, a_8, a_{10} and a_{11} only. Although not all terms in (15) are crucial, it still remains to be proven that some of these omissions in [22,23] are justified from a quantitative point of view.

B. It is worthwhile to show how pair correlations are predicted through FOTOS because of nonorthonormality.

Example: The Be $1s^2 2s2p\ ^1P^o \rightarrow \overline{1s}^2\overline{2p}^2\ ^1D$ transition.

Due to the fact that $\langle v_p|\overline{2p}\rangle \neq 0$, where v_p is a virtual for the $^1P^o$ state, pair correlations of the type $(2s2p)\rightarrow(v_s v_p)$, $(v_p v_d)$ are included in the FOTOS $^1P^o$ function. These vectors also interact directly with the other F-S configuration of the ^1D state (see Table 1 of ref. 1), the 2s3d ^1D. These correlations were not included in our old work on the Be sequence [35,40] since they correspond to all-external excitations. They are included in the approximate FOTOS calculations reported in section V.

C. The major advantage of FOTOS is the dramatic reduction in the number of correlation vectors which are employed. Below we give approximate FOTOS functions for two molecular transitions in O_2 and H_2O.

1) The O_2 $X^3\Sigma_g^- \longrightarrow {}^3\Sigma_u^-$ first and second excited states:

These important transitions have recently been examined through large CI by Buenker and Peyerimhoff [7] and Yoshimine et al [8]. The number of configurations employed were in the thousands.

Let's assume the ground state F-S configurations to be : $a|\pi_u^4\pi_g^2\rangle + b|\pi_u^2\pi_g^4\rangle$, and the upper state $a'|\overline{\pi}_u^3\,\overline{\pi}_g^3\rangle +$

$b'|\overline{\pi}_u^4\ \overline{\pi}_g\ \overline{3p\pi_u}\rangle$. $|3\sigma_u\rangle$ is also a F-S orbital. If we assume orthonormality so that no pair subshell excitations are permitted, the FOTOS wave-functions are composed of only 15 vectors for the ground state and only 10 vectors for the excited state with a few virtuals of one STO each, to be optimized according to [1]. These are:

$|{}^3\Sigma_g^-\rangle$: $3\sigma_g^2\pi_u^4\pi_g^2$, $3\sigma_g^2\pi_u^2\pi_g^4$, $3\sigma_g\pi_u^3\pi_g^3 3\sigma_u$, $3\sigma_g\pi_u^4\pi_g 3p\pi_u 3\sigma_u$,
$3\sigma_g\pi_u^4\pi_g 3p\pi_u^2$, $3\sigma_g\pi_u^3\pi_g^3 v_{\sigma_u}$, $3\sigma_g\pi_u^4\pi_g 3p\pi_u v_{\sigma_u}$, $3\sigma_g\pi_u^3\pi_g^2 3p\pi_u$,
$3\sigma_g^2\pi_u^2\pi_g^3 v_{\pi_g}$, $3\sigma_g^2\pi_u^3\pi_g 3p\pi_u v_{\pi_g}$, $3\sigma_g^2\pi_u^3\pi_g^2 3p\pi_u$, $3\sigma_g^2\pi_u^4 3p\pi_u^2$,
$3\sigma_g^2\pi_u^3\pi_g^2 v_{\pi_u}$, $3\sigma_g^2\pi_u^4 3p\pi_u v_{\pi_u}$, $3\sigma_g^2\pi_u^4\pi_g v_{\pi_g}$

$|{}^3\Sigma_u^-\rangle$: $3\sigma_g^2\pi_u^3\pi_g^3$, $3\sigma_g^2\pi_u^4\pi_g 3p\pi_u$, $3\sigma_g\pi_u^4\pi_g^2 3\sigma_u$, $3\sigma_g\pi_u^2\pi_g^4 3\sigma_u$,
$3\sigma_g\pi_u^4\pi_g^2 v_{\sigma_u}$, $3\sigma_g\pi_u^2\pi_g^4 v_{\sigma_u}$, $3\sigma_g^2\pi_u^3\pi_g^2 v_{\pi_g}$, $3\sigma_g^2\pi_u\pi_g^4 v_{\pi_g}$, $3\sigma_g^2\pi_u^4\pi_g v_{\pi_u}$,
$3\sigma_g^2\pi_u^2\pi_g^3 v_{\pi_u}$.

2) The H_2O ${}^1A_1 \longrightarrow {}^1B_1$ Transition: This line has been studied accurately by Buenker and Peyerimhoff [69] who used wave-functions of about 2000 configurations.

Assuming single configurational Fermi-Seas and orthonormality, FOTOS predicts 12 ground state and 7 excited state configurations:

$|{}^1A_1\rangle$: $2a_1^2 1b_2^2 3a_1^2 1b_1^2$, $2a_1 1b_2^2 3a_1^2 1b_1^2 4a_1$, $2a_1 1b_2^2 3a_1^2 1b_1 4a_1 2b_1$,
$2a_1 1b_2^2 3a_1^2 1b_1 4a_1 v_{b_1}$, $2a_1^2 1b_2^2 3a_1 1b_1^2 4a_1$, $2a_1^2 1b_2^2 3a_1 1b_1 4a_1 2b_1$,
$2a_1^2 1b_2^2 3a_1 1b_1 4a_1 v_{b_1}$, $2a_1^2 1b_2^2 3a_1^2 4a_1^2$, $2a_1^2 1b_2^2 3a_1^2 4a_1 5a_1$,
$2a_1^2 1b_2^2 3a_1^2 4a_1 v_{a_1}$, $2a_1^2 1b_2^2 3a_1^2 1b_1 2b_1$, $2a_1^2 1b_2^2 3a_1^2 1b_1 v_{b_1}$

$|{}^1B_1\rangle$: $2a_1^2 1b_2^2 3a_1^2 1b_1 4a_1$, $2a_1^2 1b_2^2 3a_1^2 1b_1 5a_1$, $2a_1^2 1b_2^2 3a_1^2 1b_1 v_{a_1}$,
$2a_1^2 1b_2^2 3a_1 1b_1^2 2b_1$, $2a_1^2 1b_2^2 3a_1 1b_1^2 v_{b_1}$, $2a_1 1b_2^2 3a_1^2 1b_1^2 2b_1$,
$2a_1 1b_2 3a_1^2 1b_1^2 v_{b_1}$.

The small FOTOS molecular functions derived above for O_2 and H_2O cannot, of course, compete in detail with those employed in [7,8,69]. However, they indicate clearly how drastic a reduction of the magnitude of computation of transition probabilities can be within FOTOS. Furthermore, if our experience with atoms can serve as an indicator, the accuracy of the results using these--or slightly improved--functions, is expected to be satisfactory.

D. Electric Quadrupole Transitions

FOTOS is of course applicable to higher multipole transitions. Consider for example the well-known OI $1s^2 2s^2 2p^4\ {}^1S_o \rightarrow {}^1D_2$ electric quadrupole transition [67,70]. For ^1S the FS is: $a|1s^2 2s^2 2p^4\rangle + b|1s^2 2p^6\rangle$. For ^1D, FS: $1s^2 2s^2 2p^4$. Application of FOTOS using the electric quadrupole selection rules yields:

$$|{}^1S\rangle_{FOTOS} : 1s^2 2s^2 2p^4,\ 1s^2 2p^6,\ 1s^2 2s 2p^4 v_s,\ 1s^2 2s 2p^4 v_d,\ 1s^2 2s^2 2p^3 v_p$$

$$|{}^1D\rangle_{FOTOS} : 1s^2 2s^2 2p^4,\ 1s^2 2s 2p^4 v_s,\ 1s^2 2s 2p^4 v_d,\ 1s^2 2s^2 2p^3 v_p,\ 1s^2 2s^2 2p^3 v_f,\ 1s^2 2p^5 v_p',\ 1s^2 2p^5 v_f.$$

These small functions contain all the correlation effects which were computed by us in our previous study of the effect of electron correlation on such E2 transitions [67]. Our predictions proved to be in good agreement with an accurate experiment which followed [70], and therefore FOTOS is expected to be at least as accurate.

V. QUANTITATIVE EXAMPLES

We now proceed with the presentation of numbers for a variety of transition probabilities in atoms. We note that the FOTOS calculations need not be perfect. I.e., as with every other theoretical scheme, there is always room for improvement of the implementation steps (e.g. basis sets) but within the theory. On the other hand, when compared to experimental values or other extensive calculations, the FOTOS results are consistently accurate. Depending on the amound of effort put into the calculation of a particular line, errors may range between 20% and 2% where other theoretical models can sometimes fail by factors of 2-100.

Therefore it appears that our implementation procedures are adequate. Very small oscillator strengths (~ 0.001) carry of course larger uncertainties.

Although FOTOS has not been tested in molecules yet, we see no reason why the same theoretical model shouldn't work there as well. In fact, since many molecular transitions involve closed shell initial states, small molecules may prove, on the average, easier to handle than atoms. Finally, nuclear transition probabilities can also be subjected to a FOTOS analysis at least in light nuclei where there is sufficient information to define the Fermi-Seas (e.g. in O^{16}). Just like methods such as the RPA, equations of motion, Green's functions, etc. were taken from nuclear physics and adopted to selected Quantum Chemical problems with much success, the FOTOS analysis and methodology can prove useful to nuclear physics where the shell model and symmetry are known to have a considerable physical significance as well.

The selected examples presented below are grouped separately with a brief, related commentary. Some of the FOTOS results have been published before and some have not.

A. Photoexcitation of Be and He to $^1P^o$ Rydberg series

These atoms have become something of a "must" for testing new approaches to the theory of electronic structure and dynamics. He because of its simple structure and Be because, although it is essentially a pseudo-two electron system, it is the simplest many-electron system where certain open-shell correlations become important. In Tables 1 and 2 we present our FOTOS results from calculations which are rather simple when compared to the other theories. The FS of the excited $^1P^o$ of Be states was represented by H-F functions and the remaining symmetry space by virtual orbitals optimized on the two lowest states separately.

We note that the higher Be $^1P^o$ states ($np \geqslant 3$) show extreme sensitivity to the type of one-electron functions employed (with $f_L \simeq f_V$ throughout) even though the series is unperturbed by a valence configuration. Our values there represent what we believe to be a well-balanced FOTOS calculation. The K-shell and K-L intershell correlations were not considered. The RPA results (ref. c Table 2) indicate a related error of 10%.

A noteworthy observation on Table 2 is the following: According to the findings of Oddershede et al (ref. j of Table 2), when the Time-Dependent Hartree-Fock (TDHF) theory is extended to second order, there is noticeable difference between the

Table 1

Oscillator Strengths for the Photoexcitation of He

	Hartree-Fock			FOTOS			Large CI with r_{ij} basis sets b		c		RPA (d)
	f_L	f_V	f_A	f_L	f_V	f_A	f_L	f_V	f_L	f_V	
$1s^2 \rightarrow 1s2p\ ^1P^o$	.258	.240	.319	.263	.263	.252	.276	.276	.276	.276	.252
3p	.070	.065	.084	.069	.069	.066	.0736	.0734	.0734	.0730	.070
4p	.029	.026	.034	.028	.028	.026			.030	.030	.029
5p	.015	.013	.017	.014	.014	.013					.015
6p	.0084	.0077	.0099	.0079	.0078	.0074					.0085
7p	.0053	.0048	.0062	.0048	.0048	.0046					.0054

a) This work--f_L,f_V,f_A stand for length, velocity and acceleration operators [71]. For the K-shell f_A is not as bad as it is for valence shell transitions [2,71].

b) B. Schiff and C.L. Pekeris, Phys. Rev. 134 ,A638 , (1964).

c) A.W. Weiss, J. Res. NBS 71A ,163 , (1967).

d) M. Ya. Amusia et al, Phys. Rev. A13 ,1466,(1976).

Table 2

Oscillator Strengths for the Photoexcitation of Be

	H-F[a]		FOTOS[a]		Exp.[b]	(c)	(d)	(e)	(f)		(g)		(h)		(i)	(j)		(k)		(l)
	f_L	f_V	f_L	f_V			f_L	f_L	f_L	f_V	f_L	f_V	f_L	f_V	f_L	f_L	f_V	f_L	f_V	
$2s2p\ ^1P$	1.68	1.03	1.36	1.46	1.34±.05	1.36	1.25	1.344	1.25	1.14	1.11	1.13	1.386	1.378	1.37	1.43	1.28	1.381	1.435	1.378
$3p\ ^1P^o$	.203	.062	.014	.016		.023	.016								.022	.00002	.0013			.028
$4p\ ^1P^o$	.054	.012	.00022	.00016		.001	.010								.0002					.001

a) This work. First reported at the 8th EGAS conference, Oxford July (1976).

b) I. Martinson, A. Gaupp and L.J. Curtis, J. Phys. B7, L463 (1974).

c) M. Ya. Amusia et al., Phys. Rev. A13, 1466 (1976)/RPA.

d) H.P. Kelly, Phys. Rev. 136B, 896 (1964)/MBPT.

e) J.S. Sims and R.C. Whitten, Phys. Rev. A8, 2220 (1973)/very large CI with r_{ij} basis sets.

f) C.A. Nicolaides, D.R. Beck and O. Sinanoglu, J. Phys. B6, 62 (1973)/ limited CI following the NCMET classification scheme.

g) K.E. Banyard and G.K. Taylor, Phys. Rev. A10, 1019 (1974)/ large CI following Sinanoglu's MET.

h) C.M. Moser, R.K. Nesbet and M.N. Gupta, Phys. Rev. 13A, 17 (1976)/"Variational Bethe-Goldstone" calculations.

i) G.A. Victor and C. Laughlin, Nucl. Inst. Methods 110, 189 (1973)/parametrized model potential--also quoted in ref. (l).

j) I. Oddershede, P. Jørgensen and N.H.F. Beebe, Int. J. Qu. Chem. 12, 655 (1977)/"Second Order Polarization Propagator Approximation"(SOPPA); This type of calculation goes beyond the RPA. Note the disagreement between RPA and SOPPA which, if the correct value is close to that of FOTOS, seems to indicate lack of uniform convergence in the hierarchy of such methods--provided the numerical implementation of the theory in ref. j is complete.

k) A. Hibbert, J. Phys. B7, 1417 (1974)/Large CI.

l) R.F. Stewart, J. Phys. B8, 1 (1975)/ Time Dependent Hartree-Fock (equivalent to RPA).

transition probabilities of the two methods. As Oddershede et al point out, this contradicts the suggestion of Shibuya et al [72] and seems to indicate lack of uniform convergence, at least for weak transitions (e.g. see the Be $2s^2\ {}^1S \rightarrow 2s3p\ {}^1P^o$ transition). On the other hand, given the fact that the TDHF results of Oddershede et al disagree with those of Stewart (ref. ℓ, Table 2) and Amusia et al (ref. c, Table 2), more work should be carried out to differentiate between purely theoretical and basis sets inefficiency effects.

Surprisingly, there are no experimental values for the lifetimes of the np $(n > 2)\ {}^1P^o$ states. Accurate measurements would be very helpful to theory.

B. Oscillator Strengths Along Isoelectronic Sequences: The NI $2s^22p^3\ {}^4S^o \rightarrow 2s2p^4\ {}^4P$ Sequence

Straight forward Z-dependent perturbation theory [34,73] shows that the nonrelativistic oscillator strength, f, of a particular transition can be expressed as a series of Z^{-1}:

$$f = f_0 + f_1 Z^{-1} + f_2 Z^{-2} + \ldots \qquad (16)$$

As Z becomes large, electron correlation effects other than those suggested by the hydrogenic near-degeneracies lose their critical importance and f goes smoothly to f_0, the hydrogenic value. Of course, relativistic effects introduce deviations from this uniformly converging nonrelativistic behavior [74].

The practical value of eq. 16 is that is provides the theoretical background for easy (graphical) interpolations and extrapolations in the high Z region, provided an accurate knowledge of the neutral end and an accurate number for an ionic transition exist. (E.g. in ref. [75] accurate f-values for transitions of the type $2s^22p^n \rightarrow 2s2p^{n+1}$ in Si ions were published (due to a computer input data error, the Si IX $2s^22p^2\ {}^1D \rightarrow 2s2p^3\ {}^1D^o$ f-value is wrong) with the aim to serve as standards for constructions of accurate isoelectronic f-value sequences).

The theoretical value of eq. 16 is that the coefficients $f_0, f_1, f_2, \ldots$ can be computed from first principles within Z-dependent (hydrogenic) perturbation schemes so that f-values which are quite accurate for large Z can be easily obtained [34] and be used to systematize experimental data from Beam-Foil Spectroscopy [27].

Table 3

Oscillator Strengths for the N $2s^2 2p^3\ {}^4S^o \rightarrow 2s2p^4\ {}^4P$ Isoelectronic Sequence

(Length Form)

	λ(Å)	H-F(a)	FOTOS(a)	NCMET(b)	NCMET(c)	Z-Dep. Pert. Th.(d)	Experiment
N	1134	.490	.084	.145	.035	.279	.085(e)
OII	833	.428	.235	.206		.265	.26(e)
FIII	658	.365	.233			.248	.262(e)
NeIV	542	.322	.218			.231	.167(e)
NaV	463	.292	.210			.214	.24(f)
SiVIII	319	.217	.170			.176	

(a) Ref. 10; (b) Ref. 38; (c) S.L. Davis and O. Sinanoglu, J. Chem. Phys. 62, 3664 (1975); (d) Ref. 32; (e) E.H. Pinnington et al., Can. J. Phys. 42, 1014 (1976); and in Beam Foil Spectroscopy Volume 1, edited by I. Sellin and D. Pegg, Plenum Press, 1976; (f) M. Buchet, private communication

In Table 3 we present results of FOTOS along the N $2s^2 2p^3\ {}^4S^o \rightarrow 2s2p^4\ {}^4P$ isoelectronic sequence [10] and compare them with other theories and experimental values. FOTOS calculations on ions are very simple and yet accurate. On the other hand, they are also accurate on the neutral end where other theories often fail.

C. Transitions to Excited States Exhibiting Heavy valence-Rydberg Mixing; An Effective Hamiltonian Approach

When a calculation of the oscillator strength is made for a region where a valence configuration (inner hole or doubly excited) is near or embedded in Rydberg or continuum series and mixes with them, the redistribution of oscillator strength from the IPM to the correct many-body description can be unpredictably large [6, 56]. This is because the mixing of the valence configuration with Rydberg-continuum series may be large whether it lies below or is embedded in the discrete or continuous spectrum (in which case it is autoionizing). (Examples: Li $1s2p^2 \leftrightarrow 1s^2 ns,nd$; He_2^+ $1\sigma_g 1\sigma_u^2 \leftrightarrow 1\sigma_g^2 n\sigma_g$).

The quantitative theoretical description of such important situations (e.g. avoided crossings) is very difficult to achieve accurately. This is because it is often very sensitive to the relative positions of the diagonal matrix elements and to the basis sets describing the excited states in the interacting region. As already stated [1], in such cases FOTOS employs explicitly SCF H-F functions in addition to the other vectors describing electron correlation. Furthermore, since FOTOS, being a truncated CI calculation, cannot yield the exact energy differences between the valence and Rydberg levels of interest, since the correlation which is left out is different for the valence and for the Rydberg configurations, an effective Hamiltonian approach has been adopted [6,55] which corrects this inadequacy of the truncated CI. This is accomplished by shifting downwards the diagonal energies until the best energy match between the diagonalized vectors and the experimental spectrum in the region of the perturber is achieved. This Down-Shift Model (DSM) [6,55] does not change the off-diagonal matrix elements which are expected not to be affected much if H-F basis functions are employed. (Compare this approach with the recent effective Hamiltonian approach of Nitzsche and Davidson [75] to a molecular case). The final correct adjustment of the diagonal matrix elements is expected to yield the correct mixing coefficients of the nearly degenerate configurations without having to perform extensive calculations on correlation effects which FOTOS otherwise singles out as unimportant for oscillator strengths.

Below we present results from FOTOS and other type calculations on oscillator strengths to valence-Rydberg mixed states. For BI [6] and Cl III (Tables 4 and 6) DSM was employed. Agreement with experiment is very good, in contrast to other models. In BI our FOTOS results show a redistribution of oscillator strength over the entire series. In Be I (Table 5) the FOTOS calculations are not complete and the results are given only to present a global description of the ^1D series. These approximate FOTOS calculations show that the first, valence like state loses all its oscillator strength (this value is so small that only an order of magnitude accuracy can be assigned to it) which goes mainly to the second and third states and of course to the rest of the Rydberg and continuum series.

D. Allowed Transitions in Highly Ionized Atoms: The Mg- and Zn-like resonance lines in Mo ions.

Lines in certain highly ionized metals (e.g. Mo, W) are critically important in current plasma work [27]. The main theoretical importance of such transitions is the study of relativistic effects on transition probabilities. In principle, these enter through the energy difference factor, the wave-function changes and the transition operator. It turns out [56,74,53, 84,85] that for medium degrees of ionization and for spin-allowed

Table 4

Oscillator strengths for the BI $2s^2 2p\ ^2P^o \rightarrow 2s^2 ns$, $2s2p^2\ ^2S$ transitions. According to FOTOS, the valence-Rydberg Interaction redistributes oscillator strength over the whole discrete spectrum. On the other hand, Nesbet's results indicate a disappearance of oscillator strength from the valence configuration.

	Theory										Experiment
	λ(Å)	Beck and Nicolaides [6]				Nesbet [76]	Weiss [77]		Zibinic [41]		Beam-Foil Spectroscopy
		HF		FOTOS-DSM							
		f_L	f_V	f_L	f_V	f_L	f_L	f_V	f_L	f_V	
$2p\ ^2P^o \rightarrow 3s\ ^2S$	2497	0.051	0.062	0.074	0.090	0.0695	0.067	0.074	0.062	0.068	0.087± 0.005 [78]
4s	1818	0.0087	0.0097	0.017	0.018	0.0314					
5s	1663	0.0032	0.0034	0.012	0.013	0.0038					
6s	1601	0.0015	0.0016	0.017	0.019	0.0089					
$2s2p^2$	1573	0.091	0.051	0.014	0.028	0.0004					0.035 [79]
7s	1559	0.00086	0.00091	0.0084	0.0098	0.0033					

Table 5

Oscillator Strengths for the Be 2s2p $^1P^o \rightarrow 2p^2$, 2snd ^1D Transitions(a)

	This Work				CI[40]		NCMET[35]		"Bethe-Goldstone [39]		CI [81]	Model Pot.[80]	Exp.	
	H-F		FOTOS											
	f_L	f_V	f_L	f_V	f_L	f_V	f_L	f_V	f_L	f_V			(b)	(c)
$2p^2\ ^1D$	.164	2.06	.0086	.0016	.00065	.00047	.020	.016	.0003	.0000	.001	.0084	<.0048	
2s3d	1.09	0.464	.64	.41							.41	.418	.50	.43
2s4d	.0365	4×10^{-6}	.19	.16							.173	.191	.25	.19
2s5d	.0108	8×10^{-5}	.08	.07							.081	.090	.16	
2s6d	.0047	9×10^{-5}	.05	.05							.043	.050	.07	

a) The FOTOS results are approximate and are given only to provide a global description of the ^1D series. The oscillator strength lost by the first, valence configuration, is distributed over the entire Rydberg series and, presumably, goes over to the continuum.

b) Ref. 82.

c) I. Bergstrom, J. Bromander, R. Buchta, L. Lundin, and I. Martinson, Phys. Letts. 28A, 721 (1969).

Table 6

f-values for the Cl III $3s^2 3p^3\ ^4S^o \longrightarrow 3s3p^4$, $3s^2 3p^2 3d\ ^4P$ transitions. Comparison of the HF and FOTOS results shows that the amount of strength lost by one state is gained by the other (see section VII D).

	HF		FOTOS-DSM		NCMET [37]		Exper. [83]
	f_L	f_V	f_L	f_V	f_L	f_V	
$^4S \rightarrow 3s3p^4\ ^4P$	.685	.777	.041	.026	.085	.103	0.043 ± .003
$3s^2 3p^2 3d\ ^4P$	2.9	2.0	3.47	3.47			

electric dipole transitions in the valence shell, the effect is almost exclusively through the energy difference factor provided the length form of the operator is employed. In Table 7 we present f-values in Mo ions from simple FOTOS calculations [57] where the experimental wave-lengths have been used. They are compared with recent Relativistic [53] and nonrelativistic MCHF [51] and Relativistic RPA calculations [84,85]. The agreement is very good. We note that for ions, FOTOS is expected to yield nonrelativistic f-values with an accuracy of better than 10%.

E. Radiative Autoionization: The Li $1s2p^2\ ^2P$ Lifetime

A recently revealed discrepancy between theoretical [86] and experimental [87] lifetimes of the Li $1s2p^2\ ^2P$ state led us to a FOTOS analysis of this case [26]. It turns out that this highly excited metastable state can decay to the adjacent continuum radiatively: (Li $1s2p^2\ ^2P \longrightarrow Li^+\ 1s^2 + e^-\ ^2P^o$). This decay channel was not considered before [86,87]. Such radiative autoionization (RA) deexcitation mechanisms are in principle possible in a variety of neutral and ionic atoms and molecules [26, 88,89]. In Table 8 we present the results [26] of the first complete calculation of the transition probabilities of an excited state decaying radiatively to discrete as well as to continuum channels. In this case, RA, which is a mechanism producing a continuous distribution of photons and electrons, accounts for about 10% of the total lifetime.

Table 7

Oscillator Strengths for the $^1S \rightarrow ^1P^o$ resonance lines in Mo^{12+} and Mo^{30+}. The nonrelativistic FOTOS calculations employed the length form and the experimental wavelength. The length form minimizes the relativistic effects on the transition moment and the experimental wavelength includes the effect of relativity on the transition energy. The first line of MCHF results of [51] are completely nonrelativistic calculations which take core-polarization into account and employ the theoretical wavelength. The results of the second line employ the experimental wavelength.

	FOTOS [57]	Rel. MCHF [53]		MCHF [51]		Rel. RPA [84,85]	
	f_L	f_L	f_V	f_L	f_V	f_L	f_V
Mo^{12+} $4s^2\ ^1S$-$4s4p\ ^1P^o$	1.56			1.30	1.29	1.54	
				1.43			
Mo^{30+} $3s^2\ ^1S$-$3s3p\ ^1P^o$	0.574	.540	.547			.549	.566

Table 8

Transition probabilities and theoretical lifetime of the Li $1s2p^2\ ^2P$ state from FOTOS calculations (length and velocity forms) and comparison with previous theory and experiment. Radiative autoionization contributes 8%-11% and the Rydberg series another 9%-10%. $\Delta = E(Li^+\ 1s^2\ ^1S) - E(Li\ 1s2p^2\ ^2P) = 2.07$ a.u. The integrated continuum transition probability includes a contribution of 1.2×10^8 sec^{-1} from the $1s[(2s2p)^3P^o]^2P^o$ autoionizing state and $(3\times10^{-6})\times10^8$ sec^{-1} from the $1s[(2s2p)^1P^o]^2P^o$ autoionizing state (length form).

	Transition probabilities (sec^{-1})			Lifetime of the Li $1s2p^2\ ^2P$ state (sec)		
	FOTOS [26]		Large CI [86]	FOTOS	[86]	Experiment [87]
	Length	Velocity	Length			
$1s^22p\ ^2P^o$	208×10^8	189×10^8	185×10^8	0.040×10^{-9} (length)	0.054×10^{-9}	$0.015\pm0.01\times10^{-9}$
3p	13.5	13.9		0.042×10^{-9} (velocity)		
4p	3.8	4.0				
5p	1.7	1.8				
6p	0.9	1.0				
7p	0.5	0.5				
$\Sigma_{n=8}np$	1.5	1.7				
$\int_0^\Delta dE$	19.1	26.2				
Total	249×10^8	238.1×10^8				

F. Applications to the Calculation of Polarizabilities

FOTOS can be extended straight forwardly to the economical and accurate calculation of polarizabilities [90], and, due to its resulting small and compact wave-functions, of hyperpolarizabilities. Such an application has been made to the Be dipole static polarizability. Its result is presented in Table 9 and is compared to other extensive calculations.

We note that, for certain systems where the oscillator strength distribution favors the low lying states, accurate oscillator strengths can yield very good lower bounds to polarizabilities [93].

G. A Precision Study: Be^+ $1s^22s$ $^2S \longrightarrow 1s^22p$ $^2P^0$

Recent experiments on lifetimes have claimed accuracies of better than 1% (e.g. ref. [94]). This is of course challenging to theory--although from the chemical point of view such accuracy is rather uninteresting. Nevertheless, in order to test the capability for precision of a full scale FOTOS calculation, we chose to compute the Be^+ $1s^22s$ $^2S \rightarrow 1s^22p$ $^2P^o$ f-value--currently under investigation by the Berlin group [95]. For this case we broke the K shell and formed the F-S configurations $|^2S\rangle$: $1s^22s$, $1s2s^2$, $1s2p^2$; $|^2P^o\rangle$: $1s^22p$, $1s2s2p$. The virtual space was made flexible and a total of 50 vectors for each state were employed. The results are presented in Table 10 together with previous CI [96] and Coulomb approximation [97] results as well as two experiments which, however, are not accurate to 1%. The FOTOS f_L, f_V spread is 2%.

Table 9

Static polarizabilities for the Be ground state, in $Å^3$, from three many-body methods (length form).

FOTOS [90]	5.49	^1S: 9 vectors, $^1P^o$: 34 vectors
CI-Hylleraas [91]	5.42 $\pm$ .12	^1S: 105 vectors, $^1P^o$: 53 vectors
PNO-CEPA [92]	5.61	

Table 10

Oscillator Strengths for the Be^+ $1s^22s\ ^2S \rightarrow 1s^22p\ ^2P^o$ Transition

This work							
HF		FOTOS		CI [96]		Coulomb approx. [97]	Exper.
f_L	f_V	f_L	f_V	f_L	f_V		
0.511	0.549	0.500	0.510	0.505	0.521	0.492	0.52 (a)
							0.54 ± .03 (b)

a) T. Andersen et al, Phys. Rev. 188, 76 (1969).
b) I. Martinson, quoted in [97].

VI. LENGTH, VELOCITY AND ACCELERATION ELECTRIC DIPOLE OPERATORS [71]

It is well known that formally there are at least three equivalent expressions for the nonrelativistic oscillator strengths of electric dipole transitions (we neglect summations over degeneracies in initial and final states):

the length form (in a.u.): $$f_L=\frac{2}{3}(E_f-E_i)\ |\langle\Psi_i|\sum_{j=1}^{N}\vec{r}_j|\Psi_f\rangle|^2 \quad (17a)$$

the velocity form: $$f_V=\frac{2}{3}(E_f-E_i)^{-1}\ |\langle\Psi_i|\sum_{j=1}^{N}\vec{\nabla}_j|\Psi_f\rangle|^2 \quad (17b)$$

the acceleration form: $$f_A=\frac{2}{3}(E_f-E_i)^{-3}\ |\langle\Psi_i|\sum_{j=1}^{N}\frac{\vec{r}_j}{r_j^3}|\Psi_f\rangle|^2 \quad (17c)$$

The energy-less expression is [71]:

$$\bar{f} = f_L\ (f_A/f_V)^{1/2} \quad (18)$$

In general, the use of approximate N-electron wave-functions results in answers for eqs. 17 which disagree with each other. This fact has given rise to the following question: Which form is more appropriate for the accurate prediction of electric dipole transition probabilities?

This question was examined by us some time ago [71] and we reached the conclusion that in nonrelativistic quantum mechanics, formal arguments for or against a particular form are irrelevant in practice. The physically relevant question should refer only to accuracy of results (i.e. agreement with experiment) and in this respect arguments based on the proper matching between operator and the configuration space emphasized by the wavefunctions are the most convincing.

In the same paper we also related the fact that the RPA yields $f_L = f_V$ [98,99] to its gauge invariance. This notion has recently been formalized by Lin [100].

VII. EXTRAORDINARY ABSORPTION PHENOMENA

The usual deviation of the IPM from a many-body (exact) description of many absorption oscillator strengths is no more than 50%. This may or may not alter the qualitative picture of the photoabsorption process and the proper assignment of the excited states.

However, there are cases where, due mainly to the valence-Rydberg type interactions, the true absorption properties of an atom or a molecule in a certain energy region differ considerably from those predicted by the IPM. We list them below, together with conditions for their occurence (where possible), since they may prove useful to qualitative and quantitative photochemistry and photophysics.

A. "Collective Excitations" [107]

These are cases where, contrary to the predictions of the IPM, transitions at a particular energy seem to have an unnusually large probability while the surroundings hardly absorb. Simple physical and mathematical arguments based on the conservation of oscillator strength lead to the conclusion that this occurs whenever [107]:

$$a_m = B \langle \Phi_m | D_z | \Psi_i \rangle \tag{19}$$

where a_m is the CI coefficient of the zeroth order configuration Φ_m in the excited state and Ψ_i is the initial state function. B is a fixed constant.

Conditions (19) may occur in the continuum as well as in the discrete spectrum of small, medium or large atoms and molecules.

Its satisfaction depends on the choice of the form of D and on Ψ_i. This implies that in the case of RPA-type calculations the choice of the zeroth-order potential may be crucial [29,101].

B. Nearly Constant Distribution of Oscillator Strengths

This is the case of the FOTOS results on Boron (Table 4). The absorption strength is spread over a relatively large energy region--in the discrete or continuous spectrum--in a nearly constant manner. The IPM spectrum may look completely different, having extrema or sharp increases or decreases.

C. Nearly Zero Oscillator Strength

This is the case of the Be $1s^2 2s2p\ ^1P^o \rightarrow 1s^2 2p^2\ ^1D$ transition (Table 5). It is a one particle transition with a large HF oscillator strength. One might expect a non vanishing transition probability. Yet, the mixing with the Rydberg-continuum 2snd, $\varepsilon d\ ^1D$ series apparently results in a completely destructive interference whereby the oscillator strength becomes nearly zero.

D. Localized Transfer of Oscillator Strength [9,10]

If for certain cases, such as avoided crossings, we make the assumption that <u>the mixing coefficients of two closely lying excited states with non-vanishing off-diagonal matrix elements are dictated primarily by their mutual interaction</u>, the following situations may occur:

1) There is an exchange of oscillator strength from the IPM to the many-body description of equal amounts (the usual case, e.g. see Table 6). Experimentally this can be seen in sharp increases or decreases of photoabsorption cross-sections in regions of (avoided) crossings of atomic (as a function of Z) or molecular (as a function of geometry) levels [10,102].

2) There is <u>no transfer</u> of oscillator strength if

$$c = \frac{2f_2}{(f_1 + f_2)} - 1 \tag{20a}$$

$$d = \pm 2 \frac{f_1 f_2}{(f_1 + f_2)} \tag{20b}$$

where c,d are the mixing coefficients in the two component excited state and f_1, f_2 are the corresponding zeroth order f-values.

3) There is no absorption to one state if

$$c = \pm \frac{f_2}{(f_1 + f_2)} \tag{21a}$$

and $\pm 2cd < 0$

$$d = \pm \frac{f_1}{(f_1 + f_2)} \tag{21b}$$

We note that conditions (2) and (3) cannot be met in an avoided crossing region if $f_1 \gg f_2$.

VIII. RELATIVISTIC FOTOS

For systems of high Z or for ionized atoms with large (Z/N) ratios (N equals the number of electrons, e.g. He-like Argon), relativistic effects gain importance. Experimentally, these effects manifest themselves when they induce "forbidden" transitions with high probability. The literature on this subject is growing fast and methods are being presented where a portion of correlation and a portion of relativity are taken into account [43,53,74,84,103-106].

Inclusion of relativistic effects in the FOTOS framework can take the following forms in approximately ascending order of accuracy or rigor:

A) Employ the nonrelativistic FOTOS formalism and introduce the low-Z Pauli approximation perturbation (spin-orbit, spin-spin etc.) only to correct the configurational composition of the relevant wave-functions by mixing states of different nonrelativistic symmetry. Among other things, this approximation neglects the effect of relativity on the radial parts of the Fermi-Sea functions. The standard (e.g. eqs. 17 for electric dipole) nonrelativistic operators should be used with experimental wavelengths.

B) Define quasi-relativistic Fermi-Sea wave-functions (for the definition and construction of Rel. F-S see section 5 of [107]) whose orbitals are obtained from a Hartree-Fock-Dirac calculation [108] but with coefficients from a small CI which includes only the Coulomb and not the Breit operator. The derivation of the nonrelativistic correlation effects and their computation is carried out as in nonrelativistic FOTOS with nonrelativistic operators and the major components of the FS functions. For high energy transitions where retardation effects become important, higher multipoles should be considered in deriving the "first order symmetry configurations" (see sections III B,C).

The final transition probability computation should employ the relativistic operators or, to a very good approximation, the non-relativistic length operator [43,57,103,105] , and experimental wavelengths.

C) Define the fully relativistic Fermi-Sea wave-functions by using Fock-Dirac spinor configurations and including the Breit operator in the small CI. The relativistic first-order symmetries and correlation vectors are then derived by applying the relativistic selection rules for the transition of interest. The Rel.-FOTOS wave-functions now include F-S and virtual spinors whose optimization can be carried out variationally by projecting out the positron solutions. The fully relativistic transition operators and the experimental wave-length should be used.

The efficient and correct evaluation of atomic and molecular transition probabilities within a rigorous relativistic correlation theory is certainly one of the most challenging new directions of Quantum Chemistry.

IX. SYNOPSIS

The main goal of this paper was to present a simple methodology for the systematic analysis and easy computation of transition probabilities for photoabsorption and photoemission processes in atoms and molecules. A number of heuristic and quantitative examples were given. The degree of accuracy of FOTOS is high and perhaps allows considerable optimism regarding applications to molecular and/or fully relativistic N electron systems.

REFERENCES

1. D.R. Beck and C.A. Nicolaides, "Theory of the Electronic Structure of Excited States in Small Systems with Numerical Applications to Atomic States", this volume.
2. C.A. Nicolaides and D.R. Beck, J. Chem. Phys. 66, 1982 (1977).
3. C.M. Brown, S.G. Tilford and M.L. Ginter, J. Opt. Soc. Am. 65, 1404 (1975).
4. A.W. Weiss, Phys. Rev. A9, 1524 (1974).
5. C.A. Nicolaides and D.R. Beck, J. Phys. B9, L259 (1976).
6. D.R. Beck and C.A. Nicolaides, Phys. Letts. 61A, 227 (1977).
7. R.J. Buenker and S.D. Peyerimhoff, Chem. Phys. Letts. 34, 225 (1975).
8. M. Yoshimine et al, J. Chem. Phys. 64, 2254 (1976).
9. C.A. Nicolaides and D.R. Beck, Chem. Phys. Letts. 53, 87 (1978).
10. D.R. Beck and C.A. Nicolaides, Chem. Phys. Letts. 53, 91 (1978).
11. U. Gelius, J. El. Spect. 5, 985 (1974).
12. S.P. Kowalczyk et al, Phys. Rev. B7, 4009 (1973).
13. F. Wuilleumier and M.O. Krause, Phys. Rev. A10, 242 (1974).
14. J.P. Connerade and M.W.D. Mansfield, Proc. Roy. Soc. A343, 415 (1975).
15. G.K. Wertheim and H.J. Guggenheim, Phys. Rev. Letts. 27, 479 (1971).
16. G. Wendin and M. Ohno, Phys. Scripta 14, 148 (1976).
17. P.S. Bagus and E.K. Viinikka, Phys. Rev. A15, 1486 (1977).
18. C.A. Nicolaides, Chem. Phys. Letts. 17, 436 (1972); G. Nowak, W.L. Borst and J. Fricke, Phys. Rev. A17, 1921 (1978).
19. D.R. Beck, G. Aspromallis and C.A. Nicolaides, work in progress.
20. L. Goodman and B.J. Laurenzi, Adv. Qu. Chem. 4, 141 (1968).
21. F.W. Byron and C.J. Joachain, Phys. Rev. 164, 1, (1967).
22. T.N. Chang, T. Ishihara and R.T.Poe, Phys. Rev. Letts. 27, 838 (1971).
23. S.L. Carter and H.P. Kelly, Phys. Rev. A16, 1525 (1977).
24. P.L. Altick, "Atomic Photoionization Cross-Sections" in this volume.
25. Y. Komninos, D.R. Beck and C.A. Nicolaides, work in progress.
26. C.A. Nicolaides and D.R. Beck, Phys. Rev. A17, 2116 (1978).
27. I. Martinson, "Experimental Determination of Lifetimes", this volume.
28. H.G. Berry, Phys. Scr. 12, 5 (1975).
29. P. Erman, Phys. Scr. 11, 65 (1975).
30. J.M. Esteva, G. Mehlman-Balloffet and J. Romand, J. Quant. Spect. Rad. Trans. 12, 1291 (1972).
31. P.M. Dehmer, J. Berkowitz and W.A. Chupka, J. Chem. Phys. 59, 5777 (1973).

32. J.B. West and B.V. Marr, Proc. R. Soc. A349, 397 (1976).
33. M.J. Van der Wiel and G.R. Wright, Phys. Letts. 54A, 83 (1975).
34. C. Laughlin and A. Dalgarno, Phys. Rev. A8, 39 (1973).
35. C.A. Nicolaides, D.R. Beck and O. Sinanoğlu, J. Phys. B6, 62 (1973).
36. C.A. Nicolaides, Chem. Phys. Letts. 21, 242 (1973).
37. D.R. Beck and O. Sinanoğlu, Phys. Rev. Letts. 28, 934 (1972).
38. P. Westhaus and O. Sinanoğlu, Phys. Rev. 183, 56 (1969).
39. C.M. Moser, R.K. Nesbet and M.V. Gupta, Phys. Rev. A13, 17 (1976).
40. A. Hibbert, J. Phys. B7, 1417 (1974).
41. Z. Sibincic, Phys. Rev. A5, 1150 (1972).
42. J.S. Sims and R.C. Whitten, Phys. Rev. A8, 2220 (1973).
43. U.I. Safronova and Z.B. Rudzikas, J. Phys. B10, 7 (1977).
44. M. Ya. Amusia, V.K. Ivanov, N.A. Cherepkov and L.V. Chernysheva, Phys. Letts. 40A 361 (1972); Sov. Phys. JETP 39, 752 (1974).
45. S. Garpman, Phys. Scripta 12, 295 (1975).
46. T.I. Shibuya and V. McKoy, Phys. Rev. A2, 2208 (1970).
47. L.S. Cederbaum, J. Phys. B8, 290 (1975).
48. W. Von Niessen, L.S. Cederbaum and W. Domcke, in this volume.
49. J. Oddershede, P. Jørgensen and N.H.F. Beebe, J. Phys. B11, 1 (1978).
50. J. Linderberg and Y. Ohrn, "Propagators in Quantum Chemistry", Acad. Press (1973).
51. C. Froese-Fischer and J.E. Hansen, Phys. Rev. A17, 1956 (1978).
52. D.K. Watson, A. Dalgarno, and R.F. Stewart, Phys. Rev. A17, 1928 (1978).
53. K.T. Cheng and W.R. Johnson, Phys. Rev. A15, 1046 (1977).
54. C.A. Nicolaides and D.R. Beck, Chem. Phys. Letts. 36, 79 (1975).
55. D.R. Beck and C.A. Nicolaides, Int. J. Qu. Chem. S10, 119 (1976).
56. D.R. Beck and C.A. Nicolaides, Phys. Letts. 56A, 265 (1976).
57. D.R. Beck and C.A. Nicolaides, Phys. Letts. 65A, 293 (1978).
58. B. Edlen, Encyclopaedia of Physics, S. Flügge ed., Springer Verlag, Berlin (1964), vol. 27.
59. A.W. Fliflet, R.L. Chase and H.P. Kelly, J. Phys. B: At. Mol. Phys. 7, L443 (1974).
60. L.W. Bruch and A.P. Lehnen, J. Chem. Phys. 64, 2065 (1976).
61. C.A. Nicolaides and D.R. Beck, Can. J. Phys. 53, 1224 (1975).
62. C.A. Nicolaides, Phys. Rev. A6, 2078 (1972); Nucl. Inst. Meth. 110, 231 (1973).
63. C.A. Nicolaides and D.R. Beck, "Time Dependence, Complex Scaling and the Calculation of Resonances in Many-Electron Systems", Int.J. Qu. Chem. 14, to be published.
64. T.N. Rescigno et al., J. Chem. Phys. 68, 970 (1978) and refs. therein.

65. Y. Komninos, D.R. Beck and C.A. Nicolaides, work in progress.
66. H.F. King et al, J. Chem. Phys. 47, 1936 (1967).
67. C.A. Nicolaides, O. Sinanoglu and P. Westhaus, Phys. Rev. A4, 1400 (1971).
68. M.W.D. Mansfield and J.P. Connerade, Proc. R. Soc. (Lond.) A359, 389 (1978).
69. R.J. Buenker and S.D. Peyerimhoff, Chem. Phys. Letts. 29, 253 (2974).
70. J.A. Kernahan and P.H-L Pang, Can. J. Phys. 53, 455 (1975).
71. C.A. Nicolaides and D.R. Beck, Chem. Phys. Lett. 35, 202 (1975).
72. T. Shibuya, J. Rose and V. McKoy, J. Chem. Phys. 58, 500 (1973).
73. W.L. Wiese and A.W. Weiss, Phys. Rev. 175, 50 (1968).
74. Y.-K. Kim and J.P. Desclaux, Phys. Rev. Letts. 36, 139 (1976).
75. L.E. Nitzsche and E.R. Davidson, J. Chem. Phys. 68, 3103 (1978).
76. R.K. Nesbet, Phys. Rev. A14, 1065 (1976).
77. A.W. Weiss, Phys. Rev. 188, 119 (1969).
78. J.A. Kernahan et al., Phys. Scripta 12, 319 (1975).
79. I. Martinson et al., J. Opt. Soc. Am. 60, 1213 (1970).
80. G.A. Victor and C. Laughlin, Nucl. Inst. Meth. 110, 189 (1973).
81. A.W. Weiss, quoted in ref. 80.
82. T. Andersen, et al., Phys. Scripta 4, 52 (1971).
83. S. Bashkin and I. Martinson, J. Opt. Soc. Am. 61, 1686 (1971).
84. P. Shorer and A. Dalgarno, Phys. Rev. A16, 1502 (1977).
85. P. Shorer, C.D. Lin and A. Dalgarno, Phys. Rev. A16, 1109 (1977).
86. J.L. Fox and A. Dalgarno, Phys. Rev. A16, 283 (1977).
87. J.P. Buchet et al., Phys. Rev. A7, 922 (1973).
88. T. Åberg, Phys. Rev. A4, 1735 (1971).
89. G.W.F. Drake, Astroph. J. 184, 145 (1973).
90. D.R. Beck and C.A. Nicolaides, Chem. Phys. Letts. 48, 135 (1977).
91. J.S. Sims and J.R. Rumble Jr., Phys. Rev. A8, 2231 (1973).
92. H. J. Werner and W. Meyer, Phys. Rev. A13, 13 (1976).
93. D.R. Beck and C.A. Nicolaides, Chem. Phys. Letts. 49, 357 (1977).
94. G. Astner et al., Z. Physik, A279, 1, (1976).
95. J. Andrä, private communication.
96. A.W. Weiss, Astroph. J. 138, 1262 (1963).
97. A. Lindgard and S.E. Nielsen, J. Phys. B8, 1183 (1975); value using optimal cutoff.
98. R.A. Harris, J. Chem. Phys. 50, 3947 (1969).
99. M.Ya. Amusia, N.A. Cherepkov and L.V. Shernysheva, Soviet Phys. (JETP) 33, 90 (1971).
100. D.L. Lin, Phys. Rev. A16, 600 (1977).
101. C.A. Nicolaides and D.R. Beck, J. Phys. B9, L259 (1976).
102. J. Oddershede and N. Elander, J. Chem. Phys. 65, 3495 (1976).

103. G.W.F. Drake, J. Phys. B9, L169 (1976).
104. J. Migdalek, Can. J. Phys. 54, 2272 (1976).
105. L. Armstrong Jr., W.R. Fielder and D.L. Lin, Phys. Rev. A14, 1114 (1976).
106. M. Aymar and E. Luc-Koenig, Phys. Rev. A15, 821 (1977).
107. D.R. Beck and C.A. Nicolaides, "Theory of one electron binding energies including correlation, relativistic and radiative effects: Application to free atoms and metals" in this volume.
108. J.P. Desclaux, Comp. Phys. Commun. 9, 31 (1975).

ON GREEN'S FUNCTION METHODS FOR THE STUDY OF IONIC STATES IN ATOMS AND MOLECULES[+]

W. von Niessen

Theoretische Chemie, Institut für Physikalische Chemie, Technische Universität Braunschweig, D-33 Braunschweig, W.-Germany

L.S. Cederbaum and W. Domcke

Fakultät für Physik, Universität Freiburg, D-78 Freiburg, W.-Germany

Introduction

Photoelectron spectroscopy is concerned with the study of ionic states in atoms, molecules and solids. It has developed to become a very useful tool for the understanding of the electronic structure of the system. Actually it is not just a single type of spectroscopy but represents rather a spectrum of spectroscopies. The incident radiation may be varied both in type and energy. The most frequent excitation sources are UV radiation (HeI line at 21.21eV, HeII line at 40.8eV, X-rays of various energies and nowadays also synchroton radiation as well as electrons ((e,2e)-spectroscopy) or excited atoms (Penning ionization). The electrons and/or ions formed in the process can be spectroscopically examined. There also exists the possibility of coincidence measurements which may supply additional data. The photoelectron spectrum (PES) contains a lot of information. There are the different electronic

[+]This article consists of a set of lectures given by one of the authors (WVN) at the NATO Advanced Study Institute on the Island of Cos, Greece, June 4-18, 1978

Cleanthes A. Nicolaides and Donald R. Beck (eds.), Excited States in Quantum Chemistry, 183–272.

states, each ionization process giving rise to a band. These bands show vibrational and rotational structure in the case of molecules. Whereas the vibrational structure can frequently be resolved at least for small molecules the rotational structure is in general unresolvable and only leads to a broadening of the lines.

It is generally assumed that the most prominent bands in the PES arise from transitions to electronic states that are obtained by ejection of a single electron out of a molecular orbital (MO) in the ground state. In addition to these "simple" one-electron transitions two-electron transitions can sometimes be observed, i.e. processes which are ionization of one plus simultaneous excitation of another electron. They appear as satellite bands in the PES and borrow their intensity from the simple transitions. In the outer valence region and in the core region their intensity is small compared to the main line; i.e. a one-electron picture or a quasiparticle picture of the ionization process is valid. In the inner valence region, however, this need not be the case as we are going to see. The one-electron picture of ionization can break down completely.

The assignment of a PES requires a calculation of the ionization potentials (IP's) as the fundamental quantities. The simplest and most frequently used method to calculate IP's is the use of Koopmans' theorem (1). According to Koopmans' theorem the i-th IP is given by the negative of the i-th orbital energy.

$$I_i \approx -\varepsilon_i = -\langle \varphi_i | F | \varphi_i \rangle, \qquad (1)$$

where $|\varphi_i\rangle$ is the i-th MO and F the Hartree-Fock (HF) operator. This approximation corresponds to the frozen orbital approximation (neglect of electronic reorganization) and to the neglect of electronic correlation energy. By experience it has been found that if no further approximations are made Koopmans' theorem gives in many cases a reasonable ordering of the IP's and an estimate of their values because the reorganization and correlation energies are of about the same magnitude but often of different sign and thus they tend to cancel. There can certainly be no guarantee that these two effects cancel. There are a number of well-known cases, where Koopmans' theorem fails to predict the correct ordering of states, e.g. N_2, F_2, CS, PN, P_2, HOF, O_3, OF_2, N_2H_2, C_2N_2, C_4N_2, ethylene oxide, all azabenzenes and SF_6 among others. Electronic reorganization can be included by performing separate SCF calculations for the ionic states (ΔE_{SCF} method). For core ionizations this method has been very successful because the dominant effect is the reorganization in these cases (2). For valence ionizations the ΔE_{SCF} method has not been found to be successful (3). The best-known example is supplied by the N_2 molecule (3a, 4-9). Koopmans' theorem fails as does a ΔE_{SCF} calculation (3a). The inclusion of correlation effects becomes thus necessary.

The next step in improving the calculation of the IP's must take into account the electronic correlation energy by some type of configuration interaction (CI) procedure or by equivalent methods. For a review of CI methods see ref.19, of Green's function methods in general ref.11 and of Green's function methods for the study of ionic states in particular ref.12. These methods can be divided into two groups. The conventional methods (10, 13-17) obtain the IP by taking the difference between the energies of the N-1 and N electron systems. The direct methods as their name implies calculate the IP directly (7, 12, 18-28). Because they do not compute the total correlation energy the direct methods tend to have a computational advantage.

In the present article we are concerned with the Green's function method as developed by Cederbaum (12, 27,28) for the calculation of IP's. We are going to discuss the Green's function, the Dyson equation and the diagrammatic expansion of the self-energy in chapter 1. In chapter 2 applications of this method to the calculation of IP's and electron affinities (EA's) will be presented.

In the case of the vibrational structure in a PES a direct approach can be formulated as well (12, 29). The traditional approach requires the calculation of the separate potential surfaces of the initial and final states and is only applicable to the smallest molecules. The direct approach is computationally simpler and more appropriate for interpretative purposes. It has also been extended to include nonadiabatic effects. The theory of the vibrational structure is presented together with applications in chapter 3.

In the inner valence regions of atoms and molecules the MO picture of ionization can break down. The intensity becomes distributed over numerous lines and a main line ceases to exist. A method to calculate these ionic states is the 2-particle-hole Tamm-Dancoff method to be discussed together with applications in chapter 4.

I. The one-particle Green's function and the Dyson equation (30)

The one-particle Green's function is defined as the expectation value with respect to the exact ground state wavefunction $|\Psi_o^N\rangle$ of the N-particle system of a time ordered product of an annihilation and a creation operator a_k and a_l^+

$$G_{k\ell}(t,t')=-i\langle\Psi_o^N|T\{a_k(t)a_\ell^+(t')\}|\Psi_o^N\rangle \quad . \tag{2}$$

T is Wick's time ordering operator. Applied to a product of operators T orders them in chronological order with time increasing from right to left. A multiplicative factor of -1 is added if the chronological order is an odd permutation of the original order. With the help of the unit step function this can be rewritten as

$$T\{a_k(t)a_\ell^+(t')\}=a_k(t)a_\ell^+(t')\Theta(t-t')- a_\ell^+(t')a_k(t)\Theta(t'-t). \tag{3}$$

The $a_k(t)$ ($a_k^+(t)$) are destruction (creation) operators for an electron in state k. They are operators in the Heisenberg representation.

$$a_k^{(+)}(t)=e^{iHt}\,a_k^{(+)}\,e^{-iHt};\ a_k^{(+)}=a_k^{(+)}(0) \tag{4}$$

and obey the anticommutation relations:

$$\left[a_k^{(+)}(t),\ a_\ell^{(+)}(t)\right]_+=0;\ \left[a_k(t),\ a_\ell^+(t)\right]_+=\delta_{k\ell}. \tag{5}$$

H is the full Hamiltonian of the system and is given by

$$H=\sum_i h_i a_i^+a_i+\frac{1}{2}\sum_{ijk\ell}V_{ijk\ell}a_i^+a_j^+a_\ell a_k=H_o+V \tag{6}$$

$h_i=\langle\varphi_i(1)|h(1)|\varphi_i(1)\rangle$ is the matrix element of the one-particle operator and

$$V_{ijk\ell}=\langle\varphi_i(1)\varphi_j(2)|\frac{1}{r_{12}}|\varphi_k(1)\varphi_\ell(2)\rangle$$

is the two-electron integral. The $|\varphi_i\rangle$ are the eigenfunctions of h(1) here. The perturbation may also be chosen in a different way by taking the HF operator $\sum_i \varepsilon_i a_i^+ a_i$ as the unperturbed operator:

$$H=\sum_i \varepsilon_i a_i^+ a_i + \frac{1}{2}\sum_{ijk\ell} V_{ijk\ell} a_i^+ a_j^+ a_\ell a_k - \sum_{ij}\sum_\ell V_{i\ell[j\ell]} a_i^+ a_j n_\ell . \quad (7)$$

The ε_i are the orbital energies; $V_{i\ell[j\ell]}=V_{i\ell j\ell}-V_{i\ell\ell j}$ and $n_\ell=1$ for ℓ an occupied orbital and $n_\ell=0$ for ℓ an unoccupied orbital. For later use we also define $\bar{n}_\ell=1-n_\ell$. In the second decomposition of H the perturbation is much smaller than in the first case. This and the availability of HF calculations make the latter choice the appropriate one.

If the Hamiltonian does not explicitly depend on the time, the Green's function depends on time only via the difference $\tau=t-t'$. The Green's function in energy space is obtained via the Fourier transform

$$G(\omega)=\int_{-\infty}^{\infty} G(\tau)\, e^{i\omega\tau} d\tau; \quad G(\tau)=\frac{1}{2\pi}\int_{-\infty}^{\infty} G(\omega)\, e^{-i\omega\tau} d\omega. \quad (8)$$

Using the relation (3) in the Green's function (2) and inserting in the first term the complete set of states of the (N+1)-particle system and in the second one the complete set of states of the (N-1)-particle system one obtains:

$$G_{k\ell}(t,t')=-i\Big\{\Theta(t-t')\sum_n e^{i(E_o^N-E_n^{N+1})(t-t')} \times$$

$$\times\langle\Psi_o^N|a_k|\Psi_n^{N+1}\rangle\langle\Psi_n^{N+1}|a_\ell^+|\Psi_o^N\rangle$$

$$-\Theta(t'-t)\sum_m e^{i(E_m^{N-1}-E_o^N)(t-t')} \times$$

$$\times\langle\Psi_o^N|a_\ell^+|\Psi_m^{N-1}\rangle\langle\Psi_m^{N-1}|a_k|\Psi_o^N\rangle . \quad (9)$$

The IP's and EA's are given by the energy differences in the exponentials.

$$I_m = E_n^{(N-1)} - E_o^N; \quad A_n = E_o^N - E_n^{N+1}. \tag{10}$$

With the help of the formulas

$$\int_{-\infty}^{\infty} \Theta(\tau) e^{i(\alpha+\omega)\tau} d\tau = \lim_{\eta \to +0} \frac{i}{\omega+\alpha+i\eta} \tag{11}$$

and

$$\int_{-\infty}^{\infty} \Theta(-\tau) e^{i(\alpha+\omega)\tau} d\tau = \lim_{\eta \to +0} \frac{-i}{\omega+\alpha-i\eta} \tag{12}$$

one can perform the Fourier transformation to obtain

$$G_{k\ell}(\omega) = \lim_{\eta \to +0} \left\{ \sum_n \frac{\langle \Psi_o^N | a_k | \Psi_n^{N+1} \rangle \langle \Psi_n^{N+1} | a_\ell^+ | \Psi_o^N \rangle}{\omega + A_n + i\eta} + \sum_m \frac{\langle \Psi_o^N | a_\ell^+ | \Psi_m^{N-1} \rangle \langle \Psi_m^{N-1} | a_k | \Psi_o^N \rangle}{\omega + I_m - i\eta} \right\}. \tag{13}$$

This form of G is called the spectral representation. Except for the term $\pm i\eta$ which is only needed to obtain the correct time dependence in the Fourier back transformation one notes that the poles of $G_{k\ell}(\omega)$ are the negative exact IP's and EA's. They are the vertical quantities because only the geometry of the neutral ground state enters. It is immediately apparent that the spectral representation of G is quite inappropriate to calculate its poles. Later we shall come to an equivalent formulation which is appropriate for this purpose.

The one-particle Green's function is not an approximation to physical reality; so far no approximations have been made. Let us calculate at this point the Green's function in the HF approximation (or the free Green's function). H has to be replaced by H_o and $|\Psi_o^N\rangle$ by the HF wavefunction $|\Phi_o^N\rangle$

$$H_o |\Phi_o^N\rangle = E_o^N |\Phi_o^N\rangle; \quad E_o^N = \sum_{i=1}^{N} \varepsilon_i. \tag{14}$$

With these relations one obtains

$$G^{o}_{k\ell}(t,t')=\delta_{k\ell}e^{-i\varepsilon_k(t-t')}\begin{cases}-i, t>t' & k \text{ unoccupied}\\ i, t\leq t' & k \text{ occupied}\end{cases} \tag{15}$$

or after Fourier transformation

$$G^{o}_{k\ell}(\omega)=\lim_{\eta\to+0}\frac{\delta_{k\ell}}{\omega-\varepsilon_k-i\eta a}\;, \qquad a=\begin{cases}1 & \text{for } k \text{ occupied}\\ -1 & \text{for } k \text{ unoccupied}\end{cases} \tag{16}$$

The poles of $G_{k\ell}$ are the orbital energies, i.e. the negatives of the IP's in Koopmans' approximation.

The Green's function can be expanded in the time representation by means of perturbation theory. Equivalently one could derive the equation of motion of the Green's function and uncouple the resulting hierarchy of equations. A form more appropriate for the calculation of the poles is the Dyson equation in the energy representation

$$\underline{G}=\underline{G}^{o}+\underline{G}^{o}\underline{\Sigma}\,\underline{G} \tag{17}$$

or since $\underline{G}$ and $\underline{G}^{o}$ have inverses

$$\underline{G}^{-1}=\underline{G}^{o-1}-\underline{\Sigma}(\omega)\,. \tag{18}$$

The Dyson equation connects G with G^{o} via a quantity Σ called the self energy. Σ plays the role of a potential, it is the exact potential seen by an electron due to its interactions with and in its surroundings. As G^{o} is known the IP's and EA's are given by the zeros of the equation

$$\underline{G}^{-1}=\omega\underline{1}-\underline{\varepsilon}-\underline{\Sigma}(\omega)\,. \tag{19}$$

One can physically understand this equation in a simple way. Consider the (free or) HF Green's function. If a potential is added to H_o, G^{o}_{kk} will be transformed to

$$G_{kk}=\frac{1}{\omega-\varepsilon_k-\Sigma_{kk}(\omega)}\;. \tag{20}$$

Σ_{kk} depends on the position of all particles or equivalently on time and after Fourier transformation on the energy. It is, however, not necessary that G remains diagonal. The general form of G to be expected is

$$\underline{G} = (\omega \underline{1} - \underline{\varepsilon} - \underline{\Sigma}(\omega))^{-1} \tag{21}$$

which is the Dyson equation. All difficulties are transferred now to obtain the self-energy. Σ is expanded in a perturbation series in the electronic interaction with the help of diagrammatic techniques.

As a first step the Dyson equation will be written in diagrammatic form. We represent the full Green's function by a double line with an arrow

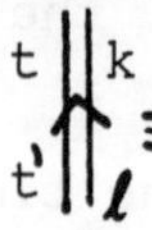

$\equiv iG_{k\ell}(t,t')$ and G^{o} by a single line

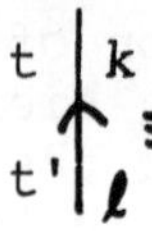

$\equiv iG^{o}_{k\ell}(t,t')$.

The self-energy is represented by

$\equiv -i\Sigma_{k\ell}(t,t')$.

Then the Dyson equation can be written in diagrammatic form

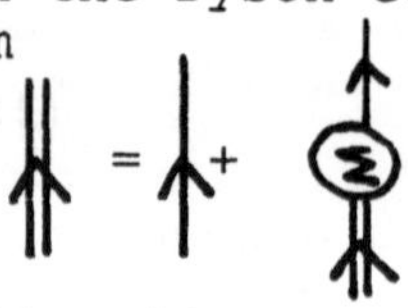
(22)

By iteration one obtains

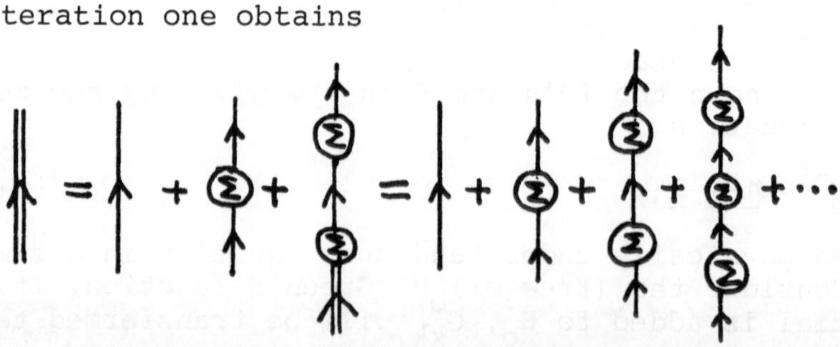
(23)

One thus obtains the full Green's function by the sum of all possible insertions of  into the G^{o} lines. As Σ will be expanded in the electronic interaction as perturbation one needs a few more definitions (we will consider only closed shell systems):

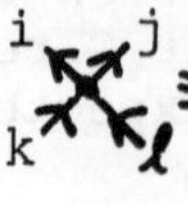

$\equiv -iV_{ij[k\ell]}$, (24)

this is the antisymmetrized interaction vertex,

$$\equiv -iV_{ijk\ell} \tag{25}$$

and

$$\equiv -iv_{k\ell}; \quad v_{k\ell} = -\sum_i V_{ki[\ell i]} n_i \tag{26}$$

The diagrams representing the n-th order of the expansion of G are obtained by drawing all topologically nonequivalent linked diagrams having n $V_{ij[k\ell]}$ points and (2n+1) G^o lines. The elements of one kind can only be connected with the elements of the other kind. Two diagrams are topologically equivalent if it is possible to transform them into one another by twisting or pulling the diagrams without breaking any G^o line. Such diagrams are called Hugenholtz or Abrikosov diagrams. If we draw the diagrams using the wiggly lines representing $V_{ijk\ell}$ one obtains the so-called Feynman diagrams. Each Abrikosov diagram contains several Feynman diagrams as can be seen in the second order diagram

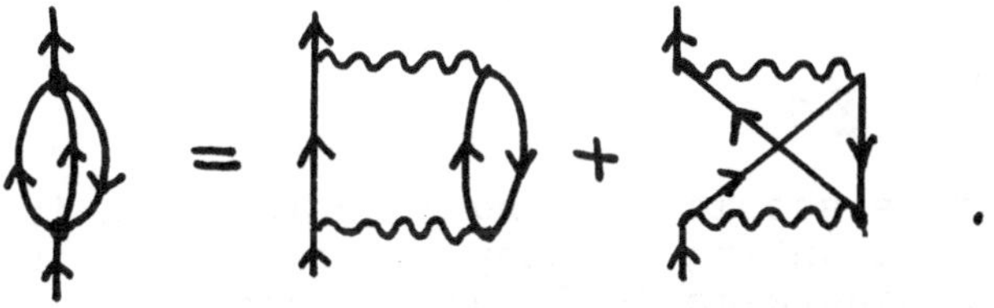

(27)

Abrikosov diagrams are thus more compact than Feynman diagrams. Only linked diagrams are to be drawn. The linked cluster theorem states that the unlinked diagrams can be factored from the linked diagrams and cancel against the normalization denominator in the perturbation expansion of the Green's function. Diagrams such as

do not appear. It is, however, necessary to include the so-called exclusion principle violating diagrams with more than one electron in the same one-electron state at the same time. In the unlinked expansion they cancel against unlinked diagrams but in the simpler linked expansion they must be kept.

From the Dyson equation and the diagrammatic expansion of G one can derive the diagrammatic expansion of Σ. Inserting one diagram of Σ into the Dyson equation implies an infinite number of diagrams in the expansion of G. Diagrams which fall apart by cutting one G^o line do not belong to Σ. They are automatically included by solving the Dyson equation. Their inclusion would lead to a double counting of diagrams. Thus Σ contains only irreducible diagrams, i.e. those which do not fall apart by cutting a single G^o line. The rules to draw the diagrams of Σ are simple: connect n $V_{ij[k\ell]}$ points with (2n-1) G^o lines (instead of (2n+1) G^o lines for G) and retain only irreducible graphs.

Until now we have not talked about the diagrams for the $v_{k\ell}$ perturbation. If one starts from free particles they do not appear as there is no such perturbation. If one starts from HF particles they do not appear either. One can use the Dyson equation to go from the free particle Green's function G^o to G^{HF}

$$G^{HF}=G^o+G^o\Sigma G^{HF}. \tag{28}$$

Σ contains as the only diagram  in this case. If one calculates now the Green's function from G^{HF}

$$G=G^{HF}+G^{HF}\Sigma G \tag{29}$$

these diagrams should no longer be included in Σ_{HF} to avoid double counting as they are contained in G^{HF} already. In diagrammatic language this finds its expression in the fact that

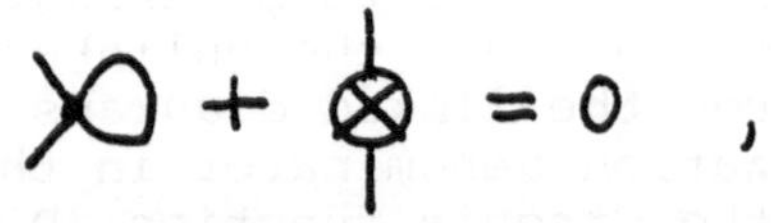 (30)

i.e. all diagrams containing insertions of  and  are to be omitted from Σ. Using the rules to evaluate diagrams which will be given below one obtains:

 $= \sum_i V_{ki[\ell i]} n_i$ (31)

and

 $= -\sum_i V_{ki[\ell i]} n_i$. (32)

This means there is no first order contribution to Σ in the case of HF particles. Koopmans' theorem is thus correct to first order in the electronic interaction.

The rules to evaluate the diagrams of Σ are given only for the ω-representation. Each G^o line points from a time t' to a time t and each $V_{ij[k\ell]}$ point is therefore associated with a fixed time on the time axis. By permuting the $V_{ij[k\ell]}$ points in a diagram of n-th order one obtains n! time-ordered diagrams. In fig.1 the two time-ordered diagrams of second order and in fig.2 the 18 time-ordered diagrams of third order are given.

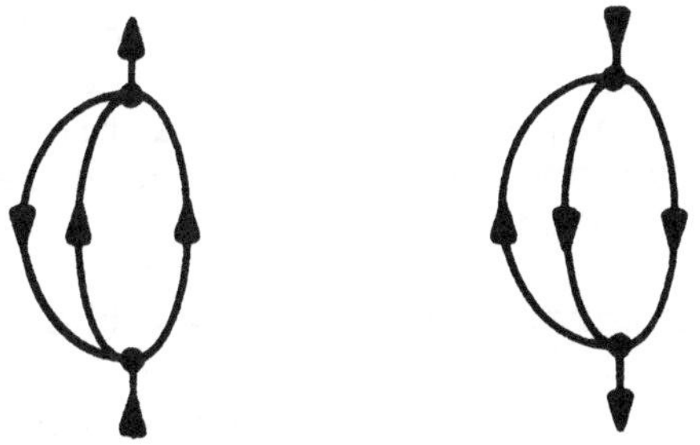

Fig.1: The time-ordered self-energy diagrams of second order

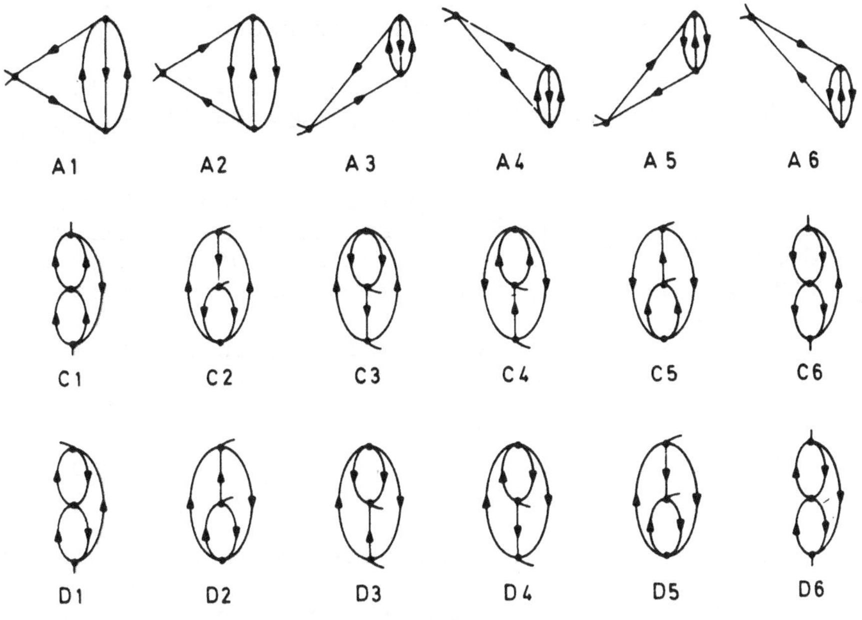

Fig.2: The notation of the time-ordered self-energy diagrams of third order

The rules for the evaluation of a given time-ordered diagram of $\Sigma_{k\ell}(t,t')$ in energy space are:

1.) join the external indices k and ℓ by a $e^{-i\omega(t'-t)}$ line

2.) draw n-1 horizontal lines between successive pairs of points. Each of these horizontal lines i is associated with a contribution A_i to the diagram

3.) each G^o line and $e^{-i\omega(t'-t)}$ line cut by a horizontal line i supplies an additive contribution to $1/A_i$, namely $+\omega(-\omega)$ when the $e^{-i\omega(t'-t)}$ line points downward (upward), $+\varepsilon_j(-\varepsilon_j)$ when the G^o_{jj} line points downward (upward). A G^o line pointing downward is called a hole line, otherwise a particle line. A hole line corresponds to j being occupied, a particle line to j being unoccupied

4.) multiply the interactions $V_{ij[k\ell]}$, the contribution of the horizontal lines and a factor $(-1)^{h+\ell}$ where h is the number of hole lines and ℓ the number of loops; then sum over the internal indices

5.) multiply the above contribution by 2^{-q} where q is the number of permutations of two G^o lines in the diagram leaving the diagram unchanged

6.) these rules 1.) to 4.) are also valid for a time-ordered Feynman diagram; only replace the $V_{ij[k\ell]}$ points by the $V_{ijk\ell}$ wiggles

7.) the sign of the $V_{ij[k\ell]}$ points and the number of loops is not uniquely determined in an Abrikosov diagram. To obtain the proper sign of the diagram compare with a Feynman diagram contained in it.

All orbitals so far are spin orbitals. In the closed shell case one can add a rule which only serves to perform the spin summation.

8.) Multiply the contribution of a Feynman diagram by 2^{ℓ} and replace orbitals by spatial orbitals.

As an example we will evaluate the second order diagram

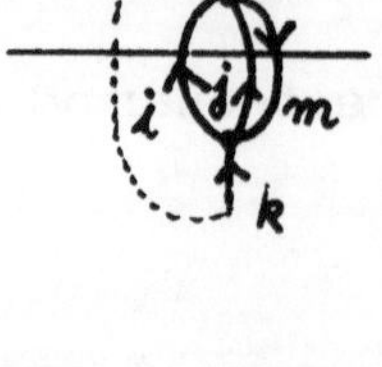

$$= \frac{1}{2}(-1)(-1)\sum_{ijm} \frac{V_{\ell m[ij]}\, V_{ij[km]}\, \bar{n}_i\, \bar{n}_j\, n_m}{\omega + \varepsilon_m - \varepsilon_i - \varepsilon_j}$$

or with spin summation

$$\sum_{ijm} \frac{(2V_{kmij}-V_{kmji})V_{\ell mij}}{\omega+\varepsilon_m-\varepsilon_i-\varepsilon_j}\bar{n}_i\bar{n}_j n_m \tag{33}$$

If the self-energy is given (up to a given order) the Dyson equation can be solved for the IP's and EA's. In fig.3 the graphical representation of the self-energy is given.

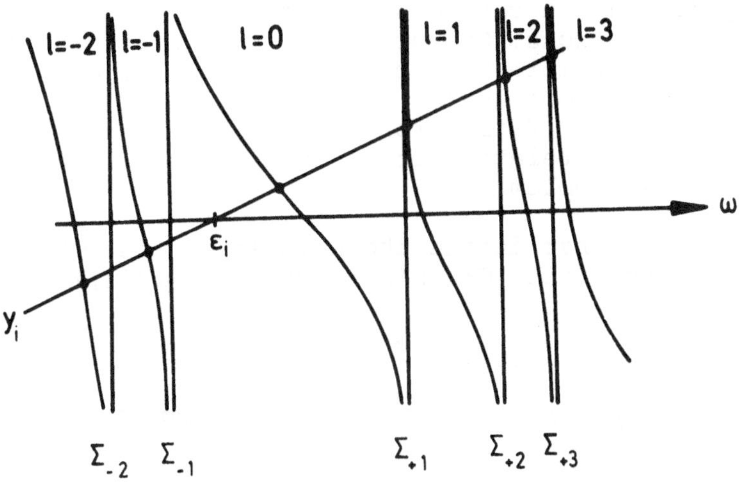

Fig.3: A schematic plot of Σ_{kk} as a function of ω and of the solution of the Dyson equation

The $-\omega$ values of the intersection points of the straight line $y=\omega-\varepsilon_k$ with $\Sigma_{kk}(\omega)$ are the IP's and EA's. Σ itself has poles and is monotonically decreasing in each interval. There always exists (for closed shell systems) a large interval free of poles of Σ of about 20 eV width. The poles start (if we neglect symmetry for the moment) at $\varepsilon_{LUMO}-2\varepsilon_{HOMO}$ and $\varepsilon_{HOMO}-2\varepsilon_{LUMO}$, where LUMO and HOMO are the lowest unoccupied and highest occupied MO's. In this interval one finds the outer valence IP's. The self-energy varies smoothly here and a detailed knowledge of the pole structure is not required. As the assignment of the closely spaced outer valence IP's represents an important and difficult problem it is worthwhile to construct an approximation specially adapted for these cases, this we will call the outer valence-type Green's function method or simply the Green's function method. If the detailed pole structure of Σ must be taken into account a different approximation should be constructed which will be called the 2-particle-hole (2ph) Tamm-Dancoff

Green's function method. The outer valence-type Green's function method has been pushed to achieve high accuracy as this is required in the outer valence region. In the expansion of Σ all terms up to and including the third order are taken into account fully. Higher order terms - called renormalization terms - which introduce an effective interaction are summed to infinite order. They can be derived by a simple procedure discussed in detail in ref.12 and 27.

From the graphical representation of the self-energy (fig.3) it can be seen that the Dyson equation has many solutions. One of these ionic states may correspond to a simple hole configuration, the other ones are of a more complicated nature. They give rise to the satellite lines. This information is only useful if one calculates the relative intensities as well. We turn now to this problem.

Let us first define the pole strengths. These are the residues of the Green's function (eq.(13)) or in shorthand notation

$$G_{k\ell}(\omega)=\lim_{\eta\to+0}\left\{\sum_s \frac{x_k^{(s)}x_\ell^{(s)*}}{\omega+I_s-i\eta}+\sum_p \frac{y_k^{(p)}\;y_\ell^{(p)*}}{\omega+A_p+i\eta}\right\} \tag{34}$$

$$P_k^{\mathrm{ion}(s)}=x_k^{(s)}x_\ell^{(s)*}$$

$$P_k^{\mathrm{aff}(p)}=y_k^{(p)}y_\ell^{(p)*} \tag{35}$$

For these pole strengths one obtains the sum rule

$$\sum_s P_k^{\mathrm{ion}(s)}+\sum_p P_k^{\mathrm{aff}(p)}=\delta_{k\ell}. \tag{36}$$

Let $D_k(\omega)$ be the k-th eigenvalue of $G(\omega)$. For the pole $\omega=-I_s$ of $D_k(\omega)$ in the interval s and for the corresponding pole strength $P_k^{(s)}$ one obtains from the Dyson equation

$$I_s=\left[\underline{\varepsilon}+\underline{\Sigma}(-I_s)\right]_k \tag{37}$$

$$P_k^{(s)}=\left\{1-\frac{\partial}{\partial\omega}\left[\underline{\varepsilon}+\underline{\Sigma}(-I_s)\right]_k\right\}^{-1}, \tag{38}$$

where the index k denotes the k-th eigenvalue. The pole strength is thus large if Σ' is small and is small if the slope of Σ is large.

In a photoelectron experiment one measures the photocurrent as a function of the kinetic energy, E_e, of the ejected electrons. This energy is equal to the difference between the photon energy ω_o and the electron binding energy I

$$E_e = \omega_o - I \quad . \tag{39}$$

The photocurrent is proportional to the intensity of the external radiation field over many orders of magnitude. The photoelectron spectrum can thus be calculated in a very good approximation with the golden-rule formula. The transition probability per unit time and unit energy at the energy ω is given by

$$P(\omega) = \frac{2\pi e^2}{m^2 c^2 \hbar} \sum_F \left| \left\langle F \left| \sum_n \vec{A}_n \cdot \vec{P}_n \right| \Psi_o^N \right\rangle \right|^2 \times$$

$$\times \delta(\omega - E_e)\, \delta(E_F - E_o^N - \omega_o)\,, \tag{40}$$

where $|F\rangle$ and $|\Psi_o^N\rangle$ are the final and initial states of the system with energies E_F and E_o^N, respectively. $\vec{A}_n$ is the vector potential at the n-th electron and $\vec{P}_n$ the momentum of this electron. The δ-functions arise from energy conservation. In second quantization $P(\omega)$ becomes

$$P(\omega) = \frac{2\pi e^2}{m^2 c^2 \hbar} \sum_F \left| \left\langle F \left| \sum_{k\ell} \tau_{k\ell}\, a_k^+ a_\ell \right| \Psi_o^N \right\rangle \right|^2 \times$$

$$\times \delta(\omega - E_e)\, \delta(E_F - E_o^N - \omega_o)$$

$$= -\frac{2e^2}{m^2 c^2 \hbar} \sum_{k\ell mn} \tau_{mn}^* \tau_{k\ell} \lim_{\eta \to +0} Im\left\{ iG_{\ell kmn}(-\omega_o) \right\} \delta(\omega - E_e) \;, \tag{41}$$

where $\tau_{ij} = \langle \varphi_i | \vec{A} \cdot \vec{P} | \varphi_j \rangle$. In this expression the two-particle Green's function appears (its particle-hole component). The calculation of these quantities is a difficult task. Further on an approximation to continuum wavefunctions is required, which for molecules is a formidable and unsolved problem. We aim at a simplified expression for $P(\omega)$ which still gives exact IP's and a fairly accurate description of the main features of the spectrum.

We approximate the final state $|F\rangle$ as an antisymmetrized product of a one-electron excited state $|e\rangle = a_e^\dagger|0\rangle$ and a (N-1)-electron state $|\Psi_s^{N-1}\rangle$, i.e. the outgoing electron does not correlate with the electrons of the ion. It does not mean that $|e\rangle$ describes a free electron; it is in general not a plane wave. Since the component of $|e\rangle$ in $|\Psi_o^N\rangle$ is almost exactly zero if E_e is sufficiently large it is necessary that a_k annihilates the electron in $|e\rangle$ in the final state. Otherwise $P(\omega)$ vanishes. One obtains

$$P(\omega) = \frac{2e^2}{m^2c^2\hbar} \lim_{\eta \to 0} \sum_{e,n,\ell} \tau^*_{en} \tau_{e\ell} \operatorname{Im} G_{\ell n}(\omega-\omega_o)\,\delta(\omega-E_e)$$

$$= \frac{2e^2}{m^2c^2\hbar} \sum_s \sum_e \delta(\omega-E_e) \left| \sum_\ell \tau_{e\ell} x_\ell^{(s)} \right|^2 \times$$

$$\times\, \delta(\omega+E_s^{N-1}-E_o^N-\omega_o)\,. \tag{42}$$

For a particular ionic state s we have for the intensity at the position $\omega=\omega_o-I$

$$P^{(s)}(\omega) = \text{const} \sum_e \left| \sum_\ell \tau_{e\ell} x_\ell^{(s)} \right|^2 \delta(\omega-E_e)\,. \tag{43}$$

This holds in general. In the case that the intensity of the line derives only from a state $|\Psi_\ell^{N-1}\rangle$ which differs from $|\Psi_o^N\rangle$ in the occupation of $|\varphi_\ell\rangle$ we have for the intensity

$$P_\ell^{(s)}(\omega) = \text{const}\ x_\ell^{(s)*} x_\ell^{(s)} \sum_e |\tau_{e\ell}|^2 \delta(\omega-E_e)$$

$$= \text{const}\ P_\ell^{(s)} \sum_e |\tau_{e\ell}|^2 \delta(\omega-E_e)\,, \tag{44}$$

where $P_\ell^{(s)}$ is the pole strength defined above. Let us now assume that $\tau_{e\ell}$ does not depend strongly on ω, so we can set it equal to a constant. We obtain then for the ratio of the transition probabilities to ionic states $|\Psi_s^{N-1}\rangle$ and $|\Psi_{s'}^{N-1}\rangle$ which both derive their intensity from a state $|\Psi_\ell^{N-1}\rangle$:

$$P_{\ell}^{(s)}(\omega)/P_{\ell}^{(s')}(\omega)=P_{\ell}^{(s)}/P_{\ell}^{(s')}, \tag{45}$$

i.e. it is equal to the ratio of the pole strengths.

Because of the relation between the slope of Σ and the pole strength (eq.38) one can deduce the pole strength from fig.3. The steeper the slope of Σ at the intersection point, the smaller is the intensity of the corresponding line in the spectrum. In the main interval and far from the poles of Σ the states will have a large pole strength (about 0.9). Satellite states borrowing intensity from these states will thus appear only with small intensity in the spectrum.

A final remark should be made on where the diagrammatic method comes from. One starts from the time-dependent Schrödinger equation. The time development operator U supplies a formal solution to this equation. For the time development operator in the interaction representation $\tilde{U}$ one derives a differential equation which transformed to an integral equation can be iterated. The Green's function can be given in terms of $\tilde{U}$. The perturbation expansion of $\tilde{U}$ leads to a perturbation expansion of the Green's function. Matrix elements are evaluated with the aid of Wick's theorem. The final expressions are translated into diagrams. For a rigorous derivation see e.g. ref.30.

II. Application of the Green's function method to the calculation of ionization potentials and electron affinities

Numerical considerations

The calculation of IP's and EA's by the Green's function method involves four steps:

1.) evaluation of one- and two-electron integrals over the basis functions. Cartesian Gaussian functions are used.

2.) solution of the SCF equations

3.) transformation of the integrals from the basis of Gaussian functions to the basis of MO's

4.) evaluation of the diagrams from the list of integrals and solution of the Dyson equation. As the energy enters the denominators of the diagrams (Brillouin-Wigner-type perturbation theory) the Dyson equation must be solved iteratively. Usually two to three steps are required where the starting value for ω is the lower order calculation. It should be noted that the lack of size consistency in the Brillouin-Wigner perturbation theory is of no relevance here as the correlation energy difference between two states of the same system is not a monotonic function of the number of electrons as is the case for the correlation energy.

The first three steps are performed with the program system MUNICH of Diercksen and Kraemer (31). All calculations to be described have been performed on an IBM 360/91 or an Amdahl 470/V6 computer in double precision algebra.

The self-energy is a matrix and the calculations should be performed with the complete matrix. The inverse Green's function matrix, however, is constructed in a basis of MO's. If the MO picture of ionization, i.e. ejection of an electron from an occupied MO is a valid one, the inverse Green's function matrix should be nearly diagonal. The calculations in second order have always been performed using a diagonal approximation to Σ and the full Σ. The results agree in most cases very closely thus justifying the diagonal approximation. In third order and in the calculation including renormalization corrections the diagonal approximation is used.

In the applications we wish to demonstrate the accuracy which can be achieved in the IP's and EA's, the size of the molecules which can be treated and what type of problems can be solved.

To achieve high accuracy in the computed IP's and EA's large basis sets must be used. They must be able to describe wavefunctions close to the HF limit both for the neutral ground state and for the various ionic states. Calculations with double-zeta basis sets cannot in general give accurate IP's. This does not preclude their usefulness and the necessity of doing such calculations in particular on larger molecules.These computations are more aimed towards an assignment of a PES, i.e. a reliable ordering of states without the claim for quantitative agreement with experiment for all IP's. The double-zeta basis appears to be a minimal (but in general reliable) basis for calculations including the effects of electronic correlation. To obtain accurate results basis sets larger than double-zeta should be used which are then supplemented by at least one polarization function per center. For the second row atoms it was, however, found that diffuse d-type functions are required as well, although this may depend on the molecule and the computed quantity. For H_2S this proved to be necessary but for SO_2 not (33). With the consequently resulting large basis sets one is tempted to look for computational simplifications. These can indeed be found. The core orbitals may be deleted from the calculations (or kept frozen) but this is in general not permitted for the C, N, O, F 2s orbitals. In a perturbation approach such as the present one the virtual orbital space may be truncated. This is a rather safe procedure with basis sets not including polarization functions. When polarization functions are included this can also be done, but it should be noted that orbitals with strong d- and f-function participation are lying energetically fairly high. If such a truncation is used a decline in the quantitative agreement with experiment is sometimes observed. The major simplification and one which is no approximation is the use of molecular symmetry. This can drastically reduce the computational expense. There are many technical points hidden in the programs which make them efficient and require an enormous amount of work for their implementation but one never talks about these, and so we will not do it either.

In the discussion of the accuracy achievable in the IP's we will discuss the three molecules N_2, H_2O and H_2S. In these calculations the basis sets have been exhausted completely.

Application to N_2

The N_2 molecule represents an ideal test case since Koopmans' theorem predicts an incorrect ordering for the $^2\pi_u$ and $^2\Sigma_g$ states (3a). This incorrect ordering persists in ΔE_{SCF} calculations and only by including the electronic correlation energy can the correct ordering of states be obtained. A substantial number of calculations with different methods and different basis sets are published in the literature. Calculations have here been performed with the (11s7p)/[5s4p] basis set (34) supplemented with 1, 2 and 3 d-type functions and with 2 d-type and 1 f-type functions. The results for the $2\sigma_g$, $1\pi_u$ and $1\sigma_u$ IP's are compiled together with literature values and the experimental IP's (35) in table 1. Note that the numbering of orbitals starts with the first valence orbital.

Table 1

IP's of N_2 with different basis sets

IP	9s5p	11s7p1d	11s7p2d	11s7p3d	11s7p2d1f	ref.9	ref.63	ref.36	exp.[35]
$2\sigma_g$	14.85	15.31	15.45	15.53	15.52	15.94	15.35	16.04	15.60
$1\pi_u$	16.34	16.80	16.76	16.75	16.83	17.20	16.63	17.11	16.98
$1\sigma_u$	18.37	19.01	18.91	18.96	18.98	19.01	18.31	19.68	18.78

Only calculations are listed if they include polarization functions. These results are very stable in particular for the basis set with more than one polarization function. These values for the IP's are the most accurate values reported in the literature for this molecule. With a double-zeta basis plus polarization functions Herman et al. obtained a maximum error of 0.34eV in the IP's (9). The CI calculations which employ the same basis set (36) do in general agree. The result for the double-zeta basis as obtained with the present method is included in the table as well. The maximum error of 0.75eV clearly demonstrates the necessity of including polarization functions for obtaining high accuracy.

Application to H_2O

Many calculations and several ones of great accuracy have been performed on the H_2O molecule and its ions. In this way a good basis for comparison with other methods is obtained. Five calculations have been performed with the (11s7p/6s1p)/[5s4p/3s1p] basis supplemented with 1, 2, 3 and 4d-type functions and with 3d-type and 1f-type functions. Table 2 lists the results of the calculations together with the values obtained by Meyer using a (11s7p4d1f) (O atom) and (5s1p) (H atom) basis set with functions added on the bonds (15a), by Chong et al. (37), and by Hubač et al. (38).

Table 2

IP's of H_2O with different basis sets

IP	9s5p	11s7p1d	11s7p2d	11s7p3d	11s7p4d	11s7p3d1f
$1b_1$ (π)	12.37	12.78	12.87	12.83	12.81	12.90
$2a_1$	14.32	14.78	14.85	14.85	14.84	14.85
$1b_2$	18.95	18.94	18.98	18.94	18.92	18.91

ref.15a	ref.37	ref.38	exp.[35]
12.48	12.42	12.79	12.78
14.68	14.73	15.12	14.83±0.11
18.85	18.97	19.19	18.72±0.22

The experimental values for the IP's (35) have been corrected for vibrational effects by Meyer (15a). The present results are again the most accurate ones available in the literature. A double-zeta basis set is not sufficient for the calculation of accurate IP's.

Application to H_2S

The following points were investigated in rather extensive calculations (32): extension of the s-p basis on the S atom and on the H atoms, number and exponential parameters of the polarization functions on the S and H atoms. Table 3 gives some results.

Table 3

IP's of H_2S with different basis sets[a,b]

IP	12s9p	12s9p1d	12s9p2d	12s9p3d	12s9p2d1f	exp.[35]
$1b_1$ (π)	9.91	10.14	10.25	10.24	10.38	10.48
$2a_1$	12.76	13.19	13.28	13.32	13.36	13.4
$1b_2$	15.72	15.67	15.59	15.57	15.58	15.5

a) centroids of the experimental bands are estimated from the spectrum in ref.35
b) the 12s9p basis set on the S atom is taken from ref.57

A double-zeta basis set leads to a maximum error of 0.64eV. The large error practically persists for the b_1 (π) IP if one or two d-type functions with standard exponential parameters are added to the S atom

(α=0.54, 2.0). It was found that diffuse d-type functions must be added to the basis set to describe correlation energy changes in the diffuse part of the charge cloud. Using one diffuse d-type function the maximum error is 0.34eV, using two d-type functions the maximum error is 0.23eV, and using three it is 0.24eV. The results appear to be stable under variation of the number of d-type functions and their exponential parameters. Further improvement is obtained by including f-type functions in the basis. The maximum error is then reduced to 0.1eV for all IP's.

To conclude the maximum errors which occurred in the other applications of the Green's function method will shortly be discussed. In general smaller basis sets with fewer polarization functions have been used than in the present work. When basis sets of double-zeta quality are used, in particular for large molecules, the errors in the low energy region are 0.1 to 0.6eV and in the high energy part 0.5 to 1eV. These larger errors are certainly in part due to basis set limitations, but in part also to the presence of intense satellite lines in the high energy range. In this case the present renormalization method is not applicable and the method of ref.28 should be used to achieve higher accuracy. The latter comment applies also to the higher energy valence IP's of molecules computed with large basis sets. The typical maximum errors in the IP's which are found in the calculations with basis sets including polarization functions are 0.1 to 0.25eV.

Larger errors were occasionally found and the following reasons are likely to be responsible for it: Intense satellite lines in the energy range (see above), difficulty in determining the maximum or centroid of the bands due to limited resolution or to predissociation problems, etc. (in the latter cases the apparent maximum or centroid of the band does not correspond to the vertical IP), incomplete exhaustion of the basis in the case of large and unsymmetrical molecules and finally limitation of basis sets if very strong d-function participation is present and an extension of the basis set is impossible due to the size of the molecule.

Application to C_2N_2

The HeI spectrum of C_2N_2 exhibits four bands which have been assigned (35) in order of increasing binding energy as $1\pi_g$, $3\sigma_g$, $2\sigma_u$ and $1\pi_u$. (The orbitals are again numbered starting with the first valence orbital). For this molecule near HF limit data are available (39). From these data and using Koopmans' theorem the following sequence of IP's is obtained: $1\pi_g$, $1\pi_u$, $3\sigma_g$, $2\sigma_u$ which is clearly in serious contradiction to the experimental result. Because of the quality of the HF calculation only many-body effects can be responsible for the interchange of the $1\pi_u$ IP with both σ-IP's.

To study the sensitivity of the results to the quality of the basis two basis sets have been employed. The first one (basis A) is of double-zeta quality (9s5p)/[4s2p]. The second basis set (basis B) has added polarization functions with exponential parameters $\alpha_d(C)=0.8$ and $\alpha_d(N)=0.85$.

The results obtained with the two basis sets are shown in table 4. The ordering of the IP's calculated in first order (Koopmans' theorem) is incorrect. In contrast to N_2 a wrong sequence is also obtained in second and third order. It is seen that both σ-orbitals are strongly shifted. The renormalization causes the σ-IP's to shift only moderately compared to the above shifts and yields the correct ordering as well as good numerical values for the IP's.

Table 4

IP's of C_2N_2 calculated with bases A and B. $IP^{(i)}$ i=1, 2, 3 stands for the IP calculated in the i-th order perturbation theory, respectively. $IP^{(R)}$ represents the final result. The P are the pole strengths corresponding to $IP^{(R)}$. All energies in eV.

	MO	$IP^{(1)}$	$IP^{(2)}$	$IP^{(3)}$	$IP^{(R)}$	P	IP(exp)[35]	IP^a
Basis A	$1\pi_g$	13.75	13.34	13.53	13.41	0.91	13.36	13.28
	$3\sigma_g$	16.79	12.70	14.62	13.94	0.90	14.49	14.24
	$2\sigma_u$	17.15	13.06	15.05	14.31	0.89	14.86	14.68
	$1\pi_u$	16.73	15.72	16.05	15.89	0.88	15.6	15.49
Basis B	$1\pi_g$	13.60	13.17	13.24	13.20	0.91	13.36	13.22
	$3\sigma_g$	16.93	13.09	15.30	14.40	0.90	14.49	14.56
	$2\sigma_u$	17.35	13.44	15.79	14.80	0.89	14.86	14.97
	$1\pi_u$	16.42	15.50	15.66	15.56	0.88	15.6	15.56

a) These values are obtained by adding the final Koopmans' defects to the near HF limit orbital energies of ref.39.

The final results obtained with basis A are in qualitative agreement with those obtained with basis B. Quantitatively the σ-IP's obtained with basis A are somewhat lower than the corresponding experimental IP's. Adding the calculated Koopmans' defects to the accurate orbital energies of McLean and Yoshimine the values in the last column of table 4 are obtained. It is seen that the results obtained with basis A are now of an accuracy that is comparable to that obtained with basis B.

The calculated PES including the vibrational structure which will be discussed later on is given in fig.4. The agreement with experiment is quantitative.

Koopmans' theorem fails to reproduce the ordering of the IP's for the N_2 molecule and it fails dramatically for the C_2N_2 molecule. In the same dramatic way it fails for the C_4N_2 molecule. It is always the change of the σ(n)- and π-IP's which occurs, where the σ-MO's are the lone pair orbitals on the N atoms. These are relatively close together in energy in the HF calculation. If one would like to connect this change of ordering to the localized character of the lone pair orbitals (the lone pair hybrid on one N atom does hardly overlap with the one on the other N atom) one should note that no change of ordering occurs in the case of the molecules C_2F_2 and C_4F_2 (41). The following simple rule has, however, emerged from a number of calculations (42, 43). If a molecule possesses low lying virtual orbitals of non-diffuse character it is an indication that considerable non-uniform many-body corrections can be expected and thus the ordering of ionic states obtained from Koopmans' theorem may not be the correct one. More precisely for a linear or planar molecule the existence of a low-lying $\pi(\sigma)$-type unoccupied orbital of non-diffuse character and of an outer valence $\pi(\sigma)$-type occupied orbital leads to large many-body corrections for outer valence IP's of $\sigma(\pi)$-type orbitals. Whether a change of ordering occurs is finally decided by the relative magnitude of matrix elements. E.g. trans-N_2H_2, cis-N_2H_2 and H_2NN are all candidates for a change of ordering but this occurs only for cis-N_2H_2 and H_2NN (44). This model cannot explain everything but it gives ar reasonable rationalization.

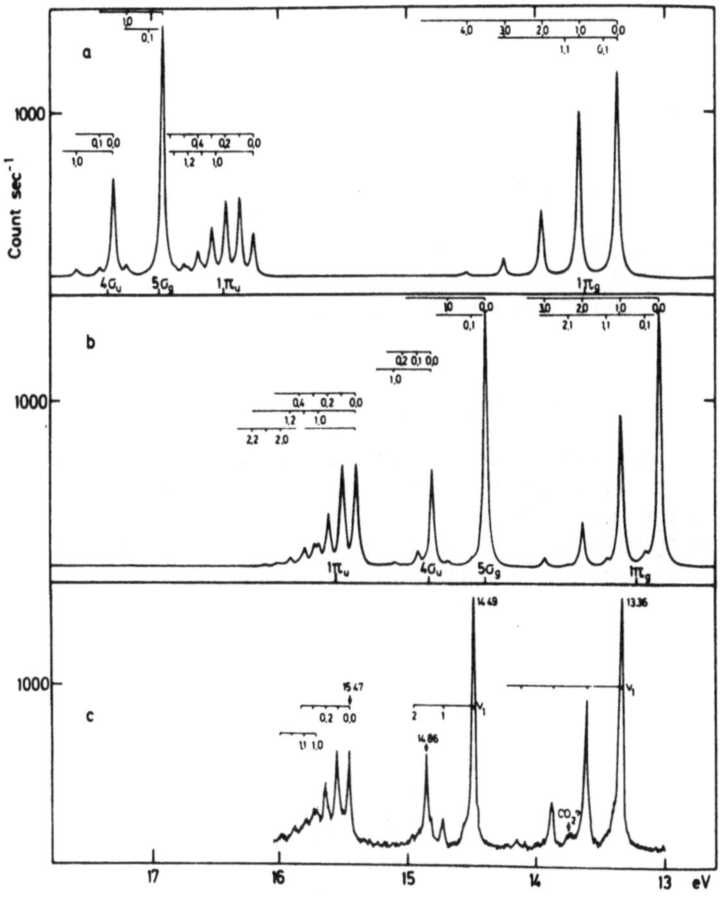

Fig.4: A comparison of the calculated and the experimental PES of C_2N_2. a) spectrum calculated on the HF level, b) spectrum calculated by including many-body effects, c) the experimental spectrum. The calculated vertical IP's are indicated in the figure.

Application to s-tetrazine

The azines constitute a family of interesting molecules where, based on the above arguments, one can expect a frequent failure of Koopmans' theorem in supplying an ordering of states. The lone pair orbitals on the N atoms and the π-IP's are close together in energy. The molecules also possess in general low lying virtual orbitals of π-symmetry. In the previous examples we could study basis set effects and eliminate them to a large extent. In the case of larger molecules this cannot be done. It is, however, interesting to investigate whether the method is also useful for studying larger molecules, where errors due to basis effects cannot be excluded. The PES of molecules with many (say about 10) atoms are usually very difficult to interpret since frequently many broad bands overlap. Without further knowledge even the question of how many bands are involved must be left unanswered. A powerful method to calculate IP's is therefore of great importance. A fair number of fairly large molecules have been computed with the present method (for references see ref.5). Here we select a very recent example, the molecule s-tetrazine.

The PES of s-tetrazine (both HeI and HeII) has been taken by Gleiter et al. (45) and by Lindholm and coworkers (46) and has been interpreted based on extended Hückel-type calculations among others. The SCF calculations were performed with a double-zeta basis set (47) (9s5p/4s)/[4s2p/2s]. The IP's have been computed by taking all valence orbitals into account (15 occupied and 37 virtual MO's). The results are given in table 5 together with the experimental IP's and the IP's obtained by Almlöf et al. (48). These authors performed SCF calculations and estimated empirically correlation corrections to the IP's.

Table 5

IP's of s-tetrazine (in eV)

Symmetry	$IP^{(1)}$	$IP^{(2)}$	$IP^{(3)}$	$IP^{(R)}$	P	$IP(exp)^{46}$	IP^{48}
$2b_{1g}$	11.48	8.18	9.94	9.24	0.90	9.7	9.6
$3b_{2u}$	14.64	10.67	12.82	11.97	0.87	12.1	12.6
$1b_{3g}(\pi)$	12.69	12.37	12.58	12.52	0.91	12.1	12.3
$4a_g$	14.57	11.55	13.26	12.64	0.90	12.8	12.8
$3b_{3u}$	15.63	11.64	13.74	12.89	0.87	12.8	13.3
$1b_{2g}(\pi)$	13.95	13.22	13.68	13.47	0.89	13.5	13.7

According to Koopmans' theorem the ordering of the first six IP's is (with respect to increasing binding energy) $2b_{1g}$, $1b_{3g}(\pi)$, $1b_{2g}(\pi)$, $4a_g$, $3b_{2u}$, $3b_{3u}$. The $4a_g$ and $3b_{2u}$ orbital energies are very close together. In the many-body calculation this ordering changes drastically. Instead of the above ordering one obtains $2b_{1g}$, $3b_{2u}$, $1b_{3g}(\pi)$, $4a_g$, $3b_{3u}$ and $1b_{2g}(\pi)$, i.e. instead of the ordering 15, 14, 13, 12, 11 and 10 one obtains 15, 11, 14, 12, 10 and 13. The σ-IP's are shifted strongly, whereas the π-IP's are shifted only moderately by many-body effects. Koopmans' theorem is thus totally useless for the interpretation of the PES of s-tetrazine. The empirical model of Almlöf et al. is, however, quite successful, it differs from the present results only in the ordering of the $3b_{2u}$ and $1b_{3g}(\pi)$ IP's which are close together.

We have so far only given the ordering of the ionic states. A detailed interpretation of the PES also requires the analysis of which ionization processes give rise to which bands. In this case we encounter a problem typical for such relatively large molecules with very closely spaced bands. According to the calculations the $2b_{1g}$ IP is assigned to the band at 9.7eV and the $3b_{2u}$, $1b_{3g}(\pi)$, $4a_g$, $3b_{3u}$ and $1b_{2g}(\pi)$ IP's to the band system between 11.9 and 13.5eV. There are prominent peaks at 12.1eV (with a shoulder at 11.9eV) at 12.8eV and at 13.5eV and five IP's are to be distributed among them. It is fairly clear that the $3b_{2u}$ IP is to be assigned to the shoulder at 11.9eV, the $1b_{3g}(\pi)$ IP to the band at 12.1eV, the $3b_{3u}$ IP to the band at 12.8eV and the $1b_{2g}(\pi)$ IP to the band at 13.5eV. But it is not clear whether the $4a_g$ IP is to be assigned to the band at 12.1eV or to the one at 12.8eV. We prefer the first interpretation as the HeI spectrum shows a shoulder on the 12.1eV band at about 12.3eV, but we cannot exclude with certainty the alternative. In such cases (a similar one occurred in the PES of para-difluorobenzene (49) three points should be considered: limited resolution in the experimental spectrum, errors due to basis set deficiencies in the calculation and inaccuracy of the molecular geometry used in the calculations. The last point should e.g. be taken into account in the case of formamide (50) where the geometry determined by microwave spectroscopy is relatively poor. Use of this geometry in the calculations leads to a different assignment of the PES than use of the geometry determined by electron diffraction. In the higher energy part of the PES of s-tetrazine the MO picture of ionization starts to break down. Intense satellite lines appear around 15eV. The $1b_{1u}(\pi)$ IP has a pole strength of only 0.69 and the $3a_g$ IP of only 0.48. Only the $2b_{2u}$ IP has a pole strength of 0.82. 2ph-Tamm-Dancoff calculations have been performed for these states, but they will not be discussed here.

Application to benzene

Owing to the fundamental importance of the benzene molecule in chemistry it has been extensively investigated. Many investigations by photoelectron spectroscopy do exist (see ref.51 for a list of references). But in spite of all this work a few points in the assignment of the spectrum remained controversial. In the beginning it was not clear how many IP's should be found in the energy range of the HeI line. The overlapping of bands, the possibility of Jahn-Teller splitting in the ionic states and the occasional observation of additional peaks in the spectrum rendered this problem as well as the assignment of the bands difficult. It was agreed that 8 IP's should lie in the energy range up to 21.21 eV and that the first IP of benzene is due to ionization from the degenerate e_{1g} (π) orbital. The next two IP's are attributed to the degenerate e_{2g} (σ) and the nondegenerate a_{2u} (π) orbitals, but the relative order could not be unambiguously established. The convergence of Rydberg series, the perfluoro effect, deuterium substitution, the vibrational structure and many other data were used to assign the spectrum, but agreement could not be reached; the assignment remained a matter of interpretation. Another controversy concerned the relative ordering of the $2a_{1g}$ and $1b_{1u}$ IP's at 15.45eV and 16.85eV (the MO's are again numbered starting with the first valence orbital). This problem was settled by Gelius using intensity arguments to interpret the ESCA spectrum (52). As the $1b_{1u}$ MO has strong C 2s character in contrast to the $2a_{1g}$ MO it should appear with higher intensity in the ESCA spectrum. This is the case for the band at 15.45eV.

Theoretical calculations on the SCF level of approximation were of little help in the clarification of these problems as they gave rather divergent results. Only the ab initio calculations employing larger basis sets gave acceptable results. But as the e_{2g} (σ) and a_{2u} (π) MO's are quite close together the inclusion of many-body effects is necessary. The calculations have been performed with the same basis sets as described for s-tetrazine. In the many-body calculations the 15 occupied valence orbitals and the 21 lowest virtual orbitals have been included (51). The results are given in table 6 and fig.5.

Table 6

IP's of benzene (in eV)

Symmetry	$IP^{(1)}$	$IP^{(2)}$	$IP^{(3)}$	$IP^{(R)}$	P	IP^{a}(exp)
$1e_{1g}$ (π)	9.31	8.68	9.08	9.10	0.92	9.3 (9.3)
$2e_{2g}$	13.47	11.12	12.13	11.95	0.92	11.4 (11.7)
$1a_{2u}$	13.79	11.84	12.36	12.26	0.83	12.1 (12.2)
$2e_{1u}$	16.14	13.60	14.62	14.46	0.90	13.8 (14.1)
$1b_{2u}$	17.02	13.72	15.11	14.83	0.88	14.7 (14.7)
$1b_{1u}$	17.42	14.96	15.87	15.75	0.88	15.4 (15.5)
$2a_{1g}$	19.47	16.60	17.63	17.48	0.85	16.9 (17.0)
$1e_{2g}$	22.43	18.92	20.20	20.01	0.83	19.2 (19.3)

a) IP's deduced from band maxima (ref.53) and (in brackets) estimated from centroids

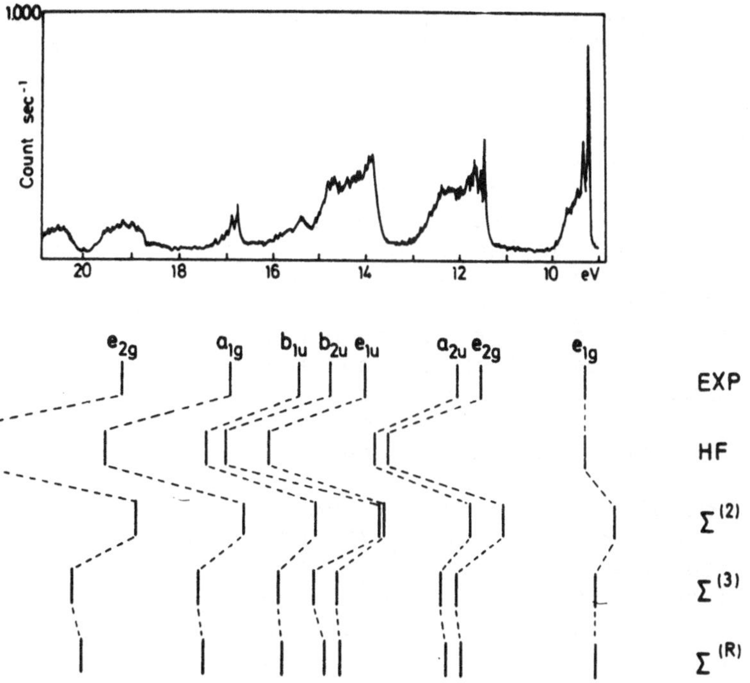

Fig.5: HeI PES of benzene from ref.35 and calculated vertical IP's

In this case there is no change of ordering in going from the Koopmans' theorem result to the second and third order and to the final results. The e_{2g} IP is smaller than the a_{2u} (π) IP and the $1b_{1u}$ IP is smaller than the $2a_{1g}$ IP to mention only the two most controversial points. Moreover these crucial IP's are shifted nearly parallelly in the different orders of the perturbation calculation. This is a very strong argument that the calculated ordering of ionic states is the correct one. The overall quantitative agreement of the computed IP's with experiment is satisfactory in view of the size of the molecule and the restricted size of the basis set which consequently had to be used.

But due to these basis set deficiencies it might still be argued that the ordering of the second and third IP's is open to some debate as they are separated by only 0.3eV. In this case the calculation of the vibrational structure of the two bands could bring the decision. Due to the high symmetry of benzene and as only totally symmetric vibrations couple in first order to the electronic motion (as will be demonstrated below) the vibrational structure in the PES can be computed. This has been done for the e_{2g} and a_{2u} bands neglecting, however, the Jahn-Teller effect in the e_{2g} band. The results are given in fig.6. Comparing with the experimental spectrum in fig.5 it becomes obvious that the onset of the second band system cannot be due to the a_{2u} band which shows strong vibrational structure. As the a_{2u} band cannot show a Jahn-Teller effect we can make a definite assignment. In a recent article Itah et al. (54) claim to have arrived at a purely experimental assignment of the bands in the second band system by a study of the Rydberg series of benzene and deuterium substituted benzenes.

Actually every chemist should have argued for the present assignment. The a_{2u} (π) MO is the lowest occupied π-orbital and must be strongly C-C bonding and thus should show strong vibrational excitation. This is not the case for the onset of the second band system. Instead, it was argued that the onset of the second band system shows weak vibrational excitation and thus the a_{2u} (π) MO is nonbonding!!

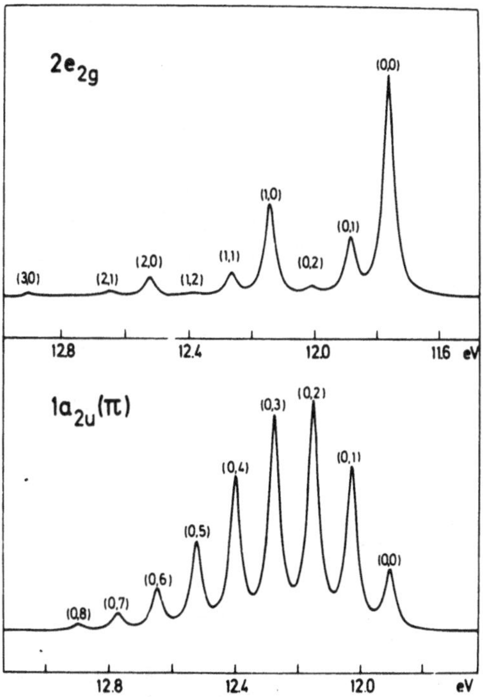

Fig.6: The calculated vibrational structure of the $2e_{2g}$ and $1a_{2u}(\pi)$ bands of benzene

Application to electron affinities

The knowledge of EA's of atoms and molecules is important for the understanding of a variety of physical and chemical processes. Unfortunately the experimental as well as the theoretical determination of accurate EA's is difficult. Experimentally there has been a breakthrough by introducing laser photodetachment techniques. These can, however, only be applied to species the negative ions of which are available to a sufficiently high density. Theoretically there has been progress in the last years as well.

For the calculation of EA's very large basis sets are usually required. This requirement derives from the diffuse character of the charge cloud of the negative ions and the small value of the EA. Already the outer valence parts of atoms and molecules are frequently not adequately described by using basis sets of double-zeta quality. These basis sets are commonly determined by the variation principle. But this is a strongly weighted variation which orients a basis towards the inner shells with at least doubtful effects on the valence electrons. For some properties this is not important but for others this is the case. To calculate EA's (especially small ones) the normal basis sets should be supplemented by diffuse functions. To describe correlation energy changes in the diffuse part of the charge cloud diffuse polarization functions should if possible be added as well.

EA's can be calculated in essentially two different ways, either as the IP's of negative ions or as EA's of the neutral system. These correspond to the two processes

$$A^- + h\nu \Rightarrow A + e^- \quad (1)$$

$$A + e^- \rightarrow A^- + h\nu \quad (2)$$

Process (2) has rarely been used to measure EA's. If we consider molecules with a positive EA process (1) corresponds to starting in the lower potential curve and making transitions to vibrationally excited states of the neutral system. Process (2) corresponds to starting in the upper potential curve and making transitions to vibrationally excited states of the negative ion. The latter process is thus more appropriate to study properties of negative ions.

For the present applications we use process (2). We start from the neutral molecules and calculate the EA directly. We shall discuss the molecules C_2, P_2 (55), O_3 and SO_2 (56). The basis sets employed and the results obtained are given in table 7 together with the experimental values. The basis sets inlcude the 11s7p basis set of Salez and Veillard (34) for first row atoms, the 9s5p basis set determined by Huzinaga (47) and the 12s9p basis set for second row atoms determined by Veillard (57). The 11s7p basis set already contains diffuse functions which may suffice for the calculation if the EA is reasonably large as is the case for C_2 and O_3. To establish the effect of diffuse functions they have been added to this or the smaller basis sets.

Table 7

EA's of C_2, P_2, O_3 and SO_2 calculated with different basis sets (in eV)

Basis	EA(HF)	vertical EA	adiabatic EA	Exp.
C_2				
(11s7p1d)/[5s4p1d] [a]	3.19	3.46	3.47	
(12s8p1d)/[6s5p1d] [b]	3.24	3.54	3.55	3.54 [k]
(11s7p2d)/[5s4p2d] [c]	3.20	3.59	3.60	
P_2				
(12s9p2d)/[7s5p2d] [d]	-0.59	0.002	0.12	0.3 ± 0.5 [l]
(12s9p3d)/[7s5p3d] [e]	-0.60	0.006	0.13	0.24 ± 0.23 [m]
(13s10p2d)/[7s5p2d] [f]	-0.38	0.176	0.30	
O_3				
(11s7p1d)/[5s4p1d] [g]	1.44	1.76	2.26	1.9 to 2.2 [n]
(10s6p1d)/[5s3p1d] [h]	1.60	1.67	2.17	
SO_2				
(12s9p1d/9s5p)/[6s4p1d/4s2p] [i]	0.73	0.84	1.08	1.097 [p]
(13s10p1d/10s6p1d)/[7s5p1d/5s3p1d] [j]	0.51	0.69	0.93	

a)α_d=0.6 b)α_s=0.016, α_p=0.01, α_d=0.6 c)α_d=0.3, 0.8 d)α_d=0.15, 0.45 d)α_d=0.15, 0.45, 1.5 f)α_s=0.04, α_p= 0.03, α_d=0.15, 0.45 g)α_d=0.8 h)α_s=0.09, α_p=0.07, α_d=0.45 i)α_d=0.15 j)α_s(S)=0.05, α_p(S)=0.035, α_d(S)= 0.15, α_s(O)=0.08, α_p(O)=0.06, α_d(O)=0.2 k) ref.58, l) ref.59 m) ref.60 n) ref.61 p) ref.62

The EA determined for C_2 is very stable in Koopmans' approximation as in the Green's function calculation and the agreement with experiment is quite satisfactory. For P_2 Koopmans' theorem predicts the incorrect sign and only many-body effects lead to a positive EA. The experimental value has so large error bounds that it cannot be stated whether P_2 has a positive EA. Diffuse basis functions are important for the calculation and to obtain more accurate results even larger basis sets should be used. The EA's of O_3 and SO_2 are in quite good agreement with experiment (always the adiabatic quantities have to be compared with the experimental values). Diffuse s- and p-type functions have an effect of about 0.1eV on the value of the EA. Basis set extension would still be desirable but the basis sets become excessively large in these cases.

The vibrational structure in the electron attachment process has been computed as well. Electron attachment to C_2 leads to practically no vibrational excitation, but it does so for P_2, O_3 and SO_2. The computed spectra of O_3 and SO_2 are given in fig.7. The stretching vibration dominates in both cases and the excitation is stronger for O_3 than for SO_2. The spectrum for O_3 is a prediction, but the electron detachment spectrum of SO_2 has been recorded by Celotta et al. (62) and is given in fig.8. The agreement with the computed spectrum is quantitative if one notes that both have to be mirror images of each other about the (0,0) line in the case one neglects the change of frequency on electron attachment or detachment and also anharmonicity effects. These effects are expected to be of greater importance for O_3 than for SO_2 because of the stronger vibrational excitation.

We have so far only talked about the EA of the C_2 molecule leading to the $^2\Sigma_g^+$ state. It is known, however, that the C_2 molecule is one of very few molecules whose negative ion has bound exited states. There is a bound $^2\Pi_u$ and a bound $^2\Sigma_u^+$ state (the $^2\Sigma_g^+ - {}^2\Sigma_u^+$ tran-

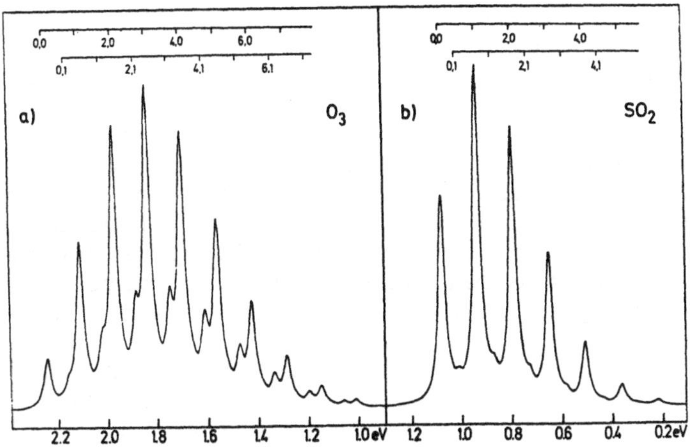

Fig.7: The calculated electron attachment spectrum of a) O_3 and b) SO_2. The energy scale represents the energy of the emitted photons for incident electrons of zero kinetic energy

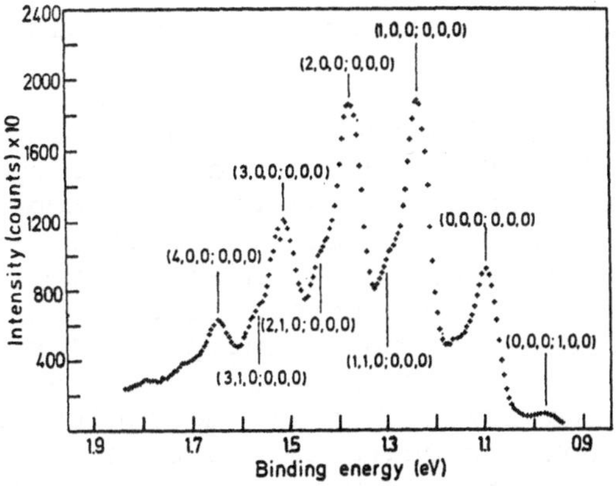

Fig.8: Photo-detachment spectrum of SO_2 as observed by Celotta et al. (67) with vibrational transitions labelled

sition is the Herzberg-Lagerquist band). Indications for further excited states exist. How do these states arise? Except for the $3\sigma_g$ MO which has a negative orbital energy C_2 possesses a number of virtual orbitals with negative or very small positive orbital energies.

The $^2\Sigma_g^+$ state of C_2 arises by direct electron attachment. The Green's function method can in the same way as for this state be used to calculate other negative ion states if they arise from a direct electron attachment. As there are low-lying orbitals of appropriate symmetry one might be tempted to conclude that direct electron attachment is the correct mechanism for the other states too. But no further bound negative ion states which correspond to direct electron attachment could be found. They all have a small but negative EA. Satellite lines arise from ionization plus electron excitation. In the same way electron capture can be accompanied by electron excitation. Thus the 2ph-Tamm-Dancoff-method to be discussed in the last chapter is appropriate to study such states. It was in fact found that the $^2\Pi_u$ and $^2\Sigma_u^+$ states arise in this rather complicated way. In addition another $^2\Sigma_g^+$ state of C_2 was calculated. As the Tamm-Dancoff-method in its presently implemented version lacks ground state correlation energy (see below) the results are not yet quantitative, but at least the calculations supply the understanding of the electronic structure of these states and for the first time give all these states as bound states.

III. The vibrational structure in molecular ionization spectra

The vibrational structure is a prominent feature of molecular PES. No discussion of these spectra can be complete without its consideration. Some bands show strong, others only very weak vibrational structure. From these observations the bonding properties of the individual electrons can be deduced. The orbitals of diatomic molecules can thus simply be classified as bonding, antibonding or nonbonding. The situation is more complicated for polyatomic molecules. Here the PES reflects the bonding character of each particular electron with respect to the various normal coordinates of the molecule.

As a result of the rapidly increasing number of normal coordinates the vibrational structure in the PES becomes exceedingly complex for larger molecules. The interpretation of the vibrational structure (i.e. the assignment of the fundamentals to the observed vibrational lines) is therefore frequently controversial, even for relatively small molecules. The analysis of the spectra becomes greatly facilitated if it can be guided by some theoretical prediction of the vibrational structure. For complex spectra, in particular it is unnecessary to calculate accurately the large number of Franck-Condon factors that determine the intensities of the individual lines. Even a qualitative prediction of the shape of the intensity distribution suffices for interpretation purposes.

A theoretical determination of the vibrational structure could also provide a useful tool for assigning the bands. If two or more bands of the spectrum lie close together, the vertical IP's must be calculated to high accuracy in order to interpret the spectrum unambiguously. Such assignment problems may be attacked more easily by using features of the vibrational structure to identify the bands. Many assignments of molecular PES are in fact based largely on arguments of this type. Theoretical calculations are thus desirable to put these arguments on a firmer basis.

The vibrational structure in molecular electronic spectra can be calculated if the initial and final state potential surfaces are known. The spectroscopic data (ie. equilibrium geometry, force constants) of the molecular ground state are frequently well determined, whereas the corresponding data for the ionic states are in general not experimentally available. In principle these data could be obtained by ab initio calculations, but in practice this approach is too expensive at least for polyatomic molecules. Furthermore the Franck-Condon factors depend mainly on the difference between the initial and final state potential functions (29). Therefore both surfaces need to be calculated to high accuracy in order to obtain good results.

To avoid these difficulties we extend the Green's function method to include the vibrational effects. It is an important advantage of this approach that only the data for the initial state are required. The Franck-Condon factors are expressed in terms of certain coupling constants which can easily be calculated on the one-particle level. These coupling constants can be corrected (renormalized) to include many-body effects. Due to its simplicity the method is applicable to fairly large polyatomic molecules yielding reasonably accurate results with only a moderate computational effort.

We will first present the Hamiltonian and subsequently derive the spectral distribution function. The derivation is restricted to the one-particle approximation. A more complete derivation is given in ref.12+29.

The molecular Hamiltonian is

$$H = T_N + V_{NN} + H_{EN} , \tag{46}$$

where T_N is the nuclear kinetic energy, V_{NN} the Coulomb repulsion of the nuclei, and H_{EN} the Hamiltonian of the electrons in the field of the nuclei, which in the one-particle approximation is

$$H_{EN} = \sum_i \varepsilon_i(R) a_i^+(R) a_i(R) . \tag{47}$$

The coordinates R are internal coordinates; translational and rotational degrees of freedom are assumed to be separated. The orbital energies and the electronic operators depend on these coordinates, but the dependence of the operators will from now on not be explicitly indicated. In a later section where this dependence becomes important we will come back to this point.

We subtract from H_{EN} the electronic ground state energy and add it to V_{NN}

$$V_o = V_{NN} + \sum_i \varepsilon_i n_i \quad . \tag{48}$$

Thus

$$H = T_N + V_o + \sum_i \varepsilon_i(R)(a_i^+ a_i - n_i) \quad . \tag{49}$$

We now make the harmonic approximation for the ground state potential energy

$$V_o(R) = V_o(0) + \frac{1}{2} \sum_{s,s'} F_{ss'} R_s R_{s'} \tag{50}$$

$$T_N = -\frac{1}{2} \sum_{s,s'} G_{ss'} \frac{\partial}{\partial R_s} \frac{\partial}{\partial R_{s'}} , \tag{51}$$

where $F_{ss'}$ is the force constant matrix and $G_{ss'}$ the kinematic matrix.

We transform to normal coordinates which diagonalize the potential and kinetic energy

$$\underline{Q} = \underline{\omega}^{1/2} \underline{L}^{-1} \underline{R} . \tag{52}$$

These normal coordinates are dimensionless due to the frequency factor $\omega^{1/2}$. The L-matrix is the solution of the equation

$$\underline{G}\,\underline{F}\,\underline{L} = \underline{L}\omega^2 , \quad \underline{L}\,\underline{L}^+ = \underline{G} . \tag{53}$$

Boson creation and annihilation operators are then introduced by the relations

$$Q_s = \frac{1}{\sqrt{2}} (b_s + b_s^+) \; ; \quad \frac{\partial}{\partial Q_s} = \frac{1}{\sqrt{2}} (b_s - b_s^+) , \tag{54}$$

where b_s, b_s^+ obey the commutation relations

$$\left[b_s^{(+)}, b_{s'}^{(+)}\right] = 0 \; ; \quad \left[b_s, b_{s'}^+\right] = \delta_{ss'} . \tag{55}$$

In the harmonic approximation we obtain then

$$V_o = V_o(0) + \frac{1}{2}\sum_s \omega_s Q_s^2$$

$$T_N = -\frac{1}{2}\sum_s \omega_s \frac{\partial^2}{\partial Q_s^2} , \tag{56}$$

where ω_s are the normal frequencies corresponding to the Q_s. We substitute these expressions in the Hamiltonian and leave all anharmonic terms away to obtain

$$H = V_o(0) + \sum_s \omega_s (b_s^+ b_s + \tfrac{1}{2}) + \sum_i \varepsilon_i(Q)(a_i^+ a_i - n_i) . \tag{57}$$

$\varepsilon_i(Q)$ is now expanded in a Taylor series

$$\varepsilon_i(Q) = \varepsilon_i(0) + \sum_s \left(\frac{\partial \varepsilon_i}{\partial Q_s}\right)_o Q_s + \ldots\ldots . \tag{58}$$

The second and higher order terms introduce frequency shifts and will not be considered here. For their inclusion see ref.12, 29. The "o" denotes the equilibrium configuration of the molecular ground state. Introducing the first order coupling constants

$$\kappa_s(i) = -\frac{1}{\sqrt{2}} \left(\frac{\partial \varepsilon_i}{\partial Q_s}\right)_o \tag{59}$$

we obtain finally for the Hamiltonian

$$H = V_{NN}(o) + \sum_s \omega_s (b_s^+ b_s + \tfrac{1}{2}) + \sum_i \varepsilon_i(0) a_i^+ a_i$$

$$- \sum_i \sum_s \kappa_s(i)(a_i^+ a_i - n_i)(b_s + b_s^+) \quad . \tag{60}$$

The spectrum is given by the transition probability per unit time and unit energy, i.e. $P(\omega)$. We derive $P(\omega)$ in the adiabatic, harmonic, Franck-Condon and one-particle approximation. Many-body effects will be discussed at the end. In the adiabatic approximation the electronic operators are considered to depend only parametrically on the nuclear coordinates Q, that is the nuclear kinetic energy operator is assumed to commute with a_i and a_i^+. The electronic operators then commute with the boson operators b_s and b_s^+. The transition probability decomposes into a sum of terms one for each orbital j.

$$P(\omega) = \sum_j P_j(\omega)$$

$$P_j(\omega) = \int_{-\infty}^{\infty} dt e^{i\omega t} \langle \phi_o | T_j^+ T_j(t) | \phi_o \rangle \tag{61}$$

with

$$T_j = \tau_{ej} a_j \quad , \quad T_j(t) = e^{iHt} \, T_j \, e^{-iHt} \quad . \tag{62}$$

For $|\phi_o\rangle$ we make in the adiabatic approximation a product ansatz $|\phi_o\rangle = |\psi_o\rangle|0\rangle$ where $|\psi_o\rangle$ is the electronic ground state and $|0\rangle$ the nuclear ground state ($b_s|0\rangle = 0$ for all s, i.e. no vibrations are excited). In the one-particle approximation $|\psi_o\rangle$ is the HF wavefunction.

We are going to introduce a pure boson operator $\tilde{H}$ (we neglect the constant term in the Hamiltonian in the following).

$$H\, a_j|\phi_o\rangle = \sum_s \omega_s (b_s^+ b_s + \tfrac{1}{2}) a_j|\phi_o\rangle + \sum_i \varepsilon_i(0) a_i^+ a_i a_j|\phi_o\rangle$$

$$- \sum_i \sum_s \kappa_s(i)(a_i^+ a_i - n_i)(b_s + b_s^+) a_j|\phi_o\rangle \quad . \tag{63}$$

For $i \neq j$ we have $a_i^+ a_i a_j |\phi_o\rangle = a_j a_i^+ a_i |\phi_o\rangle = a_j n_i |\phi_o\rangle$ and for $i = j$ we have $a_i^+ a_i a_i |\phi_o\rangle = 0$. Thus

$$Ha_j|\phi_o\rangle = \sum_s \omega_s (b_s^+ b_s + \tfrac{1}{2}) a_j |\phi_o\rangle + \sum_{i(\neq j)} \mathcal{E}_i n_i a_j |\phi_o\rangle + \sum_s K_s(j) n_j a_j (b_s + b_s^+) |\phi_o\rangle \tag{64}$$

$$\begin{aligned} Ha_j|\phi_o\rangle &= \hat{H} a_j|\phi_o\rangle + (E_o - \mathcal{E}_j) a_j |\phi_o\rangle \\ &= \tilde{H} a_j |\phi_o\rangle . \end{aligned} \tag{65}$$

$$\hat{H} = \sum_s \omega_s (b_s^+ b_s + \tfrac{1}{2}) + \sum_s K_s(j)(b_s + b_s^+) \tag{66}$$

is a pure boson operator. From this it is easily seen that

$$\begin{aligned} H^n a_j |\phi_o\rangle &= \tilde{H}^n a_j |\phi_o\rangle \\ e^{iHt} a_j |\phi_o\rangle &= e^{i\tilde{H}t} a_j |\phi_o\rangle \end{aligned} \tag{67}$$

which we can use to simplify $P_j(\omega)$.

$$\begin{aligned} P_j(\omega) &= \int dt e^{i\omega t} \langle \phi_o | T_j^+ e^{i\tilde{H}t} T_j e^{-i(E_o + \sum_s \frac{\omega_s}{2})t} |\phi_o\rangle \\ &= \int dt e^{i\omega t} \langle \phi_o | \tau_{ej}^* a_j^+ e^{i\hat{H}t} \tau_{ej} a_j e^{-i(\mathcal{E}_j + \sum_s \frac{\omega_s}{2})t} |\phi_o\rangle \\ &= |\tau_{ej}|^2 \int dt e^{i(\omega - \mathcal{E}_j - \sum_s \frac{\omega_s}{2})t} \langle \phi_o | a_j^+ e^{i\hat{H}t} a_j |\phi_o\rangle \end{aligned} \tag{68}$$

and as $\hat{H}$ commutes with the $a_j^{(+)}$

$$P_j(\omega) = |\tau_{ej}|^2 \int dt e^{i(\omega - \mathcal{E}_j - \sum_s \frac{\omega_s}{2})t} \langle 0 | e^{i\hat{H}t} | 0 \rangle \tag{69}$$

for j an occupied orbital.

$\hat{H}$ is diagonalized by a unitary transformation U, where

$$U = \exp\left(-\sum_s \frac{K_s}{\omega_s}(b_s - b_s^+)\right) \quad , \quad U^+U = 1 \quad . \tag{70}$$

For this purpose we need the relation

$$e^A B e^{-A} = B + [A,B] + \frac{1}{2}[A,[A,B]] \tag{71}$$

which holds if A and B are bosonic operators. Since

$$[b+b^+, b-b^+] = -2 \quad , \quad [b^+b, b-b^+] = -(b+b^+) \tag{72}$$

we obtain with $B = \omega b^+b + K(b+b^+)$ and $A = \alpha(b-b^+)$

$$[A,B] = \alpha\omega(b+b^+) + 2\alpha K; \quad [A,[A,B]] = 2\alpha^2\omega$$

$$[A,[A,[A,B]]] = 0 \quad \text{as all higher commutators.}$$

Thus

$$U^+\hat{H}U = \sum_s \omega_s(b_s^+b_s + \frac{1}{2}) - \sum_s \frac{K_s^2}{\omega_s} \quad . \tag{73}$$

For $\langle 0|e^{i\hat{H}t}|0\rangle$ we obtain

$$\langle 0|e^{i\hat{H}t}|0\rangle = \langle 0|U\underbrace{U^+e^{i\hat{H}t}U}_{\text{diagonal}}U^+|0\rangle \tag{74}$$

$$= \sum_{n_1,n_2\ldots} |\langle n_1 n_2 \ldots |U^+|0\rangle|^2 \times$$

$$\times e^{i\left(\sum_s (n_s+\frac{1}{2})\omega_s - \sum_s \frac{K_s^2}{\omega_s}\right)t} \tag{75}$$

$$P_j(\omega) = |\tilde{\tau}_{ej}|^2 \sum_{n_1,n_2\ldots} |\langle n_1 n_2 \ldots |U^+|0\rangle|^2 \times$$

$$\times 2\pi\delta\left(\omega - \varepsilon_j(0) - \sum_s \frac{K_s^2}{\omega_s} + \sum_s n_s\omega_s\right) \tag{76}$$

We are left with having to evaluate terms of the form

$\langle n| e^{\frac{K}{\omega}(b-b^+)} | 0\rangle$. We use the Baker-Hausdorff formula.

$$e^{x+y} = e^x e^y e^{-\frac{1}{2}[x,y]} \tag{77}$$

which is valid if $[x,y]$ is not an operator.

$$\langle n| e^{\frac{K}{\omega}(b-b^+)} | 0\rangle = \langle n| e^{-\frac{K}{\omega}b^+} e^{\frac{K}{\omega}b} | 0\rangle e^{-\frac{1}{2}(\frac{K}{\omega})^2} \tag{78}$$

As $e^{\frac{K}{\omega}b} | 0\rangle = (1+\frac{K}{\omega}b+\frac{1}{2}(\frac{K}{\omega})^2 b^2+\ldots.) | 0\rangle = | 0\rangle$

we obtain

$$\langle n| e^{\frac{K}{\omega}(b-b^+)} | 0\rangle = \langle n| e^{-\frac{K}{\omega}b^+} | 0\rangle e^{-\frac{1}{2}(\frac{K}{\omega})^2}$$

$$= \langle n| \sum_m (-\frac{K}{\omega})^m \frac{1}{m!} b^{+m} | 0\rangle e^{-\frac{1}{2}(\frac{K}{\omega})^2}$$

$$= \sum_m (-\frac{K}{\omega})^m \frac{1}{\sqrt{m!}} \delta_{nm} e^{-\frac{1}{2}(\frac{K}{\omega})^2} \tag{79}$$

where the relation $b^{+n}|0\rangle = \sqrt{n!}\,|n\rangle$ has been used.

$$|\langle n|U^+|0\rangle|^2 = \left(\frac{K^2}{\omega^2}\right)^n e^{-\frac{K^2}{\omega^2}} \frac{1}{n!} = \frac{a^n}{n!} e^{-a} \tag{80}$$

and

$$P_j(\omega) = |\tau_{ej}|^2 \sum_{n_1,n_2\ldots} \prod_s \frac{a_s^{n_s}}{n_s!} e^{-a_s} \times$$

$$\times\ \delta(\omega - \varepsilon_j(0) - \sum_s a_s \omega_s + \sum_s n_s \omega_s)\ . \tag{81}$$

a_s is called coupling parameter. The intensity distribution is thus a product of Poisson distributions one for each normal mode. The quantity $-\varepsilon_j(0)-\sum_s a_s\omega_s$ is the adiabatic IP ((0,0) transition) and by a moment analysis it can be shown that the vertical IP, i.e. $-\varepsilon_j(0)$ corresponds to the centroid of the band. (12,29).

Many-body effects can be included in a rather simple way. The adiabatic hypothesis tells us that we can think in terms of well defined potential energy surfaces also when the electrons interact (beyond the HF approximation). It follows that the entire effect of the electron interaction terms is to replace the bosons by "renormalized" bosons, $-\varepsilon_j(0)$ by the exact vertical IP and the coupling constants by renormalized coupling constants. Thus all the results obtained can be taken over if the ω_s are identified with the molecular frequencies in the exact state and if $-\varepsilon_j(0)$, $\kappa_s(j)$ etc. are replaced by the corresponding renormalized quantities. Since the spectroscopic constants of the electronic ground state are well known experimentally for most molecules of interest it is only necessary to calculate the renormalized coupling constants in order to obtain the vibrational structure.

Before proceeding to the applications let us briefly discuss the content of the method presented above. For diatomic molecules rather accurate Franck-Condon factors can be calculated by using potential functions constructed by the Rydberg-Klein-Rees method from spectroscopic data (64) or by fitting Morse potentials to the spectroscopic data of the initial and final states (65). The calculation of Franck-Condon factors for polyatomic molecules, on the other hand, has been in general confined to the harmonic approximation, although the importance of anharmonic effects is known. It should be noted that this traditional harmonic approximation and the harmonic approximation discussed above are not identical. It can be shown that a major part of the anharmonic effects neglected in the traditional approach is taken into account in the present harmonic approximation (12, 29).

The Hamiltonian (60) has been obtained by expanding all Q-dependent terms (with the exception of the electronic operators) about the equilibrium geometry of the electronic ground state. Calculating $P(\omega)$ with this Hamiltonian implies that the ground state and the ionic state potential energy surfaces are expanded about the ground state equilibrium geometry. It must be emphasized at this point that the traditional approach to calculate Franck-Condon factors within the harmonic approximation is to expand both potential surfaces up to second order about their respective minima. The drawback of the traditional approach is easily understood from an inspection of fig.9 which shows schematically the initial and final state potential surfaces. The initial state vibrational wavefunction can be assumed to be well described within the harmonic approximation. In the ionic state, however, the nuclei perform large amplitude vibrations in the case of strong coupling.

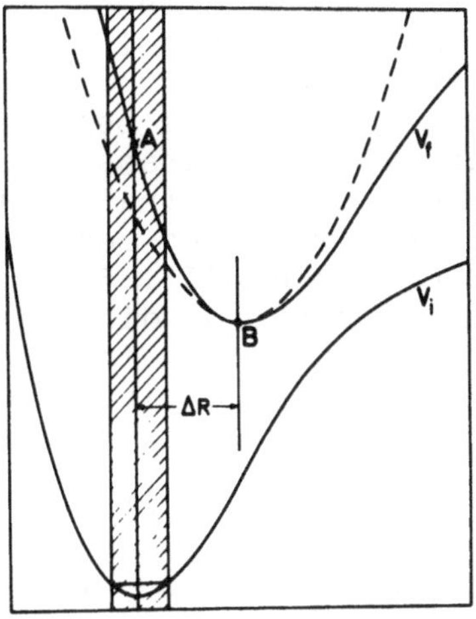

Fig.9: A schematic one-dimensional drawing of the initial (i) and final (f) potential energy surfaces. The Franck-Condon region is indicated by the shaded area. The harmonic expansion of V_f about its equilibrium geometry is represented by a broken line.

In particular the minimum of the upper potential surface may lie considerably outside the Franck-Condon region as indicated by the shaded area in fig.9. It appears reasonable and it can be shown in detail (12, 29) that the overall shape of the spectrum depends only on the behaviour of the final state potential surface within the Franck-Condon region. It is of advantage therefore to expand the final state potential surface about a point within the Franck-Condon region, i.e. the ground state equilibrium configuration, point A in fig. 9. An expansion about the final state equilibrium configuration (point B) will give a poor description of the final state energy surface within the Franck-Condon region if the coupling is strong. An expansion about point B is of advantage if one is interested in an accurate description of the first few vibrational energy levels in the final state. In the strong coupling case, however, these levels are excited with negligible intensity and do not contribute to the observed vibrational structure.

It is known from the literature that the calculation of the vibrational structure of some Rydberg transitions and of the first band in the PES of NH_3 presents a problem (for a discussion and references see ref.66). These transitions show an extended progression in the bending mode ν_2. The intensity maximum occurs at n=6 in the UV absorption spectrum and at n=7 in the PES. The strong excitation of the bending mode is in qualitative agreement with the transition from a pyramidal ground state to a planar final state. Calculations in the traditional way have been performed within the harmonic approximation (67, 68) and employing anharmonic potentials for the bending vibration (69). In the harmonic approximation the intensity maximum was found to occur for n=4 in the excitation spectrum in rather poor agreement with experiment. The use of anharmonic potentials did surprisingly not improve the result. However, if the vibrational structure in the PES is calculated according to the procedure outlined above very good agreement with experiment is obtained as can be seen from fig.10. The intensity maximum is calculated to occur at the n=7 line of the bending progression as is indeed observed in the high resolution spectrum of Rabalais et al. (70). The calculation presented here is an absolute one including the calculation of the position of the band on the energy scale. The potential in the case of NH_3^+ is rather anharmonic. But as the Franck-Condon zone is narrow, a second order Taylor expansion about the center of the zone gives a fairly accurate description of the final state potential surface within the zone. In the traditional approach this part of the potential surface is represented poorly and a high order Taylor expansion would be required to obtain an accurate description here.

In contrast to the intensities the vibrational energy levels in the final state depend on the potential surface as a whole. Therefore its harmonic expansion about the initial state equilibrium geometry does not necessarily lead to accurate line spacings in the calculated spectrum. It is our opinion, however, that for interpretative purposes the accurate calculation

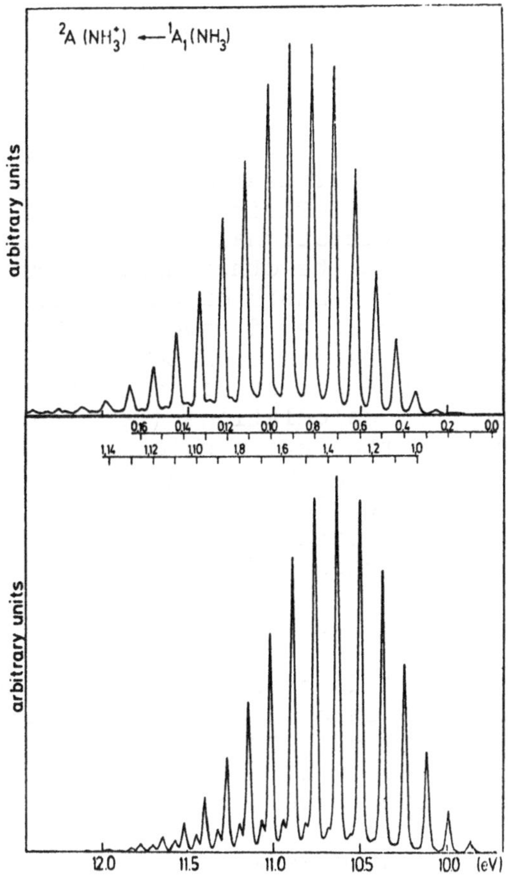

Fig.10: The first band in the PES of NH_3 as recorded by Rabalais et al. (70) (upper part) and the calculated spectrum (lower part)

of intensities is more valuable than the accurate calculation of line spacings. It is important to note that the calculated intensities are independent of the line spacings in the final state. In the case of simple well resolved spectra the line spacings may be taken from experiment. In the case of complex and unresolved spectra one is in general not interested in the intensities and energies of the numerous vibrational lines but in a prediction of the shape of the spectral distribution. As shown in ref.(12, 29) the low moments which represent the overall shape of the spectrum are obtained rather accurately with the present method, whereas this is not the case when the traditional harmonic expansion is used.

Application to formaldehyde

The formaldehyde molecule is well studied by photoelectron spectroscopy. The HeI spectrum exhibits four bands. There have been many discussions in the literature about the assignment of the third and the fourth band. According to the original proposal of Baker et al. (71, 35) the third IP corresponds to the $1b_2$ MO and the fourth IP to the $3a_1$ MO. Their argument was based on the analysis of the vibrational structure of both bands. Brundle et al. (72) on the other hand favoured the reverse assignment on the basis of an ab initio calculation. Recently a X_α-scattered wave calculation employing overlapping spheres has been performed (73) and the ordering proposed by Baker et al. was obtained. Previous Green's function calculations (74) as well as Rayleigh-Schrödinger perturbation calculations (75) predicted the ordering given by Koopmans' theorem to be the correct one. To settle this problem definitely the IP's have been recalculated with the Green's function method employing an extensive basis set (29, 76).

The interpretation of the vibrational structure in the PES of H_2CO has proved to be difficult as well owing to the accidental degeneracy of vibrational frequencies in some of the ionic states. Especially for the second band ($1b_1$ (π)) of H_2CO and D_2CO no assignment of the vibrations involved could be given, although the vibrational structure is well resolved and apparently simple. For these reasons the vibrational structure of the bands was calculated as well.

The basis set used in the calculation is (11s7p1d/6s1p)/[5s3p1d/3s1p] (34). The final results for the IP's are given in table 8.

Table 8

IP's of formaldehyde (in eV)

Symmetry	IP	IP(exp) [a]
$2b_2$	10.84	10.9
$1b_1$	14.29	14.5
$3a_1$	16.36	16.2
$1b_2$	17.13	~17.0

a) estimated centroid of the band in the PES of H_2CO (35

The computed values are within 0.2eV of the experimental ones. Because of this agreement it can be concluded that the third band has to be assigned as $3a_1$ and the fourth one as $1b_2$. Koopmans' theorem thus supplies for this molecule the correct ordering of states. The ordering deduced from the overlapping spheres X_α-scattered wave calculation is incorrect.

In calculating the vibrational structure only the three totally symmetric normal vibrations ν_1, ν_2 and ν_3 are taken into account. The coupling to the non-totally symmetric vibrations occurs only through the quadratic coupling constants and is expected to be very weak. The appropriate internal symmetry coordinates are

$$s_1 = \frac{1}{\sqrt{2}}(r_1+r_2), \quad S_2=R, \quad s_3=\sqrt{\frac{3}{2}}\, r_o \phi \quad ,$$

where r_1 and r_2 denote the change of the two C-H bond distances, R the change of the C-O bond length and ϕ the deviation of the HCH angle from its equilibrium value. The force field of Shimanouchi and Suzuki (77) is used to construct the normal coordinates of H_2CO and D_2CO in their electronic ground states.

The vibrational coupling parameters in the HF approximation $a_s^{(o)}$ and the renormalized coupling parameters a_s are listed in table 9.

Table 9

Vibrational coupling parameters for valence ionization of H_2CO and D_2CO. The $a_s^{(o)}$ are calculated on the HF level, the a_s with inclusion of many-body effects

		$a_1^{(o)}$	$a_2^{(o)}$	$a_3^{(o)}$	a_1	a_2	a_3
$2b_2$	H_2CO	0.217	0.375	0.087	0.033	0.129	0.126
	D_2CO	0.456	0.229	0.028	0.092	0.122	0.090
$1b_1$	H_2CO	0.128	3.555	0.310	0.036	2.792	0.270
	D_2CO	0.778	3.125	0.008	0.414	2.611	0.011
$3a_1$	H_2CO	0.003	2.176	0.147	0.004	1.156	0.301
	D_2CO	0.168	1.660	0.731	0.098	0.765	0.855
$1b_2$	H_2CO	0.886	0.201	1.439	0.757	0.195	1.533
	D_2CO	0.765	0.995	1.420	0.635	0.950	1.534

We show the calculated vibrational structure together with the corresponding band in the experimental spectrum in figs. 11 to 14. The lines are drawn as Gaussians for the first band and as Lorentzians for the others (half width of 0.03eV) as these forms fit best the experimental line shapes which are of instrumental origin. H_2CO and D_2CO represent complicated examples since the structure of several bands is the result of a superposition of two nearly degenerate normal modes. It is therefore essential to employ correct ionic state vibrational frequencies in drawing the spectra. Having calculated only linear vibrational coupling constants we are not in a position to predict the frequency change due to ionization. Therefore the experimental ionic frequencies are used to draw the calculated vibrational structure. Although the ionic frequencies are not unambiguously known for most of the ionic states of H_2CO and D_2CO it is found that they can easily be determined from the structure observed in the spectrum with the help of the calculated Franck-Condon factors. An exception is only the fourth band because of its low intensity, the very complex vibrational structure and the severe overlap with the third band. However, the vibrational structure calculated for this band is very complex and does not critically depend on the frequencies chosen in drawing the spectra.

The first band (fig.11) corresponds to a nonbonding electron. The calculated vibrational structure is in satisfactory agreement with experiment showing that all three normal modes are weakly excited.

The second band corresponds to ionization of the $1b_1$ CO π-bonding electron and exhibits the expected strong excitation of the C-O stretching mode ν_2 (fig. 12). The higher resolution spectra, however, reveal that the lines are doublets for both molecules. Thus at least one other mode must be excited. The assignment of this mode has caused difficulties. According to Turner et al. (35) the structure is due to ν_2 together with one quantum of the C-H stretching mode ν_1. It is then difficult to explain, however, the lack of isotope effects. One would expect ν_1 to be considerably reduced for D_2CO and thus the accidental near degeneracy of both frequencies in H_2CO should be removed for D_2CO. The problem is resolved by the calculations. For H_2CO the vibrational structure is due to excitation of ν_2 and ν_3, for D_2CO, on the other hand, the structure is due to ν_1 and ν_2. Thus the coupling is completely different for these two isotopic species.

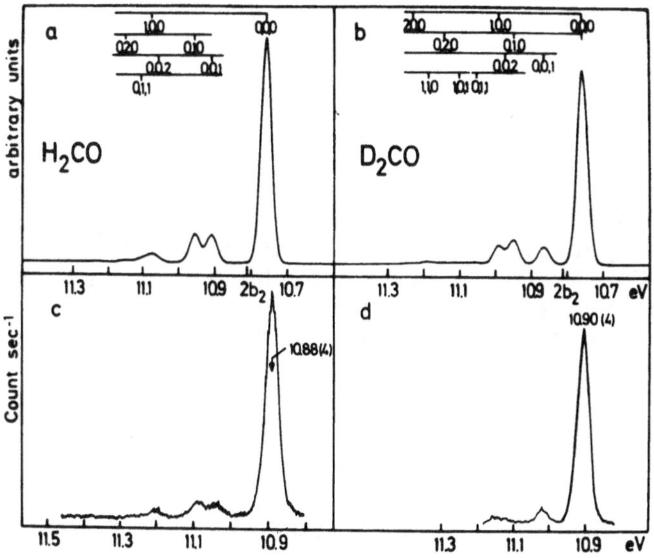

Fig.11: The calculated (a), (b) and observed (c), (d) first band in the spectrum of H_2CO and D_2CO

This result is already obtained on the HF level of approximation. The calculated vibrational structure is in nearly quantitative agreement with experiment.

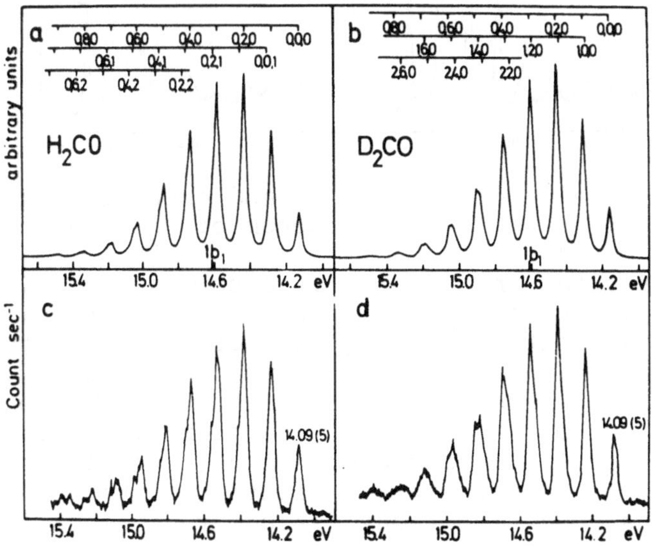

Fig.12: The calculated (a), (b) and observed (c), (d) second band in the spectrum of H_2CO and D_2CO

The third band in the spectrum of H_2CO consists of a single series of narrow lines (fig.13). From a comparison with the third band of D_2CO which shows a considerably more complex vibrational structure due to strong excitation of both ν_2 and ν_3 Turner et al. (35) concluded that the simplicity of the third band of H_2CO was due to a degeneracy of the frequencies ν_2 and ν_3 in the ionic state. This interpretation is confirmed by the present calculations. We find strong excitation of ν_2 accompanied by weak excitation of ν_3 in the case of H_2CO. For D_2CO, on the other hand, both ν_2 and ν_3 are calculated to couple strongly, the coupling of ν_3 being even stronger than that of ν_2.

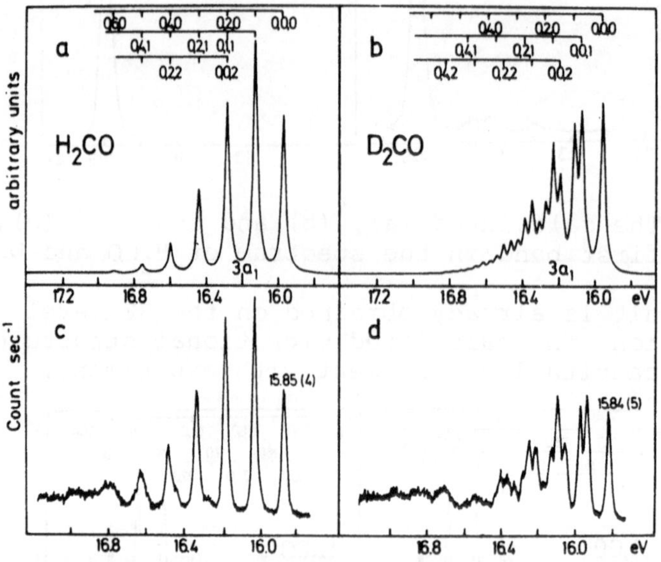

Fig.13: The calculated (a), (b) and observed (c), (d) third band in the spectrum of H_2CO and D_2CO

The fourth band appears as a complex and diffuse structure in the spectra of both H_2CO and D_2CO (fig.14). From the calculated coupling parameters we see that all three normal vibrations are strongly excited in the case of D_2CO, whereas for H_2CO only ν_1 and ν_3 couple strongly. In both cases a very complex vibrational structure results in agreement with the experimental situation. It is seen that excitation of three normal modes leads to a vibrational structure of such a complexity that it cannot be resolved with present-day spectrometers. From the computed vibrational structure it is now obvious that the third band of H_2CO has to be assigned as $3a_1$ and the fourth one as $1b_2$. The vibrational mode ν_1 is not excited in the third band of D_2CO but should appear in the fourth band. In the experimental spectrum there is clearly no indication of the ν_1 mode in the third band.

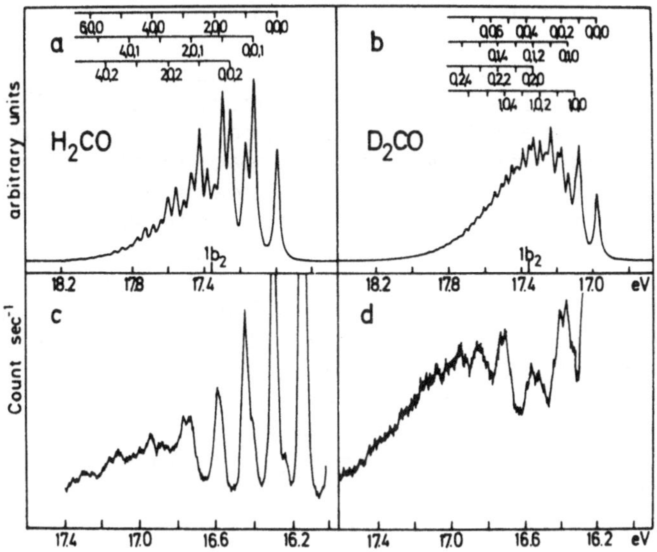

Fig.14: The calculated (a), (b) and observed (c), (d) fourth band in the spectrum of H_2CO and D_2CO. Note that the intense lines at lower energies in parts (c) and (d) belong to the third band. The fourth band is the diffuse structure centered at about 17.0eV. Note also that the experimental spectrum of H_2CO (c) has been recorded with less amplification than that of D_2CO (d)

The HF coupling parameters reproduce all trends discussed above, but the quantitative agreement is better when the renormalized coupling parameters are used.

The calculations reveal very pronounced isotope effects in the second, third and fourth bands of these molecules. These results demonstrate that electronic bonding properties derived from so-called overlap populations are of a limited value for interpreting the vibrational structure in the PES of polyatomic molecules. It is the bonding strength of a particular electron with respect to a particular normal coordinate which is reflected in the spectrum and not the bonding strength with respect to any two nuclei in the molecule. Therefore the normal coordinate derivatives of the orbital energy or of the pole of the Green's function are the appropriate quantities to discuss the vibrational structure in a PES. Isotope effects on the vibrational structure are then automatically taken into account in the construction of the ground state normal coordinates.

Nondiabatic effects in the spectrum of butatriene (80)

Until now we have based our considerations on the adiabatic approximation. This approximation relies on the fact that the energy difference between electronic states is large compared to the spacing of the rotational and vibrational energies. A breakdown of the adiabatic approximation may therefore occur when two electronic states become degenerate or nearly degenerate. Most important is the case of symmetry induced degeneracy leading to the well-known Jahn-Teller and (for linear molecules) Renner-Teller effects. In particular the Jahn-Teller effect manifests itself in the PES. Molecules of high symmetry may possess spatially degenerate orbitals which according to the Jahn-Teller theorem (78) can couple linearly to vibrations of suitable symmetry. If the coupling is sufficiently strong a very complex vibrational structure and a splitting of the band is observed. The Jahn-Teller effect in the second band of NH_3 has been treated in ref.79. Here we will discuss the strong vibronic coupling effects in the PES of butatriene (C_4H_4) (80).

The PES of butatriene has been recorded and discussed by Brogli et al. (81). The low energy part of the spectrum shows three bands which are well separated from the fourth band in the spectrum. Theoretical and experimental investigations have definitely shown that only two of the bands in this lower part can be attributed to direct ionization from orbitals occupied in the molecular ground state. Therefore one of the bands 1, 1' and 2 in fig.15 must be either a satellite line or/and due to a vibrational phenomenon. By comparing with spectra of similar molecules it is evident that band 1' is the band in question. In table 10 we give the computed and experimental valence IP's of C_4H_4 which show that one band in the low energy part of the spectrum is too many. The basis set used in these calculations is (9s5p1d/4s1p)/[4s2p1d/2s1p].

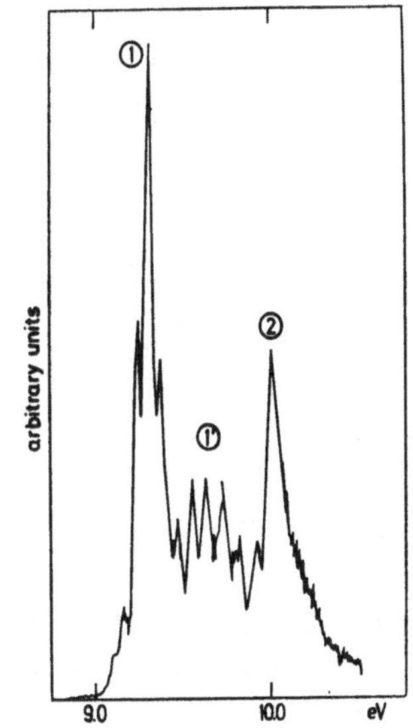

Fig.15: The first band system in the PES of butatriene (81). The band 1' is referred to as the "mystery" band

Table 10

IP's of butatriene (in eV)

Symmetry	IP	IP(exp) (81)
$1b_{3g}(\pi)$	8.95	9.30
		9.63
$2b_{3u}$	9.67	9.98
$1b_{2u}$	12.07	11.78
$1b_{2g}$	14.66	14.2
$1b_{3u}$	15.10	15.0
$3a_g$	15.76	15.5
$2b_{1u}$	16.76	16.8

A calculation of the satellite lines by a many-body Tamm-Dancoff method showed that no satellite lines appear in this low energy part of the spectrum. The first satellite lines appear near 13.8eV and can clearly be detected in the spectrum of ref.81. We will demonstrate here that this mystery band arises from vibronic coupling effects.

The molecular Hamiltonian is again given by eq's (46) and (47). We need to maintain here that all quantities, also the electronic operators, depend on the nuclear coordinates. Rearranging the Hamiltonian as has been done above leads to

$$
\begin{aligned}
H &= H_{NN} + H_{EN} \\
H_{NN} &= T_N + V_o(R) \\
H_{EN} &= \sum_i \varepsilon_i(R)\left[a_i^+(R)\,a_i(R) - n_i\right] \\
V_o(R) &= V_{NN}(R) + \sum_i \varepsilon_i(R)\,n_i \;.
\end{aligned}
\tag{82}
$$

$V_o(R)$ will be approximated by a harmonic potential where we expand about the molecular equilibrium geometry R_o. $R-R_o$ is then replaced by dimensionless normal coordinates. Introducing boson destruction and creation operators the pure nuclear Hamiltonian reads

$$
H_{NN} = V_o(0) + \sum_s \omega_s (b_s^+ b_s + \tfrac{1}{2})
\tag{83}
$$

with ω_s the ground state vibrational frequencies. We then expand the orbital energies about R_o up to first order and obtain

$$
H_{EN} = V_{NN}(0) - V_o(0) + \sum_i \varepsilon_i(0)\,a_i^+ a_i - \sum_{i,s} \kappa_s^{(o)}(i)\left[a_i^+ a_i - n_i\right](b_s + b_s^+) \;.
\tag{84}
$$

The $\kappa_s^{(o)}(i)$ are the electron-vibrational coupling constants given by

$$
\kappa_s^{(o)}(i) = -\frac{1}{\sqrt{2}} \left(\frac{\partial \varepsilon_i}{\partial Q_s}\right)_o \;.
\tag{85}
$$

Non-degenerate electronic states $|\varphi_i\rangle$ couple to totally symmetric vibrational modes only because $(\partial\varepsilon_i/\partial Q_s)_o$ vanishes for non-totally symmetric modes. The fermion operators a_i still depend on the normal coordinates. This means that the commutator of a_i with the boson operator b_s does not vanish. Putting this commutator to zero means that the a_i depend only parametrically on the nuclear coordinates which is equivalent to the adiabatic approximation as already discussed above.

The adiabatic approximation may loose its validity when the energy difference between electronic states is not large compared to the vibrational spacings. In these cases one cannot neglect the dependence of fermion operators on the nuclear coordinates. There is, however, a simple way to avoid the calculation of complicated commutators by transforming the $a_i(Q)$ to a Q-independent electronic basis

$$a_i(Q) = \sum_j \langle \varphi_i(Q) | \varphi_j(0) \rangle a_j(0) \ . \tag{86}$$

We insert this expression in eq. (84) for H_{EN}, expand $|\varphi_i(Q)\rangle$ in a Taylor series and collect all terms linear in Q

$$\begin{aligned} H = & \sum_s \omega_s (b_s^+ b_s + \tfrac{1}{2}) + \sum_i \varepsilon_i(0) a_i^+ a_i \\ & - \sum_{s,i} \kappa_s^{(0)}(i) \left[a_i^+ a_i - n_i \right] (b_s + b_s^+) \\ & + \sum_{i<j} \sum_s \lambda_s^{(0)}(i,j) \left[a_i a_j^+ + a_j a_i^+ \right] (b_s + b_s^+) \ . \end{aligned} \tag{87}$$

Here the fermion operators do not depend on the nuclear coordinates. The constant term has been set equal to zero and

$$\lambda_s^{(0)}(i,j) = \lim_{Q\to 0} \frac{1}{\sqrt{2}} (\varepsilon_i(Q) - \varepsilon_j(0)) \times \times \langle \varphi_j(Q) | \frac{\partial}{\partial Q_s} | \varphi_i(Q) \rangle \ . \tag{88}$$

By restricting the orbital space to $\{\varphi_1, \varphi_2\}$ we can obtain another form for λ as will be seen below

$$\lambda_s^{(0)}(i,j) = \frac{1}{4} \left\{ \left(\frac{\partial^2 (\varepsilon_1(Q) - \varepsilon_2(Q))^2}{\partial Q_s^2} \right)_0 \right\}^{1/2} \ . \tag{89}$$

The above Hamiltonian contains adiabatic and nonadiabatic coupling terms. Many of the coupling constants vanish on symmetry grounds. The selection rule for nonvanishing adiabatic coupling constants, κ_s, is

$$\Gamma_i \times \Gamma_i \times \Gamma_s \supset \Gamma_A \tag{90}$$

and for the nonadiabatic ones (λ_s)

$$\Gamma_i \times \Gamma_j \times \Gamma_s \supset \Gamma_A \quad , \tag{91}$$

where Γ_i is the representation of MO $|\varphi_i\rangle$, Γ_s the representation of Q_s and Γ_A the totally symmetric representation.

Up to now we have given the Hamiltonian in the one-particle approximation. As discussed above many-body effects can be taken into account by using renormalized coupling constants.

$$\kappa_s(i) = -\frac{1}{\sqrt{2}}\left(\frac{\partial E_i}{\partial Q_s}\right)_o$$

$$\lambda_s(1,2) = \frac{1}{4}\left\{\left(\frac{\partial^2 (E_1(Q)-E_2(Q))}{\partial Q_s^2}\right)_o\right\}^{1/2} \quad . \tag{92}$$

The E_i are the poles of the one-particle Green's function.

In many cases it is sufficient to restrict oneself to two close lying ionic states of different symmetry which we denote by $|\varphi_1\rangle$ and $|\varphi_2\rangle$. If in addition only one totally symmetric vibration α and one non-totally symmetric vibration β couple to the electronic motion then the Hamiltonian takes the form

$$\begin{aligned} H = {} & \omega_\alpha (b_\alpha^+ b_\alpha + \tfrac{1}{2}) + \omega_\beta (b_\beta^+ b_\beta + \tfrac{1}{2}) + E_1 a_1^+ a_1 + E_2 a_2^+ a_2 \\ & + (\kappa_1 a_1 a_1^+ + \kappa_2 a_2 a_2^+)(b_\alpha + b_\alpha^+) \\ & + \lambda (a_1 a_2^+ + a_2 a_1^+)(b_\beta + b_\beta^+) \quad . \end{aligned} \tag{93}$$

This Hamiltonian underlies the calculation of the vibrational structure of C_4H_4.

We now have to evaluate $P(\omega)$ for this case.

$$P(\omega) = \int dt e^{i\omega t} \langle \Phi_o | T^+ T(t) | \Phi_o \rangle \, , \tag{94}$$

where $T = \sum_i \tau_{ei} a_i$. The a_i are taken in the Q independent basis. The dipole matrix elements τ_{ei} between a continuum and an occupied MO are independent of Q as well. In our specific case:

$$T = \tau_{e1} a_1 + \tau_{e2} a_2 \tag{95}$$

The ground state $|\Phi_o\rangle$ of $\underline{H}$ is a product of an electronic ground state $|\Psi_o\rangle = a_2^+ a_1^+ |\text{vacuum}\rangle$ and a nuclear ground state $|0\rangle$ of two harmonic oscillators with frequencies ω_α and ω_β. We now wish to introduce pure bosonic operators. This can be done by the same procedure used above. Due to the term involving λ, which is nondiagonal in the electronic operators, one obtains, however, a matrix boson operator.

$$\begin{aligned} H a_1 |\Phi_o\rangle &= \tilde{\mathcal{H}}_{11} a_1 |\Phi_o\rangle + \tilde{\mathcal{H}}_{12} a_2 |\Phi_o\rangle \\ H a_2 |\Phi_o\rangle &= \tilde{\mathcal{H}}_{21} a_1 |\Phi_o\rangle + \tilde{\mathcal{H}}_{22} a_2 |\Phi_o\rangle \end{aligned} \tag{96}$$

or

$$H \begin{pmatrix} a_1 \\ a_2 \end{pmatrix} |\Phi_o\rangle = \underline{\tilde{\mathcal{H}}} \begin{pmatrix} a_1 \\ a_2 \end{pmatrix} |\Phi_o\rangle \tag{97}$$

and

$$e^{iHt} \begin{pmatrix} a_1 \\ a_2 \end{pmatrix} |\Phi_o\rangle = e^{i\underline{\tilde{\mathcal{H}}}t} \begin{pmatrix} a_1 \\ a_2 \end{pmatrix} |\Phi_o\rangle \, , \tag{98}$$

where $\underline{\tilde{\mathcal{H}}} = \underline{\mathcal{H}} + (E_1+E_2)\underline{1}$ and

$$\underline{\mathcal{H}} = \left[\omega_\alpha (b_\alpha^+ b_\alpha + \tfrac{1}{2}) + \omega_\beta (b_\beta^+ b_\beta + \tfrac{1}{2})\right] \underline{1} + \begin{bmatrix} -E_1 + \kappa_1 (b_\alpha + b_\alpha^+) & -\lambda (b_\beta + b_\beta^+) \\ -\lambda (b_\beta + b_\beta^+) & -E_2 + \kappa_2 (b_\alpha + b_\alpha^+) \end{bmatrix} . \tag{99}$$

This can be used to obtain a compact form for the transition probability

$$P(\omega) = \int dt e^{i\omega' t} \underline{\tau}^{+} \langle \underline{0} | e^{i\mathcal{H}t} | \underline{0} \rangle \underline{\tau}, \quad (100)$$

where $\underline{\tau} = \begin{pmatrix} \tau_{e1} \\ \tau_{e2} \end{pmatrix}$ and $|\underline{0}\rangle = \begin{pmatrix} |0\rangle & 0 \\ 0 & |0\rangle \end{pmatrix}$

and $\omega' = \omega - \frac{1}{2}\omega_\alpha - \frac{1}{2}\omega_\beta$.

The Hamiltonian $\mathcal{H}$ describes the nuclear motion in the ionic states. The adiabatic potential surfaces V_a and V_b associated with the ionic states are obtained by diagonalizing $\mathcal{H}-T_N$:

$$2V_{a,b} = (\omega_\alpha Q_\alpha^2 + \omega_\beta Q_\beta^2) - (E_1+E_2) + \sqrt{2}(\kappa_1+\kappa_2)Q_\alpha$$

$$\pm\left\{\left[(E_1-E_2) - \sqrt{2}(\kappa_1-\kappa_2)Q_\alpha\right]^2 + 8\lambda^2 Q_\beta^2\right\}^{1/2}. \quad (101)$$

From the difference $V_a - V_b$ the equation (92) for λ can be derived.

For a fixed value of Q_α, say $Q_\alpha=0$, the upper potential curve $V_a(Q_\beta)$ has a single minimum at $Q_\beta=0$. The lower potential curve $V_b(Q_\beta)$, on the other hand, has two minima at

$$Q_\beta^2 = \frac{1}{8}\lambda^{-2}\left[16\lambda^4/\omega_\beta^2 - (E_1-E_2)^2\right] \quad (102)$$

if the relation $4\lambda^2/\omega_\beta > |E_1-E_2|$ is fulfilled. The double minimum character of V_b indicates that the vibrational structure will be rather complicated. In addition one should note that for this range of the coupling constant it is a poor approximation to solve separately the Schrödinger equations for the Hamiltonians

$$H_{a,b} = \frac{1}{2}\omega_\alpha \frac{\partial^2}{\partial Q^2} + \frac{1}{2}\omega_\beta \frac{\partial^2}{\partial Q^2} + V_{a,b} \quad (103)$$

and to calculate the intensities via the Franck-Condon principle. The eigenvectors which diagonalize $\mathcal{H}-T_N$ do not commute with the kinetic energies in $\mathcal{H}$. This leads to a complicated non-diagonal kinetic energy matrix operator. The normal modes decouple only if $\kappa_1 = \kappa_2$. The strongest mixing of the modes is found if κ_1 and κ_2 are equal in magnitude but have different sign. This turns out to be the case for butatriene.

$P(\omega)$ cannot be solved for analytically. We therefore solve for it numerically. We calculate the eigenvalues and eigenvectors of $\mathcal{H}$ in a basis of unperturbed harmonic oscillator states $\{|n_\alpha m_\beta\rangle\}$. One first obtains a supermatrix with 2x2 matrices $\mathcal{H}_{n_\alpha m_\beta, n'_\alpha m'_\beta}$ as elements. The elements are obtained via

$$\mathcal{H}_{n_\alpha m_\beta, n'_\alpha m'_\beta} = \begin{pmatrix} \langle n_\alpha m_\beta| & 0 \\ 0 & \langle n_\alpha m_\beta| \end{pmatrix} \mathcal{H} \begin{pmatrix} |n'_\alpha m'_\beta\rangle & 0 \\ 0 & |n'_\alpha m'_\beta\rangle \end{pmatrix} \tag{104}$$

The supermatrix decouples into two submatrices $\mathcal{H}_1$ and $\mathcal{H}_2$. $\mathcal{H}_2$ is obtained from $\mathcal{H}_1$ by interchanging all indices 1 and 2. The eigenvalues of $\mathcal{H}_1$ and $\mathcal{H}_2$ give the positions of the vibrational lines in the spectrum and as we expand the vibrational wavefunctions in ground state harmonic oscillator wavefunctions the intensity of a line in the spectrum is obtained by multiplying the square of the first element of the corresponding eigenvector by $|\tau_{e1}|^2$ or $|\tau_{e2}|^2$ for a line emerging from $\mathcal{H}_1$ or $\mathcal{H}_2$, respectively.

Butatriene belongs to the D_{2h} symmetry group. The two lowest ionic states are definitely the $^2B_{3g}$ ground state and the $^2B_{3u}$ first excited state. The next ionic state $^2B_{2u}$ is energetically well separated and is thus excluded from the calculation. In order to study the vibrational structure in the low energy part of the PES we have to include the four totally symmetric vibrational modes as well as the mode of symmetry species A_u which couples the ionic states $^2B_{3u}$ and $^2B_{3g}$ ($B_{3u} \times B_{3g} \times A_u = A_g$). This problem would be of defeating complexity if all coupling constants were nonvanishing.

The coupling constants are evaluated first in internal symmetry coordinates S_m and then transformed to normal coordinates Q_m

$$-\kappa_m(i) = \frac{1}{\sqrt{2}}\left(\frac{\partial E_i}{\partial Q_m}\right)_o = \left(\frac{\hbar}{2\omega_m}\right)^{1/2} \sum_n L_{nm}\left(\frac{\partial E_i}{\partial S_n}\right)_o . \tag{105}$$

The matrix $\underline{L}$ is obtained from the force field of butatriene. As this information is unavailable it was pieced together from the force field of ethylene and the frequencies of C_4H_4. Details are given in ref.80. HF calculations of the energies and coupling constants showed that only the mode ν_2 couples strongly, all other adiabatic coupling constants are negligible: $E(^2B_{3g})=-9.00eV$, $E(^2B_{3u})=-10.31eV$, $\lambda(A_u)=0.26eV$, $\kappa_2(^2B_{3g})=-0.26eV$, $\kappa_2(^2B_{3u})=0.29eV$. These quantities will be abbreviated by E_1, E_2, λ, κ_1, κ_2, respectively. We are faced with the situation described by the simplified Hamiltonian (93). Many-body calculations on energies and coupling constants have also been performed. They reduce the energy split E_1-E_2 as well as the absolute magnitude of the coupling constants. The energies and coupling constants which enter the calculation of the spectrum are listed in table 11. There are four sets of calculations.

Table 11

The energies and coupling constants used in the calculations of the PES of butatriene HF=Hartree-Fock, MB= many-body, I=freely chosen energies E_1 and E_2 and the coupling constants from MB, II=best fit. All quantities in eV if not otherwise stated

	HF	MB	I	II
$-E_1$	9.00	8.95	9.4	9.45
$-E_2$	10.31	9.67	9.9	9.85
κ_1	-0.26	-0.19	-0.19	-0.15
κ_2	0.29	0.23	0.23	0.18
λ	0.26	0.26	0.26	0.225

$\omega_\alpha = 2079\ cm^{-1}$ (83), $\omega_\beta = 736\ cm^{-1}$ (83)

To be able to compare with experiment we have drawn the calculated spectra in fig.16 A-D representing the vibrational lines by Lorentzians of width (hwhm) 0.025 eV and have assumed $|\tau_{e1}|^2 = |\tau_{e2}|^2$. The spectra in figs. A and B have been computed using the HF and many-body results, respectively. It is seen that the vibronic coupling effects influence the first band only little, i.e. the individual vibrational lines of mode α show only little structure due to interaction with mode β. This is due to the large energy difference E_1-E_2 obtained in the HF approximation. By including many-body effects this difference becomes smaller, the modes α and β interact stronger and the more complex structure shown in fig.B results. The distance between the most intense peaks is still too large in the calculated spectrum. This is mainly due to basis set deficiencies as discussed in ref.(82). In particular, E_1 will be lowered by using extended basis sets. In fig.C we have thus chosen E_1=-9.4eV and E_2=-9.9eV and otherwise taken the many-body results. The fine structure of the first band in the experimental spectrum is well reproduced and it is seen that a third band emerges between the two bands. This band clearly represents the mystery band.

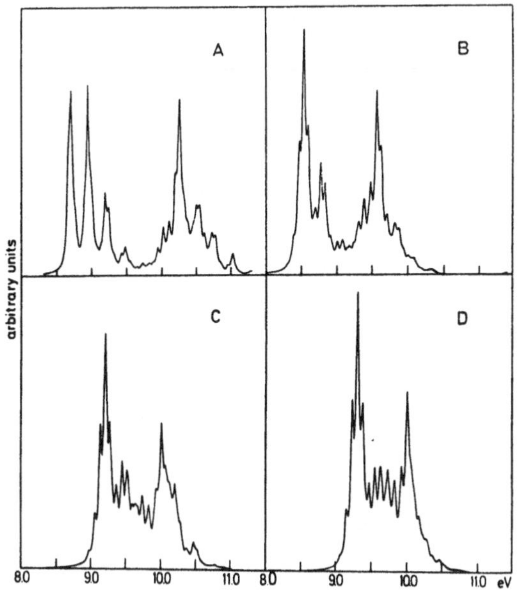

Fig.16: The calculated PES of butatriene (see text)

In order to find out how accurately the experimental spectrum can be reproduced with the Hamiltonian (93) we have varied the energies and coupling constants to obtain the best fit. This best result is plotted in fig.D and the best fit parameters are listed in table 11. The result is very satisfactory. The number of peaks in bands 1 and 1' is accurately reproduced as is also the onset of band 2 and the peculiar diffuse structure of this band.

The best fit coupling constants are quite close to those obtained from the many-body calculation. The calculated spectrum is very sensitive to changes in the coupling constants and especially in the energy difference.

To gain some further insight into the coupling mechanism we have performed three more calculations whose results are plotted in fig.17 A to C.

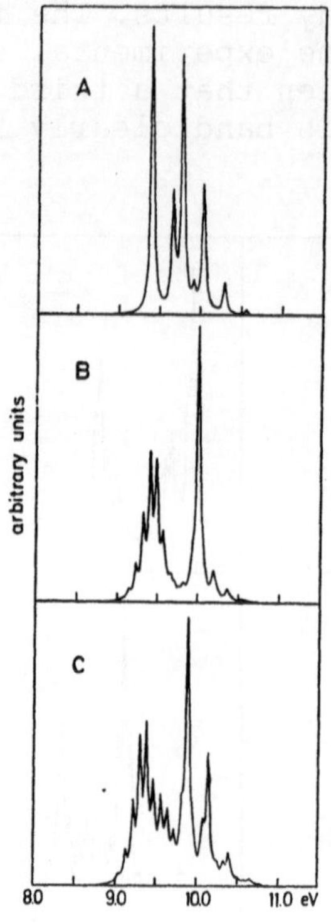

Fig.17: Investigation of the influence of the individual coupling constants on the first band system of butatriene

In the first calculation the best fit parameters are used, but λ is set to zero, i.e. no vibronic coupling is allowed for. Neglecting κ_1 and κ_2 but including λ leads to the spectrum in fig.B. Both spectra have little in common with the experimental one. In the last calculation the sign of κ_1 is reversed (fig.C). Again a rather different spectrum results. These results underline the fact that the mystery band arises from a rather subtle interplay between the totally and non-totally symmetric vibrations.

IV. The 2-particle-hole Tamm-Dancoff approximation (12, 28)

In the first chapters we have presented the Green's function method as it can be applied to the calculation of outer valence IP's, i.e. to IP's which are far from the poles of the self-energy. These IP's have pole strengths close to unity. A consequence is that there is one line in the PES for each MO. Satellite lines will accompany these main lines, but they have only small relative intensities. The corresponding electronic states are of a complicated nature. These solutions lie in the pole region of the self-energy. The inner valence IP's will lie in this region too. To calculate these IP's and the satellite lines a method is required which takes the pole structure of the self-energy into account. Quite new phenomena may occur in this energy range. We are going to see that in the inner valence region the one-particle picture of ionization may break down completely. The familiar concepts useful in the outer valence region and in the core are without validity here. The method used for these calculations is the 2-particle-hole Tamm-Dancoff approximation (2ph-TDA).

The exact self-energy has a constant and an energy-dependent term which we denote by $M(\omega)$

$$\Sigma(\omega) = \Sigma(\infty) + M(\omega) \quad . \tag{106}$$

As $\Sigma(\infty)$ can be obtained from $M(\omega)$ (see ref.12) we investigate only $M(\omega)$ here. The straightforward perturbation expansion of the self-energy is certainly limited to low orders because of the rapidly increasing effort involved in calculating higher orders. The lowest order energy-dependent term is the second order term $M^{(2)}(\omega)$

$$M^{(2)}_{pq}(\omega) = \frac{1}{2} \sum_{jk\ell} V_{pj[k\ell]} V_{qj[k\ell]} \times$$

$$\times \left\{ \frac{\bar{n}_j n_k n_\ell}{\omega+\varepsilon_j-\varepsilon_k-\varepsilon_\ell-i\eta} + \frac{n_j \bar{n}_k \bar{n}_\ell}{\omega+\varepsilon_j-\varepsilon_k-\varepsilon_\ell+i\eta} \right\}$$

$$= M^{I}_{pq} + M^{II}_{pq} \quad . \tag{107}$$

This expansion up to second order has the analytical structure of the exact $M(\omega)$. Unfortunately $M^{(2)}$ has, in general, proved to provide only a poor approximation (27). The extension to the third order or any higher finite order will destroy the simple spectral structure of the second order expression, since quadratic poles are added already by $M^{(3)}(\omega)$. It is thus clear that an expansion up to a finite order will completely fail to describe $M(\omega)$ in the neighbourhood of its poles, whereas it might be useful if a region far from the poles is only of interest. The latter is the case for the outer valence IP's of closed shell systems.

We have to look for a structure conserving approximation, that is, an approximation for $M(\omega)$ exhibiting the spectral form reflected by the exact self-energy. Such an approximation can only be obtained by some kind of infinite partial summation. Well-known examples of partial summations of diagrams are the summation of all ring diagrams (RPA approximation) and of all ladder diagrams. Unfortunately neither the RPA nor the ladder self-energy parts are satisfactory for finite electronic systems. A simple addition of the RPA and ladder series indeed contains all third order terms of $M(\omega)$ that have been found to be essential, but leads to negative intensities (28).

Let us consider the second order expression eq. 107 (see also the diagrams given in chapter I). The poles of M^{I} describe ionic states where one electron is removed from the occupied MO $k(\ell)$ and another one is excited from the occupied MO $\ell(k)$ into the unoccupied orbital j. Analogously the poles M^{II} correspond to an electron attachment process to an unoccupied orbital accompanied by a simultaneous particle-hole excitation. The corresponding states have the configuration $k^{-1}\ell^{-1}j$ and $j^{-1}k\ell$, respectively. By solving the Dyson equation with $M^{(2)}$ these configurations are allowed to interact with the one-hole and one-particle configurations p^{-1} and p which are described by G^{o}. The resulting ionic states now include correlation effects.

The mechanism leading to the breakdown to the MO picture of ionization can also be understood based on these considerations. A configuration $k^{-1}\ell^{-1}j$ where k and ℓ specify occupied outer valence orbitals and j a low lying unoccupied orbital can be nearly degenerate with the configuration p^{-1} of a more deeply lying valence orbital p. In this case strong configuration interaction can take place (this depends on the value of $V^2_{pjk\ell}$) and the two resulting ionic states will have a considerable contribution of the p^{-1} configuration. The lines in the spectrum will then have comparable intensities. In actual calculations we often find several poles of M^{I}_{pp} to be nearly degenerate with the one-particle energy ε_p. The intensity originally corresponding to the p orbital (and thus to a single line in the MO picture) is about equally distributed over many lines. It is then not possible to identify any of these lines as the "main" line representing the ionization of the orbital p and the remaining ones as satellite lines. At this place we wish to mention that in some cases two (or more) one-hole configurations p^{-1} and p'^{-1} interact indirectly via a configuration of the type $k^{-1}\ell^{-1}j$. As a consequence the Green's function becomes nondiagonal and the resulting line corresponding to the ionic state which develops from the $k^{-1}\ell^{-1}j$ configuration cannot be thought of as arising from the ionization of a specific MO.

As mentioned before the second order expression for M is in general not adequate to obtain accurate results. The 2h1p(2p1h) excitations are free and their energies are thus too large by about 10 to 20eV to describe adequately the energies of ionization with simultaneous excitation. In higher orders additional interactions come into play which can be considered to renormalize these excitations. Such processes can be taken into account to infinite order in the electronic interaction by summing certain classes of diagrams. We consider the 2h1p case. The 2h1p diagrams arise from the second order diagram by successively introducing further electronic interaction points. Thereby one can connect either lines 1 and 2 or 2 and 3 or 1 and 3. This construction principle leads to a simple recursion formula. We separate $M_{pq}(\omega)$ into its external vertices and the remainder.

$$M_{pq}(\omega) = \frac{1}{4} \sum_{\substack{jk\ell \\ j'k'\ell'}} V_{qj[k\ell]} \Gamma_{jk\ell,j'k'\ell'}(\omega) V_{pj'[k'\ell']} \qquad (108)$$

thus defining the kernel of the self-energy, Γ. As Γ is antisymmetric in k and ℓ and in k' and ℓ'

$$\Gamma_{jk\ell,j'k'\ell'} = -\Gamma_{j\ell k,j'k'\ell'} = -\Gamma_{jkl,j'\ell'k'}$$
$$= \Gamma_{j\ell k,j'\ell'k'} \qquad (109)$$

it follows that

$$M_{pq}(\omega) = \sum_{\substack{jk\ell \\ j'k'\ell'}} V_{qjk\ell}\,\Gamma_{jk\ell,j'k'\ell'}(\omega)\,V_{pj'k'\ell'} \tag{110}$$

or in matrix notation

$$M_{pq}(\omega) = \underline{V}_q^T\underline{\Gamma}(\omega)\,\underline{V}_p \quad . \tag{111}$$

From the definition of Γ it is clear that it also has a spectral representation similar to $M(\omega)$ and is subject to an expansion in powers of $V_{ij[k\ell]}$, where, however, the n-th order term $\Gamma^{(n)}$ corresponds to the (n+2) -th term of the self-energy. The zero-th order term is obtained to be

$$\Gamma^{(o)}_{jk\ell,j'k'\ell'} = \gamma^{o}_{jk\ell,j'k'\ell'}\left[\frac{\bar{n}_j n_k n_\ell}{\omega+\varepsilon_j-\varepsilon_k-\varepsilon_\ell-i\eta} + \frac{n_j\bar{n}_k\bar{n}_\ell}{\omega+\varepsilon_j-\varepsilon_k-\varepsilon_\ell+i\eta}\right] \quad , \tag{112}$$

where

$$\gamma^{o}_{jk\ell,j'k'\ell'} = \delta_{jj'}(\delta_{kk'}\delta_{\ell\ell'}-\delta_{k\ell'}\delta_{\ell k'}) \quad . \tag{113}$$

We shall omit the infinitesimal quantity η in what follows and restrict ourselves to the 2h1p and 2p1h space in the whole development. We further introduce the diagonal matrix

$$(\omega\underline{1}-\underline{K})_{jk\ell,j'k'\ell'} = \delta_{jj'}\delta_{kk'}\delta_{\ell\ell'}(\omega+\varepsilon_j-\varepsilon_k-\varepsilon_\ell) . \tag{114}$$

$\Gamma^{(o)}$ can then be written as

$$\Gamma^{(o)}(\omega) = (\omega\underline{1}-\underline{K})^{-1}\underline{\gamma}^{o} \tag{115}$$

or in diagrammatic form (for the 2h1p case)

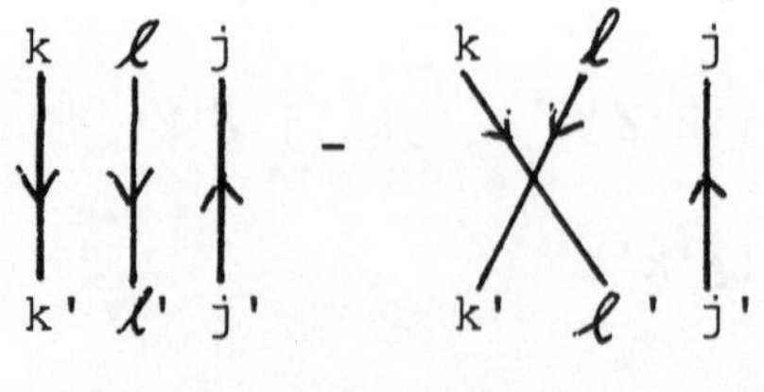

(116)

where the second term is the exchange part. Evaluation by the diagram rules leads immediately to the algebraic form given above. In the next order one obtains in the 2h1p case six diagrams. By connecting e.g. lines k and ℓ one obtains the two diagrams

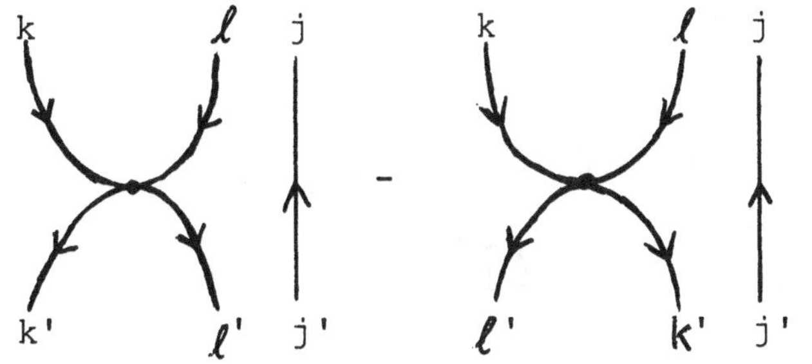

(117)

By the diagram rules one obtains for these two diagrams

$$(\omega+\varepsilon_j-\varepsilon_k-\varepsilon_\ell)^{-1} V_{k\ell[k'\ell']} (\omega+\varepsilon_{j'}-\varepsilon_{k'}-\varepsilon_{\ell'})^{-1} \gamma^o . \quad (118)$$

Including the other terms as well one obtains instead of the V term a matrix C which contains combinations of the V terms. Thus

$$\Gamma^{(1)} = (\omega-K)^{-1} C \Gamma^{(o)} , \quad (119)$$

where

$$C_{jk\ell,j'k'\ell'} = -\delta_{jj'} V_{k\ell[k'\ell']} + \delta_{kk'} V_{j\ell'[j'\ell]} + \delta_{\ell\ell'} V_{jk'[j'k]} \quad (120)$$

from which the recursion formula for Γ can easily be derived

$$\Gamma^{(n+1)} = (\omega-K)^{-1} C \Gamma^{(n)} . \quad (121)$$

In the same way one can obtain a Dyson like equation for Γ. Let $\Gamma_{j''k''\ell'',j'k'\ell'}(\omega)$ be represented by the diagram

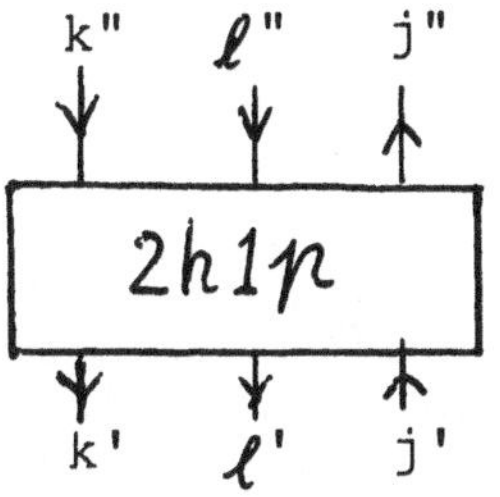

(122)

We can add an interaction point in three possible ways, e.g.

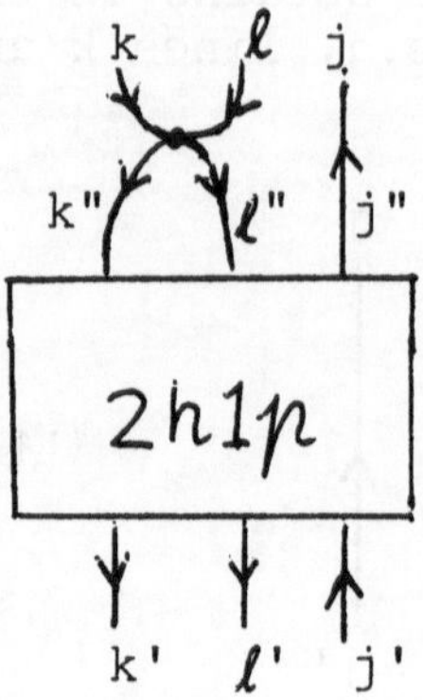

$$= (\omega+\varepsilon_j-\varepsilon_k-\varepsilon_\ell)^{-1}\delta_{jj"} \times$$

$$\times V_{k\ell[k"\ell"]}\Gamma_{j"k"\ell",j'k'\ell'} \qquad (123)$$

and by adding the other possible diagrams one obtains

$$(\omega-K)^{-1}C\Gamma. \qquad (124)$$

With each added interaction point a factor $(\omega-K)^{-1}C$ is added. Continuing in this way one obtains a geometric series in $\Gamma^{(o)}C$. As Γ is an infinite summation with first term $\Gamma^{(o)}$ we have

$$\Gamma = \Gamma^{(o)} + \Gamma^{(o)}C\Gamma = (\omega-K)^{-1}\gamma^{o} + (\omega-K)^{-1}C\Gamma \qquad (125)$$

or in diagrammatic form

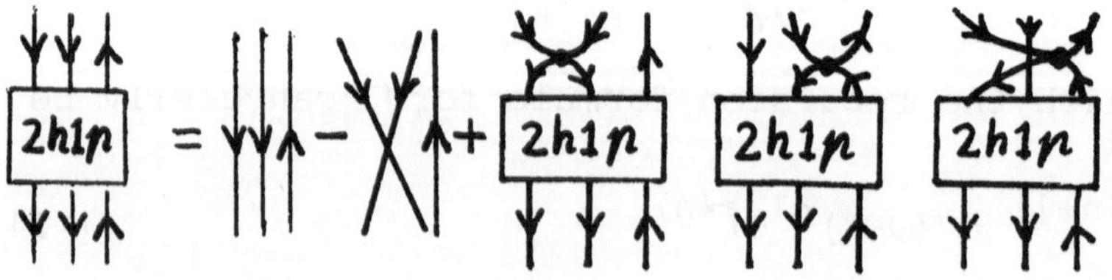

(126)

This is the Dyson like equation for Γ. This equation can be solved to give

$$\Gamma(\omega) = (\omega-K-C)^{-1}\gamma^{o} \quad . \qquad (127)$$

Analogous equations result in the 2p1h case. The two parts of Γ may be denoted as Γ^{I} and Γ^{II} with corresponding matrices C^{I} and C^{II} for the 2h1p and 2p1h case, respectively. (For a more complete derivation and the general RPA equation see ref.12)

Taking these two parts of Γ together and adding the external vertices to form $M(\omega)$ we have summed over all diagrams with two hole and one particle line and with two particle and one hole line between any two interaction points. The summation includes all ladder and RPA type diagrams to infinite order and is the simplest approximation which gives the correct analytical structure of the self-energy. Having obtained the self-energy via an infinite summation we are left with having to solve the Dyson equation. By solving the Dyson equation the 1h and 1p processes are coupled to the 2h1p and 2p1h processes. In this way we obtain meaningful excitations in G.

There exists a connection between the present approach and a wavefunction approach. If we leave the 2p1h part of the self-energy away the Dyson equation is equivalent to CI with single excitations, i.e. no ground state correlation is included. More precisely, the exact $(N\pm 1)$ electron wavefunctions can be expanded according to

$$|\Psi_s^{N-1}\rangle = \sum_i \alpha_i^s a_i |\Psi_o^N\rangle + \sum_{ijk} \alpha_{ijk}^s a_i^+ a_j a_k |\Psi_o^N\rangle + \dots \tag{128}$$

$$|\Psi_s^{N+1}\rangle = \sum_i \beta_i^s a_i^+ |\Psi_o^N\rangle + \sum_{ijk} \beta_{ijk}^s a_i^+ a_j^+ a_k |\Psi_o^N\rangle + \dots .$$

In the 2ph-TDA the ground state $|\Psi_o^N\rangle$ is replaced by the unperturbed ground state $|\Phi_o^N\rangle$. Since one is only interested in such ionic states characterized by a set of three indices $s=\{ijk\}$ the α_i^s (β_i^s) and $\alpha_{ijk\ell m}^s$ $(\beta_{ijk\ell m}^s)$ and higher coefficients are set equal to zero in the TDA self-energy part. Considering the α_i^s, α_{ijk}^s and β_i^s, β_{ijk}^s as variational parameters the energies $E_s^{N-1}-E_o^N(\text{HF})$ and $E_s^{N+1}-E_o^N(\text{HF})$ are obtained as roots of secular equations equivalent to the Dyson equation without the 2p1h or 2h1p self-energy parts, respectively. Solving the Dyson equation with both self-energy parts one takes account of the α_i^s (β_i^s) and higher excitations and of correlation in $|\Psi_0^N\rangle$, although the ground state correlation will still be included to an insufficient degree in general. This will result in the outer valence IP's being too small by 0 to 1.5eV. Ionic state correlation effects, however, dominate the processes in which we are interested and thus the 2ph-TDA should be adequate to obtain at least a qualitatively correct ionization spectrum over the whole energy range.

Some technical considerations

Let us start from a self-energy of the form

$$\Sigma_{pq}(\omega) = A_{pq} + \sum_n \frac{\chi_{pn}\chi^*_{qn}}{\omega-\alpha_n-i\eta} + \sum_m \frac{\kappa_{pm}\kappa^*_{qm}}{\omega-\beta_m+i\eta} . \qquad (129)$$

With the aid of partitioning it is easily shown that solving the Dyson equation is equivalent to solving the eigenvalue equation for the matrix

$$\underline{Y} = \begin{pmatrix} \underline{\varepsilon} + \underline{A} & \underline{\chi} & \underline{\kappa} \\ \underline{\chi}^+ & \underline{\alpha} & \underline{0} \\ \underline{\kappa}^+ & \underline{0} & \underline{\beta} \end{pmatrix} \qquad (130)$$

where $\underline{A}$, $\underline{\chi}$, $\underline{\kappa}$ are matrices with elements A_{pq}, χ_{pn} and κ_{pm}, respectively and $\underline{\varepsilon}$, $\underline{\alpha}$, $\underline{\beta}$ are diagonal matrices with the orbital energies and the poles of the self-energy as elements. The eigenvalues of $\underline{Y}$ are the poles of the Green's function and the p-th component of the corresponding eigenvectors are the amplitudes of G. More generally $G_{pq}=(\omega 1-Y)^{-1}_{pq}$. In the 2ph-TDA the α_n and β_m are the eigenvalues of $K+C^{I}$ and $K+C^{II}$, respectively. In this case one can combine the equations $\Gamma^{I,II}=(\omega-K-C^{I,II})\gamma^{o}$ and eq.(130) to give $G_{pq}=(\omega\underline{1}-\underline{Z})^{-1}_{pq}$ with

$$\underline{Z} = \begin{pmatrix} \underline{\varepsilon} & \underline{V}^{I} & \underline{V}^{II} \\ \underline{V}^{I+} & \underline{K}+\underline{C}^{I} & \underline{0} \\ \underline{V}^{II+} & \underline{0} & \underline{K}+\underline{C}^{II} \end{pmatrix} \qquad (131)$$

where $\underline{V}^{I}$ and $\underline{V}^{II}$ are matrices with elements $V_{pj[k\ell]}$ with $\bar{n}_j n_k n_\ell=1$ and $n_j \bar{n}_k \bar{n}_\ell=1$, respectively.

We now have three different ways to obtain the Green's function. The first method is to calculate the self-energy first and then to solve directly the Dyson equation using e.g. the pole search procedure. The other two methods involve diagonalization of matrices either by calculating the self-energy first followed by the diagonalization of $\underline{Y}$ or by diagonalizing $\underline{Z}$ directly. At a first glance one may prefer the last method because only one diagonalization is required. This is, however, misleading. The matrices C^{I} and C^{II} are usually of large dimensions which renders a direct diagonalization of $\underline{Z}$ difficult if many orbitals are to be included. The first two methods take advantage of the fact that M^{I} and M^{II} can be calculated separately in the TDA and that M^{II} which involves the usually larger matrix C^{II} has much less influence on the calculated IP's than M^{I}. The latter fact is especially important. One can reduce the dimensions of $\underline{Y}$ considerably by replacing $\underline{\beta}$ and $\underline{\kappa}$ by a few effective poles and amplitudes which approximate M^{II} well in the energy region of ionization. Compared to the others the first method is found to be the fastest. It has, however, two disadvantages. One may miss solutions due to numerical problems because of the strange shape $M(\omega)$ may have and it does not provide the complete eigenvectors of $\underline{Y}$ which are important for interpreting the computational results.

It should be noted that if the Green's function matrix is diagonal, the self-energy is also diagonal and one obtains a matrix $\underline{Y}$ as in eq. (130) for each orbital p. $\underline{\varepsilon}$ and $\underline{A}$ are then reduced to one element each: ε_p and A_{pp}. In this case the Dyson equation can be written for each orbital separately

$$G^{-1}_{pp}(\omega) = \omega - \varepsilon_p - \Sigma_{pp}(\omega) \quad . \tag{132}$$

We have performed the TDA calculations both in the full and diagonal versions of the Green's function. The Green's function is in general well approximated by its diagonal form and we will discuss only the results obtained in this way. A notice on what may happen if the full form of the Green's function is used has been made above already.

Application to CS (84)

As the first application we will discuss the CS molecule. The CS molecule is selected because the effects we wish to investigate occur already at fairly low energies and because its three valence σ-type IP's exhibit all effects which can occur. The calculations have been performed with the basis set (12s9p2d/11s7p 1d)/[7s5p2d/5s3p1d]. The basis set has been completely exhausted in the Green's function calculation, whereas in the TDA calculation the five occupied valence and the 15 lowest virtual orbitals have included. The PES of CS exhibits a curious phenomenon (fig.18) (85).

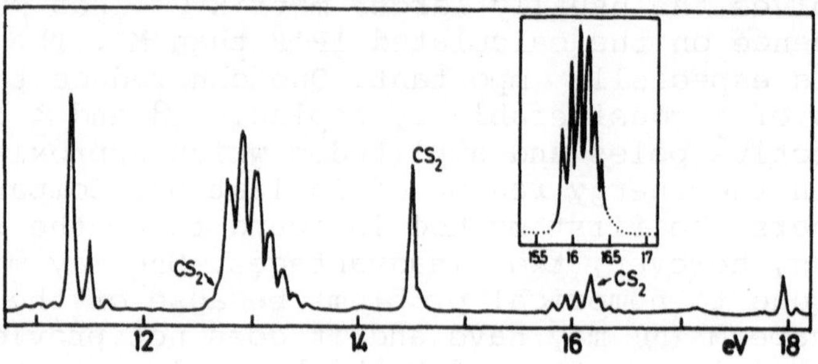

Fig.18: The PES of CS (85)

In the low energy region up to 21.21eV there should be 3 bands, but four are actually found (84). The first two IP's are 7σ and 2π, whereas Koopmans' theorem predicts the reverse ordering. The Green's function as well as the TDA calculations predicts the correct ordering. The nature of the next two bands is not clear from the experiment. Let us first discuss the 7σ orbital. The self-energy $\Sigma_{7\sigma 7\sigma}$ is graphically presented in fig.19. The first pole of Σ corresponds to the configuration $7\sigma^{-1} 2\pi^{-1} 3\pi$. The intersection of Σ with the line $Y=\omega-\varepsilon_{7\sigma}$ occurs in the main interval and leads to a large pole strength. There will be thus one main line and only unimportant satellite lines in the PES. One of these satellite lines corresponds to the configuration $7\sigma^{-1} 2\pi^{-1} 3\pi$ with a small admixture of $7\sigma^{-1}$. The situation is already different for the 6σ MO (fig.20). The intersection of $\Sigma_{6\sigma 6\sigma}$ with $Y=\omega-\varepsilon_{6\sigma}$ occurs in the main interval but close to the pole region of the self-energy and one finds actually two solutions with roughly equal pole strengths. The other satellite lines are unimportant. The main line is thus split into two lines of roughly equal intensity, thus explaining the experimentally observed phenomenon. Both ionic states have

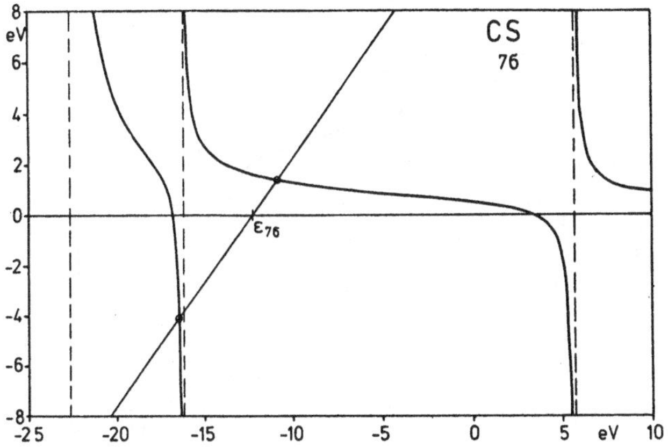

Fig.19: Schematic plot of $\Sigma_{7\sigma7\sigma}$ as a function of ω for CS

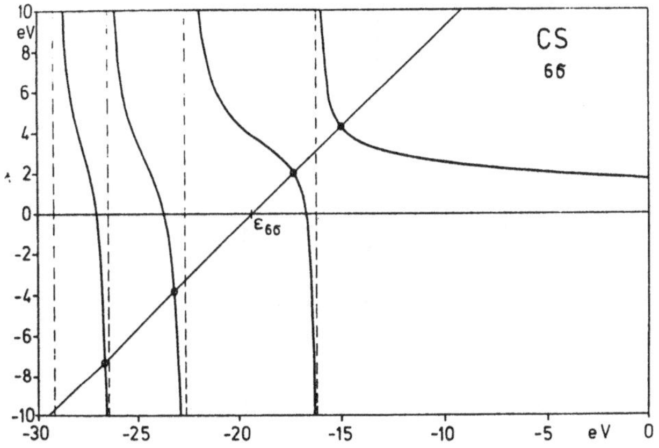

Fig.20: Schematic plot of $\Sigma_{6\sigma6\sigma}$ as a function of ω for CS

configurations which are mixtures of the configurations $7\sigma^{-1}$ and $7\sigma^{-1}2\pi^{-1}3\pi$. A one-particle picture of ionization is inappropriate. The next higher excitations are already mixtures of different 2h1p configurations.

The 5σ MO is a typical inner valence MO. Its energy lies amidst the poles of the self-energy (fig. 21) resulting in a strong interaction of the $5\sigma^{-1}$ configuration with various 2h1p configurations. The result is a complete breakdown of the MO picture of ionization. The intensity is distributed about evenly over many lines. In fig.22 we summarize the results

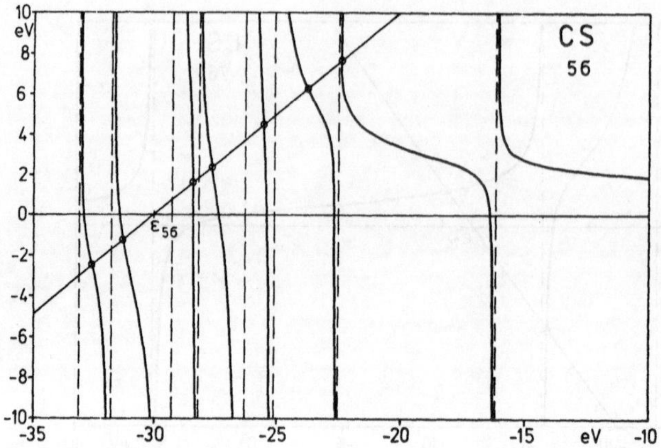

Fig.21: Schematic plot of $\Sigma_{5\sigma 5\sigma}$ as a function of ω for CS

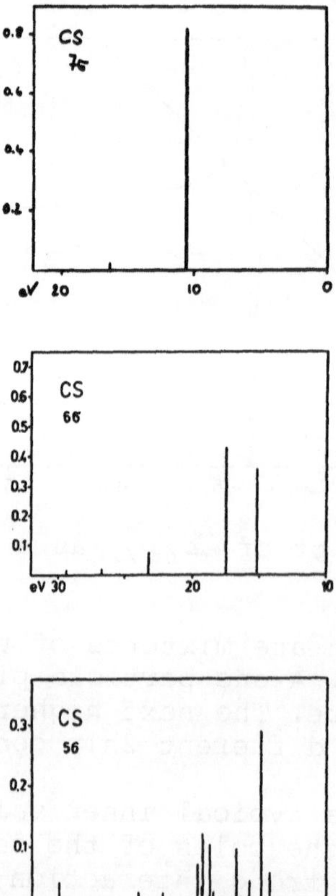

Fig.22: The calculated ionization spectra of CS (7σ, 6σ and 5σ MO's). Note the different ordinate scale for the 5σ spectrum

obtained for the valence σ ionization of CS in the form of a line spectrum. The height of the lines is given by the pole strengths. Note the different scale in the last spectrum corresponding to the 5σ MO.

Application to N_2 and CO

The first three IP's lie in the main interval of the self-energy and the MO picture of ionization applies. Here we are mainly interested in the $2\sigma_g$ MO of N_2 and the 3σ MO of CO. The calculations have been performed with two basis sets (9s5p)/[4s2p] (47) and (11s 7p)/[5s3p] (34). The results are qualitatively the same in both calculations and only the results obtained with the larger basis will be discussed (86). It should be noted that a further enlargement of the basis (e.g. by including polarization functions) may change the relative intensities and energies of the lines, but will not change the qualitative aspect, i.e. the breakdown of the MO picture. The results for the 3σ line of CO is shown together with the experimental spectrum (87) in fig.23 and for the "$2\sigma_g$ line" of N_2 in fig.24. (Experimental spectrum from ref.88). The calculated Franck-Condon envelops of the lines are drawn in the figures for a better comparison with experiment. For CO we find a main line at 39.9eV which carries most of the intensity (45%). It is accompanied by several more or less weak satellite lines. It is clear that the intense line at 39.9eV represents the ionization of an electron from the 3σ MO and that the satellite lines have borrowed intensity from the main line. It should be noted that the satellite lines appear on both sides of the main line. The ones at lower energy can obtain their intensity only via correlation effects. Recently Bagus and Viinikka (89) have performed CI calculations for the valence and inner valence region of CO in a limited model space of 5 occupied and 3 unoccupied orbitals. Apart from the absence of some intense satellite lines especially on the high energy side of the 3σ peak their results are qualitatively similar to the present ones. Their calculated inner valence IP's are 5 to 7eV higher than the experimental ones. This discrepancy is probably due to the missing of a sufficient number of virtual orbitals in the basis leading to a strong underestimation of relaxation effects. The errors in the present calculations are about 1 to 2eV.

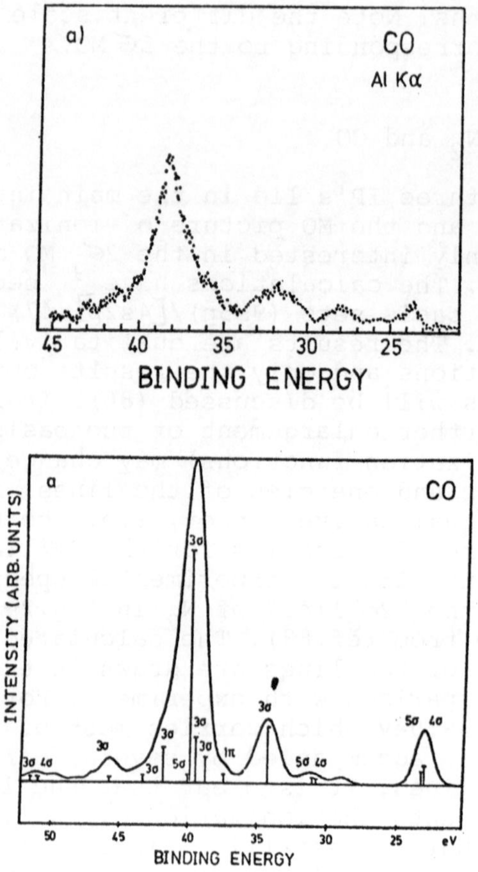

Fig.23: Calculated and observed inner valence spectrum of CO (87)

The situation is somewhat different for the N_2 molecule (fig.24). The main part of the intensity is distributed about evenly over three lines at 37.9, 39.9 and 41.6eV (86). This explains the large width and strange shape of the band found in the experiment. It is no longer possible to identify any one of these lines as the main line corresponding to the $2\sigma_g$ orbital. We cannot exclude with certainty that the 3σ line which survived in CO in the present calculation is also split into several closely spaced lines when the basis set is further enlarged. There are a number of satellite lines on both sides of the lines around 40eV. The major ones are indicated in the spectrum. It is also indicated from which MO's they derive their intensity.

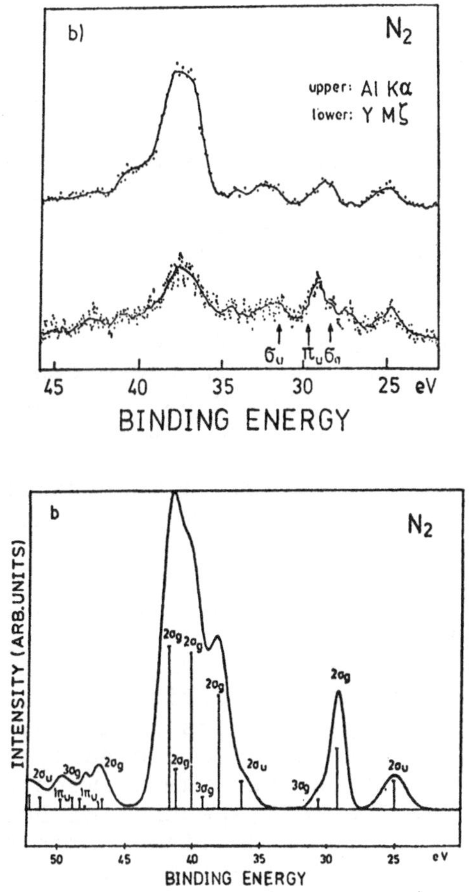

Fig.24: Calculated and observed inner valence spectrum of N_2 (88)

In the diagonal approximation to the self-energy the satellite at 28.9eV of N_2 derives its intensity from the $2\sigma_g$ orbital, but in the calculation using the full self-energy it obtains its intensity from both the $2\sigma_g$ and the $3\sigma_g$ orbital. As a consequence one can expect that the intensity of this line relative to the lines of pure $2\sigma_g$ character varies with incident photon energy as is in fact observed. It is seen from fig.5 of ref.88 that the cross section for $3\sigma_g$ ionization increases markedly when going from AlKα to YMζ radiation. This explains why the intensity of the line near 29eV increases relative to the intensity of the

$2\sigma_g$ lines when going from AlK_α to YM_ζ radiation. The same effect is observed in CO.

Application to formic acid (90)

The breakdown of the one-particle picture of ionization becomes even more prominent in the case of larger molecules. Unfortunately there are few spectra in the relevant energy range which are free of intensity, background and impurity problems. For formic acid, however, we could find a HeII PES (91) which at least partly covers the relevant energy range so that the theoretical predictions can be checked. The calculations on formic acid were performed with the basis set (9s5p1d/4s)/[4s2p1d/2s] (47). In the Green's function calculation the basis set was exhausted and in the TDA calculation only the 9 occupied orbitals and the 7 unoccupied orbitals lowest in energy were taken into account. This is due to the large dimension of the matrices which have to be completely diagonalized.

For the 10a', 2a", 9a', 1a", 8a' and 7a' outer valence orbitals of HCOOH the MO picture of ionization is retained. There is one intense line carrying about 90% of the spectral intensity. More interesting are the results for the inner valence orbitals. For the 6a', 5a' and 4a' orbitals there exists no longer a "main line". To exhibit the results more clearly the calculated vertical ionization spectrum between 20 and 40eV is shown in fig.25. The HeII spectrum of Potts et al. (91) is given in fig. 26. Each line represents an ionization process, the relative intensity being given by the pole strength. Satellite lines originating from the outer valence orbitals are assigned in the figure. For the 6a' orbital we find a splitting into several very closely spaced lines. The correlation effects are most spectacular, however, for the 5a' and 4a' orbitals. As shown in fig.25 the 5a' and 4a' strength is spread over a large number of lines. No individual line carries more than about 10% of the total strength associated with one orbital. The lines extend over more than 5eV for both orbitals. The experimental spectrum shows broad peaks at 22.0, 30.7 and 33.0eV. The predicted splitting of the 6a' band can of course not be resolved experimentally due to the vibrational broadening of the lines. The calculated ionization spectrum shows that the remaining two bands in the HeII spectrum arise mainly from the 5a' orbital, although the band at higher energy also contains lines originating from the 4a' orbital. The observed peak separation of 2.3eV is in excellent agreement with the calculated spacing of 2.4eV. A further intense peak is to be expected near 38eV originating mainly from the 4a' orbital.

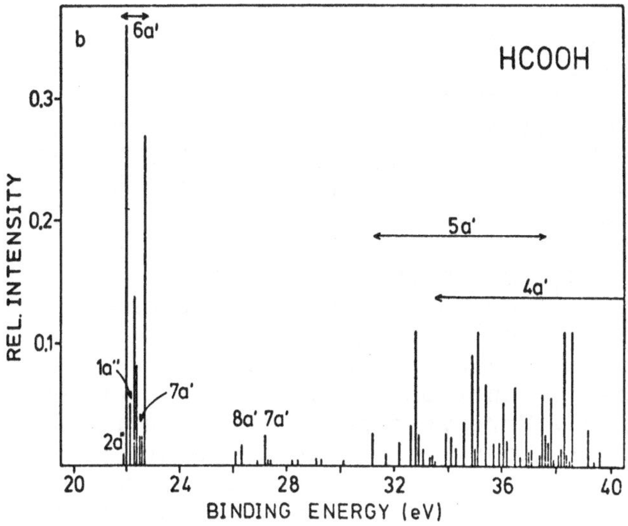

Fig.25: Calculated vertical ionization spectrum of HCOOH between 20 and 40eV

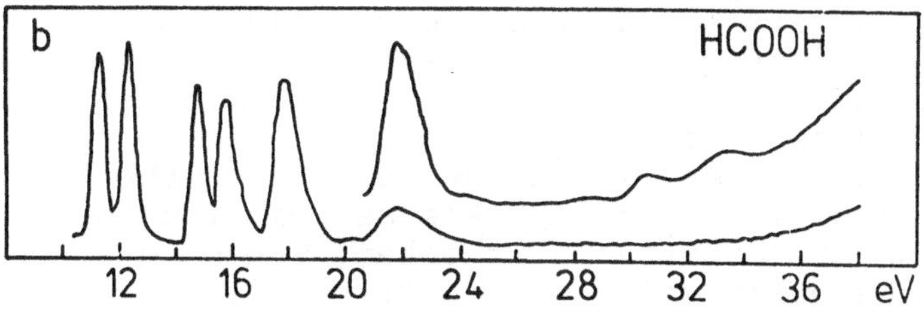

Fig.26: HeII PES of HCOOH as recorded by Potts et al. (91)

Application to N_2O_4 (92)

For molecules containing first row atoms the one-particle picture of ionization may break down for the inner valence orbitals of mainly 2s character. For molecules containing second row atoms this breakdown starts at much lower energies. But where does this breakdown start? This question can only be answered by a calculation for each molecule. One can only say that if there are low lying non-diffuse virtual orbitals of appropriate symmetry which lead to low lying poles of the self-energy the breakdown tends to occur at lower energies. With the statement that for molecules containing first row atoms the breakdown occurs for the 2s-type orbitals one may be on rather shaky grounds. We wish to demonstrate this for the case of the N_2O_4 molecule. Several authors have recently been able to extract the PES of N_2O_4 from a mixture of NO_2 with N_2O_4 and to measure the IP's of the latter molecule (for a discussion and references see ref.92 where the N_2O_4 molecule is studied in detail). In fig.27 we give the PES of N_2O_4 as recorded by Gan et al. (93). This is the best resolved spectrum. The assignment of the spectrum proved to be quite difficult. The basis set used for the calculation is of double zeta quality. The Green's function and TDA are presented in detail in ref.92. Here we discuss only the results of the TDA calculations. For these calculations the 17 occupied valence orbitals and the 8 unoccupied orbitals lowest in orbital energy have been included. The ionization spectrum of N_2O_4 computed in this way is shown in fig.28 (10-21eV range) and fig.29 (21 to 51eV range).

The experimentally observed separation of the spectrum into a group of lines below 15eV and a group of lines between 15 and 21eV is clearly reproduced by the calculations. The five lines below 15eV have pole strengths between 0.8 and 0.9. The ordering of the IP's is the same as obtained by the outer valence type calculation, i.e. there is an interchange of the $3b_{2g}$ IP with both the $1b_{1g}$ and $1a_u$ IP's as compared to the prediction of Koopmans' theorem. The good agreement between the outer valence type Green's function results and the TDA results leaves no doubt as to the assignment of the first five IP's of N_2O_4. Fig.28 shows furthermore that below 15eV the ionization spectrum of N_2O_4 is completely free of satellite lines. For the group of lines between 15 and 21eV very strong correlation effects are predicted by the TDA calculation. All lines in this energy range have pole strengths of 0.7 or less. The $3a_g$ line near 19eV is split into closely spaced components with intensities of 41% and 25%, respectively. The remaining intensity appears above 22eV. A similar situation is found for the $2b_{3u}$ orbital. For the $1b_{2u}$ orbital we find a line near

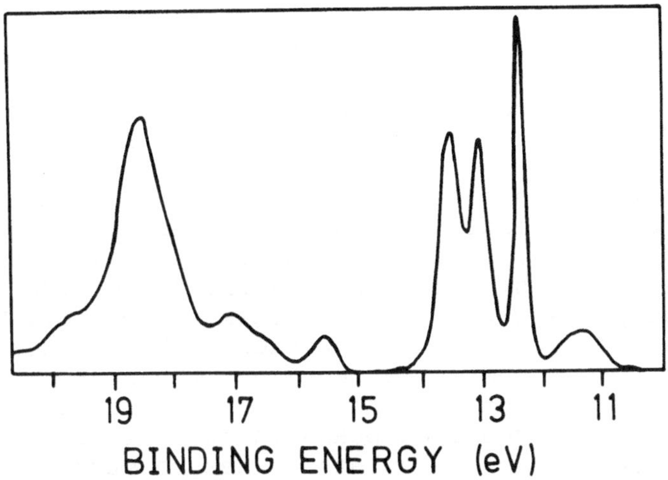

Fig.27: HeI PES of N_2O_4 as recorded by Gan et al. (93)

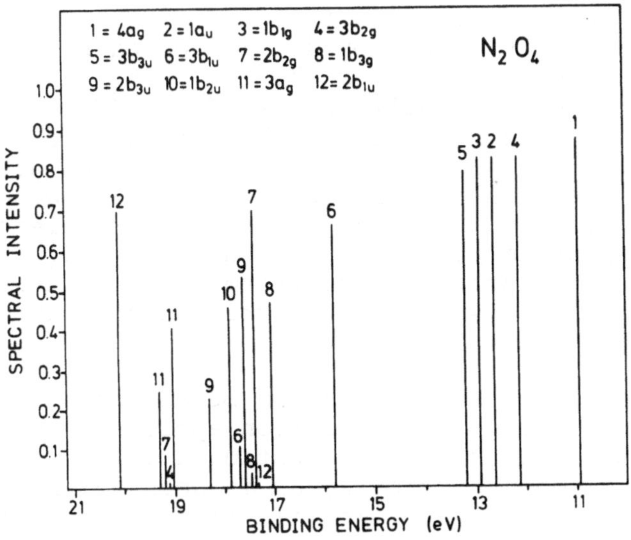

Fig.28: The ionization spectrum of N_2O_4 up to 21eV binding energy obtained by the 2ph-TDA Green's function method

18eV which carries, however, only 46% of the total intensity. A group of three lines originating from the $1b_{2u}$ orbital and carrying altogether 28% of the intensity appears near 23eV separated by 5eV from the lowest energy $1b_{2u}$ line. A similar situation is found for the $1b_{3g}$ orbital with intense lines at 17.1eV(P=0.47) and 22.7eV(P=0.23). For the $3b_{1u}$ orbital we note that a strong (P=0.11) satellite line appears at 17.7eV, less than 2eV higher than the $3b_{1u}$ main line. For the $2b_{1u}$ orbital, on the other hand, a weak satellite line is found 2.7eV below $2b_{1u}$ main line. All these features show that for N_2O_4 the familiar MO picture of ionization breaks down above 15eV ionization energy. A distinction between main lines and satellite lines is no longer possible there. The self-energy has poles with large residues above 15eV. The outer valence type Green's function method is not applicable in this energy range. Since there are more lines in the theoretical ionization spectrum between 15 and 21eV than resolved peaks in the HeI PES a detailed assignment of calculated lines to structures in the experimental spectrum is hardly possible except for the isolated band observed at 15.6eV which corresponds to the calculated $3b_{1u}$ line at 15.8eV. Above 17eV the concept of orbital ordering becomes questionable. The $1b_{2u}$ line near 18eV e.g. is situated between two of the lines originating from the $2b_{3u}$ orbital. The question how many IP's should be observed below 21eV finds no unique answer. For the $1b_{3g}$, $2b_{3u}$ and $1b_{2u}$ orbitals roughly one half of the intensity appears below 21eV, the second half above 21eV. Apart from such difficulties the calculated ordering of the IP's is in agreement with the conclusions of Gan et al. The high energy part of the valence ionization spectrum of N_2O_4 is shown in fig.29. There is a dense line structure covering nearly the whole range from 21 to 44eV. The 2s orbitals of O and N have dissolved into numerous lines.

One question should finally be answered. Why is the orbital picture of ionization applicable to the core orbitals? It should be mentioned immediately that it can break down completely as it does for heavier atoms (e.g. Xe, ref.94). We will consider here only molecules with first row atoms. Correlation effects between simple hole and 2h1p configurations can lead to the breakdown of the MO picture of ionization. The core hole energies, however, are found to be much higher than can be reached by 2h1p valence type excitations. The self-energy exhibits only poles which lie at higher energies than the core IP. In addition the self-energy has a cut in the core region. But as this continuum contributes mainly to the lifetimes of the states the situation is quite similar to the outer valence region.

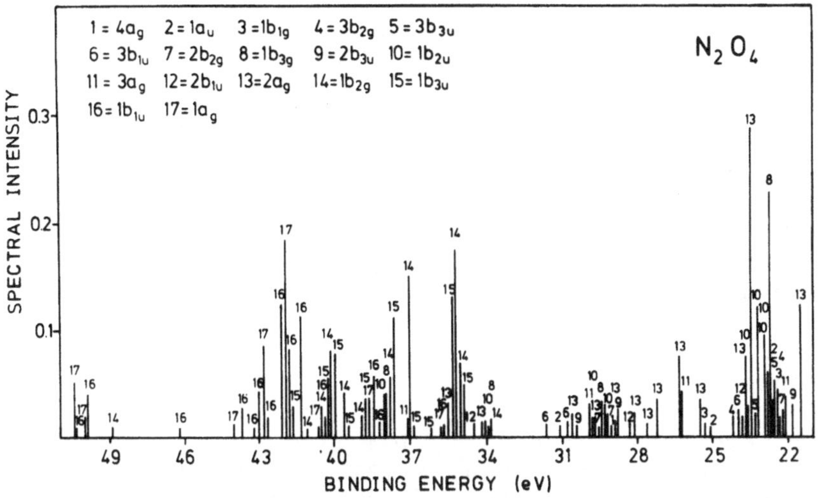

Fig.29: The continuation of fig.28 up to 51eV binding energy. Note that the ordinate scale is different from that in fig.28

There is a large interval free of poles of Σ. In principle larger molecules with several atoms of the same type but in different chemical surroundings might show in the core a breakdown of the MO picture of ionization as simple hole configurations may have about the same energy as configurations with one core hole and a valence type excitation. The residues of the self-energy poles must be large enough for strong correlation effects to occur. For ionization of orbital i these residues are proportional to $V^2_{ij[k\ell]}$. Whether they can be large enough remains to be seen. In discussing these effects it should be noted that the localization characteristics of core valence and unoccupied orbitals are very different. In fact recent calculations demonstrate that this breakdown does occur, e.g. in para-nitroaniline (95). Experiment gives one N1s peak as simple, the other one as a doublet (96). The calculations predict in fact a triplet.

Acknowledgment

The authors gratefully acknowledge the opportunity to use the MUNICH program system written by Dr. Diercksen and Dr. Kraemer with which a significant part of the calculations was performed. They also gratefully acknowledge many helpful discussions and the fruitful cooperation with Dr's Diercksen, Kraemer and Schirmer. Without this cooperation much of the work presented here would not have been done. Financial support from the Fonds der Chemischen Industrie is gratefully acknowledged.

References

1.) T. Koopmans, Physica 1, 104 (1933)
2.) P.S. Bagus, Phys. Rev. A 139, 619 (1965); P.S. Bagus and H.F. Schaefer, J. Chem. Phys. 55, 1474 (1971); 56, 224 (1972)
3.) a) P.E. Cade, K.D. Sales and A.C. Wahl, J. Chem. Phys. 44, 1973 (1966); At.Data 13, 339 (1974); b) A.E. Smolyar, A.I. Boldyrev, O.P. Charkin, N.M. Klimenko and V.I. Avdeev, J. Struct. Chem. 17, 188 (1976); c) M.F. Guest and V.R. Saunders, Mol. Phys. 29, 873 (1975)
4.) L.S. Cederbaum, G. Hohlneicher and W. von Niessen, Chem. Phys. Lett. 18, 503 (1973); L.S. Cederbaum and W. von Niessen, J. Chem. Phys. 62, 3824 (1975)
5.) W. von Niessen, G.H.F. Diercksen and L.S. Cederbaum, J. Chem. Phys. 67, 4124 (1977)
6.) T.T. Chen, W.D. Smith and J. Simons, Chem. Phys. Lett. 26, 296 (1974) (for corrections see ref.9)
7.) G.D. Purvis and Y. Öhrn, J. Chem. Phys. 60, 4063 (1974); H. Kurtz, G.D. Purvis and Y. Öhrn, Int. J. Quantum Chem. Symp. 10, 331 (1977)
8.) D.P. Chong, F.G. Herring and D. McWilliams, J. Chem. Phys. 61, 3567 (1974)
9.) M.F. Herman, D.L. Yeager, K.F. Freed and V. McKoy, Chem. Phys. Lett. 46, 1 (1977)
10.) Modern Theoretical Chemistry, edited by H.F. Schaefer (Plenum, New York, 1977) Vol.'s 3 and 4
11.) G. Csanak, H.S. Taylor and R. Yaris, Adv. At. Mol. Phys. 7, 287 (1971)
12.) L.S. Cederbaum and W. Domcke, Adv. Chem. Phys. 36, 205 (1977)
13.) A.C. Wahl and G. Das, Adv. Quantum Chem. 5, 261 (1970); F. Grein and T.C. Chang, Chem. Phys. Lett. 12, 44 (1971), J. Chem. Phys. 57, 5270 (1972);

J. Hinze, J. Chem. Phys. 59, 6424 (1973); H. Basch, J. Chem. Phys. 55, 1700 (1971), Chem. Phys. Lett. 19, 323 (1973)

14.) F.P. Billingsley, J. Chem. Phys. 62, 864 (1975); 63 2267 (1975); M. Hackmeyer, Int. J. Quantum Chem. 8, 783 (1974)

15.) a) W. Meyer, Int. J. Quantum Chem. Symp. 5, 341 (1971);
b) J. Chem. Phys. 58, 1017 (1973);
c) R. Ahlrichs, H. Lischka, V. Staemmler and W. Kutzlnigg, J. Chem. Phys. 62, 1235 (1975)

16.) W.J. Hunt, P.J. Hay and W.A. Goddard, J. Chem. Phys. 57, 738 (1972); W.A. Goddard and R.C. Ladner, J. Am. Chem. Soc. 93, 6750 (1971)

17.) D.P. Chong, F.G. Herring and D. McWilliams, J. Chem. Phys. 61, 78 (1974)

18.) D.W. Smith and O.W. Day, J. Chem. Phys. 62, 113 (1975); O.W. Day, D.W. Smith and R.C. Morrison, ibid. 62, 115 (1975); M.M. Morell, R.G. Parr and M. Levy, J. Chem. Phys. 62, 549 (1975)

19.) I. Hubač, V. Kvasnička and A. Holubec, Chem. Phys. Lett. 23, 381 (1973)

20.) J. Simons and W.D. Smith, J. Chem. Phys. 58, 4899 (1973); J. Simons, Chem. Phys. Lett. 25, 122 (1974)

21.) B.T. Pickup and O. Goscinski, Mol. Phys. 26, 1013 (1973)

22.) G. Hohlneicher, F. Ecker and L.S. Cederbaum, Electron Spectroscopy, edited by D.H. Shirley (North-Holland, Amsterdam, 1972), p. 647

23.) F. Ecker and G. Hohlneicher, Theor. Chim. Acta 25, 289 (1972)

24.) J.D. Doll and W.P. Reinhardt, J. Chem. Phys. 57, 1169 (1972)

25.) B. Schneider and H.S. Taylor, Phys. Rev. A1, 855 (1970)

26.) B.S. Yarlagadda, G. Csanak, H.S. Taylor, B. Schneider and R. Yaris, Phys. Rev. A7, 146 (1973)

27.) L.S. Cederbaum, Theor. Chim. Acta 31, 239 (1973); J. Phys. B8, 290 (1975)

28.) L.S. Cederbaum, J. Chem. Phys. 62, 2160 (1975); J. Schirmer and L.S. Cederbaum, J. Phys. B, in press

29.) L.S. Cederbaum and W. Domcke, J. Chem. Phys. 60, 2878 (1974), 64, 603, 612 (1976)

30.) A.L. Fetter and J.D. Walecka, Quantum Theory of Many-Particle Systems, McGraw-Hill, New York (1971)

31.) G.H.F. Diercksen and W.P. Kraemer, MUNICH, Molecular Program System Reference Manual, Special Technical Report, Max-Planck-Institut für Physik

und Astrophysik (to be published); G.H.F. Diercksen, Theor. Chim. Acta 33, 1 (1974)
32.) W. von Niessen, L.S. Cederbaum, W. Domcke and G.H.F. Diercksen, J. Chem. Phys. 66, 4893 (1977)
33.) L.S. Cederbaum, W. Domcke, W. von Niessen and W.P. Kraemer, Mol. Phys. 34, 381 (1977)
34.) C. Salez and A. Veillard, Theor. Chim. Acta 11, 441 (1968)
35.) D.W. Turner, C. Baker, A.D. Baker and C.R. Brundle, Molecular Photoelectron Spectroscopy (Wiley-Interscience, New York, 1970)
36.) W. Ermler and A.C. Wahl, unpublished results cited in ref.9
37.) D.P. Chong, F.G. Herring and D. McWilliams, J. Chem. Phys. 61, 958 (1974)
38.) I. Hubač and M. Urban, Theor. Chim. Acta 45, 185 (1977)
39.) A.D. McLean and M. Yoshimine, IBM J. Res. Develop. Suppl. (1967)
40.) L.S. Cederbaum, W. Domcke and W. von Niessen, Chem. Phys. 10, 459 (1975)
41.) G. Bieri, E. Heilbronner, J.-P. Stadelmann, J. Vogt and W. von Niessen, J. Am. Chem. Soc. 99, 6832 (1977)
42.) L.S. Cederbaum, Chem. Phys. Lett. 25, 562 (1974)
43.) D.P. Chong, F.G. Herring and D. McWilliams, J. Electron. Spectroscopy 7, 445 (1975)
44.) W. von Niessen, W. Domcke, L.S. Cederbaum and W.P. Kraemer, J. Chem. Phys. 67, 44 (1977)
45.) R. Gleiter, E. Heilbronner and V. Hornung, Helv. Chim. Acta 55, 255 (1972)
46.) C. Fridh, L. Åsbrink, B.-Ö. Jonsson and E. Lindholm, Int. J. Mass. Spectrom. Ion. Phys. 9, 485 (1972)
47.) S. Huzinaga, J. Chem. Phys. 42, 1293 (1965)
48.) J. Almlöf, B. Roos, U. Wahlgren and J. Johansen, J. Electron. Spectrosc. 2, 51 (1973)
49.) W. von Niessen, G.H.F. Diercksen and L.S. Cederbaum, Chem. Phys. Letters 45, 295 (1977)
50.) W. von Niessen, unpublished results
51.) W. von Niessen, L.S. Cederbaum and W.P. Kraemer, J. Chem. Phys. 65, 1378 (1976)
52.) U. Gelius, J. Electron. Spectrosc. 5, 985 (1974)
53.) B.-Ö. Jonsson and E. Lindholm, Ark. Fys. 39, 65 (1969)
54.) J. Itah, B. Katz and B. Scharf, Chem. Phys. Lett. 52, 92 (1977)
55.) L.S. Cederbaum, W. Domcke and W. von Niessen, J. Phys. B10, 2963 (1977)

56.) L.S. Cederbaum, W. Domcke and W. von Niessen, Mol. Phys. 33, 1399 (1977)
57.) A. Veillard, Theor. Chim. Acta 12, 405 (1968)
58.) D. Feldmann, Z. Naturforschung 25a, 621 (1970)
59.) J.D. Carette and L. Kerwin, Canad. J. Phys. 39, 1300 (1961)
60.) S.L. Bennell, J.L. Margrave and J.L. Franklin, J. Chem. Phys. 61, 1647 (1974)
61.) R.H. Wood and L.A. D'Orazio, J. Phys. Chem. 69, 2562 (1965); J. Berkowitz, W.A. Chupka and D. Gutman, J. Chem. Phys. 55, 2733 (1971); R. Byerly and E.C. Beatly, J. Geophys. Res. 76, 4596 (1971); S.F. Wong, T.V. Vorburger and S.B. Woo, Phys. Rev. A5, 2598 (1972); G. Sinnott and E.G. Beaty, Electronic and Atomic collisions (North Holland), 1971
62.) R.J. Celotta, R.A. Bennett and J.L. Hall, J. Chem. Phys. 60, 1740 (1974)
63.) S.-K. Shih, W. Butscher, R.J. Buenker and S.D. Peyerimhoff, Chem. Phys. 29, 241 (1978)
64.) W. Benesch, J.T. Vanderslice, S.G. Tilford and P.G. Wilkonson, Astrophys. J. 143, 236 (1966)
65.) R.W. Nicholls, J. Quant. Spectr. Radiat. Transfer 2, 433 (1962)
66.) W. Domcke, L.S. Cederbaum, H. Köppel and W. von Niessen, Mol. Phys. 34, 1759 (1977)
67.) W.L. Smith and P.A. Warsop, Trans. Faraday Soc. 64, 1265 (1968)
68.) R. Botter and H.M. Rosenstock, Adv. Mass. Spectrometr. 4, 579 (1968). The calculated FC factors are compared with the PES in: H.J. Weiss and G.M. Lawrence, J. Chem. Phys. 53, 214 (1970)
69.) W.R. Harshberger, J. Chem. Phys. 53, 903 (1970); J. Chem. Phys. 56, 177 (1972); S. Durmaz, J.N. Murrell, J.M. Taylor and R. Suffolk, Molec.Phys. 19, 533 (1970)
70.) J.W. Rabalais, L. Karlsson, L.O. Werme, T. Bergmark and K. Siegbahn, J. Chem. Phys. 58, 3370 (1973)
71.) A.D. Baker, C.R. Brundle and D.W. Turner, Int. J. Mass. Spectr. Ion Phys. 1, 285 (1968)
72.) C.R. Brundle, M.B. Robin, N.A. Kuebler and H. Basch, J. Am. Chem. Soc. 94, 1451 (1972)
73.) I.P. Batra and O. Robaux, Chem. Phys. Lett. 28, 529 (1974)
74.) L.S. Cederbaum, G. Hohlneicher and S. Peyerimhoff, Chem. Phys. Lett. 11, 421 (1971); L.S. Cederbaum, G. Hohlneicher and W. von Niessen, Molec. Phys. 26, 1405 (1973)
75.) D.P. Chong, F.G. Herring and D. McWilliams, J. Chem. Phys. 61, 958 (1974)

76.) L.S. Cederbaum, W. Domcke and W. von Niessen, Chem. Phys. Lett. 34, 60 (1975)
77.) T. Shimanouchi and I. Suzuki, J. Chem. Phys. 42, 296 (1965)
78.) H.A. Jahn and E. Teller, Proc. Roy. Soc. (London), A 161, 220 (1977)
79.) L.S. Cederbaum, W. Domcke, H. Köppel and W. von Niessen, Molec. Phys., in press
80.) L.S. Cederbaum, W. Domcke, H. Köppel and W. von Niessen, Chem. Phys. 26, 169 (1977)
81.) F. Brogli, E. Heilbronner, E. Kloster-Jensen, A. Schmelzer, A.S. Manocha, J.A. Pople and L. Radom, Chem. Phys. 4, 107 (1974)
82.) W. von Niessen, G.H.F. Diercksen, L.S. Cederbaum and W. Domcke, Chem. Phys. 18, 469 (1976)
83.) F.A. Miller and I. Matsubara, Spectrochim. Acta 22, 173 (1966)
84.) J. Schirmer, W. Domcke, L.S. Cederbaum and W. von Niessen, J. Phys. B in press
85.) N. Jonathan, A. Morris, M. Okuda, D.J. Smith and K.J. Ross, Chem. Phys. Letters, 13, 334 (1972)
86.) J. Schirmer, L.S. Cederbaum, W. Domcke and W. von Niessen, Chem. Phys. 26, 149 (1977)
87.) U. Gelius, E. Basilier, S. Svensson, T. Bergmark and K. Siegbahn, J. Electron. Spectrosc. 2, 405 (1974)
88.) R. Nilsson, R. Nyholm, A. Berndtsson, J. Hedman and C. Nordling, J. Electron. Spectrosc. 9, 337 (1976)
89.) P.S. Bagus and E.K. Viinikka, Phys. Rev. A15, 1486 (1977)
90.) J. Schirmer, L.S. Cederbaum, W. Domcke and W. von Niessen, to be published
91.) A.W. Potts, T.A. Williams and W.C. Price, Faraday Disc. 54, 104 (1972)
92.) W. von Niessen, W. Domcke, L.S. Cederbaum and J. Schirmer, to be published
93.) T.H. Gan, J.B. Peel and G.D. Willett, J. Chem. Soc. Faraday Trans. II 73, 1459 (1977)
94.) G. Wendin and M. Ohno, Phys. Scripta, 14, 148 (1976)
95.) W. von Niessen, unpublished results
96.) S. Pignataro and G. Distenfano, J. Electron. Spectrosc. 2, 171 (1973); S. Tsuchiya and M. Senō, Chem. Phys. Lett. 54, 132 (1978)

AUGER ELECTRON SPECTRA FROM FREE ATOMS AND MOLECULES

Hans Siegbahn

Institute of Physics, University of Uppsala

1. INTRODUCTION AND GENERAL ASPECTS OF AUGER ELECTRON SPECTRA

A system excited to a discrete, quasistationary state above the first ionization potential may decay by the emission of an electron. In current terminology, electron emission from neutral systems is generally referred to as autoionization, while the term Auger electron emission is usually reserved to designate decay from initially ionic states. The latter processes were first observed by P. Auger in 1925 /1/. He studied X-ray absorption by means of cloud chamber experiments and found that besides the photoelectron tracks, other tracks appeared, whose lengths were independent of the X-ray photon energy. He concluded thus that these were due to internal conversion processes.

The kinetic energy of an emitted Auger electron is given by:

$$E_{Aug} = E_{Tot}(X) - E_{Tot}(YZ) \tag{1}$$

where $E_{Tot}(X)$ is the total energy of the initial state with a vacancy in shell X and $E_{Tot}(YZ)$ that of the final state with two vacancies in shells Y and Z. The transition is referred to as an XYZ transition (KLL, LMM etc.). Special types of transitions are those where Y or Z is equal to X, which are generally termed Coster-Kronig transitions. Super Coster-Kronig processes are defined by Y = Z = X.

Eq. (1) involves the assumption that the production of Auger electrons is a two-step process, where the creation of the initial vacancy does not affect the subsequent Auger decay. This may not be entirely true, if the excitation energy is close to

Cleanthes A. Nicolaides and Donald R. Beck (eds.), Excited States in Quantum Chemistry, 273–295.

the ionization threshold. Schmidt et al. /2/ have recently employed synchrotron radiation to excite Auger electrons from xenon. It was then found, that a shift occurred in the position of the $N_5O_{2,3}O_{2,3}(^1S_0)$ Auger line, when the photon energy was only slightly higher than the N_5 threshold. A postcollision interaction (PCI) thus occurs between the photoelectron and the Auger electron. In the following, we shall consider Auger electron spectra excited with energies well above threshold, such that PCI effects on the energies are absent.

The rate of an Auger transition is given by the matrix element coupling the discrete hole state to the continuum:

$$P_{XYZ} \propto |\langle\Psi(X) \mid H \mid \Psi(YZ)\rangle|^2 \qquad (2)$$

where H is the (N -1)-particle Hamiltonian and Ψ(YZ) contains a free Auger-electron continuum part /3 - 7/. For a non-zero rate, Ψ(X) and Ψ(YZ) should have the same symmetry. This leads to selection rules in the Auger spectra. In the atomic case, the main line KLL Auger spectrum contains seven possible final ionic states for pure LS-coupling ($(2s)^{-2}(^1S_0)$, $2s^{-1}2p^{-1}(^1P_1$ and $^3P_{0,1,2})$ and $(2p)^{-2}(^1S_0$ and $^1D_2)$). In jj-coupling the number of states is six ($L_1L_1, L_1L_2, L_2L_2, L_1L_3, L_2L_3$ and L_3L_3) and the full intermediate coupling case contains nine states ($(2s)^{-2}(^1S_0)$, $2s^{-1}2p^{-1}(^1P_1$ and $^3P_{0,1,2})$ and $(2p)^{-2}(^1S_0, {}^1D_2$ and $^3P_{0,2})$).

In addition to the main Auger electron lines, satellite lines are generally observed in the spectra. These may arise through a number of mechanisms. First, the initial state of the Auger process may be excited above the hole ground state or multiply ionized. Such states may be populated by means of multielectron transitions in the primary ionization (shake-up, shake-off) or by Auger electron decay from lower lying hole states (e.g. an L_1 hole in Ar, which decays by a Coster-Kronig transition to an $L_{2,3}M_{2,3}$ double hole state) second, multielectron processes in the Auger process itself analogous to those in the primary ionization may occur (shake-up and shake-off (double Auger process)). Third, transitions may occur from excited states of the neutral species - autoionization. The intensities of these latter transitions is highly dependent on the mode of excitation. Thus, autoionization satellites are generally much more intense if a source continuous in energy is used, such as electrons, than if e.g. characteristic X-rays are employed.

The first category of satellites mentioned above may have lower or higher energy than the main lines, depending on whether the initial state shake-up or shake-off excitation participates in the Auger decay or not. The second type (final state shake-up, shake-off) leads to low-energy lines and autoionization peaks occur on the high-energy side of the main lines.

The free-atom Auger spectra motly studied so far have been the noble gas spectra /8 - 17/. For these systems, naturally, the most advanced theoretical calculations have been made. Since continuum functions can be generated by numerical integration of the radial Schrödinger equation for these cases, transition rates can be calculated to a high degree of sophistication, including also interchannel coupling /16, 18 - 20/. These systems furnish thus critical tests of theory. The most accurately studied spectrum so far in this respect is the NeKLL spectrum / 8, 9, 13, 15, 18 - 20/.

Auger transitions in molecules and solids are appropriately classified according to the character of the final vacancies. If these occur in shells which can be considered core-like, the main line Auger electron spectrum will be very similar to a free-atom spectrum, both with regard to relative intensities and relative energies. The entire spectrum may shift with respect to changes in the chemical state of the element, however. A chemical dependence of the satellite peaks will in general also be present, when these involve valence electron excitations.

When any of the final vacancies of the Auger transition occur in the valence shell of a molecule or a solid, the spectrum will in general lose its atomic character. The analysis of such spectra are often complicated, since the effects of final state couplings and electron polarization must be considered for multicenter hole-state wavefunctions. A first approcimation to Auger spectra in solids, where the final vacancies are both in the outermost electron band, is a selfconvolution of the occupied density of states. This simple consideration would give an Auger spectrum with approximately twice the width of the band. Deviations from such predictions have been found, however /21 - 24/.

Molecular Auger spectra with vacancies among the valence electrons is interpretationally simpler than the solid band structure case, but requires extensive calculations for a detailed account. A number of small molecules have been investigated, where it has proved possible to assign most of the observed structures /9, 25 - 34/. In the case of the water molecule /28, 33/ the energies of the transitions were calculated by open-shell Hartree-Fock procedures, followed by a limited CI. The intensities of the transitions were estimated on the basis of a quasiatomic model, in which the rates were given by linear combinations of atomic transition rates. By this approach structures due to the main transitions could be identified. In general, the origin of satellite structures in molecular and solid state Auger spectra is difficult to assign in any reliable fashion.

I shall consider below some K Auger spectra, which have been studied in our laboratory. These will serve to illustrate the concepts mentioned above and some of the methods used in the interpretations of these types of spectra.

2. EXPERIMENTAL CONDITIONS

The recording of Auger electron spectra requires basically three units: an excitation source, an electron energy analyzer and an electron detector. The excitation source most commonly used is an electron gun, producing a beam of electrons of energies of up to say, 10 keV. Electrons provide effective agents in producing the primary vacancies and intense Auger signals are thus achieved. A special use of electron excitation has been made in solid surface investigations, where Auger electron spectroscopy is a versatile investigative method, due to its high surface sensitivity (the Auger electrons are sampled from the first few atomic layers of the sample). This type of electron excitation, Scanning Auger Electron Spectroscopy (SAES), uses a fine-focus electron gun which sweeps over the sample surface. Since the Auger electron spectra provide elemental identification, one can obtain a picture of the atomic distribution on the surface in this way. The lateral resolution is of the order of 0.5 μm.

Ion or photon beams are alternative modes of exciting Auger spectra. As previously indicated, the various excitation modes will lead to the same relative intensities among the main lines. However, the ratio between the satellite and the main line intensities may be affected. This furnishes a possibility of assignment of satellite structure. In particular, it is expected that final state shake-up, shake-off satellites will be independent of excitation mode, whereas the other two types mentioned above will not. In this context the possibility of using synchrotron radiation is an attractive alternative. Selective excitation to auto-ionizing states with synchrotron radiation in conjunction with Auger electron spectroscopy has been studied in krypton and xenon /35/.

The electron energy analyzers that are used for Auger spectroscopy are generally either of retarding grid type or deflection type, hemispherical capacitor and cylindrical mirror spectrometers. The first type is a nonimaging analyzer, in which the electron energies are analyzed by means of retarding potentials between spherical grids. The deflection type analyzers create an electron-optical image of the electron source, whose position is dependent on the energy. The retarding grid type is generally used at much lower resolution and signal to background than the deflection type analyzers. Its main use has been in solid surface investigations, where it is employed in combined LEED/Auger systems.

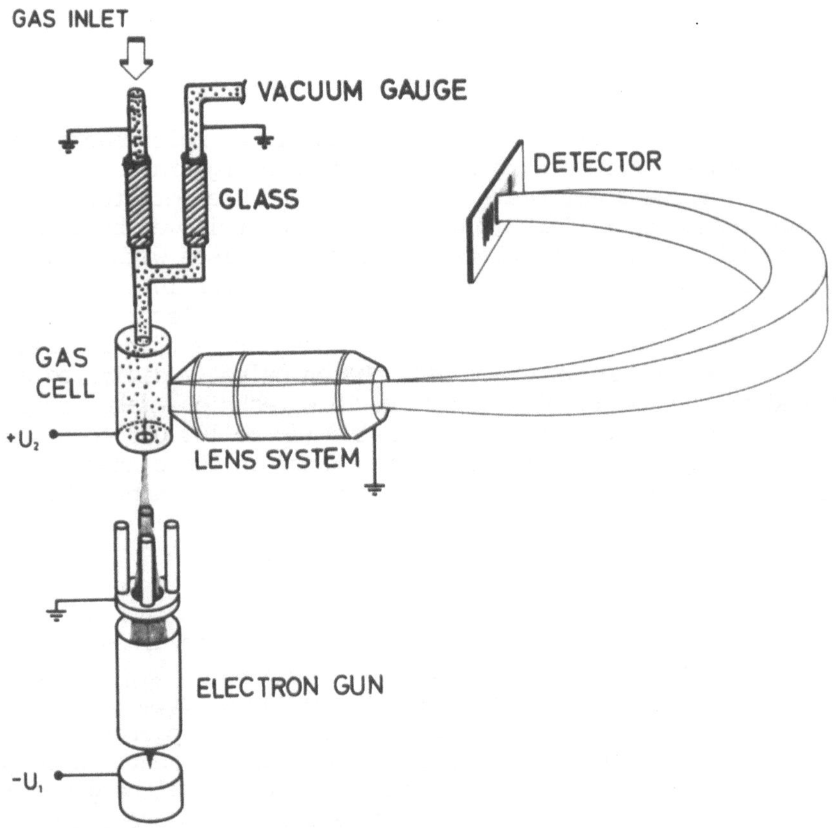

Fig. 1. Schematic view of experimental arrangement for recording of Auger electron spectra by electron impact.

The experimental arrangement used to record the spectra presented below is shown in Fig. 1. The electrons used for excitation are generated in an electron gun with a small focus (50 µm at 5 kV and 5 µA). By means of an electrostatic quadrupole lens the beam is adjusted to enter the sample gas cell through a small hole. The gas cell is an aluminum tube, the inside of which is coated with graphite to avoid local charging. The upper part of the cell is connected to the gas inlet system. Since the Auger electrons that are analyzed by this spectrometer may have high energies (E_{Aug} >2000 eV), a preretardation lens system is incorporated. The purpose of this device is to enhance the total resolution of the system ($\Delta E_{kin}/E_{kin}$). With a normal resolution of the spectrometer of say, $5 \cdot 10^{-4}$ the instrumental contribution to the line widths would become unacceptably large (~1 eV) without retardation. The electrostatic lens system, schematically shown in Fig. 1, was designed to operate in the magnetic field of the electron energy analyzer. It allows a retardation ratio of 1:5. The gas cell is held at a high potential (U_2) which is the retarding voltage of the outgoing electrons.

The detector unit in this case is of multichannel type which allows the recording of an extended energy range of the spectrum at one spectrometer field setting. The detector consists of two channeltron plates coupled in series followed by a phosphor screen. The channeltron plates are circular discs with a diameter of 2.5 cm, each containing small holes (diameter 10 - 50 μm). The holes through the two plates act as indivicual electron multipliers with a high gain. The light from the phosphor is then focused on to a photodiode array, where each photon pulse is digitized and stored in the memory of a computer. This technique makes it possible to record very high counting rates (up to 100 000 counts per second per channel). This is of importance in the context of Auger electron spectra, where large intensities may often be achieved.

3. THE KLL AUGER ELECTRON SPECTRUM OF ARGON

The KLL Auger spectrum of argon is shown in Fig. 2 and numerical data are contained in Table 1. The main lines arising from the decay of a single vacancy in the K shell are denoted by the LS- and jj-coupling notations. Since the spectrometer contribution to the line widths has been reduced by preretardation, the $L_2L_3(^1D_2)$ main line has nearly the natural Lorentzian shape. It has a width of 1.19 eV and a curve fit to this line gives a Voigt function with 0.56 eV Gaussian and 0.92 eV Lorentzian contributions. This natural linewidth is dominated by the lifetime of the K-shell hole state.

A number of satellite peaks can be observed in the spectrum. In order to gain insight into the formation of these satellites, relativistic MCSCF calculations were performed. The program used is based on expansion of the wave function in jj-states, for which the coefficients and one-electron radial functions are optimized by means of the variational principle /36/. The initial states which were calculated had K, KM and KL holes and the final states LL, LM, LLL and $L_{2,3}L_{2,3}M$ holes. A total of 87 hole states were thus considered /37/.

For the case of shake-off in the M shell at the ionization of the K shell, one expects to find a spectrum similar to the normal KLL spectrum wtih regard to the intensity ratios but shifted towards lower kinetic energies. This shift can be estimated from optical data by considering it as a difference between two binding energies:

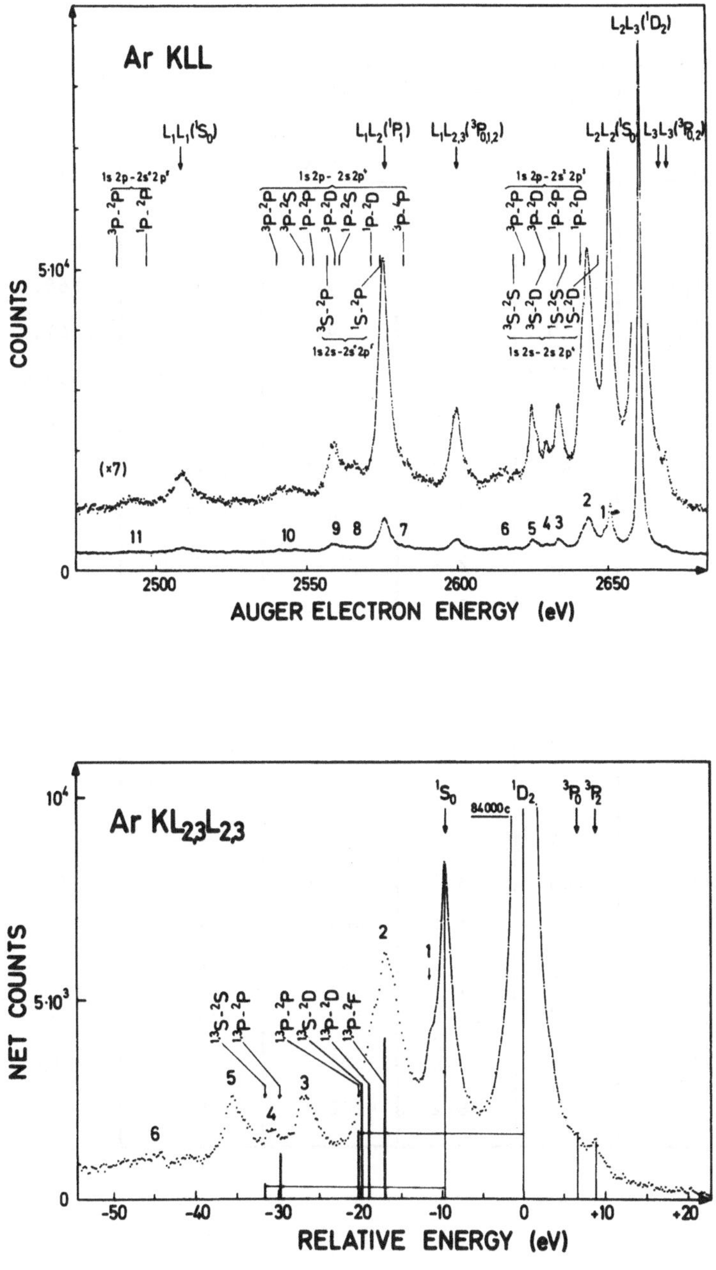

Fig. 2. KLL Auger electron spectrum of argon excited by electron impact. Top figure: Full spectrum. Bottom figure: $KL_{2,3}L_{2,3}$ part of the spectrum on an expanded energy scale.

Table 1: Experimental and theoretical data for Ar KLL spectrum. Energies in eV.

LINE/ FINAL STATE	EXPERIMENTAL				CALCULATION
	INT.	FWHM	RELATIVE ENERGY	ENERGY	ENERGY
$L_3L_3(^3P_2)$	3	1.4	8.6	2669.1	2670.9
$L_3L_3(^3P_0)$	1	1.2	6.4	2666.9	2668.4
$L_2L_3(^1D_2)$	540	1.19	0	2660.51	2661.9
$L_2L_2(^1S_0)$	73	1.79	-9.60	2650.91	2653.0
1	19	2.0	-11.3	2649.2	-
2	97	4.8	-17.0	2643.5	-
3	15	2.7	-26.7	2633.8	-
4	5	2.2	-30.7	2629.8	-
5	19	3.4	-35.3	2625.2	-
6	3	3.7	-44.7	2615.8	-
$L_1L_3(^3P_2)$					2601.7
$L_1L_3(^3P_1)$	34	4.1	-60.8	2599.7	2600.2
$L_1L_2(^3P_0)$					2599.4
7	8	3.7	-76.9	2583.6	-
$L_1L_2(^1P_1)$	100	3.7	-84.8	2575.7	2575.5
8	13	5.1	-94.5	2566.0	-
9	23	4.8	-101.6	2558.9	-
10	5	10.9	-117	2544	-
$L_1L_1(^1S_0)$	26	6.6	-151.6	2508.9	2508.3
11	6	8.2	-168	2492	-

$$\Delta = E_{Aug} - E^{M}_{Aug} =$$
$$= (E_{Tot}(LLM) - E_{Tot}(LL)) - (E_{Tot}(KM) - E_{Tot}(K)) \quad (3)$$

The second bracket binding energy can be approximated as the ionization potential of KII. Similarly, the binding for the M electron in the doubly (LL) ionized atom can be taken as the ionization potential of CaIII. Thus:

$$\Delta \approx I_p(\text{CaIII}) - I_p(\text{KII}) \quad (4)$$

Optical data for these quantities gives $\Delta \approx 19.4$ eV. This is in approximate agreement with the observed value of 17.0 eV for the satellite.

The results obtained from the calculations for the M-shake-off satellites are indicated by bars in the bottom part of Fig. 1, which shows the energy region around the $^1S_0$ and $^1D_2$ lines on an expanded energy scale. The theoretical energies of the transitions have been normalized to the main $^1D_2$ transition, i.e. they have been corrected for the difference between experiment and theory (-1.4 eV) for this transition. The heights of the bars are proportional to the product of the intensity of the corresponding main line ($^1D_2$ or $^1S_0$) and the statistical weights in the final state. The calculations show that the shift between the satellite and main line is the same for the $^1S_0$ and $^1D_2$ transitions.

As mentioned, a similar M-shake-off satellite structure is expected to appear at each main line. This is also found to be the case and the expected satellites to the $L_1L_1(^1S_0)$, $L_1L_2(^1P)$ and $L_1L_{2,3}(^3P)$ lines occur at a distance of about 17 eV from the corresponding main line (peaks no. 11, 9 and 7, respectively).

The fractional intensity of the M-shake-off satellite (defined as $I_{SAT}/(I_{SAT} + I_{MAIN})$) gives the probability for shake-off in the primary ionization. It is found to be 16% from the present spectrum.

In the case of shake-off in the L shell at the ionization of the K shell (which has a substantially lower probability than M shell shake-off) satellite peaks will also be produced in the Auger spectrum. The possible terms of the initial states, 1s2s or 1s2p hole configurations, are $^{1,3}S$ and $^{1,3}P$, respectively. In the calculations, the splitting between the ^1P and ^3P terms was calculated to be 12.0 eV and between the ^1S and ^3S terms 17.0 eV. The transitions that are allowed in LS coupling are indicated by bars in Fig. 2. They are clustered in three different groups. An initial shake-off in 2p leads to two possible transitions to a final state with two vacancies in 2s. These transitions are calculated to lie in the energy region around the

$L_1L_1(^1S_0)$ main line. The next group corresponds to $K\rightarrow L_1L_{2,3}$ transitions and the calculated energies of these cover an energy range from peak 7 to peak 10 in the spectrum. From previous studies on the NeKLL spectrum /13/, only three of these are expected to have appreciable intensity. The first of the transitions with $^{1,3}P$ initial state ($^3P\rightarrow{}^2P$) could be associated with peak number 10. The $^1P\rightarrow{}^2P$ transition is expected to fall into the M-shake-off peak no. 9 and the $^3P\rightarrow{}^2D$ is probably manifested in peak no. 8.

The third group of L-satellites are those coming from the $K\rightarrow L_{2,3}L_{2,3}$ transitions with a vacancy in the L_1 or $L_{2,3}$ subshell. These are contained in the satellites with the numbers from 3 to 6 in the spectrum.

4. KLM AUGER ELECTRON SPECTRUM OF ARGON

The recorded spectrum of the KLM Auger transitions is shown in Fig. 3 and the experimental and theoretical data are given in Table 2. To get the total intensity of the KLM transitions compared to the intensity of the KLL transitions, the strongest peak in the KLM spectrum was recorded simultaneously with the $^1D_2$ peak in the KLL spectrum. The intensities of the two lines were then compared to each other and to the total intensity of the spectra. This gave a ratio of 1:7 for the probabilities of the KLM to the KLL transitions. The KMM spectrum is weak and does not contribute by more than 1% to the total K Auger transition rate.

The kinetic energies derived from the calculations are marked by bars in Fig. 3. In the $KL_{2,3}M_{2,3}$ transitions the energies corresponding to the $^{1,3}P$ final states have been omitted, since these are forbidden in LS coupling. The spin orbit splittings of the S and D states in the same group of transitions have been indicated.

In assigning this spectrum, use is made of the fact that the transition probability to a triplet state is expected to be appreciably lower than to the corresponding singlet state of that configuration /28, 33/. The strongest peak in the spectrum is thus assigned to the $^1D_2$ final state of the $L_{2,3}M_{2,3}$ configuration. The bump observed on the low energy side of this peak fits energetically well with the $^1S_0$ transition, obtained from the calculation. The peak on the high energy side of $^1D_2$ is assigned to the ^3D transition and the ^3S peak is probably hidden under the $^1D_2$ intensity. The two peaks at 12 eV lower energy than the main $^1D_2$ line are the $K\rightarrow L_{2,3}M_1$ transitions with possible final states ^1P and ^3P. The intensities of the two observed peaks are almost equal, which is probably due to a satellite under the ^3P peak. The $L_1M_{2,3}$ peak consists mainly of the $^1P_1$ state with a small contribution from the ^3P state on the high energy side. The L_1M_1

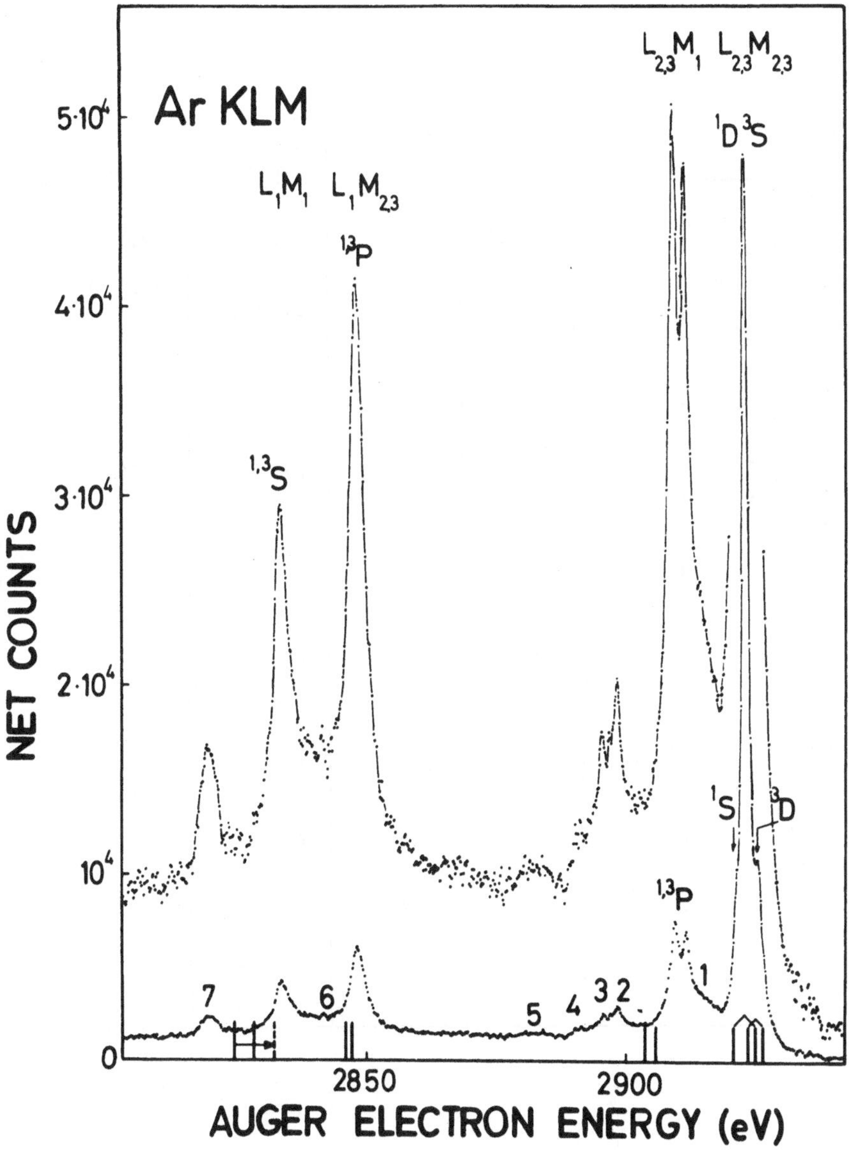

Fig. 3. KLM Auger electron spectrum of argon. Full bars indicate transition energies obtained by relativistic DF-calculations. The dashed bar indicates the transition energy obtained for the L_1M_1 state in an MCSCF calculation including ($\underline{L}_1 3s^2 3p^4 3d$) and ($\underline{L}_1 3s^2 3p^4 4s$) configurations.

Table 2: Experimental and theoretical data for the Ar KLM spectrum. Energies in eV.

LINE/ FINAL STATE	EXPERIMENTAL INT.	FWHM	RELATIVE ENERGY	ENERGY	CALCULATION ENERGY
$L_{2,3}M_{2,3}(^3D)$	19	2.0	2.5	2925.8	2929.1*
$L_{2,3}M_{2,3}(^3S_1)$	100	1.46	0	2923.35	2927.6
$L_{2,3}M_{2,3}(^1D_2)$					2926.0
$L_{2,3}M_{2,3}(^1S_0)$	10	1.5	-1.8	2921.6	2923.1
1	9	6.2	-6.8	2916.5	
$L_{2,3}M_1(^3P)$	16	2.0	-11.5	2911.8	2908.2* a
$L_{2,3}M_1(^1P_1)$	32	2.7	-13.9	2909.5	2906.1
2	10	3.7	-24.9	2898.5	
3	2	2.4	-27.8	2895.5	
4	3	4.4	-32.1	2891.2	
5	3	8.0	-39.6	2883.7	
$L_1M_{2,3}(^3P)$	26	3.7	-75.1	2848.2	2849.7*
$L_1M_{2,3}(^1P_1)$					2848.5
6	4	7.4	-83.1	2840.2	
$L_1M_1(^3S_1)$	16	3.5	-89.9	2833.5	2830.6
$L_1M_1(^1S_0)$					2826.6 b
7	6	4.4	-103.8	2819.5	

* weighted average

a. Inclusion of ($\underline{L}_{2,3}3s^23p^43d$) and ($\underline{L}_{2,3}3s^23p^44s$) configurations shifts the $^3P_0$ transition energy from 2906.5 eV to 2913.4 eV.

b. Inclusion of ($\underline{L}_1 3s^23p^43d$) and ($\underline{L}_1 3s^23p^44s$) configurations shifts this energy to 2834.5 eV.

peak is composed of ^3S and ^1S lines, the singlet being the stronger.

Fig. 3 and Table 2 show that the $L_1M_{2,3}$ and $L_{2,3}M_{2,3}$ energies are well reproduced by the relativistic calculations. However, the L_1M_1 and $L_{2,3}M_1$ transitions are off from the observed peaks by more than 5 eV. This is much larger than any of the observed discrepancies between experiment and theory in both the KLL and the KLM spectra. We therefore extended the calculations for the ^1S state of the L_1M_1 configuration and the ^3P state of the $L_{2,3}M_1$ configuration by adding additional configurations. These were of the type ($\underline{L}_1 3s^2 3p^4 3d$), ($\underline{L}_1 3s^2 3p^4 4s$) and ($\underline{L}_{2,3} 3s^2 3p^4 3d$), ($\underline{L}_{2,3} 3s^2 3p^4 4s$), respectively. It was found, that the transition energies shifted by as much as 7.9 eV and 6.9 eV, respectively. These MCSCF results are in good agreement with the observed peaks (cf. Fig. 3). The main contributing configurations mixing with the L_1M_1, $L_{2,3}M_1$ hole states were the d-excited states. This type of CI for inner valence holes has been found also in a large number of other cases /39 - 46/. In particular, in the noble gas photoelectron spectra, energy shifts as well as prominent satellites have been observed for Ar3s, Kr4s and Xe5s /39, 40, 46/. In all of these cases, interactions between the s hole state and p^4nd configurations have been found to be of considerable importance. In addition, for Ar, effects pertaining to this type of CI has been observed also in the LMM Auger spectrum /14, 16, 17/.

The satellites appearing in the KLM spectrum are partly due to shake-off in the M-shell in the primary ionization. It is reasonable to assume that they will have total intensities comparable to those found in the KLL spectrum. The positions of these satellites in the KLM spectrum will be different compared to the KLL case, however, since the final state will have two holes in the M shell. In addition to these satellites, one expects also lines due to the final state CI mentioned above. Further investigations of the satellite structure is being carried out.

5. MOLECULAR KLL AUGER SPECTRA; H_2S, SO_2 and SF_6

Since sulphur is a second row element, the S KLL Auger spectra will have atomic character. The observed spectra for the three molecules H_2S, SO_2 and SF_6 are very similar to the Ar KLL spectrum with respect to the appearance of the main lines. In Tables 3 - 5 the measured data for these three molecules are given. The main lines are denoted by the same atomic notation as for the Ar case. Calculations were performed for H_2S and SO_2 and these results are also given in Tables 3 and 4.

The calculations were performed for the main transitions and for satellite transitions, which involve an initial state valence shake-off hole. The initial K-shell hole state was computed by SCF's followed by a limited CI. The CI calculations were

Table 3: Experimental and theoretical data for the S KLL spectrum of H_2S. Satellite peaks are indicated by X in column 3. For the theoretical values on the satellites, the mean energy and spread of several components are given /55/. Energies in eV.

LINE			EXPERIMENTAL				CALCULATION	
No	FINAL STATE	SAT.	INT.	ENERGY	RELATIVE ENERGY	FWHM	RELATIVE ENERGY	SPREAD
A			4	2108.5	10.1	1.5	-	-
	$L_{2,3}^2(^3P_{0,2})$		2	2105.4	7.0	1.5	6.6	-
	-"-	x	-	-	-	-	-6.7	3.4
	$L_{2,3}^2(^1D_2)$		454	2098.42	0	1.35	0	-
2	-"-	x	128	2086.3	-12.1	3.7	-13.1	1.6
1			61	2091.7	-6.7	4.6	-	-
	$L_{2,3}^2(^1S_0)$		41	2089.94	-8.48	1.5	-7.1	-
3	-"-	x	13	2077.6	-20.8	4.0	-19.6	1.1
4			20	2073.2	-25.2	4.0	-	-
	$L_1L_{2,3}(^3P_{0,1,2})$		29	2047.4	-51.0	3.1	-49.6	-
5	-"-	x	7	2035.7	-62.6	4.0	-63.4	4.2
	$L_1L_{2,3}(^1P_1)$		100	2026.9	-71.5	3.0	-71.4	-
6	-"-	x	27	2014.9	-83.5	4.5	-84.2	1.2
7			8	2003.2	-95.2	7.0	-	-
	$L_1L_1(^1S_0)$		27	1971.4	-127.4	5.6	-126.0	-
8	-"-	x	9	1961.1	-137.3	7.0	-139.1	1.0

Table 4: Data for the S KLL spectrum of SO_2 /55/. Energies in eV.

LINE			EXPERIMENTAL				CALCULATION	
No	FINAL STATE	SAT.	INT.	ENERGY	RELATIVE ENERGY	FWHM	RELATIVE ENERGY	SPREAD
A			10	2105.8	10.4	1.8	-	-
	$L_{2,3}^2(^3P_{0,2})$		5	2103.0	7.6	1.9	6.7	-
	-"-	x	-	-	-	-	-4.7	3.3
	$L_{2,3}^2(^1D_2)$		493	2095.40	0	1.38	0	-
2	-"-	x	50	2085.7	-9.7	1.8	-11.5	3.5
1			64	2087.9	-7.5	3.8	-	-
	$L_{2,3}^2(^1S_0)$		54	2087.0	-8.4	1.5	-6.7	-
3	-"-	x	22	2077.7	-17.7	5.0	-17.7	2.1
4			8	2068.5	-26.9	3.8	-	-
	$L_1L_{2,3}(^3P_{0,1,2})$		32	2044.4	-51.0	3.1	-49.4	-
5	-"-	x	6	2034.9	-60.5	4.8	-60.9	3.1
	$L_1L_{2,3}(^1P_1)$		100	2024.0	-71.4	3.0	-71.3	-
6	-"-	x	16	2014.6	-80.8	3.5	-82.3	2.0
	$L_1L_1(^1S_0)$		26	1969.0	-126.4	5.9	-126.2	-
8	-"-	x	9	1960.6	-134.8	6.4	-137.2	1.6

Table 5: Data for the S KLL spectrum of SF_6. /55/ Energies in eV.

LINE			EXPERIMENTAL			
No	FINAL STATE	SAT.	INTENSITY	ENERGY	REL. ENERGY	FWHM
A			18	2102.2	9.7	1.6
	$L_{2,3}^2(^3P_{0,2})$		2	2099.6	7.1	1.6
X			46	2095.0	2.5	3.6
	$L_{2,3}^2(^1D_2)$		441	2092.52	0	1.22
2	-"-	x	53	2085.7	-6.8	2.4
1			28	2088.2	-4.4	3.1
	$L_{2,3}^2(^1S_0)$		34	2084.8	-7.7	1.4
3	-"-	x	20	2078.9	-13.6	4.8
4a			30	2068.9	-23.6	6.2
4b			15	2062.5	-30.0	4.0
	$L_1L_{2,3}(^3P_{0,1,2})$		23	2042.0	-50.5	3.0
5	-"-	x	9	2027.6	-64.9	3.8
	$L_1L_{2,3}(^1P_1)$		100	2021.8	-70.7	5.5
6	-"-	x	20	2015.3	-70.2	6.2
	$L_1L_1(^1S_0)$		61	1968.5	-124.3	10.4

partly made to correct the OS-SCF results, which is necessary when the state has more than hole per symmetry (singlet coupled) /33/. The CI's for the main line final states involved a redistribution of the two holes within the internal orbitals (those orbitals occupied in the ground state of the neutral species). For H_2S test computations were made which showed that differences in the Auger transition energies of less than 0.35 eV were induced by limiting the CI to the L-shell instead of to the full internal space.

The satellite transition energies were computed by a ΔSCF-CI procedure. For the initial states a sum-of-states SCF was performed, in which the $1a_1$ was singly occupied and another hole was distributed over the $5a_1$, $2b_1$ and $2b_2$ (with weight 1/3 each) in H_2S and over the $5a_1$, $2b_1$ and $4b_2$ in SO_2. The energies of the initial states were evaluated from these vectors by using either the SCF or the CI program. Similarly, final state vectors were obtained from an SCF having the $2a_1$ and $3a_1$ open in H_2S and the $3a_1$ and $4a_1$ open in SO_2, and having a valence hole distributed as for the initial states. Limited CI's were then performed involving a redistribution of the shake-off hole among the valence orbitals. All theoretical results in Tables 3 and 4 are normalized to the $L_{2,3}L_{2,3}(^1D_2)$ transition energy (the discrepancy between theory and experiment for the $^1D_2$ transition in H_2S is 4.8 eV, principally due to relativistic effects).

In Fig. 4 the energy region around the main $^1D_2$ and $^1S_0$ peaks are shown with the $^1D_2$ peaks normalized to the same energy. The satellite structure is seen to have roughly the same intensity as was observed in the Ar KLL case. In these cases, however, a shift in the position is found when going from H_2S to SF_6. This shift is reproduced by the calculations (cf. Tables 3 and 4). There is thus little doublt as to the origin of these satellite peaks as due to initial state shake-off. Moreover, a satellite peak of similar relative intensity is found for each main line in the spectra.

The shift in energy of the satellite from H_2S to SF_6 is seen to be accompanied by a decrease in relative intensity. In discussing the origin of such an effect the intensities of satellites due to shake-up observers should also be considered. Such lines will appear between the shake-off satellite and the main line. These shake-up lines will compete with the shake-off lines for intensity and the ratio between the two could well vary from system to system. It is also possible for the total shake-up + shake-off intensity to vary. Some of the relevant intensities were thus estimated from the calculated wavefunctions. In the sudden approximation the relative intensity of monopole transitions is given by:

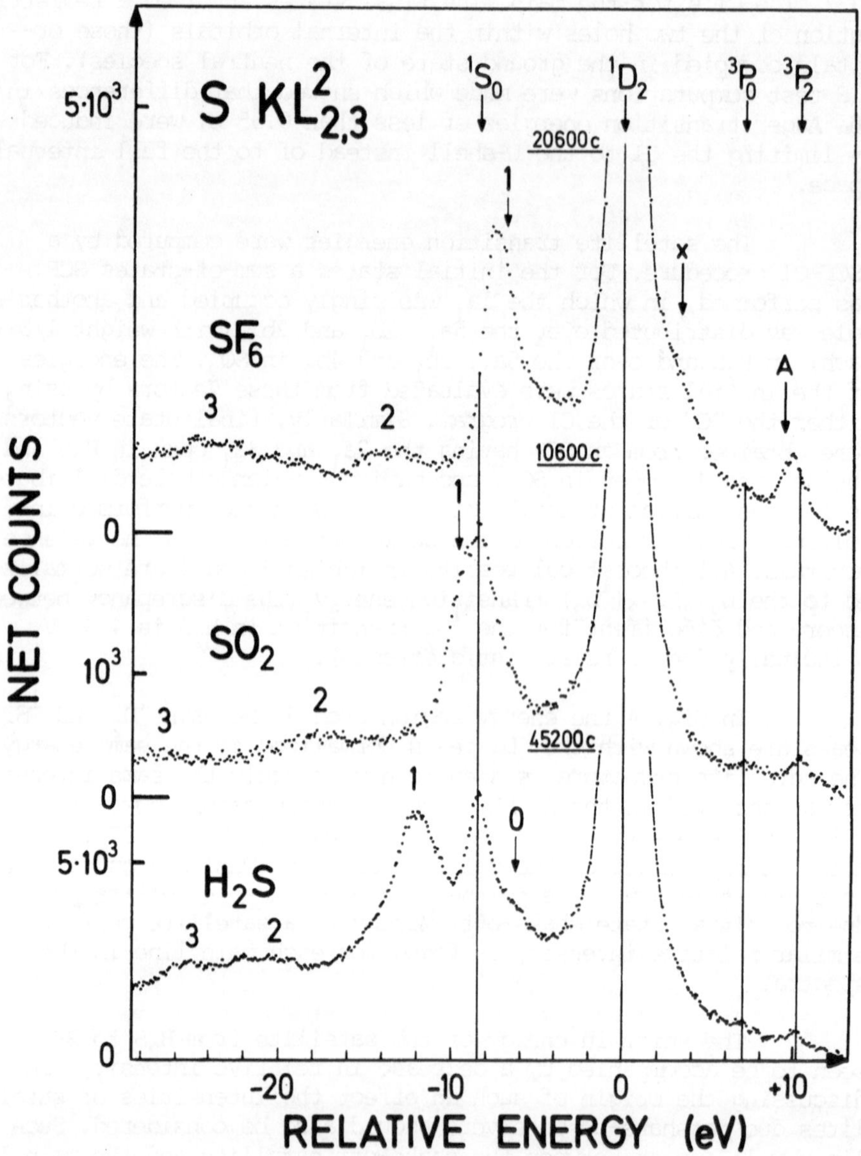

Fig. 4. $KL_{2,3}L_{2,3}$ part of the S KLL Auger electron spectra for H_2S, SO_2 and SF_6. The peak marked X for the SF_6 spectrum refers to an impurity.

$$P_{k,1} = |\langle Na_k \Psi_G | \Psi_1 \rangle|^2 \qquad (4)$$

where Ψ_G is the ground state, a_k is the annihilator of the kth electron, Ψ_1 is an eigenfunction for the ionized system and N is a normalizing factor. Thus, the total relative shake-up + shake-off intensity accompanying the initial $1a_1$ ionization is:

$$P_s = 1 - |\langle Na_{1a_1} \Psi_G | \Psi(1a_1) \rangle|^2 \qquad (5)$$

The total shake-up intensities for H_2S and SO_2 were estimated by summing over the intensities of the lowest shake-up states. It was found that the total shake-up + shake-off intensity was 25% for both H_2S and SO_2. The ratio of total shake-up to total shake-off intensity was also found to be the same for these two molecules, roughly 1:4. Our calculations were thus not able to point to any major effects which could account for the observed differences between the satellite intensities for these two molecules. Since only a very small number of shake-up states were considered in these calculations, however, it could well be that SO_2 has more shake-up intensity than H_2S, due to the much larger number of available transitions in the former case.

An interesting peak (marked A in Fig. 4) appears at the high energy side of the $^1D_2$ transition. This peak has nearly constant energy shift with respect to the $^1D_2$ line. The possibility of this line as originating from an impurity can be ruled out. Further, our calculations on these systems, as well as on atomic S and Ar, clearly shows that it cannot be due to a main line transition (a possibility would be the $^3P_2$ transition, but this occurs at a lower energy). This latter interpretation is ruled out also on intensity grounds, since the relative intensity of the line is not constant for the three molecules. It is therefore proposed that this peak is due to the $^1D_2$ Auger transition from an auto-ionizing K-excited neutral state. This type of excitation is allowed in the present case, since electron bombardment is used to produce the spectra.

6. CONCLUDING REMARKS

The methods for recording and interpreting Auger electron spectra from atoms, molecules and solids are in a continuous development. For atoms, theory is at present capable of predictions of Auger spectral features beyond one-particle models both with regard to transition rates and energies. On the experimental side, free atom spectra from metal vapors can now be studied by means of high temperature techniques /47/. The use of synchrotron radiation in the excitation of Auger spectra is being explored and, as demonstrated above, the use of preretardation of the Auger electrons for high energy spectra allows substantial improvements in resolution. A large number of problems still lack a satisfactory treatment, however. For molecules and solids, this concerns for instance transition rate calculations, where the final vacancies occur among the valence electrons, as well as satellite transitions Another area of particular interest, not explicitly discussed above, is the chemical shifts of Auger lines and their relation to the electronic structure. A number of investigations indicate that these shifts are related to the core electron binding energy shifts via the relaxation energy of the photoionization process. Other quantities will also contribute, however, and the approximations involved in discussing these effects deserve further study /48 - 54/.

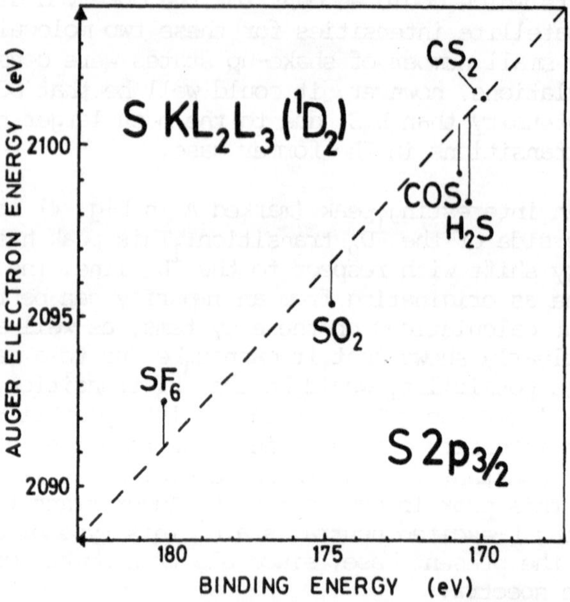

Fig. 5. Shift of $KL_2L_3(^1D_2)$ Auger line vs. S $2p_{3/2}$ binding energy shift for a number of sulphur-containing molecules.

REFERENCES

1. P. Auger, J. Phys. Radium 6, 205 (1925)
2. V. Schmidt, N. Sandner, W. Mehlhorn, M.Y. Adam and F. Wuilleumier, Phys. Rev. Lett. 38, 63 (1977)
3. G. Wentzel, Z. Phys. 43, 524 (1927)
4. U. Fano, Phys. Rev. 124, 1866 (1961)
5. U. Fano and J.W. Cooper, Rev. Mod. Phys. 40, 441 (1968)
6. F. Mies, Phys. Rev. 175, 164 (1968)
7. C.A. Nicolaides, Phys. Rev. A6, 2078 (1972)
8. K. Siegbahn et al., ESCA-Atomic, Molecular and Solid State Structure Studied by Means of Electron Spectroscopy, Almqvist-Wiksells, Uppsala (1967)
9. K. Siegbahn et al., ESCA Applied to Free Molecules, North-Holland, Amsterdam (1969)
10. W. Mehlhorn, Z. Physik 187, 21 (1965)
11. W. Mehlhorn, W. Schmitz and D. Stalherm, Z. Physik 252, 399 (1972)
12. H. Körber and W. Mehlhorn, Z. Physik 191, 217 (1966)
13. V. Schmidt in Inner Shell Ionization Phenomena (Ed. R.W. Fink et al.) U.S.A.E.C.R., CONF-720404, 538 (1973)
14. L.O. Werme, T. Bergmark and K. Siegbahn, Phys. Scr. 6, 141 (1972) and 8, 149 (1973)
15. M.O. Krause, T.A. Carlson and W.E. Moddeman, J. Phys. (Paris) Colloq. C4, 139 (1971)
16. E.J. McGuire, Phys. Rev. A11, 17 and 1880 (1975)
17. D. Ridder, Thesis, Freie Universität (Berlin) (1977)
18. H.P. Kelly, Phys. Rev. A11, 556 (1975)
19. G. Howat, T. Åberg and O. Goscinski, 2nd Internat. Conf. Inner Shell Ion. Phen. (Extended Abstracts), Freiburg (April 1976) p. 126

20. G. Howat, T. Åberg, O. Goscinski, S.C. Soong, C.P. Bhalla and M. Ahmed, Phys. Lett. 60A, 404 (1977)

21. C.J. Powell, Phys. Rev. Lett. 30, 1179 (1973)

22. G. Schön, Surf. Sci. 35, 96 (1973)

23. P.J. Bassett, T.E. Gallon, J.A.D. Matthew and M. Prutton, Surf. Sci. 35, 63 (1973)

24. L.I. Yin, I. Adler, T. Tsang, M.H. Chen, D.A. Ringers and B. Crasemann, Phys. Rev. B9, 1070 (1974)

25. D. Stalherm, B. Cleff, H. Hillig and W. Mehlhorn, Z. Naturforsch. A24, 1728 (1969)

26. W.E. Moddeman, T.A. Carlson, M.O. Krause, B.P. Pullen, W.E. Bull, G.K. Schweitzer, J. Chem. Phys. 55, 2317 (1971)

27. L. Karlsson, L.O. Werme, T. Bergmark and K. Siegbahn, J. Electron Spectr. 3, 181 (1974)

28. H. Siegbahn, L. Asplund and P. Kelfve, Chem. Phys. Lett. 35, 330 (1975)

29. J.A. Connor, I.H. Hillier, J. Kendrick, M. Barber and A. Barrie, J. Chem. Phys. 64, 3325 (1976)

30. M.T. Okland, K. Faegri Jr. and R. Manne, Chem. Phys. Lett. 40, 185 (1976)

31. R.W. Shaw Jr., J.S. Jen and T.D. Thomas, J. Electron Spectr. 11, 91 (1977)

32. I.B. Ortenburger and P.S. Bagus, Phys. Rev. A11, 1501 (1975)

33. H. Ågren, S. Svensson and U.I. Wahlgren, Chem. Phys. Lett. 35, 336 (1975)

34. R.H.A. Eade, M.A. Robb, G. Theodorakopoulos and I. Csizmadia, Chem. Phys. Lett. 52, 526 (1977)

35. W. Eberhardt, G. Kalkoffen and C. Kunz, DESY-Report-78/02 (March 1978)

36. J.P. Deslaux, Comp. Phys. Comm. 9, 31 (1975)

37. L. Asplund, P. Kelfve, B. Blomster, H. Siegbahn and K. Siegbahn, Phys. Scr. 16, 268 (1977)

38. W.N. Asaad and E.H.S. Burhop, Proc. Phys. Soc. 71, 369 (1958)

39. U. Gelius, J. Electron Spectr. 5, 985 (1974)

40. G. Wendin, Phys. Scr. 16, 296 (1977)

41. N. Mårtensson, P.-Å. Malmquist, S. Svensson, E. Basilier, J.J. Pireaux, U. Gelius and K. Siegbahn, Nov. J. de Chim. 1, 191 (1977)

42. J. Schirmer, L.S. Cederbaum, W. Domcke and W. von Niessen, Chem. Phys. 26, 149 (1977)

43. R.L. Martin and E.R. Davidson, Chem. Phys. Lett. 51, 237 (1977)

44. B. Kjöllerström, N.H. Möller and H. Svensson, Ark. Fys. 29, 167 (1965)

45. J.W. Cooper and R.E. Lavilla, Phys. Rev. Lett. 25, 1745 (1970)

46. D.P. Spears, H.J. Fischbeck and T.A. Carlson, Phys. Rev. A9, 1603 (1974)

47. W. Mehlhorn, B. Breuckmann and D. Hausamann, Phys. Scr. 16, 177 (1977)

48. J.A.D. Matthew, Surf. Sci. 40, 451 (1970)

49. S.P. Kowalczyk, L. Ley, F.R. McFeely, R.A. Pollak and D.A. Shirley, Phys. Rev. B9, 381 (1974)

50. C.D. Wagner and P. Biloen, Surf. Sci. 35, 82 (1973)

51. H. Siegbahn and O. Goscinski, Phys. Scr. 13, 225 (1976)

52. L. Asplund, P. Kelfve, H. Siegbahn, O. Goscinski, H. Fellner-Feldegg, K. Hamrin, B. Blomster and K. Siegbahn, Chem. Phys. Lett. 40, 353 (1976)

53. O. Keski-Rahkonen and M.O. Krause, J. Electr. Spectr. 9, 371 (1976)

54. D.B. Adams, J. Electr. Spectr. 10, 247 (1977)

55. L. Asplund, P. Kelfve, B. Blomster, H. Siegbahn, K. Siegbahn, R.L. Lozes and U.I. Wahlgren, Phys. Scr. 16, 273 (1977)

MANY BODY PERTURBATION METHODS FOR THE CALCULATION OF EXCITED STATES

Michael A. Robb, Dermot Hegarty and Sally Prime*

Department of Chemistry,
Queen Elizabeth College,
London, W8 7AH.

These lecture notes are intended to provide an elementary introduction to the use of the methods of diagrammatic Many Body Perturbation Theory (MBPT) for calculations on excited states of molecules. They represent neither a review nor an original article. Consequently the bibliography is limited to a few essential references. Limitations of space permit only a brief discussion of the main ideas. We shall refer the reader to the literature for numerical results where appropriate.

INTRODUCTION

The configuration interaction method (CI) [1] when used with a multi-dimensional reference space is the most general method for obtaining wavefunctions for excited states of molecules. However, this method is linear in the numbers of configurations and is not "size consistent" (i.e. as the number of electrons increases, many electron effects as manifested in the unlinked cluster excitations are not included). In contrast, many body methods (quasi-degenerate many body perturbation theory(QD-MBPT) [2-4] or many body Greens Functions (MB-GF) [5,6]) are non-linear. The non-linearity of many body methods arises through diagram factorizations which lead to "exclusion principle violating" (EPV) diagrams where the intermediate states do not correspond to physical states.

In these notes we will be concerned with the application of many body methods to the calculation of a manifold of excited states from a single energy independent effective Hamiltonian.

* Kenyatta University College, P. O. Box 43844, Nairobi, Kenya.

Cleanthes A. Nicolaides and Donald R. Beck (eds.), Excited States in Quantum Chemistry, 297–316.

Our treatment is adapted from that of Brandow [2,3]. However, in order to keep the discussion at an elementary level we shall try to formulate the problem as far as possible in the language of the CI method and matrix calculus.

Many-body perturbation theory can thus be derived from a Rayleigh Schrödinger perturbation theory formulation of the CI method. One simply replaces summations over determinantal states by summations over spin-orbitals. The expressions over determinants reduce to expressions over the one and two electron integrals in the spin-orbital basis. The terms that remain after performing order by order cancellations are enumerated using diagrams.

There are two formulations of QD-MBPT that we wish to discuss the valence/core expansion of Brandow [2] and the energy independent propagator [5] method. The valence/core expansion yields an effective Hamiltonian which gives the eigenvalues of the valence states relative to a core (in which the valence electrons are absent). This method is suitable for an open shell ground state. The energy independent propagator type formulation is suitable for calculation of excitation energies relative to a closed shell reference function.

We shall illustrate these methods by first describing quasi-degenerate Rayleigh-Schrödinger perturbation theory (QD-RSPT) as an algorithm for CI with a partitioned Hamiltonian. The reduction of QD-RSPT to the valence/core expansion of QD-MBPT will then be illustrated using the example of the π - Hamiltonian of ethylene. Finally, we shall demonstrate the propagator type form of QD-MBPT using the single particle propagator (ionization potentials) as an example.

QUASI-DEGENERATE RAYLEIGH-SCHRODINGER PERTURBATION THEORY (QD-MBPT).

Our objective is to compute an effective Hamiltonian operator $\tilde{H}$ which operates on a model or reference space $\{\phi_R\}$ so that we have

$$\tilde{H} A_\alpha = E_\alpha A_\alpha \tag{1}$$

with

$$\psi_\alpha = \sum_R A_{\alpha R} \phi_R \tag{2}$$

The eigenvalues E_α span the region of the spectrum of experimental interest and ψ_α is the projection of the complete wavefunction onto the model space ϕ_R . The operator $\widetilde{H}$ will be obtained as a perturbation expansion that is not dependent on the particular eigenvalue E_α

Let us consider the partitioning of the full space on which the Hamiltonian is allowed to act, into a reference space and a complimentry secondary space. The matrix representation of the Schrodinger equation can now be written in partitioned form as

$$\left(\begin{array}{c|c} H^0+V^1 & Z^+ \\ \hline Z & W \end{array}\right)\left(\begin{array}{c} A_\alpha \\ \hline B_\alpha \end{array}\right) = E_\alpha \left(\begin{array}{c} A_\alpha \\ \hline B_\alpha \end{array}\right) \quad (3)$$

H^0 and V^1 are the diagonal and off diagonal part of the matrix representation of the Hamiltonian in the reference space. W is the representative in the secondary space, while Z^+ and Z contain the matrix elements between the reference and secondary spaces. The coefficient vectors A_α and B_α give the wave-function projection on to the reference and secondary spaces.

If we now eliminate B_α from eq. 3 we have

$$\widetilde{H} A_\alpha = E_\alpha A_\alpha \quad (4)$$

with

$$\widetilde{H}(E_\alpha) = H^0 + V^1 + Z^+(E_\alpha - W)^{-1} Z \quad (5)$$

If we now expand the E_α dependence out of the inverse matrix in eq. 5 we obtain (see ref [2-4,7])

$$\widetilde{H} = H^0 + V^1 + \sum_A^{\oplus} Z_A^+ \sum_B^{\oplus} (E_B - W)^{-1} Z_B$$

$$+ \text{ FOLDED DIAGRAMS} \quad (6)$$

The symbol $\sum^{\oplus}$ is used here to denote a direct sum, Z_A^+ (Z_A) denotes the A th row (column) of $Z^+(Z)$ and the addition in simple matrix multiplication is thus replaced by a direct sum. The fourth term (FOLDED DIAGRAMS) in eq. 6 represents the energy renormalization terms. The algebraic structure of these terms is rather complicated and we shall discuss them subsequently using diagrammatic methods. Finally E_B is just the diagonal element of H^0

The most important features of the effective Hamiltonian defined by eq. 6 are

a) it is energy independent so that all eigenvalues are obtained from a single diagonalization and

b) it is non-hermitian.

One may generate the usual RSPT perturbation series using an expansion of the form

$$(A-B)^{-1} = A^{-1} + A^{-1}BA^{-1} + \cdots \qquad (7)$$

where A is some part of $E_B - W$ which is diagonal. Finally, one should observe that one obtains perturbation series for <u>each</u> matrix element of $\bar{H}$

TRANSCRIPTION OF QD-RSPT TO THE LINKED VALENCE EXPANSION OF QD-MBPT.

The transcription of QD-RSPT to diagrammatic QD MBPT is most easily understood by considering an example [8] that is familiar to most quantum chemists - the Pariser - Parr effective π-Hamiltonian for ethylene.

The reference space consists of all possible occupancies of the π atomic orbitals ρ_1 and ρ_2 :

$$\begin{aligned} \phi_1 &= |\alpha\beta \cdots \rho_1 \bar{\rho}_1| & \phi_2 &= |\alpha\beta \cdots \rho_1 \bar{\rho}_2| \\ \phi_3 &= |\alpha\beta \cdots \bar{\rho}_1 \rho_2| & \phi_4 &= |\alpha\beta \cdots \rho_2 \bar{\rho}_2| \end{aligned} \qquad (8)$$

where we use $\alpha\beta\cdots$ to denote the frozen core orbitals. Thus the reference space corresponds to an orthogonal valence bond calculation with a frozen core of σ orbitals. The matrix elements of $H^0 + V^1$ in this reference space (using the ZDO approximation) can be written using the usual Pariser - Parr parameters as

$$H^0+V^1 = \begin{pmatrix} 2\alpha+\gamma_{11} & \beta & 0 & 0 \\ \beta & 2\alpha+\gamma_{12} & 0 & \beta \\ \beta & 0 & 2\alpha+\gamma_{12} & \beta \\ 0 & \beta & \beta & 2\alpha+\gamma_{11} \end{pmatrix} \qquad (9)$$

where

$$\begin{aligned} \alpha &= \langle \rho_i | h^{core} | \rho_i \rangle \\ \beta &= \langle \rho_i | h^{core} | \rho_j \rangle \\ \gamma_{11} &= \langle \rho_i \rho_i | \rho_i \rho_i \rangle \\ \gamma_{12} &= \langle \rho_i \rho_i | \rho_j \rho_j \rangle \end{aligned} \qquad (10)$$

The secondary space consists of all possible excitation of both core σ and valence(π) orbitals. For purposes of discussion we shall truncate the secondary space at double excitations from this manifold. If we use symbols $k\, \ell\, m \cdots$ etc. to denote virtual orbitals it is convenient to group the double substitutions into various classes.

VALENCE EXCITATIONS	$\lvert \alpha \beta \cdots k \ell \rvert$	
VALENCE-CORE EXCITATIONS	$\lvert k \beta \cdots \rho_i \ell \rvert$	(11)
	$\lvert \rho_j \beta \cdots \rho_i \ell \rvert$	
CORE-EXCITATIONS	$\lvert k \ell \cdots \rho_i \rho_j \rvert$	
	$\lvert \rho_i \rho_j \cdots \rho_k \rho_\ell \rvert$	

Let us now look at an example of the transcription to diagrammatic form of one of the contributions to the third term in eq. 6.

$$\langle \phi_1 | \tilde{H} | \phi_2 \rangle = \beta + \sum_R \frac{\langle \phi_1 | H | \phi_R \rangle \langle \phi_R | H | \phi_2 \rangle}{E_2 - E_R} \tag{12}$$

where $\langle \phi_1 | H | \phi_R \rangle$ is an element of Z^+ and E_R is obtained from the diagonal part of W. Let us look at the term where

$$\phi_R = \lvert \ell \beta \cdots \rho_1 k \rvert \tag{13}$$

Considering the energy denominators first, it is convenient to compute our energies relative to E_{core}, the energy obtained if all the π electrons have been removed.
Thus

$$E_{CORE} = \sum_\gamma \varepsilon_\gamma$$

where ε_γ is the orbital energy of orbital γ. The energy denominators relative to E_{core} one are thus given as

$$E_2 = \varepsilon_1 + \varepsilon_{\bar{1}}$$

$$E_R = \varepsilon_k + \varepsilon_\ell + \varepsilon_1 - \varepsilon_\alpha$$

Then, using the rules for matrix elements our determinantal wavefunctions to evaluate the numerator we have

$$\frac{\langle \phi_1 | H | \phi_R \rangle \langle \phi_R | H | \phi_2 \rangle}{E_2 - E_R} = \frac{\langle \bar{1} \alpha | k \ell \rangle \langle k \ell | \bar{1} \alpha \rangle}{\varepsilon_1 + \varepsilon_{\bar{1}} - (\varepsilon_k + \varepsilon_\ell + \varepsilon_1 - \varepsilon_\alpha)} \tag{14}$$

We can represent the right hand side of eq. 14 diagrammatically as shown in fig. 1.

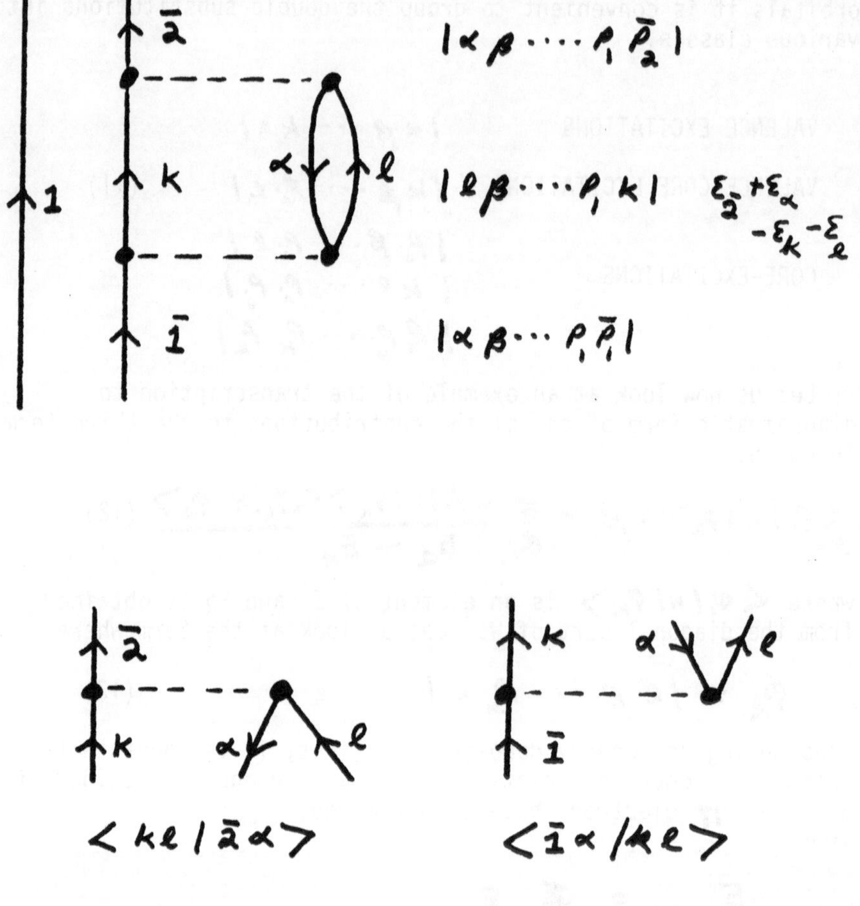

Figure 1. A second order valence-core diagram.

The rules for evaluating this diagram are summarized in fig. 2 and we shall consider fig. 1 in some detail.

There are two essential diagram elements; vertical oriented lines and the horizontal dashed lines which represent interaction vertices. The upward oriented lines run over the valence and virtual orbitals while the downward lines carry the labels of the core orbitals. The states involved in the perturbation expression can be read off from the three levels of the diagram from bottom to top: Φ_1 Φ_R Φ_2 The numerator of

RULE 1 - INTERACTION LINES

HOLE LINE CORE ORBITAL α

PARTICLE LINE k

VALENCE PARTICLE μ

$$\langle AB|CD\rangle = \iint \frac{A(1)\,B(2)\,[1-P_{12}]\,C(1)\,D(2)}{r_{12}}$$

$$\langle A|h|B\rangle$$

RULE 2 - DENOMINATOR

$$\mathcal{E}_C + \mathcal{E}_E - \mathcal{E}_B - \mathcal{E}_D$$

RULE 3 - SIGN

$$(-1)^{h+\ell}$$

h = NO. OF HOLE LINE

ℓ = NO. OF CLOSED LOOP

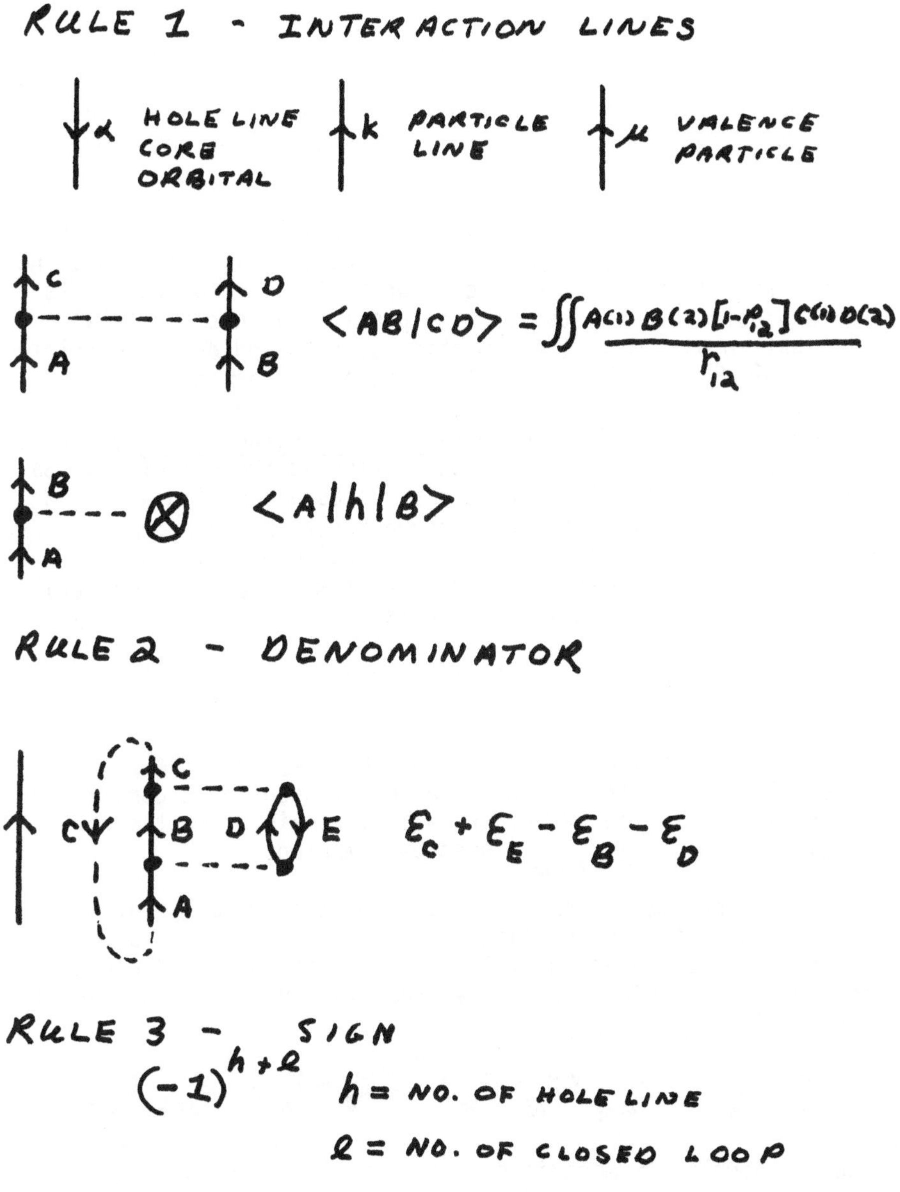

Fig. 2. Rules for the evaluation of the diagrams of QD-MBPT.

the perturbation expression is computed from the interaction vertices according to rule 2 in fig. 2. For the evaluation of the energy denominator (rule 3 fig. 2) one first closes the diagram from top to bottom, discarding the "bottom" labels. The denominator is then given as the sum of the orbital energies of the down going lines minus the orbital energies of the upward lines.

One may proceed by enumerating all possible types of diagram that arise from the configurations of eq. 11. In doing this one encounters diagrams of the type shown in fig. 3c (which arises from a double core excitation with passive valence orbitals). Note that in dia. 3c, the summation over k and ℓ is restricted so that $P_i P_j \neq k\ell$ since otherwise the intermediate state would violate the Pauli exclusion principle. However, the correlation energy of the core (valence electrons removed but valence orbitals retained) can be written as the sum of dia. 3a and dia. 3b. Thus we may write dia. 3c as a sum of dia. 3d, 3e, 3f etc. The new dia. 3e and 3f correct for the overcounting in 3d. However, if we are interested in the relative energies of the valence states, we need never evaluate ΔE_{CORE} (ie dia. 3d.) since it only contributes to the reference zero of the energy scale. However, we can no longer interpret dia. 3e and 3f in the usual fashion since the intermediate states violate the exclusion principle. (EPV diagrams).

Thus we have introduced some non-linearity into the problem. The double excitations of the core do not need to be considered explicitly. Only the EPV diagrams that remain after the partial cancellation of ΔE_{CORE} need be computed.

There is one more element of diagrammatic QD-MBPT that we must discuss - the folded diagrams of eq. 6. These diagrams result from the expansion of the explicit eigenvalue dependence from eq. 5. A typical energy renormalization term in QD-RSPT is given at third order as

$$\sum_R \sum_B - Z^{+}_{AR} (E_B - E_R)^{-1} (E_C - E_R)^{-1} Z_{RB} V^{1}_{BC}$$

Note the highly assymmetric nature of the expression. This term is enumerated using "folded diagrams" as shown in fig. 4.

In fig. 4a we show the diagram corresponding to one of the terms in eq. 15. The levels of the diagram are labelled with energies of the states ϕ_A ϕ_R ϕ_B ϕ_C . Using the usual diagram rules one obtains the correct expression for the numerator in the perturbation expression from fig. 4a. However, the diagram must be kinked or "folded" along the circled lines to give fig. 4b so that the ususal rules for the denominators

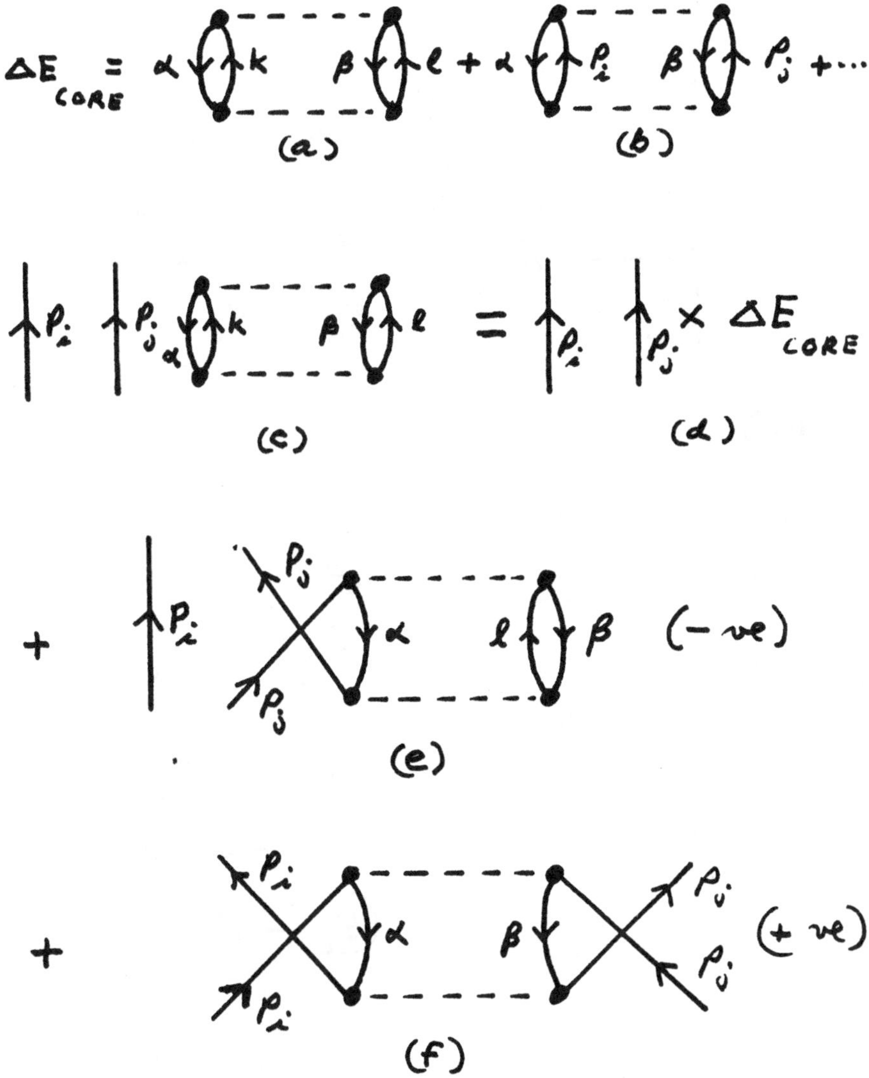

Figure 3. Factorization of the core correlation energy (dia. a & b)to give EPV dia. e and f.

$$E_C \qquad \varepsilon_{P_i} + \varepsilon_{P_j}$$

$$E_B \qquad \varepsilon_{P_\ell} + \varepsilon_{P_k}$$

$$E_R \qquad \varepsilon_\ell + \varepsilon_k + \varepsilon_{P_\ell} - \varepsilon_\alpha$$

$$E_A \qquad \varepsilon_{P_\ell} + \varepsilon_{P_m}$$

(a)

$$\left(E_B - E_R \right)$$

$$\varepsilon_{P_k} + \varepsilon_\alpha - \varepsilon_k - \varepsilon_\ell$$

$$\varepsilon_{P_i} + \varepsilon_{P_j} + \varepsilon_\alpha$$

$$-\varepsilon_k - \varepsilon_\ell - \varepsilon_{P_\ell}$$

$$\left(E_C - E_R \right)$$

(b)

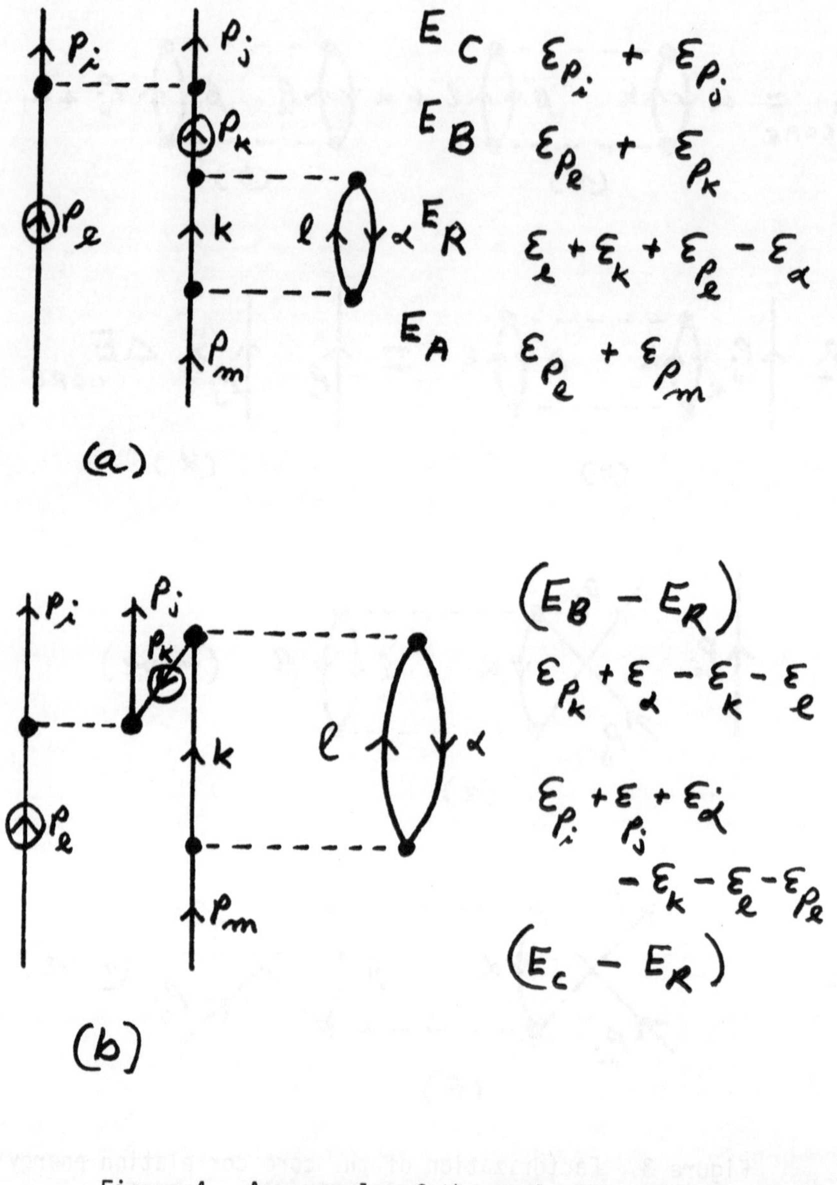

Figure 4.   An example of the evaluation of a folded
            diagram. Dia. a is folded to give dia. b
            which may be evaluated by the rules given in
            fig. 2.

apply. (Note that because we evaluate our diagrams from bottom to top we must fold downwards. If one discards the "top" labels then one must fold upwards). In fig. 4b we illustrate the application of the rules for the denominators for the folded diagrams.

We should point out at this stage that in the special case where $\rho_\ell = \rho_i$ and $\rho_k = \rho_j$ in fig. 4b one may sum the folded diagrams to infinite order (using a geometric series). The result of this procedure is that one may include the "diagonal" folded diagrams in the diagram of figure 2 as a denominator shift. Thus the denominator of the diagram in fig. 2 becomes

$$\text{dia. } 2 = \varepsilon_{\bar{2}} + \varepsilon_1 + \Delta E_3 - \varepsilon_k - \varepsilon_\ell \tag{16}$$

where

$$\Delta E_3 = \langle \phi_3 | \tilde{H} | \phi_3 \rangle - E_3 \tag{17}$$

The energy shift ΔE_3 has a simple interpretation. It is the eigenvalue shift evaluated as though all the other reference states were absent.

Finally, in fig. 5 we have given the diagrams (to second order) that contribute to the effective π Hamiltonian. Diagrams a, d, g. represent the "bare" one and two electrons integrals, diagrams b, e, i, account for σ polarization and $\sigma - \pi$ correlation, diagrams c f h represent σ-σ correlation and diagram j accounts for π-π correlation. Clearly this analysis could be regarded as a scheme for the "ab-initio" calculation of semi-empherical parameters (see reference [8] for a detailed discussion). However, the significant point for the present discussion is that one has a scheme for the construction of an energy independent effective Hamiltonian which gives the relative energies of the valence (ie π - electron) states. Further, the explicit calculation of the correlation energy of the σ electrons is not required.

The above formulation of QD-MBPT is probably the most general method for the calculation of excited state wavefunctions since the ground state need not be a closed shell. Thus to conclude this section we should make a few brief comments on the practical aspects of the method.

In our own work [7] we have examined the numerical stability of QD-RSPT in comparison with variational CI as a prelude to performining QD-MBPT calculations. Limitations of space preclude the presentation of the details of this work so we shall limit discussion to the main conclusions.

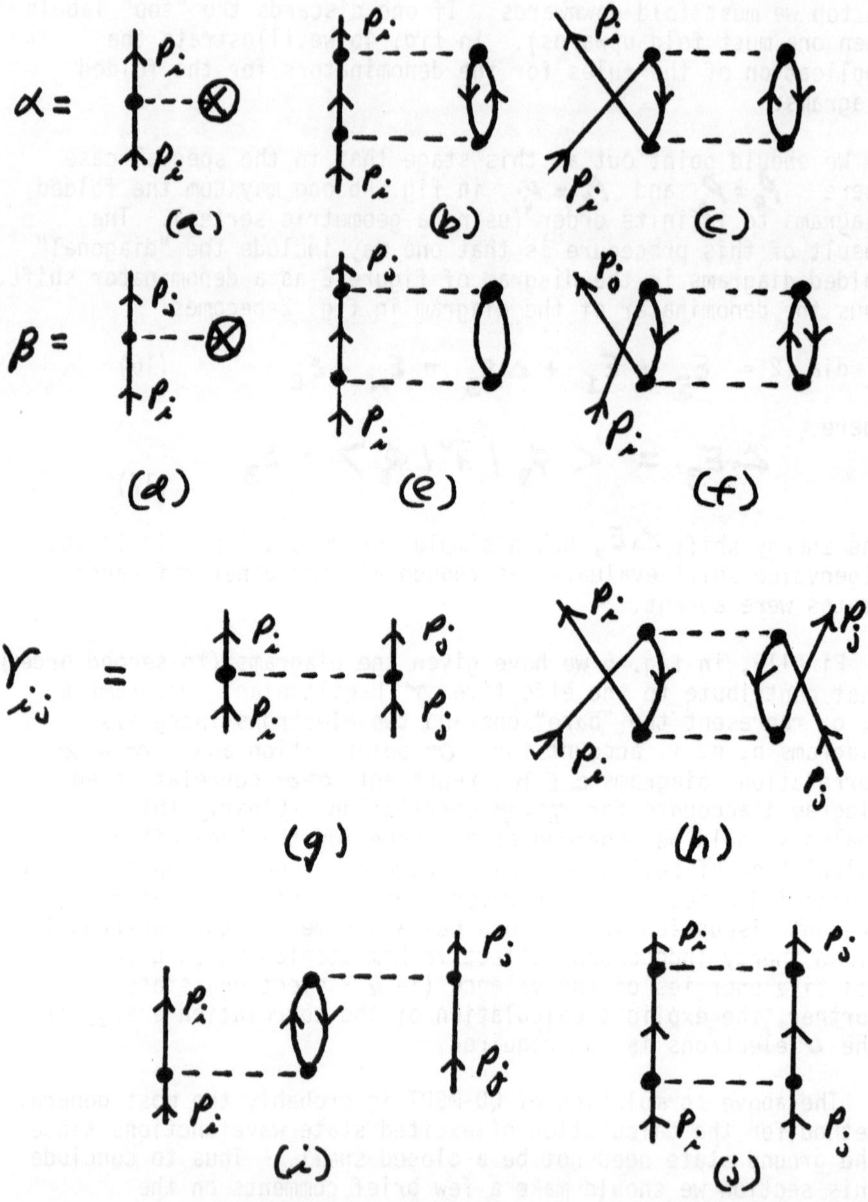

Figure 5. Diagrams for the Pariser - Parr effective π Hamiltonian.

Initially, the main practical difficulties were associated with the summation of the folded diagrams. Certainly, we have found it essential to shift denominators with the diagonal folded diagrams; however, we have found that one need only keep the off-diagonal folded diagrams to third order.

The central technical problem in our opinion is the so-called "intruder state problem". If one of the secondary states lies embedded in or very close to the reference manifold then the perturbation expansion diverges. From a practical point of view the solution to the problem is simply to include the intruder state in the reference manifold. However, in the case where the valence state is in fact embedded in the continuum the problem will become intractable as the basis set is extended to completeness.

PROPAGATOR OR EQUATIONS OF MOTION TYPE THEORIES IN QD-MBPT.

In addition to the core/valence form of QD-MBPT just discussed, it is possible to formulate the propagator or equation of motion [5] methods for excitation energies in terms of an energy independent effective Hamiltonian. As an example we shall consider the single particle many-body Greens Function(MB-GF) method as implemented by Cederbaum and co-workers[9, 10] A detailed examination of the problem is clearly not possible in the limited space available so we shall concentrate on the diagrammatic representation of Cederbaum[9] in lowest orders only.

For the problem of the computation of ionization potentials, the reference space consists of all single hole states

$$\{h_1\} = |\alpha\rangle \tag{18}$$

while the secondary space contains all single and double excitations relative to the single hole states:

$$\{h_3\} = |\alpha^{k}_{\beta}\rangle \tag{19a}$$

$$\{h_5\} = |\alpha^{k\ell}_{\beta\gamma}\rangle \tag{19b}$$

where for example $|\alpha^{k\ell}_{\beta\gamma}\rangle$ represents the state where the occupied spin orbital α has been removed and the spin orbitals β and γ have been replaced by virtual spin orbitals k and ℓ . The second order QD-MBPT diagrams are given in fig. 6.

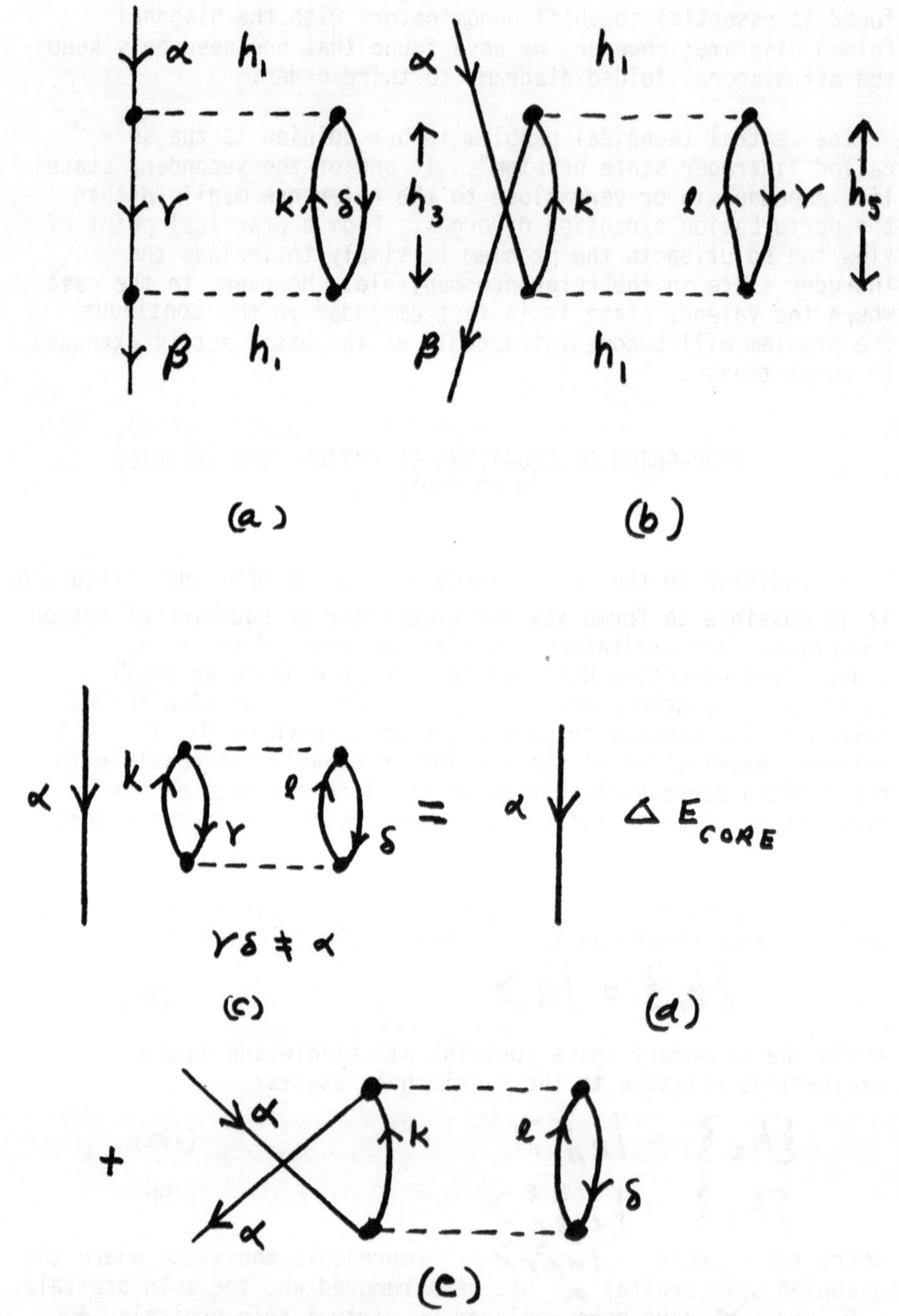

Figure 6. Second order diagrams for single hole excitation in QD-MBPT.

The diagrams are evaluated using the same rules as given in fig. 2 except that we no longer have any valence lines. Diagram 6a involves the intermediate state h_3 while dia. 6b involves the intermediate state h_5 . In fig. 6c we show again the factorization of ΔE_{CORE}; however, now ΔE_{CORE} is the correlation energy of the closed shell ground state. Again we have EPV diagrams which give rise to the non-linearity of the problem.

In the MB-GF method of Cederbaum [9] one observes a simple one - to one correspondence between the diagrams through third order for the self-energy and the diagrams for the effective Hamiltonian of QD-MBPT. However, the energy denominators are different at second order. For example the energy denominators for dia. 6a and 6b are

$$\text{dia 6a(QD-MBPT)} = -\varepsilon_\alpha + \Delta E_\alpha + \varepsilon_\gamma + \varepsilon_\delta - \varepsilon_k \tag{20a}$$

$$\text{dia 6a(MB-GF)} = -\omega + \varepsilon_\gamma + \varepsilon_\delta - \varepsilon_k \tag{20b}$$

$$\text{dia 6b(QD-MBPT)} = \varepsilon_\beta + \Delta E_\alpha + \varepsilon_\gamma - \varepsilon_k - \varepsilon_\ell \tag{21a}$$

$$\text{dia 6b(MB-GF)} = \omega + \varepsilon_\gamma - \varepsilon_k - \varepsilon_\ell \tag{21b}$$

In eq. 20a and 21a the term ΔE_α is a denominator shift from the folded diagrams and in equations (20b) and (21b) ω is the ionization potential. The relationship between the two sets of denominators can be established by observing that for near degeneracy we have

$$\omega \approx \varepsilon_\alpha - \Delta E_\alpha \approx \varepsilon_\beta - \Delta E_\beta \approx \cdots \tag{22}$$

Thus dia. 6a is almost the same in the two methods. The essential difference is that in QD-MBPT only part of the energy dependence (see eq. 16 and 17) has been resummed back into the denominators. On the other hand, for dia. 6b the two theories differ by $2\Delta E_\alpha$ in the denominators. Brandow [2] has shown how this difference may be accounted for and it is particularly instructive to illustrate the essential feature of this argument as given in fig. 7.

Consider the 4th order diagrams shown in fig. 7a and 7b. The numerators of the perturbation expressions for both diagrams are identical. However, because the two diagrams differ only in the relative order of the lowest interaction lines the sum factorizes. If we denote the denominators of the left hand and right hand parts of each diagram as E_L and E_R. Then the sum may be written as

$$\left[E_R\,(E_L+E_R)\,E_R\right]^{-1} + \left[E_R\,(E_L+E_R)\,E_L\right]^{-1} = \left[E_R\,E_L\,E_R\right]^{-1}$$

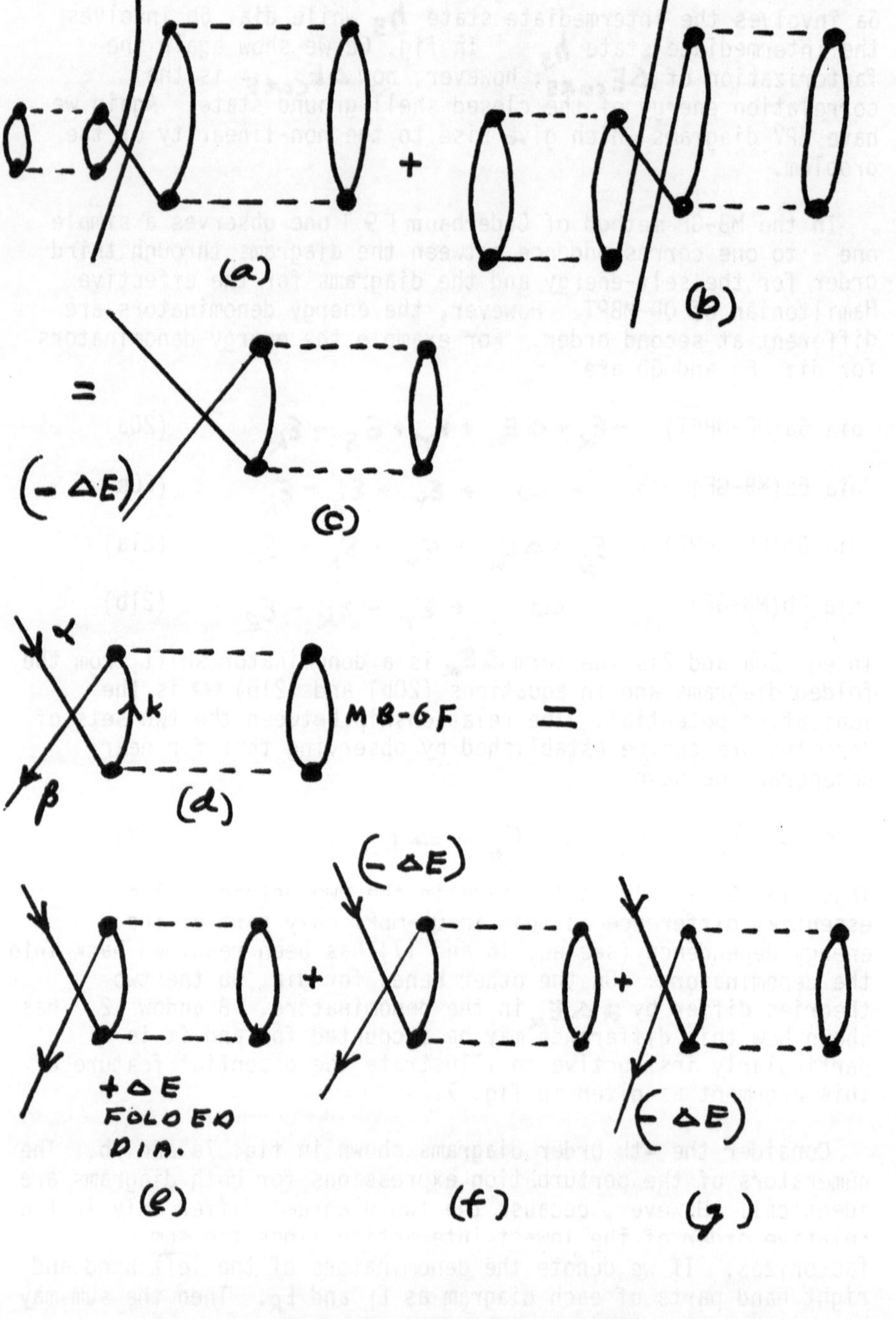

Figure 7. Expansion of the self energy diagrams of MB-GF methods in terms of QD-MBPT diagrams.

Thus we may think of the left hand part of the diagram as though it was inserted into the right. If the insertion is diagonal then it may be summed to produce a denominator shift of $-\Delta E$ as indicated in diagram 7c. One may have a similar insertion on the other external line so that one can represent the MB-GF dia 7d as the sum of the QD-MBPT dia. 7e, 7f and 7g. Thus the two $-\Delta E$ terms partly cancel the ΔE from the folded diagrams. Thus one may think of the MB-GF diagrams as being a higly summed form of QD-MBPT.

We must emphasize that the relationship between MB-GF and QD-MBPT just discussed is not exact. The ΔE_α from the folded diagrams contains many terms that are not cancelled by similar terms that result from diagram factorizations of the type just discussed. Beyond lowest orders of perturbation theory QD-MBPT and MB-GF differ significantly. The difference arises because the MB-GF theory has a different reference manifold that is constructed by allowing complete sets of second quantized operators to operate on the exact correlated ground state wavefunction [11]. As a consequence the MB-GF cannot be interpreted in terms of simple CI type language. However, for calculation of ionization potentials the numerical results through third order of perturbation theory should be very similar since the folded diagrams do not begin to occur until 4th order of perturbation theory.

It is possible to carry out the same analysis for the particle -hole propagator(electronic excitation energies). Again the analysis is simple for the Random Phase Approximation (see reference [2]); however, beyond this the theories are quite different owing to the use of a different reference manifold.

We believe that there are some advantages in the QD-MBPT approach as compared to the MB-GF method. Firstly, one obtains all the eigenvalues from a single matrix diagonalization and dependence of the MB-GF method. Secondly, the intruder state problem is more easily dealt with in QD-MBPT. A consideration of the calculation of the shake up states (the h_3 states in eq. 19a) illustrates this point. In the MB-GF theory [9] these states have been treated using a 2 particle-hole propagator. In QD-MBPT these states will occur in a natural way as intruder states in the reference space of the single hole states and do not require any special techniques.

Finally, let us comment very briefly on practical aspects of QD-MBPT methods for propagator type theories. The major problem is to decide which diagrams beyond second order are to be summed. In our own work [12] we have summed the "ladder" diagrams which do not contain non-diagonal hole line interactions. The accuracy obtainable is similar to the 3rd order MB-GF theory [10]. The most important practical consideration within this type of calculation is the extent to which the effects of ground state correlation can be taken as independent of the particular reference state. As an example, consider the third order diagrams shown in figure 8.

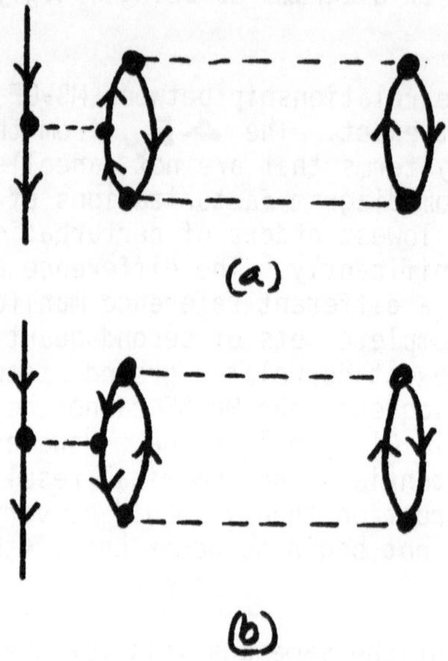

Figure 8. Modification of the effects of ground state correlation energy by an external hole line.

Here the effects of ground state correlation energy (see for example diagram 6b) is modified by the external hole line. If this sort of effect has to be included to high orders of perturbation theory then the advantage of the non-linearity of the method becomes lost. Cederbaum's calculations [9] indicate that each of these diagrams is large but their sum is very small.

CONCLUSIONS

In these lectures we have attempted to indicate the main features of QD-MBPT for the calculation of excited states. From a practical point of view these methods are still in the early stages of development. However, we have tried to show that the generality of the method, particularly in the form of the valence/core expansion, may lead to applicability in a wide range of excited state calculations.

ACKNOWLEDGEMENTS

Two of us (S. P. and D. H.) wish to acknowledge the SRC for the provision of studentships.

REFERENCES

1. I. Shavitt, Application of Electronic Structure Theory vol. 4 ed. H. F. Schaefer III(Plenum, New York, 1977)

2. B. H. Brandow, Rev. Mod. Phys. 39, 771 (1967).

3. B. H. Brandow, Adv. in Quantum Chemistry 10, 187 (1977).

4. V. Kvasnicka, Adv. Chem. Phys. 26, 345, (1977).

5. J. Linderberg and Y. Ohrn, Propagators in Quantum Chemistry (Academic Press, 1973).

6. Gy. Csanak, H. S. Taylor and R. Yaris, Advan. Atomic Mol. Phys. 7, 287 (1971)

7. D. Hegarty and M. A. Robb (to be published).

8. B. H. Brandow, Quantum Theory of Polymers ed. J. M. Andre and J. Ladik (Plenum, 1977).

9. L. S. Cederbaum and W. Domcke, Adv. Chem. Phys. 26, 205(1977)

10. W. von Niessen, G. H. F. Diercksen and L. S. Cederbaum, J. Chem. Phys. 67, 4124 (1977).

11. B. Pickup and O. Goscinski, Mol. Phys. 36, 1013 (1973).

12. S. Prime and M. A. Robb, Chem. Phys. Lett. 47, 527, (1977).

THE CALCULATION OF ATOMIC AND MOLECULAR ELECTRON BINDING ENERGIES

Yngve Öhrn

Department of Chemistry, University of Florida
Gainesville, Florida 32611

I. INTRODUCTION

Interest in accurate theoretical determination of electron binding energies has grown with the experimental advances in the field of photoelectron spectrometry. Much early work [1] on ionization potentials, photoionization cross sections, chemical shifts etc. relied on semiempirical models. It was also recognized early [2] that ground state Hartree-Fock ab-initio orbital energies (according to Koopmans' theorem) can with some success be used as measures of valence electron binding energies, while core ionization energies are rather poorly represented in this manner.

Detailed studies have been made of the nature of the correction terms which are necessary in order to go beyond the results of Koopmans' theorem. Pickup and Goscinski [3] have given an erudite analysis of the error terms through second order in perturbation theory and they stress the propagator or Green's function equations [4] as the best direct method to obtain electron binding energies. The work of Cederbaum et al. [5], Purvis, and Öhrn [6], and Simons et al. [7] indicates clearly that when a Møller-Plesset [8] partitioning of the many-electron Hamiltonian is used, terms at least through third order in perturbation theory are needed to produce a predictive theory.

We separate the correction terms to Koopmans' theorem results into two classes. The first we call relaxation terms [3] and the second class is denoted correlation terms. There are sometimes reasons to consider further subdivisions of the correction terms but for our purposes this will suffice. The relaxation terms can

Cleanthes A. Nicolaides and Donald R. Beck (eds.), Excited States in Quantum Chemistry, 317–328.

be obtained through all orders by the ΔE_{SCF} procedure [9]. This consists of separate Hartree-Fock calculations on the N-electron ground state and the (N-1)-electron state (or N+1)-electron state considered to be the final ion state of an electron detachment (or attachment) process. Subtraction of the resulting total energies then yields the appropriate electron binding energy with full account taken for relaxation. The correlation terms require further analysis and we will repeat some of the recent results of Born, Kurtz, and Öhrn [10]. The electron correlation yields complicated terms in third order and there is a great need for simplified expressions which yield adequate numerical results for electron binding energies from the innermost core region to the outer valence region for a variety of atomic and molecular systems. The relative importance of the relaxation and the correlation terms varies a great deal from the core to the valence region (see Fig. 1) and there is a delicate balance between these two kinds of terms which makes the problem of calculating electron binding energies a challanging one.

In the next section I define some of the notation used and introduce the basic ideas of the theory of the electron propagator.

II. THE ELECTRON PROPAGATOR

The importance of the electron propagator is readily appreciated from its spectral form as discussed by Linderberg and Öhrn [4]. In a spin orbital basis this spectral form is a matrix with elements

$$G_{ij}(E) = \sum_n \frac{\langle N|a_i|N+1,n\rangle\langle N+1,n|a_j^\dagger|N\rangle}{E - E_n(N+1) + E_0(N) + i\eta} + \frac{\langle N|a_j^\dagger|N-1,n\rangle\langle N-1,n|a_i|N\rangle}{E + E_n(N-1) - E_0(N) - i\eta}, \tag{1}$$

which have only simple poles in the energy plane when the spin orbital basis is discrete and finite. When the convergence parameter $\eta > 0$ approaches 0 these poles are all on the real energy axis and equal to the difference between the stationary state total energies of the N-electron ground state and the appropriate final ion states. The overlap amplitude [11] between such states

$$f_n(j) = \langle N+1,n|a_j^\dagger|N\rangle, \quad g_n(i) = \langle N-1,n|a_i|N\rangle \tag{2}$$

involve the electron field operators a_i, and $a_i^\dagger$ corresponding to the spin orbital basis [4]. Normally this would be the ground state Hartree-Fock spin orbital basis, which would be divided into occupied and unoccupied spin orbitals. Throughout this paper we use the notation a,b,c,... for occupied spin orbitals, p,q,r,... for unoccupied, and i,j,k,... for unspecified spin orbitals. These overlap amplitudes are important for the calculation of intensities in photoelectron spectrometry for principal as well as satellite structures [11].

The many-electron Hamiltonian can be expressed as

$$H = \sum_{ij}\sum (i|h|j)a_i^\dagger a_j + \tfrac{1}{4}\sum_{ijkl}\sum\sum\sum \langle ij||kl\rangle a_i^\dagger a_j^\dagger a_l a_k, \tag{3}$$

where the antisymmetric two-electron integrals

$$\langle ij||kl\rangle = \int i^*(1)j^*(2)r_{12}^{-1}(1-P_{12})k(1)l(2)d(1)d(2) \tag{4}$$

are used.

This Hamiltonian operates in the Fock space spanned by all the independent eigenstates of the number operator of the total number of electrons, which can be constructed within the chosen spin orbital basis.

When we use a spin orbital basis which diagonalizes the Fock operator

$$F_{ij} = (i|h|j) + \sum_{kl}\sum \langle ik||jl\rangle\langle a_k^\dagger a_l\rangle = \varepsilon_i\delta_{ij} \quad , \tag{5}$$

we can naturally partition the Hamiltonian as [8]

$$H = H_0 + (H-H_0), \tag{6}$$

where

$$H_0 = \sum_i \varepsilon_i a_i^\dagger a_i . \tag{7}$$

The electron propagator consistent with the uncorrelated electron dynamics described by H_0 can be denoted $\underline{G}_0$ and its matrix elements in the spin orbital basis expressed as in Eq. (1) for the full electron propagator. For $\underline{G}_0$, however, we now interpret |N as the Hartree-Fock ground state

$$|N\rangle = a_1^\dagger a_2^\dagger \ldots a_N^\dagger|vac.\rangle = |HF\rangle \tag{8}$$

where $|vac.\rangle$ is the nondegenerate eigenstate of the electron number operator corresponding to the eigenvalue zero. Similarly we have

$$E_0(N) = \sum_a \varepsilon_a \tag{9}$$

and $E_b(N-1) = E_0(N) - \varepsilon_b$, $E_p(N+1) = E_0(N) + \varepsilon_p$, (10) with

$$|N-1,b\rangle = \prod_{a=b} a_a^\dagger|\text{vac.}\rangle, \text{ and } |N+1,p\rangle = \prod_a a_a^\dagger a_p^\dagger|\text{vac.}\rangle , \tag{11}$$

respectively. Normalized determinantal states, and the anti-commutation relations

$$[a_i,a_j]_+ = [a_i^\dagger,a_j^\dagger]_+ = [a_i,a_j^\dagger]_+ - \delta_{ij} = 0 \tag{12}$$

yield $f_p(j) = \langle N+1,p|a_j^\dagger|N\rangle = \pm\delta_{pj}$, and (13)

$$g_b(i) = \langle N-1,b|a_i|N\rangle = \pm\delta_{bi} . \tag{14}$$

We see from Eq. (1) that the full electron propagator $G_{ij}(E)$ depends on the electron field operators a_i, and $a_j^\dagger$, which justifies the notation [4]:

$$G_{ij}(E) = \langle\langle a_i;a_j^\dagger\rangle\rangle_E \tag{15}$$

The expression of $G_{ij}(E)$ in Eq. (1) can now be manipulated using the identities

$$E(E - E_n(N+1) + E_o(N))^{-1}$$

$$= 1 + (E_n(N+1) - E_o(N))(E - E_n(N+1) + E_o(N))^{-1}; \tag{16}$$

$$\langle N|a_i|N+1,n\rangle(E_n(N+1) - E_o(N)) = \langle N|[a_i,H]|N+1,n\rangle , \tag{17}$$

and similar expressions for the second term inside the summation of Eq. (1). This results in the following equation

$$E\langle\langle a_i;a_j^\dagger\rangle\rangle_E = \langle N|[a_i,a_j^\dagger]_+|N\rangle + \langle\langle [a_i,H];a_j^\dagger\rangle\rangle_E , \tag{18}$$

which is the equation of motion of the electron propagator or rather a chain of equations since the more complicated propagator $\langle\langle [a_i,H];a_j^\dagger\rangle\rangle_E$ has an analogous definition to the one of $\langle\langle a_i; a_j^\dagger\rangle\rangle_E$ given in Eq. (1) with the commutator $[a_i,H]$ replacing a_i. The iteration of Eq. (18) yields

$$\langle\langle a_i;a_j^\dagger\rangle\rangle_E = E\ \langle [a_i,a_j^\dagger]_+\rangle + E\ \langle[[a_i,H],a_j^\dagger]_+\rangle$$

$$+ E^{-3}\ \langle[[[a_i,H],H],a_j^\dagger]_+\rangle+\ldots \tag{19}$$

where the average $\langle\ldots\rangle$ is taken with respect to the ground state $|N\rangle$. In order to be able to employ standard matrix and vector space techniques, Eq. (19) can be recast in terms of superoperators acting on a space of electron field operators. One introduces [12] a linear space with elements [13]

$$\{ a_a, a_p \; ; \; a_p^\dagger a_a a_b, \; a_a^\dagger a_p a_q \quad a<b, \; p<q \; \ldots \}. \tag{20}$$

For X and Y being general elements of this linear space of electron field operators we define a superoperator identity $\hat{I}$, and a superoperator Hamiltonian $\hat{H}$ such that

$$\hat{I}X = X \; , \; \hat{H}X = [X,H], \tag{21}$$

and a scalar product

$$(X|Y) = \langle [Y,X^\dagger]_+ \rangle. \tag{22}$$

The identification

$$\langle\langle a_i;a_j^\dagger \rangle\rangle_E = (a_j|(E\hat{I}-\hat{H})^{-1}a_i) \tag{23}$$

is now readily made from the expansion in Eq. (19). Introducing vector arrays of electron field operators we can write the matrix relation

$$\underline{G}(E) = \langle\langle \underline{a};\underline{a}^\dagger \rangle\rangle_E = (\underline{a}|E\hat{I}-\hat{H})^{-1}\underline{a}) \tag{24}$$

The techniques of inner projections and partitioning [14] can now be used for systematic approximate treatments of the electron propagator [15] by turning the superoperator inverse into a matrix inverse [3]:

$$\underline{G}(E) = (\underline{a}|\underline{h})(\underline{h}|(E\hat{I}-\hat{H})\underline{h})^{-1}(\underline{h}|\underline{a}), \tag{25}$$

where $\underline{h}$ is a manifold of elements from the linear space of electron field operators. When $\underline{h}$ is partitioned as $\underline{h} = \{\underline{a};\underline{f}\}$ with the orthogonality condition $(\underline{a}|\underline{f}) = \underline{0}$ [15] one readily obtains [3,15]:

$$\underline{G}^{-1}(E) = (\underline{a}|E\hat{I}-\hat{H})\underline{a}) - (\underline{a}|\hat{H}\underline{f})(\underline{f}|(E\hat{I}-\hat{H})\underline{f})^{-1}(\underline{f}|\hat{H}\underline{a}). \tag{26}$$

Different approximations are attained by truncation of the inner projection manifold $\underline{f}$ and choice of approximate ground states for the average defining the scalar product in Eq. (22). This can be systematized by perturbation theory.

Choosing the same partitioning of $\hat{H}$ as we did for H in Eqs. (6) and (7):

$$\hat{H} = \hat{H}_0 + (\hat{H}-\hat{H}_0), \tag{27}$$

we can easily implement the details of Raleigh-Schrödinger perturbation theory and identify

$$\underline{G}_0^{-1}(E) = (\underline{a}|(E\hat{I}-\hat{H}_0)\underline{a}) \tag{28}$$

with the average over $|HF\rangle$ of Eq. (8). When the approximate ground state is expressed as

$$|N\rangle = |HF\rangle + |\text{correlation}\rangle \quad , \tag{29}$$

we can write Eq. (26) as

$$\underline{G}^{-1}(E) = \underline{G}_0^{-1}(E) - \underline{\Sigma}(E) \quad , \tag{30}$$

collecting all the contributions from the correlated motion of the electrons into the "self-energy" matrix $\underline{\Sigma}(E)$.

III. APPROXIMATIONS

The electron propagator contains information of electron binding energies, photoionization cross sections, total energies, and allows the calculation of average values of arbitrary one-electron quantum mechanical operators. We will here concentrate on the electron binding energies and in particular on single-electron ionizations and electron attachment for atomic and molecular systems. The discussion is also for all practical purposes limited to vertical processes.

The matrix elements of Eq. (3) satisfy

$$(\underline{G}^{-1})_{ij} = (E - \varepsilon_i)\delta_{ij} - \Sigma_{ij}(E) \quad , \tag{31}$$

where we have utilized Eqs. (9)-(14) for the elements of $\underline{G}_0^{-1}$. From Eq. (1) we can see that $G_{ij}(E)$ has simple poles at values of the energy parameter corresponding to electron binding energies. This means that the matrix elements of $\underline{G}^{-1}(E)$ vanishes at such values of the energy parameter. Thus electron binding energies E satisfy the relation

$$0 = (E - \varepsilon_i)\delta_{ij} - \Sigma_{ij}(E) \tag{32}$$

or

$$E = \varepsilon_i + \Sigma_{ii}(E) \ . \tag{33}$$

When the part $\underline{f}$ of the inner projection manifold is neglected and we do not consider corrections to the ground state beyond $|HF\rangle$, the self energy, as defined here, vanishes and the electron binding energies are given as the orbital energies ε_i (or the negative thereof). This is contained in the Koopmans' theorem

[2] and is the so-called frozen orbital approximation, i.e. it considers neither relaxation nor correlation effects.

The corrections to the Koopmans' theorem result are contained in $\Sigma_{ii}(E)$ and this quantity is conveniently analyzed in terms of perturbation theory [6,10]. Through second order in electron interaction we can write [4,10]

$$\Sigma_{ij}^{(2)}(E) = \tfrac{1}{2}\sum_{apq} \frac{\langle ia||pq\rangle\langle pq||ia\rangle}{E+\varepsilon_a-\varepsilon_p-\varepsilon_q} + \tfrac{1}{2}\sum_{abp} \frac{\langle ip||ab\rangle\langle ab||jp\rangle}{E+\varepsilon_p-\varepsilon_a-\varepsilon_b} . \quad (34)$$

When we are considering an ionization energy, spin orbital i labelling ε_i, and Σ_{ii} in Eq. (33) will be an occupied orbital in $|HF\rangle$. One can then isolate the terms in the second sum of Eq. (34), which result when the summation index a, or b equals i. This yields

$$\Sigma_{ii}^{R(2)} = - \sum_{ap} \frac{|\langle ai||pi\rangle|^2}{\varepsilon_a - \varepsilon_p} , \quad (35)$$

and it is precisely the second-order contribution to the ionization energy calculated as the difference

$-I_i(\Delta SCF)=E_{HF}(N)-E_{HF}^{i}(N-1)$, with $E_{HF}(N)=\sum_a\varepsilon_a - \tfrac{1}{2}\sum_{ab}\langle ab||ab\rangle$, and

where $E_{HF}^{i}(N-1)$ is the total energy obtained from a separate SCF calculation on the state characterized by the spin orbital i being removed from the electron configuration of the Hartree-Fock N-electron ground state. This is the so called ΔSCF [16] approximation to the electron binding energy and is defined as the relaxation corrections. The expression (35) thus is the second-order relaxation terms and similarly we can identify the relaxation corrections in any order of perturbation theory [10].

Diagrammatic rules have been given by Born, Kurtz, and Öhrn [10] so that, through any order in perturbation theory, we can separate $\Sigma_{ii}(E)$ into a relaxation part Σ_{ii}^{R}, and a correlation part (the rest) Σ_{ii}^{C}. All the diagrams or terms in the perturbation theory expansion of -I (ΔSCF) are most easily summed to all orders in perturbation theory by simply performing the ΔSCF calculation. We can then write Eq. (33) as

$$E = -I_i(\Delta SCF) + \Sigma_{ii}^{C}(E)$$

and focus the approximate treatment on the correlation part Σ_{ii}^{C}.

The relative importance of Σ_{ii}^{R}, and Σ_{ii}^{C} varies from the core to the outer valence region and the delicate balance between these two kinds of error is quite different in the core region as compared with the outer valence electrons. In Figure 1 the relaxation error and the correlation error are displayed for the furan molecule for all the principal ionization energies (those being identifiable with a single occupied spin orbital). The trend shown here is typical for all molecular systems which have been studied so far. In the deep core the relaxation error dominates and just illustrates the success of ΔSCF or approximations thereof as the Transition Operator Method [17] for the calculation of core ionization energies. The outer region is characterized by the relaxation and the correlation errors having opposite sign offering possibilities for cancellation and realistic results with Koopmans' theorem. The deep valence region is more complicated since both kinds of errors are equally important and reinforces each other. This region is usually characterized by strong configuration mixing and photoelectron spectra, when available for this region, usually display a rich satellite structure. When several atoms of the same kind occur as the four carbon atoms in the furan molecule another consideration comes into play for the core ionizations. This can be labeled as a localization effect. When the ΔSCF or TOM calculations are carried out with the full molecular symmetry taken into account (C_{2v} for furan) the carbon core orbitals are delocalized and the resulting relaxation errors are large but only half of what they become when the calculations are carried out with reduced symmetry (C_s for furan). The correlation corrections become correspondingly less important upon localization of the core orbitals.

The treatment of $\Sigma_{ii}(E)$ can then be as ambitious as algebraic skill and computer budget will allow. In Table 1 we list some results from Kurtz, and Öhrn [18] for the water molecule using different simple approximations and a good quality basis (14 CGTO combination of Dunning's [19] oxygen basis and Huzinaga's [20] hydrogen basis augmented with a unit exponent d-orbital on oxygen and a p-orbital on each hydrogen). It is extremely important to saturate the basis set for this type of calculations before considering going to higher orders in perturbation theory, which is often leading to a different order of magnitude in complexity and computing effort. The different results listed correspond to the following expressions:

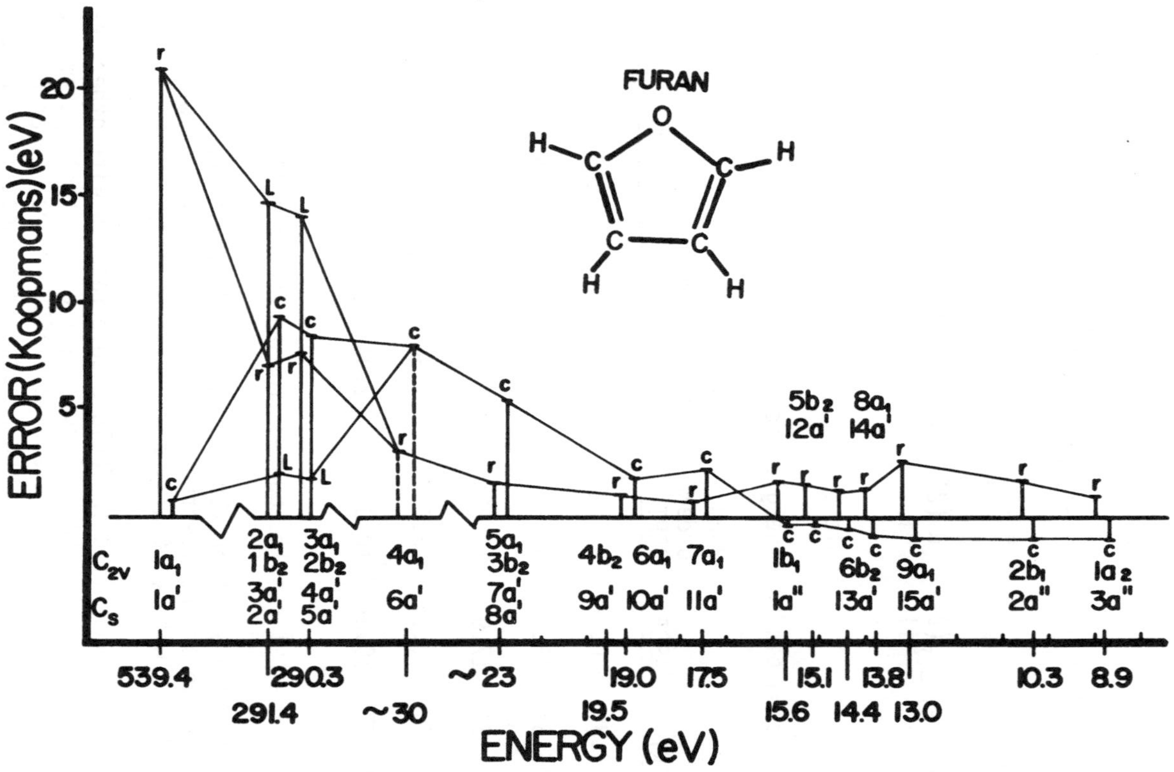

Figure 1. The relaxation error (r), the correlation error (c), and the localization error (L) in calculated electron binding energies for furan as defined in the text.

KT	$E = \varepsilon_i$
ΔSCF	$E = -I_i(\Delta SCF)$
2nd order RSPT	$E = \varepsilon_i + \Sigma_{ii}^{(2)}(\varepsilon_i)$
2nd corr. ΔSCF	$E = -I_i(\Delta SCF) + \Sigma_{ii}^{C(2)}(\varepsilon_i)$
ΔSCF q-part.	$E = -I_i(\Delta SCF) + \Sigma_{ii}^{C(2)}(E)$ (iterated)

where $\Sigma_{ii}^{C(2)}(E) = \Sigma_{ii}^{(2)} - \Sigma_{ii}^{R\ (2)}$.

Table 1

Electron binding energies for H_2O (eV)

state	KT	ΔSCF	2-nd order RSPT	2-nd corr. ΔSCF	ΔSCF q-part.	Exp.[a]
$1a_1$	559.39	540.49	529.72	540.69	540.72	540.2
$2a_1$	36.62	33.86	31.81	32.78	32.31[b]	32.2
$1b_2$	19.34	17.93	17.98	19.14	18.72	18.6
$3a_1$	15.66	13.15	13.28	14.30	14.33	14.7
$1b_2$	13.67	10.89	10.87	12.06	12.09	12.6

[a]These experimental results quoted from reference [21].

[b]There is another solution (branch) at 33.66 eV with somewhat smaller residue of the electron propagator, and still another one at 36.69 eV with virtually zero residue. [18]

For comparison I list in Table 2 some results with other related approaches. The water molecule is a common test system and it is interesting to compare the simple ΔSCF q-part. (second-order) results with those obtained using second-order (Ced2) and third-order (Ced3) Green's function theory (Cederbaum, Holneicher, and von Niessen [22]) and those using straight Raleigh-Schrödinger perturbation theory through second (HU2) and third order (HU3) (Hubac, and Urban [23]) with about the same quality basis sets.

Table 2

Electron binding energies for H_2O (eV)

with different methods

state	Ced2	Ced3	HU2	Hu3	KO(14)[a]	KO(14+pol)[b]
$2a_1$	32.93	35.10	33.38	35.22	32.8	32.31
$1b_2$	17.70	19.22	17.93	19.42	18.7	18.72
$3a_1$	13.18	15.18	12.62	14.74	13.8	14.33
$1b_1$	10.92	13.03	10.48	12.75	11.8	12.09

[a] These results are obtained as ΔSCF q-part with a 14 CGTO Dunning Huzinaga basis. The difference with the last column is a basis effect. [19]

[b] These results are repeated from Table 1.

The simple ΔSCF q-part. formula for calculating electron energies has also been used by Kurtz, and Öhrn [19] to calculate electron affinities for small molecules with encouraging results. More studies are underway and it is a hope that this simple formula or improvements of it will provide efficient ways to calculate vertical electron affinities for molecules which otherwise are too large for accurate treatments.

ACKNOWLEDGEMENTS

It is indeed a pleasure to thank Dr. Michael Hehenberger for the data used to prepare Figure 1. Support from the U.S. Air Force Office of Scientific Research (AFOSR) for the work reported here is also gratefully acknowledged.

REFERENCES

[1] K. Siegbahn, C. Nordling, G. Johansson, J. Hedman, P. F. Heden, K. Hamrin, U. Gelius, T. Bergmark, L. O. Werme, R. Manne, and Y. Baer, ESCA Applied to Free Molecules, (North-Holland Publishing Co., Amsterdam, 1969).
[2] T. A. Koopmans, Physica (The Hague) 1, 104 (1933).
[3] B. T. Pickup, and O. Goscinski, Mol. Phys. 26, 1013 (1973).
[4] J. Linderberg, and Y. Öhrn, Propagators in Quantum Chemistry, (Academic, London, 1973).

[5] L. S. Cederbaum, and W. Domcke, Adv. Chem. Phys. 36, 205, (1977), and references therein.
[6] Y. Öhrn, in The New World of Quantum Chemistry, Proceedings from the 2nd International Congress on Quantum Chemistry, (B. Pullman, and R. Parr, Eds. Reidel, Boston, 1976), p 57; and references therein.
[7] J. Simons, and W. D. Smith, J. Chem. Phys. 58, 4899 (1973); J. Simons, Chem. Phys. Lett., 25, 122, (1974); K. M. Griffing, and J. Simons, J. Chem. Phys. 62, 535, (1975); J. Kenney, and J. Simons, J. Chem. Phys. 62, 592 (1975).
[8] C. Møller, and M. S. Plesset, Phys. Rev. 46, 618 (1934).
[9] P. S. Bagus, Phys. Rev. 139, 619 (1965).
[10] G. Born, H. A. Kurtz, and Y. Öhrn, J. Chem. Phys. 68, 74 (1978).
[11] G. D. Purvis, and Y. Öhrn, J. Chem. Phys. 62, 2045 (1975).
[12] O. Goscinski, and B. Lukman, Chem. Phys. Lett., 7 573, (1970).
[13] L. Tyner Redmon, G. D. Purvis, and Y. Öhrn, J. Chem. Phys. 63, 5011 (1975).
[14] P. O. Löwdin, Phys. Rev. 139, A357 (1965); Int. J. Quantum Chem. S4, 231 (1971).
[15] C. Nehrkorn, G. D. Purvis, and Y. Öhrn, J. Chem. Phys. 64, 1752 (1976).
[16] See references [9], and [10].
[17] O. Goscinski, B. T. Pickup, and G. D. Purvis, Chem. Phys. Lett. 22, 167 (1973), O. Goscinski, M. Hehenberger, B. Roos, and P. Siegbahn, Chem. Phys. Lett. 33, 427 (1975).
[18] H. A. Kurtz, and Y. Öhrn, J. Chem. Phys. 00, 0000 (1978).
[19] T. H. Dunning, J. Chem. Phys. 53, 2823 (1970).
[20] S. J. Huzinaga, J. Chem. Phys. 42, 1293 (1965).
[21] J. Almlöf, University of Stockholm Inst. of Physics Report 72-09 (1972).
[22] L. S. Cederbaum, G. Holneicher, W. von Niessen, Chem. Phys. Lett. 18, 503 (1973).
[23] I. Hubac, and M. Urban, Theoret. Chim. Acta (Berl.) 45, 187 (1977).

THEORY OF ONE ELECTRON BINDING ENERGIES INCLUDING CORRELATION, RELATIVISTIC AND RADIATIVE EFFECTS: APPLICATION TO FREE ATOMS AND METALS

DONALD R. BECK and CLEANTHES A. NICOLAIDES

Theoretical Chemistry Institute
National Hellenic Research Foundation
Athens 501/1 Greece

ABSTRACT.

The recent emphasis on non rare-gas core Binding Energies (BE) measurements and the continued use of these BE to probe bulk and surface properties makes this a timely subject to review. The wide range of atoms treated and the accuracies required place stringent demands on any comprehensive theory, forcing it to consider effects of relaxation, relativity, radiation, correlation, nuclear size and the bulk. To date, there is no theory capable of treating these simultaneously, although we suggest a means by which relativistic and correlation effects may eventually be unified. This would leave proper treatment of bulk effects as the prominent challenge.

The existing additive theory is reviewed and extended to include a way of making a priori determinations of when and what correlation effects are large. This allows assessment of experimental observations and of independent particle model results and eases computation of correlation effects. A simple method of including the significant radiative effects is also put forth. This theory has been applied to the 1s BE of C, O, O^-, F, F^-, Ne, Na, Na^+ in excellent agreement with experiment ($\sim$0.1-0.2 eV), to all core subshells of K, (accuracies $\sim$0.1-0.5 eV), and to selected subshells of Cs. Modifying Shirley's ad hoc model for "extra atomic relaxation", which we suggest has considerable utility, we produce metallic BE which agree with measured values to within 1-2 eV.

I. INTRODUCTION

Since the mid 1960's a large successful effort using the methods of photoelectron spectroscopy (PES), initiated by the development of high resolution electron spectrometers and convenient line sources, has yielded results for binding and excitation energies, cross sections, electron decay probabilities (Auger effects) of excited and ionized states in atoms, molecules and

Cleanthes A. Nicolaides and Donald R. Beck (eds.), Excited States in Quantum Chemistry, 329–359.

the solid state [1]. The indirect use of these methods is even more impressive--they have probed surface reactions, determined charge distributions in molecules, investigated effects of electron correlation, detected trace elements, made molecular structure determinations, etc. Accuracies range from a few tenths to one electron volt.

The exploitation of the indirect applications has been greatly facilitated by noting that binding (BE) and Auger (AE) energies associated with similar species (e.g. for the same atom residing in different molecular environments or between a free and metallic atom) exhibit differences (chemical shifts) whose gross features at least are often explained by simple ad hoc models.

In this work, we will be concerned with BE of free atoms, and when they can form a metal, we predict standard state core BE as well to within 1-2 eV, using a model for "extra-atomic relaxation". BE are of particular interest in very heavy atomic species because they allow us to identify a transient species by the characteristic X-rays they emit. Furthermore BE exhibit some rather interesting correlation effects.

Since we will impose no restriction on the type of atom considered, we will have to include the effects of correlation, relativity, radiation, relaxation, and take nuclear size into account. In Table 1, we illustrate that all these effects can be important.

At present, there exists no theory capable of simultaneously treating all these effects, which will be treated additively in this work. In Section V we suggest a method by which all of them, except radiative ones can be treated properly.

Existing experimental core atomic BE are generally limited to the rare gases [1a ,2-3] and metallic elements [4-8] (some recent work [7] using fast ion beam projectile Auger spectroscopy shows promise of yielding core BE for non-metallic elements to within a few tenths of an eV). Often, BE have been extracted from standard state (e.g. thin film, bulk) X-ray energies by combining them with atomic UV measurements, and applying some ad hoc model [9,10]. These models, which can also be used to probe the bulk (e.g. the charge state) are thus of considerable interest and are discussed for metals in Section VI. However, since they only seem capable of at most 1-2 eV accuracy, they do not suffice to predict "experimental" values for atomic BE.

In Section II we discuss the definition of the binding energy and observe that it is non-unique. In the following section (III) we review existing additive theories of binding energies, considering one in detail. There, we propose a comprehensive a priori

Table 1

Comparison of the Different Contributions to the Binding Energy (eV) Throughout the Periodic Table

Type	Species 1s Ne[(a)]	1s K[(a)]	4s Cs[(a)]	1s Hg[(b)]	1s Fm[(c)]	3s Fm[(c)]
Non-relativistic[(d)] orbital eigenvalue	891.7	3633.4	236.6			
Non-relativistic[(d)] relaxation	-23.2	-32.8	-6.1	-92.8	-117	-45
One particle[(d)]	1.2	17.0	16.7	83556.3	142929	7250
Two-particle relativistic	-0.3	-2.8	-0.1	-303.3	-674	-18
Radiative	-0.1	-1.2	-0.1	-154.6	-302	-19
Non-relativistic correlation	1.1	0.9	-6.6	~1.	~1.	~ -(1-3)
Relativistic correlation[(e)]	small	?	?	?	?	?
Nuclear Size	small	small	?	-54[(f)]	± 8[(g)]	?
Total (rows 1-7)	870.4	3614.4	240.4	83099.7	141953[(h)]	7213[(h)]
Precision[(i)]	~0.2	~0.5	~2-3	~4	~30	~15

(a) This work.
(b) A.M. Desiderio and W.R. Johnson, Phys. Rev. A3, 1267 (1971).
(c) B. Fricke, J.P. Desclaux and J.T. Waber, Phys. Rev. Letts 28, 714 (1972).
(d) If non-relativistic ΔSCF results are available (Ne, K, Cs) then all three entries are given. Otherwise (Hg, Fm) relativistic ΔSCF is compared with the relativistic orbital eigenvalue.
(e) No theory exists which would yield this quantity. It is probably no larger than the non-relativistic correlation energy for larger species, and negligible for smaller species (see Section V).
(f) Difference between a point and extended nuclear model (already included in (d)).
(g) This is the uncertainty in the extended nuclear model.
(h) No correlation included.
(i) Approximately the accuracy of the theoretical result.

scheme which allows the determination of when correlation effects are important, and what the principle constituents are (in the configuration sense). In particular, we suggest a mechanism to account for the anomalous behavior [11] of the 5d BE in the actinides. In section IV, we apply the additive method to predict BE of C, O, O^-, F, F^-, Ne, Na, Na^+, K and Cs which are in excellent agreement with experiment, where available. Section V outlines a method for the simultaneous inclusion of relativistic and correlation effects which we have under development. In the final section, we discuss a semi-localized exciton model for metallic BE which we use to "correct" our free atom BE, achieving results which agree with standard state experiment to 1-2 eV.

II. DEFINITION OF THE BINDING ENERGY

In photoelectron spectroscopy one knows the energy of the photon beam and measures the energy of the electrons which emerge after the interaction between atom and beam (assumed to be instantaneous and energy independent). Energy conservation requires:

$$E_i = E_f \tag{1}$$

The initial energy E_i is defined well (for not a too intense photon beam which would induce an initial state width),

$$E_i = h\nu + E_G(N) \tag{2}$$

where $E_G(N)$ is the ground state energy of the N-electron atom or molecule. It satisfies:

$$H(N)\,\Psi_G(N) = E_G(N)\,\Psi(N) \tag{3}$$

For the final state, both experimentalists and theoreticians make use of the formal separation:

$$E_f = X(N-1) + K.E. \tag{4}$$

where K.E. is what is measured as kinetic energy of the ejected electron and X is a property of the (N-1) electron system (we assume corrections for recoil have been accounted for).

Experimentally, the quantity

$$X(N-1) - E_G(N) = h\nu - K.E. \tag{5}$$

is designated as the Binding Energy of the observed electron:

$$B.E._{Exp} \equiv h\nu - K.E. \tag{6}$$

and involves only measurable quantities.

On the other hand, X(N-1) is identified as $E_{Ion}(N-1)$ satisfying the eigenvalue equation:

$$H(N-1)\,\Psi_{Ion}(N-1) = E_{Ion}(N-1)\,\Psi_{Ion}(N-1) \qquad (7)$$

in which case the theoretical Binding Energy is defined as:

$$BE_{Th} = E_{Ion}(N-1) - E_G(N) \qquad (8)$$

Presently, the equivalence of eqs. 6 and 8 is well accepted. A variety of approximations, such as Koopmans' theorem, "frozen core", relaxation model, adiabatic model, ΔSCF, or more advanced many-body methods which include relativistic, electron correlation and radiative effects have been applied to the calculation of eq. 8 for a variety of systems and are discussed in the following sections. (We note that for inner hole states, in principle the calculation of $E_{Ion}(N-1)$ should involve the consideration of an energy shift which is energy independent only in lowest order and is due to self-energy modifications particular to nonstationary states. Currently, such corrections are usually omitted.)

However, beyond the complexities which characterize the rigorous computation of the total energy $E_{Ion}(N-1)$, it appears that the measured B.E. (eq. 6) is not identical to the theoretical B.E. (eq. 8) - contrary to what has been thought all along. This conclusion, which was stated in one sentence in 1975 [12], can be reached in the following way:

The interactions among the N-electrons in the final state can be written in Hamiltonian form to a good approximation which includes relativistic effects,

$$H(N) = \Sigma_{i=1}^{N}\, h(i) \;+\; \Sigma_{i<j}^{N} g(ij) \qquad (9)$$

where h(i) and g(ij) are one - and two - electron operators (see following sections). The corresponding Schrödinger equation is $H(N)\,\Psi(N) = E(N)\,\Psi(N)$ where $E(N) = E_f$.

A measurement of the total E(N) would yield E_f of equation 1. Such a measurement should in principle involve the simultaneous recording of the energies of the final products.

However, if the N-electron system is conceptually divided according to eqs. 4 and 7, the separation

$$E_f \equiv E(N) = E_{Ion}(N-1) + K.E. \qquad (10)$$

where $E_{Ion}(N-1)$ is the unperturbed solution of eq. 7 and K.E. is the measured free electron energy, implies the separation of the N-electron Hamiltonian and energy into two parts without any "self-energy" type corrections. (The word "self-energy" is chosen in order to show some analogy with the Lamb shift corrections which arise from the interaction of the isolated atomic system with the radiation field, to which, in fact, it is coupled continuously.)

Prediction of magnitudes of the effects mentioned above is, of course, beyond the scope of this paper. Their calculation would have to involve some assumption about the energy dependence (time dependence) of the photon-atom-ion-electron interaction process. However, we point out that according to eq. 8 the B.E. of electrons in N-electron systems should be independent of the photon energy. Therefore, small variations of observed B.E. (eq. 6) as a function of photon energy would indicate a certain anomaly in the accepted definitions. (Accurate measurements of B.E. as a function of photon energy over a large range are about to become possible using synchrotron radiation.) In fact, such variations have recently been observed in Auger spectroscopy for small free electron energies and have been attributed to "post-collision interactions" [13,14].

In conclusion, our point is that the quantity which is called Binding Energy and is measured from the relationship:

$$B.E._{Exp} \equiv h\nu - K.E. \qquad (6)$$

is not identical to the (in principle) calculable quantity that has up to now been defined as:

$$B.E._{Th} = E_{Ion}(N-1) - E_G(N) \qquad (8)$$

and is therefore independent of the photon energy. Having made this point we proceed with the theory for the accurate calculation of eq. 8 in many-electron atoms assuming $BE_{Exp} = BE_{Th}$. Modification of this definition for solids is postponed until Section VI.

III. THEORY OF FREE ATOM BINDING ENERGIES

Given the current experimental accuracies obtainable for both gas and standard state core BE (tenths of an eV), and the relatively small size of the chemical shift between atom and standard state ($\gtrsim$1eV), we need a theory capable of yielding accuracies $\sim$0.1eV. This requires the careful consideration of relativistic, correlation, radiative, nuclear size and bulk effects (standard state).

For the purposes of discussion, we will choose to divide the total energy of an atom up as follows:

$$E_{tot} = \varepsilon_{NR\text{-}SCF} + R_{NR\text{-}SCF} + E_{Dirac\text{-}SCF} + E_{Breit\text{-}SCF} + E_{NR\text{-}Corr} + E_{R\text{-}Corr} + E_{Rad} \quad (11)$$

This division closely patterns the methods used to evaluate atomic BE. The first two quantities, the non-relativistic orbital eigenvalue, $\varepsilon_{NR\text{-}SCF}$, and the non-relativistic relaxation energy, $R_{NR\text{-}SCF}$, constitute the non-relativistic SCF energy. The third is the one-body relativistic SCF correction to the first two--or equivalently, the first three terms are the one-particle-relativistic SCF energy (a different decomposition into the sum of the relativistic orbital energy and the relativistic relaxation energy is sometimes useful (see Table 1)). The next term, $E_{Breit\text{-}SCF}$ contains the relativistic two particle corrections (through order $(Z\alpha)^2$, where α is the fine structure constant), evaluated at the SCF level. $E_{NR\text{-}Corr}$ is the non-relativistic correlation energy obtained using the non-relativistic SCF solution as a zeroth order function, and the sum $E_{NR\text{-}Corr} + E_{R\text{-}Corr}$ is its relativistic equivalent. The radiative effects, E_{Rad}, here will always be added on to the result to form E_{tot}. Effects of nuclear size are accounted for within the Hamiltonian.

A. The Independent Particle Model (IPM) or SCF Solution

Most BE for Z >18 have been evaluated at the IPM level with or without inclusion of relativistic effects. For inner electrons, consideration of relativity is clearly mandatory even for the lightest species. For these, however, this can be done in a perturbative way [15]. Relativistic effects are incorporated by taking as a Hamiltonian (in a.u.)

$$H_D = \Sigma_{i=1}^{N} c\underset{\sim}{\alpha}_i \cdot \underset{\sim}{p}_i + \beta_i c^2 + V(r_i) + \Sigma_{i<j} 1/r_{ij} \quad (12)$$

where the single sum is over one-electron Dirac Hamiltonians [16]. The $\underset{\sim}{\alpha}_i$ are 4 x 4 matrices built from Pauli spin-matrices, $\underset{\sim}{p}_i$ is the linear momentum, and β_i is a 4 x 4 matrix built from 2 x 2 identity matrices. The first term is the kinetic energy, the second the rest energy, and the third the electron-nuclear potential energy.

It is computationally possible [17] to express $V(r_i)$ in terms of a point potential

$$V_{pt}(r_i) = -Z/r_i \quad (13a)$$

or a constant charge within a sphere of radius R_n where $R_n = 2.2677 \times 10^{-5} \sqrt{AM}$, with AM being the atomic mass (13b)

or a Fermi charge distribution $\rho_N=\rho_o/[1+\exp(b(r-R_n))]$ with b as the thickness parameter. (13c)

Potentials can be constructed for (13b) directly, and for (13c) by solving a differential equation. The use of a finite nuclear model has been found to be essential in the study of trans-uranic elements, with the effect largest for subshells having a high density near the nucleus (e.g. s electrons). Even in Hg, these effects change the 3s orbital eigenvalue by 3 eV [18]. They obviously will have a significant effect on hyperfine structure for these species.

IPM calculations for use in binding energy studies are usually implemented within the restricted Hartree-Fock (RHF) or Hartree-Fock-Slater (HFS) method. The latter is based on the average energy and also uses a local approximation for the exchange [19], which is inappropriate when very high precision is desired (1 part in 5000 for deep core levels of Fm) or when exchange effects (outermost subshells) are large. We should point out that most existing ΔSCF results for BE of heavy atoms use the relativistic HFS method [1a,20], and they are often quite useful.

Codes for implementing the SCF (RHF) calculation within the non-relativistic approximation have been designed by Froese-Fischer [21], and relativistically by Desclaux [17]. These are both based on the pioneering work of Hartree [22a]. Others [22b-22d] have also contributed to these developments.

Once the SCF result for the ground state is obtained, IPM BE calculations can be done in one of two ways. In the first, a second SCF calculation is carried out for the hole state, and the results subtracted to yield a ΔSCF result, which includes the effect of relaxation. For atoms, such procedures are quite tractable, providing one uses due care to make sure that SCF tolerances are high enough. This is particularly important for valence electrons in heavy atoms. Molecular systems however, apart from additional conceptual problems related to delocalization of inner holes [23], are still rather difficult to handle this way. Solids present other difficulties--as the energy of the model solid is infinite.

Methods avoiding this difficulty are principally based on corrections (for relaxation)to BE obtained from ground state orbital eigenvalues (loosely, Koopmans' theorem). These have been developed by a number of authors [24], and also form an integral part of propagator [25] and Green's functions methods [26] (which include some correlation effects). When relaxation effects are large (see Table 1), their proper inclusion forces consideration of third and higher order perturbation theory, which requires further developments of the above techniques. A more complete survey of calculations at the IPM level may be found in the work of Larkins[27].

B. $E_{Breit\text{-}SCF}$

The Hamiltonian of equation 12 may be corrected to include (to order $(Z\alpha)^2$), two-particle relativistic effects, through the use of the Breit operator [16], to yield:

$$H_{rel} = H_D + \Sigma_{i<j}\ B(i,j),\ \text{where}$$

$$B(i,j) = (-1/2r_{ij})\left[\underset{\sim}{\alpha}_i\cdot\underset{\sim}{\alpha}_j + (\underset{\sim}{\alpha}_i\cdot\underset{\sim}{r}_{ij})(\underset{\sim}{\alpha}_j\cdot\underset{\sim}{r}_{ij})/r_{ij}^2\right] \qquad (14)$$

This contains the effects of retardation. Several other forms of these corrections have been proposed [28-30], some of which appear to be more accurate, but judging from the Fm results, their inclusion can be postponed.

Most present treatments of B(i,j) use the SCF functions created from equation 12 within the context of zeroth order perturbation theory. Higher order effects are therefore postponed to $E_{Rel\text{-}Corr}$. There are two reasons for this restrictive treatment. The first is formal--both H_{rel} and H_D admit electron and positron solutions, the latter being of lower energy. If we evaluated B(i,j) in higher order perturbation theory using complete sets (positron and electron functions), the correction is the wrong order of magnitude and too large [16]. On the other hand, it may be possible (see Section V) in practice to avoid such difficulties, although treatment of B(i,j) at the SCF level is an expensive proposition and is probably unnecessary (large numbers of new two particle magnetic radial integrals appear. A precursor of the present-day relativistic codes actually allowed for such a possibility [22d]). B(i,j) may have to be included at the variational C.I. level if electron correlation effects are large or when fine structure effects significantly (beyond first order) affect wavefunction determination. (B(i,j) contains terms which serve to screen the bare nucleus spin-orbit contribution, and these screening effects have been found to be important [31]). For all core electrons, the effect of B(i,j) should be included, although for smaller species this may be done using perturbation theory [15].

C. Non-Relativistic Correlation

a. General: It is recognized that for core properties of

medium to high Z atoms relativistic effects should no longer be treated perturbatively. At the same time, as we have seen (Table 1), correlation effects can in many cases be important as well. Given the absence of a combined relativistic-correlation theory, we have no recourse but to treat such effects non-relativistically and add them on to relativistic SCF results, a procedure we call the additive theory of BE. It is not so well appreciated that a combined theory will also be required for valence properties such as molecular bonds, excitation energies, transition probabilities, etc. for the heavier atoms, but the evidence for this is mounting [32-35].

b. Review of Existing Non-Relativistic Correlation Theories There are a whole host of substantially successful correlation theories which have been applied to small ($Z \leq 18$) atoms. These include Rayleigh-Schrödinger [36] and Many Body Perturbation Theories (MBPT) [37-39], Brueckner-Goldstone equations [40], Propagator Methods [25], Green's functions [26], Equations of Motion [41], Random Phase Approximation (RPA) [42-43], Multi-Configurational methods [44], First Order methods [45], perturbation theory directed-variational C.I. implemented methods [46-48], superposition of configurations [49], etc. Some of these appear to have difficult-to-remove restrictions even for small systems (e.g. only ground or single determinantal states may be treated with ease, or only a limited number of configurations may be considered, or there is no configuration selection mechanism, etc.) which limit their general applicability. Very few have been applied to larger species, due in part to the computational complexities involved. Among the exceptions is found work using RPA methods [42] (e.g. Xe).

In the next section, we will discuss in some detail one of these methods [47-48] which at present has been used to treat rather larger species (e.g. Cs) with the practical computational limit currently being the consideration of active d electrons.

c. A Perturbation Theory Directed Variational C.I. Implemented Non-Relativistic Correlation Procedure This is an approach which we have discussed in Paper I in this volume [50]. Here we only outline what is essential for application to BE. The language used throughout is that of C.I. which remains, in our belief, the most general implementation procedure available for atoms and small molecules (certain aspects of correlation in extended systems is more appropriately described by other methods [50]).

For perturbation theory to be effective, a proper zeroth order (or reference) function Φ must be found. A satisfactory solution is to use the few (usually one) most dominant configuration(s) which are selected on the basis of past computational and physical (e.g. spectroscopic) experience. Such a function is determined self-consistently, i.e. by RHF means. The correlation function χ is then given by:

$$\Psi \equiv \Phi + \chi; \text{ where } \langle\Phi|\Phi\rangle = \langle\Phi|\Psi\rangle = 1 \tag{15}$$

and the correlation energy is the difference between the exact (non-relativistic) and reference energies:

$$E_{NR\text{-}Corr} = \frac{\langle\Psi|H|\Psi\rangle}{\langle\Psi|\Psi\rangle} - \langle\Phi|H|\Phi\rangle = \langle\Phi|H|\Psi\rangle \tag{16}$$

First order perturbation theory is then used to restrict χ to a tractable <u>form</u> i.e. we include only those configurations in χ (call them $\mathcal{C}_\chi$) which have a non vanishing matrix element with Φ, viz:

$$\langle\Phi|H|\mathcal{C}_\chi\rangle \neq 0 \tag{17}$$

Within the <u>form</u>, the solutions are obtained variationally to all orders. Failure to do this falsely emphasizes certain higher order effects which would not be compensated for. This is discussed further in Paper I. Because H contains only one and two particle operators, for a given configuration in $\Phi, \mathcal{C}_\Phi$, only single and double subshell excitations must be kept. The emphasis on subshells rather than spin-orbitals results in a more accurate tractable procedure [50].

Schematically, we have in subshell terms

$$\begin{aligned} &s_\Phi \rightarrow u_\chi \\ &s_\Phi\ s'_\Phi \rightarrow u_\chi\ \ u'_\chi \end{aligned} \tag{18}$$

where s_Φ, s'_Φ are occupied subshells in C_Φ and u_χ, u'_χ are vacant and orthogonal to all subshells in C_Φ.

In principle, this substitution must be carried over all subshell(s) (pairs) in C_Φ and all C_Φ. Such a function, providing the u_χ are variationally optimized is found to give excellent results for nearly all properties of atoms and small molecules (small electron affinities can be an exception).

A further separation of each u_χ is then made to exhibit near degeneracies directly, giving them special weight. This is helpful both for the purpose of analysis and computation [50]. Thus

$$u_\chi = s_\chi + v_\chi \tag{19}$$

where the collection of all s_Φ and s_χ (i.e. all the nearly degenerate subshells) is known as the Fermi sea (FS), all of whose members are to be determined at the same level of computational approximation. v_χ is called a virtual subshell function, orthogonal to all members of the FS, and is determined by variational C.I. methods [50]. Equation (11) then expands to:

$$s_\Phi \rightarrow s_\chi \quad \text{(internal polarization)} \tag{20a}$$

$$s_\Phi \rightarrow v_\chi \quad \text{(virtual polarization)} \tag{20b}$$

$$s_\Phi\, s'_\Phi \rightarrow s_\chi\, s'_\chi \quad \text{(internal Fermi-Sea)} \tag{20c}$$

$$s_\Phi\, s'_\Phi \rightarrow s_\chi\, v_\chi \quad \text{(hole-virtual)} \tag{20d}$$

$$s_\Phi\, s'_\Phi \rightarrow v_\chi\, v'_\chi \quad \text{(bi-virtual)} \tag{20e}$$

This grouping conveniently corresponds for the most part to the role these excitations play for different properties. For example, hole-virtual, internal, and bi-virtual correlation are largest for BE. When energetically allowed, the first corresponds to Coster-Kronig events, and the second to super Coster-Kronig events [51]. With the reference function, (20a) + (20c) form the FS configurations.

d. Examples (i) Consider the ground state of the Ne atom. Here $\Phi = 1s^2\, 2s^2\, 2p^6$ and since all nearby degeneracies have been accounted for, u=v and only (20e) survives. The correlation function is then created as follows:

$$1s^2 \rightarrow v_s^{\,2} + v_p^{\,2} + v_d^{\,2} \ldots$$

$$1s\;2s \rightarrow \hat{v}_s^{\,2} + \hat{v}_p^{\,2} + \hat{v}_s\hat{v}_s' + \hat{v}_p\,\hat{v}_p' + \hat{v}_d\hat{v}_d' + \hat{v}_d^{\,2} \ldots.$$

$$1s\;2p \rightarrow \tilde{v}_s\tilde{v}_p + \tilde{v}_p\tilde{v}_d + \ldots$$

$$2p^2 \rightarrow \bar{v}_s^{\,2} + \bar{v}_p^{\,2} + \bar{v}_s\bar{v}_d + \bar{v}_d^{\,2} \ldots.. \tag{21}$$

In the above we have limited ourselves to one radial function particular to each subshell pair (e.g. $v_s \neq \hat{v}_s \neq \tilde{v}_s \neq \bar{v}_s$) except for 1s 2s where two are needed to form the ^3S coupling. The subscript refers to the azimuthal symmetry (ℓ) which although it has no formal cut-off, unlike (20a)-(20d), in practice the expansion is found to converge rapidly with ℓ.

(ii) Consider the ground state of the Be atom, and assume that $\Phi = 1s^2\, 2s^2$ suffices (low Z). Here we have a nearby degeneracy (2s and 2p) so

$$u_\chi(p) = 2p + v_p \quad \text{and} \quad u_\chi(\ell) = v_\ell\,,\ \ell \neq p \tag{22}$$

Suppose also that we are interested in valence subshell properties, so the $1s^2$ core is considered frozen. We then have only the configurations: $1s^2$ 2s v_s (20b), $1s^2$ $2p^2$ (20c), 2p v_p (20d) appearing with (20e) (see the Ne example for these). The first four structures do have symmetry cutoffs ($\ell \leq 1$).

In carrying out our BE calculations, which are obtained by subtracting the result of two separate variational C.I. calculations for the upper and lower states, we make use of the approximate decoupling [50] of the sections of χ in the following form: We do Φ + (20a) - (20d) and Φ + (20e) independently and add the results together.

e. Which are the Important Configurations? To be included, a configuration must make a significant contribution ($\geq$0.05 eV) to the energy and to the BE. Configurations contributing only to the former are left out in the spirit of approaches such as those based on Green's functions, propagators, etc. Most of these belong to type (20e).

f. Sections (20a)-(20d) Whenever valence internal correlations, (20c), are present they should be included. They depend on symmetry and affect the multiplet structure as well as intensities. For example, consider the 2s ionization in S. The final multiplets will be $1s^2$ 2s $2p^6$ $3s^2$ $3p^4$ $^{4,2}P$, ^2D, ^2S. Out of these, only the ^2S symmetry allows the important $3s^2 \rightarrow 3p^2$ near-degeneracy correlation. Thus, internal correlation in the valence shells will have an effect on the photoelectron spectrum (energies and intensities) of an inner shell excitation. The first accurate correlation study of multiplet structure for inner electron excitation of an open shell system was done in 1973 [68].

The most important remaining correlations, (20a)-(20d), are associated with the final state, and are either of type (20c) or (20d). Past experience drawn for K,L, M subshells from optical spectroscopy, Auger spectroscopy (e.g. (super) Coster-Kronig events), computational results, suggests the following classification of the important (20c) and (20d) types for the entire periodic table (see also reference 47). (See Table 2.)

The estimates of the sizes of these effects can depend substantially on the species and subshell involved due to the following factors:

a) the structure of the off-diagonal matrix element, i.e. the group theoretical coefficients

b) how many subshells which have the replacement symmetry of the non-hole filling electron are fully occupied. The fewer there are, the larger the effect.

c) how deep the core hole is. The deeper it is, the more nearly degenerate are the nl subshells, enhancing the effect. This

Table 2

The Most Important Internal and Hole-Virtual Correlation Contributions to the Final State for Binding Energy Studies

Subshells removed from C_Φ	New Subshells in C_χ	When	Size (eV)[(a)]	Where
	s Holes			
nd^2	ns g	$n \geq 4$		
np nf	ns g	$n \geq 4$		
nf^2	ns g	$n \geq 4$		
np nd	ns f	$n \geq 3$	-7	4s in Cs; f=4f
np^2	ns d	$n \geq 2$	-2	2s in K; $d=v_d+3d$
			-7	3s in K; $d=3d+V_d$
	p Holes			
ns nf	np g	$n \geq 4$		
nd nf	np g	$n \geq 4$		
nd^2	np f	$n \geq 3$	-10	4p in Cs; f=4f
			-2.2[(b)]	3p in Sr; $f = v_f$
ns nd	np f	$n \geq 3$	-1	4p in Cs; f=4f
	d Holes			
np^2	nd g	$n \geq 4$	Weak?	
nf^2	nd g	$n \geq 4$	Large	Actinides (see text)
np nf	nd g	$n \geq 4$		
ns np	nd f	$n \geq 3$	Weak?	
	f Holes			
ns np	nf g	$n \geq 4$	Weak?	
np nd	nf g	$n \geq 4$		

(a) This work.

(b) The addition of this correlation brings Dirac-HF results into nearly ($\lesssim$ 0.1 eV) perfect agreement with recent experimental values [4]. The remaining ,weaker, correlation effects-internal valence and bi-virtual-apparently cancel out in this case.

usually competes with (b).
d) whether the replacement is energetically allowed (i.e. (super) Coster-Kronig?). If the hole state is embedded in the hole-virtual or internal continuum, one must use approaches like that of Fano and Altick [52-53] (C.I. in the continuum) to calculate the shift which may be positive or negative. For example, we have found a 3.3 eV shift to lower energies due to $2p^2 \rightarrow 2s\ v_d$ correlation for the 2s $2p^6$ ^2S hole state of F I [54]. When energetically forbidden, the BE is always reduced by this effect.

The largest contributions for s-holes and p-holes have a considerable literature associated with them. A particular case of the $p^2 \rightarrow s\ d$ substitution was suggested by Bacher [55] (in 1933!) as responsible for the anomalous term splitting observed in Mg I ^1D, which was later confirmed by Zare [56]. Related behavior in Mn was also explained by Bagus et al [57] in this way. The hole-virtual analog in the first row ($2p^2 \rightarrow 2s\ v_d$) is primarily responsible for the departure of valence properties (e.g. f-values) from their HF values [47].

More recently it was suggested that the anomalous behavior observed [2-3] for 4p BE was due [3,47] to the $4d^2 \rightarrow 4p\ f$ substitution which has now been confirmed by recent calculations [42].

Here we propose that the anomolous behavior observed for 5d BE in the actinide region [11] may be principally due to the $5f^2 \rightarrow 5d\ g$ replacement.

g. Section (20e) In the 1960's, it was shown [58] that bi-virtual correlation configurations associated with a given spin-orbital pair are decoupled to first order (and so may be computed independently of one another), and that the total bi-virtual energy is given as a sum of symmetry adapted (in practice this means the two virtuals are coupled to form a parent of pure S, L symmetry) pair energies, ε, i.e.

$$E_{vv'} = \Sigma_\beta\ a_\beta\ \varepsilon(n\ell_\beta\ \overline{n\ell}_\beta; S_\beta L_\beta) \tag{23}$$

where the a_β are purely group theoretical constants which can be computed using equations 19-20 of Paper I.

Moreover it was demonstrated, semi-empirically [59] at first for small atoms, and later in an ab-initio manner, that the bi-virtual pair energies are roughly independent of Z,N and state, viz, transferable.

Thus in our BE work, bi-virtual pairs common to both states (same a_β) can be excluded. Inter-shell pair energies ($\overline{n} \neq n$) are found to be individually smaller than intra-shell pair energies ($\overline{n} = n$) and provided there is not a considerably larger number of them (which can be the case in molecules and medium to large atoms) these can be neglected. Furthermore, the more distinct the shell structure is, the less important are these inter-shell pairs in the agregate. Consequently their neglect in larger atoms is more rigorous for inner shell BE.

h. Example: Bi-virtual Contribution to the 1s BE of Ne

From equation 21 of Paper I, the bi-virtual energy of the ground state $E_{vv'}$(g.s.) is given by:

$$E_{vv'}(\text{g.s.}) = \varepsilon(1s^2) + 3\,\varepsilon(1s\,2s\,{}^3S) + \varepsilon(1s\,2s\,{}^1S) + 3\,\varepsilon(1s\,2p\,{}^1P^o) + 9\,\varepsilon(1s\,2p\,{}^3P^o) + \varepsilon(2s^2) + 3\,\varepsilon(2s\,2p\,{}^1P^o) + 9\,\varepsilon(2s\,2p\,{}^3P^o) + \varepsilon(2p^2\,{}^1S) + 9\,\varepsilon(2p^2\,{}^3P) + 5\,\varepsilon(2p^2\,{}^1D) \qquad (24)$$

and for the 1s hole state, assuming transferability of the ε, we have for $E_{vv'}(\underline{1s})$:

$$E_{vv'}(\underline{1s}) \cong E_{vv'}(\text{g.s}) - \varepsilon(1s^2) - 1.5\,\varepsilon(1s\,2s\,{}^3S) - 0.5\,\varepsilon(1s\,2s\,{}^1S) - 1.5\,\varepsilon(1s\,2p\,{}^1P^o) - 4.5\,\varepsilon(1s\,2p\,{}^3P^o) \qquad (25)$$

Dropping the small inter-shell pairs (which was not done for our calculations on small atoms), the non-relativistic bi-virtual correlation energy's contribution to the 1s BE of Ne, is;

$$BE^{vv'}_{NR\text{-}Corr}(\underline{1s}) = -\varepsilon(1s^2) \qquad (26)$$

or about a 1 eV increase in the BE. Bi-virtual correlation will always increase the BE since all ε's are negative and there are more of them in the ground state than in the hole state. For the 2s BE of K, this increase is about 0.5 eV and for the np BE of K about 1.5 eV (this is larger essentially because more pairs are "broken" in the hole state).

Bi-virtual correlation, while important, is secondary to the larger correlations of Table 2, providing they are allowed. For 1s BE those of Table 2 are smaller because a FS subshell change is involved.

D. $E_{R\text{-}Corr}$

Since no satisfactory treatment of relativistic correlation

effects exists at this time, we must take this contribution to be zero. We may hope that such effects will usually (but not always) be only a fraction of $E_{NR-Corr}$.

In the next section, we complete the discussion of the "additive" theory of BE, by introducing the radiative effects.

E. Radiative Effects

The Dirac theory for hydrogenic atoms predicts that the energy depends only on n and j and not on ℓ. In 1947, Lamb and Retherford [60] experimentally determined that the $2s_{1/2}$ level in H was 1058 MHz higher than the $2p_{1/2}$ level in contradiction to the Dirac theory.

This deviation, now known as the Lamb shift, was explained first by Bethe [61], who included a substantial part of the correction using conventional non-relativistic quantum mechanics. This was soon afterwards improved using quantum electrodynamic techniques. Since that time a considerable amount of work on hydrogenic atoms ($1s_{1/2}$ and $2s_{1/2}$ levels) has been done to develop the higher order terms [62-63] which has been the subject of a recent review [64].

The two main radiative corrections (and the only ones of concern here) are the self-energy (SE) and vacuum polarization (VP) terms. The former arises from the interaction of the electron with an external potential (the nucleus) which spreads the electron out, diminishing the point charge Coulomb binding energy. This pushes $s_{1/2}$ levels higher than $p_{1/2}$ in hydrogenic atoms. Levels of higher n are also affected (they may go up or down) but to a much lesser extent and we ignore it in this work (this effect is important in Fm however).

The other significant term arises because the effective potential seen by the electron is modified by the vacuum polarization due to virtual electron-positron pairs. This correction lowers the $s_{1/2}$ levels and is usually no more than 20% of the self-energy term. In the independent particle model, if an ns electron is removed then that much less radiative energy is present. It is clear from Table 1 that even for small species such as K, these effects can be quite substantial ($\sim$1 eV for the 1s BE) and they must be included. Radiative effects grow as $Z^4\alpha^3 \ln\alpha$, much faster than non-relativistic effects (Z^2) and only $\alpha \ln\alpha$ smaller than relativistic effects.

Fortunately, they can be treated for our purposes in a relatively straight-forward way. We will neglect effects associated with two or more electrons as they are small [16], and only acknowledge the existence of the other electrons by modifying the nuclear charge (screening). Finally, only the deeper ns BE must include this contribution. For a given ℓ , radiative effects fall off faster than n^{-3} with increasing n, due to the additional screening present in outer shells.

Much of the work in this area [62,65] has concentrated on 1s shifts for low to medium Z for which expansions in $Z\alpha$ (including higher order terms) were developed. Unfortunately, these do not converge sufficiently rapidly for the species of concern here ($Z\alpha \simeq 0.2$-0.4). On the other hand, Desiderio and Johnson [66], following the work of Brown et al [67] have developed a method for evaluating the self-energy for K electrons which avoids the above expansion, and published results for selected values of Z ($70 \leq Z \leq 90$). In part of this work, Dirac-Slater SCF wavefunctions were used, so the effect of screening was included at the IPM level. This was found to be about 2% ($Z \simeq 70$-80) for K electrons and grew with N as one might expect. This approach is not entirely satisfactory for our purposes, since we must depend on published results, which would force us to extrapolate from Z=70, and to provide an estimatory process for other ns BE ($n > 1$).

However, Erickson has recently established a result [63] valid for all Z for the self energy of hydrogenic ns electrons which combines an analytic and graphical result (Figure 2, ref. 63). Specifically, the self-energy is given by (in a.u.):

$$E_{SE}(Z) = \frac{4(Z\alpha)^4}{3\pi n^3\alpha} F_n(Z) \qquad (27)$$

where F_n is a slightly n-dependent dimensionless function, which varies (with Z) from 1 to 2 for the species of interest. We ignore the n dependence here.

The leading term in the vacuum polarization correction is [65] (in a.u.)

$$E_{VP}(Z) = \frac{-4(Z\alpha)^4}{15\pi n^3\alpha} \qquad (28)$$

The net effect of corrections to this term seem small [65] and so will be neglected.

To represent the many-electron atom, we choose to let Z be an effective Z throughout equations (27) and (28). Since these Z's appear hydrogenically through the value of the ns radial function at the origin [61], $R_{ns}(0)$, the effective Z should be determined by matching the SCF result (non-relativistic for small atoms, relativistic for large ones) to a screened hydrogenic solution for the above quantity. Screening effects can in fact be substantial for ns electrons ($n > 1$). For example the radiative contribution to the 3s BE of Hg is reduced 0.3 a.u. by screening.

Finally, the radiative contribution to the BE, BE_{Rad}, is:

$$BE_{Rad} = E_{SE}(Z_h) + E_{VP}(Z_h) - 2\left[E_{SE}(Z_i) + E_{VP}(Z_i)\right] \qquad (29)$$

where Z_h, Z_i, are the final and initial state effective Z's. This always reduces the BE for ns electrons. We have tested equation (29) on the 1s BE of Hg and found it to be in substantial agreement with other results.

IV. GAS PHASE BINDING ENERGIES

In Table 3 we present our (this work and references 47-48, 68-70) gas phase results for binding energies. All calculations were done by performing a ΔSCF relativistic Hartree-Fock calculation. Radiative effects were added using equation (29). For the small atoms (C,O,O^-,F,F^-,Ne,Na,Na^+) non-relativistic correlation was added essentially by using the full correlation function (eqns. 20a-20e). These results are in excellent (0.1-0.2 eV) agreement with experiment [1a, 4,6,8,71-74] where available.

Our calculations of $E_{NR-Corr}$ for K and Cs are more exploratory due to their large size (Table 3 contains corrections to earlier published [48] K values). For K, we have included all the non-cancelling intra-shell bi-virtual energy, the important hole-virtual correlation, and the symmetry changing virtual polarizations. BE accuracies appear to average 0.3-0.5 eV., although the 3s BE (which is subject to large correlation effects) discrepancy remains a puzzle.

For Cs, only the important hole-virtual correlation has been added to the 3p,3d and 4s BE which were chosen for investigation as they are important for solid state studies [75] and complement existing free atom experimental values [76]. Errors for these appear to be $\lesssim 0.4$ eV.

V. TOWARDS A COMBINED RELATIVISTIC-CORRELATION THEORY

Our goal is a theory which simultaneously treats both correlation and relativisitic effects and which is capable of dealing with <u>any</u> property. This will take on a simplified form for BE, much as did the non-relativistic theory.

As noted earlier, such a theory does not exist. There has been some work in this direction. Beck [77-78] used the low Z

Table 3

BE (in eV)

Subshell/Species	Gas Phase Theory[(a,b)]	Gas Phase Expt.	Standard State Theory[(a)] Shift	Standard State Theory[(a)] Total	Standard State Expt.
1s Li		64.41 [71]	7.9	56.6	57.2 [10]
1s Be		123.6 [7]	9.6	114.0	115.6 [10]
1s B		200.8 [7]	11.0	189.8	191 [10]
1s C	296.3	296.2 [7]	15.2	281.1	289.3 [10]
1s O	544.5		non-metallic		
1s O^-	528.6		non-metallic		
1s F	698.0		non-metallic		
1s F^-	679.8		non-metallic		
1s Ne	870.4	870.31 [72]	non-metallic		
1s Na	1079.3	1079.0 [73] 1079.1 [6]	6.5	1072.8	1074.0 [10].
1s Na^+	1083.6				
1s K	3614.4 (.15)		4.7	3609.6	3610 [1a]
2s K	384.6 (.24)		4.9	379.7	379 [1a]
$2p_{1/2}$ K	303.7	303.7 [8,4]	5.0	298.6	299 [1a]
$2p_{3/2}$ K	300.9 (.07)	301.2 [8] 300.9 [4]	5.0	295.7	296 [1a]
3s K	39.6 (.67)		4.7	34.9	36 [1a]
$3p_{1/2}$ K	24.4	24.8 [74]	4.8	19.6	20.0 [1a]
$3p_{3/2}$ K	24.2 (.24)	24.5 [74]	4.8	19.4	
$3p_{1/2}$ Cs	1073.0 (.04)		4.2	1068.7	1068.9 [75]
$3d_{3/2}$ Cs	745.8 (.02)	745.8 [c]	4.2	741.6	742.9 [75]
$3d_{5/2}$ Cs	732.0 (.02)	731.8 [c]	4.2	727.8	727.9 [75]
4s Cs	238.5		4.2	234.3	233.9 [75]

(a) This work. Between the two terms of lowest energy ($Z \lesssim 11$), or between the levels of unique J.

(b) Approximate multiplet (level) splittings are given in parentheses.

(c) M.S. Banna, B. Wallbank, D.C. Frost, C.A. McDowell, and J.S.H.Q. Perera, J. Chem. Phys. 68, 5459 (1978).

Pauli approximation [16] (the Hamiltonian is formed by expanding the Dirac and Breit Hamiltonians and keeping terms through order $(Z\alpha)^2$) and incorporated correlation effects via C.I. While this was successful for low and medium Z systems (the inversion of the ^2D fine structure for the Na I isoelectronic sequence was explained [78] in this way), it was pointed out that the $Z\alpha$ expansion did not converge rapidly enough to justify its use for high Z.

Ivanov et al [79] have considered few electron systems using the Gell-Mann and Low formula within the context of perturbation theory, based on hydrogenic Dirac functions. Johnson and Lin [80] have initiated a relativistic variant of the Random Phase Approximation and applied it to the He isoelectronic sequence. We expect that this will have similar strengths and weaknesses as its non-relativistic counterpart.

Finally, there are the relativistic multi-configurational approaches [32,81]. Their principle difficulties are that only a few configurations can be treated simultaneously and there is no configuration selection mechanism. Within this context, they are quite powerful.

A. A Proposed Relativistic Pair Correlation Procedure

The essential features of this proposal will parallel those of the non-relativistic approach, viz: (1) there will be a relativistic reference function Φ_R which describes the system "well", (2) creation of a relativistic Fermi sea (RFS) and virtual space, (3) generation of the form of the correlation function χ_R by applying first order perturbation theory, (4) determination of χ_R by variational C.I. methods.

1. Relativistic Φ_R An easy way to create this function is to think non-relativistically and adjust the results. Thus, a given state is usually described well even for high Z systems by a few "non-relativistic" configurations (C_Φ). The relativistic forms are generated by the replacement $(n\ell)^q$ to $(n\ell_{-(1/2)})^{q-m}$ x $(n\ell_{+(1/2)})^m$ where $m \leq 2\ell + 2$, $q-m \leq 2\ell$ for each subshell present. For closed subshells ($q = 4\ell + 2$) only the equality obtains.

In general then, Φ_R takes on a multi-configurational character, except in the limit of complete j-j coupling, even if Φ is single configurational. This is the "natural" relativistic coupling, and all the algebra-first enumerated by Swirles [82] for many electron systems, later modernized and extended by Grant [18] - has been developed within this scheme.

On the other hand, the "physics" may dictate another coupling-for example the valence electrons may be nearly LS coupled, so expanding in LSJ eigenstates would be more desirable. The absence of a convenient means to accomplish this has impeded the use of relativistic procedures, particularly in the lighter atoms. Here, we suggest a way by which this can be overcome. We recall that any complete orthonormal set (within the configuration space) may be used. Let us then ignore the minor component (it prohibits the construction of exact L^2, S^2 eigenstates) and diagonalize L^2 and S^2 using the remaining function. This fixes all the coefficients arising from a single non-relativistic configuration (if parents are necessary to characterize the state as well, these can be treated in a similar manner). If we have near LS coupling, one vector will dominate Φ_R and we can use the fixed coefficients within the relativistic SCF process to determine just the radials, i.e. the calculation becomes single configurational (which speeds it up, and removes some possible convergence problems). The remaining vectors, if needed, can be added at the C.I. level. Only when we are far from LS or jj coupling would a true relativistic multi-configurational procedure become necessary.

2. The Relativistic Fermi Sea (RFS) and Virtual Space All spinors appearing in Φ_R and all those nearly degenerate with them form the relativistic Fermi sea (RFS). These may be generated from the non-relativistic FS (which we have given for ground states of all atoms [47]) by making the replacement $n\ell \rightarrow n\ell_{\ell-(1/2)}$ $(\ell \neq 0)$ and $n\ell_{\ell+(1/2)}$ with the proviso if $n\ell_j$ is present, $n\bar{\ell}_j$ $(\ell, \bar{\ell} = j \pm (1/2), \ \ell \neq \bar{\ell})$ must be present as well--for these two are relativistically degenerate in the high Z hydrogenic limit. All RFS spinors are to be generated at the (separate) SCF level.

To complete the space, we add a set of spinors orthogonal to all RFS spinors, which are determined at the variational C.I. level. Prior to orthonormalization, these virtuals will probably have the form:

$$\text{virtual} \sim r^g \, e^{-ar}$$

where both a and g are non-linear parameters to be chosen variationally. As in the non-relativistic case, these are truncated (relativistic) hydrogenic functions.

3. First Order Form of the Relativistic Correlation Function X_R We again select the form by use of first order perturbation theory, keeping only configurations C_{X_R} for which the matrix element:

$$\langle \Phi_R | \Sigma_{i<j} \; r_{ij}^{-1} + B(i,j) \; | \; C_{X_R} \rangle \neq 0$$

is non-zero. The configurations can be classified according to equations (20). Symmetry considerations (parity, total angular momentum) will serve to truncate (20a)-(20d) to a finite number of configurations.

When constructing C_{χ_R}, it may be advantageous (certainly for deep core excitations) to jj couple part of the configuration, and use a different coupling (perhaps LS) for the outer electrons [83]. This would most nearly diagonalize the correlation configuration and allow the decoupling ideas of Paper I to be applied here (in a somewhat modified form).

To illustrate the method, consider the ground (J=0) state of a Be-like atom, and take the 1s spinors as fixed (forming a core). The only configurations which can be generated solely from the RFS (which includes only $2s_{1/2}$ $2p_{1/2}$ $2p_{3/2}$) are: $2s_{1/2}{}^2$, $2p_{1/2}{}^2$, $2p_{3/2}{}^2$. The latter two can be coupled (ignoring the minor components) to form ${}^1S_0$ and ${}^3P_0$, and depending on Z, we may choose them to have one ($2s^2{}_{1/2}$) two ($+p^2\ {}^1S_0$) or three ($+p^2\ {}^3P_0$) vectors to form Φ_R (the others appear in (20c)).

Hole-virtual and virtual polarization configurations are: $2s_{1/2}vs_{1/2}$, $2p_{1/2}vp_{1/2}$, $2p_{3/2}vp_{3/2}$. These demonstrate the rapid symmetry cutoff. Bi-virtual configurations (allowing one virtual per symmetry and cutting the expansion off at f-symmetries) are: $vs_{1/2}{}^2 + vp_{1/2}{}^2$, $vp_{3/2}{}^2 + vd_{3/2}{}^2$, $vd_{5/2}{}^2 + vf_{5/2}{}^2$, $vf_{7/2}{}^2$. Identification of just what belongs in $E_{R\text{-}Corr}$ is difficult. Certainly ${}^3P_0$ contributions involving virtuals will always be found there. More importantly, the differences in the strengths of the off diagonal interactions with (20a)-(20e) (in first order) between the non-relativistic and relativistic treatments also belong there. Eventually, one will simply compare the correlation results of Section III with those of this method and assign the difference to $E_{R\text{-}Corr}$.

4. Variational C.I. It may be noted that we have used the Breit operator to produce the configurations in χ_R. We propose to also include B in the Hamiltonian during the variational C.I. process because, as we argued earlier, it may significantly affect the wavefunctions in some cases.

Yet, there may be some formal difficulty (Section III). How can we avoid this?

Both the SCF functional and the C.I. functional admit (with either equation (12) or equation (14)) positron as well as electron solutions (these correspond to different stationary points). At the SCF level, where we do not propose to use B (see Section III), positron solutions are avoided by simply choosing the proper (i.e. electron) input. This means major components are truly larger than minor components, radial functions have the proper number of nodes, etc. (see also the work of Kim [84] for a further discussion on this point). We should note that a number of relativistic SCF calculations have now been done and nothing catastrophic has occured.

In the C.I. calculations, we will insist that virtuals exhibit a similar major component dominance and that all correlation configurations remain small (a large correlation function could indicate we are trying to "correct" our electron solution by introducing positron components. This would certainly lower the energy, but it would be incorrect.)

Further work on this proposal is in progress.

VI. CORE BE IN METALS

Core BE of metals are not only of interest in themselves (e.g. for surface and catalytic studies), but because in the past most experimental core BE were only available from standard state measurements. There is thus considerable interest in finding a simple relationship between the free and metallic BE.

Standard state BE^S are measured relative to the Fermi level ($BE_F{}^S$) and to refer them to the common vacuum level $BE_V{}^S$, we must introduce the work function φ, viz.

$$BE_V{}^S = BE_F{}^S + \varphi \tag{30}$$

where φ is the energy necessary to remove the least bound conduction band electron from the solid. This quantity is known for most metals [85].

A series of comparisons of core BE for free atoms (BE^A) and the standard state $BE_V{}^S$ (not just metals) has yielded the observations [1a,10,86,89]:

(a) $BE^A = BE_V{}^S + S_e$; S_e= 4-15 eV

(b) within an atom, core level shifts are nearly independent of the hole position (to a few tenths of an eV)

This latter behavior has been used to combine optical BE for valence subshells of free atoms with X-ray standard state core BE [1a,9,10,87]. However, the results are quite uncertain, principally because of the different shift experienced by the core vs. the valence BE. Given these differences, even theories which accurately account for standard state valence properties (e.g. the work function) are of no direct concern to this work and will not be considered further. Present theoretical and experimental precisions would also suggest that even the difference of core BE should not be transferred from solid to atom unless essential.

The models used to account for the difference between the metal and gas phase for core BE fall into two categories: those which add the correction to the free atom perturbatively, and those which deal with the metal directly. Both approaches have had their successes and failures. For the latter, we have the image potential method of Gadzuk [88], and a density functional approach for rare gases in noble metals by Citrin and Hamann [89]. If a completely ab initio approach of this nature is desired, some way of including the local atomic-like effects discussed in Section III will have to be found.

In this section, we will concentrate on the other approach. Its principle proponents are Ley, Shirley et al [90] and Watson et al [91]. In molecules, a "potential" model [92] similar in spirit to these has been used to correlate chemical shifts among different compounds containing the atomic species of interest.

There appear to be three standard state effects contributing to the shift for which some estimates have been made. The first, and the only one examined in detail here, has been called "extra-atomic relaxation" and represents the screening of the final state hole by the electrons around the atom being probed. The second arises from the presence of a dipole sheet on the surface, and seems to vary from a few tenths to one electron volt. The final effect investigated comes from renormalization of the atomic functions due to their confinement in the unit cell. This has been estimated by Watson et al [91] to be about as important as the extra-atomic relaxation effects at least for transition metals. More work should be done on this term, for it essentially destroys the good agreement (1-2 eV) obtained using just the first effect, and also seems not to yield shifts which are independent of the core hole--an observed effect.

A. Extra-Atomic Relaxation

The essential features of the model discussed here were first put forward by Shirley, Ley et al [90,93]. In it, the excess charge in the hole state is assumed screened by the itinerant electrons forming a semi-localized exciton by the dropping down of a conduction band below E_F. This is in the spirit of a model first proposed by Friedel [94]. One further assumes that the process can be made atomic-like by assuming the exciton wavefunction is found only in the neighborhood of the hole state and that it has the symmetry of the lowest unbound state in the conduction band. These two restrictions will mean that our extra-atomic relaxation shifts are too large. Ley et al [90] then go on to make further approximations within an atomic framework, some of which give rise to rather large (3-4 eV) un-necessary (see below and ref. 95) errors particularly for light atoms.

Our interpretation of this model is as follows: Let the atomic configuration characteristic of the $n\ell$ hole state in the solid be

$$\ldots (n\ell)^{4\ell+1} \ldots (m\bar{\ell})^{q} (S,L)_{min} \quad (q < 4\bar{\ell}+2)$$

Here $m\bar{\ell}$ is the valence (in atomic notation) subshell of lowest energy which is not completely filled. The free atomic approximation to the excitonic state is then:

$$\ldots (n\ell)^{4\ell+1} \ldots (m\bar{\ell})^{q+1} (\tilde{S},\tilde{L})_{min}$$

So we subtract from the atomic BE

$$S_t = E(\ldots(n\ell)^{4\ell+1}\ldots(m\bar{\ell})^{q}(S,L)_{min}) - E(\ldots(n\ell)^{4\ell+1}\ldots (m\bar{\ell})^{q+1}(\tilde{S},\tilde{L})_{min}) \quad (31)$$

to generate a vacuum standard state BE. Two remarks are in order here. First, there can be a configuration change upon going from the atom to the solid, and if this happens (it doesn't for the species of Table 3), the free atom BE must be precorrected to account for it. Watson et al [94] found such corrections removed much of the theoretical experimental discrepancies observed in the transition metals (these are particularly sensitive to the 3d-4s degeneracies). In that work and here, it was also implicitly assumed that formation of the exciton does not induce a further configuration change. The second feature is that one computes for the BE the difference between the two terms of minimum energy.

The standard state results of Table 3 were obtained by adding a 2s (Li), a 2p (Be,B,C) and 3s (Na), 4s (K) and a 6s (Cs) electron to the hole state configuration. For all cases, S_t was evaluated from separate ΔSCF calculations- relativistic ones for $Z>10$ and non-relativistic ones for $Z \leq 10$. For the light atoms, we also included correlation effects by using the internal, internal polarization, virtual polarization and hole-virtual correlation of the species in which $n\ell$ (here 1s) is closed and Z is increased by one unit--an application of the equivalent cores idea (see below). The uncancelled bi-virtual energy was obtained by using equation 19 of Paper I to generate the group theoretical coefficients and the pair energies taken from reference [59].

The results shown in Table III are all in agreement with the standard state experiment [10] to within 1-2 eV. The C(1s) anamoly may arise from the fact that it (graphite) is partially co-valent and the assumption of a pure $^5S^o$ excitonic state may be inappropriate (a calculation based on the average energy lowers the shift by 4.3 eV).

It can be seen that the theoretical shift is quite constant for core BE belonging to thé same species. This may be understood at the SCF level as follows. The two configurations differ only in having one extra electron in the outermost subshell. The ionization potential of the extra electron may be given reasonably well by the orbital eigenvalue. If we assume this is independent of the position of the hole, the shift is constant.

In the above, we have made use of an equivalent cores approximation to obtain a portion of the correlation energy. This has been employed in a variety of contexts, and with care can be quite useful. This does not obtain when exchange effects involving the core-hole are large as for first row atoms (2eV errors result then). For larger species, it seems to be quite good (see also the work of Firsht and McWeeny [97]), and represents an extremely simple way of evaluating equation (31) for the shift is then nothing more than the first ionization potential (known optically) of the species with nuclear charge Z + 1.

In summary, we have found a rather simple model for extra-atomic relaxation predicts core BE shifts for the metallic species considered to within 1-2 eV. It does not however tell us much about "what is happening" in the solid (charge state information is acessible though). Furthermore it should not be applied to valence hole BE - these can have markedly different shifts and we do not expect a model which predicts constant shifts to account for these effects. On the other hand, it was the relatively simple localized nature of core holes that led to our original [68] interest in them (O,O^-).

Further improvement will have to involve direct consideration of solid state aspects. Perhaps this can be done by combining the local or orbital methods of Kunz and Adams [98] with the methods of the preceeding sections.

VII. ACKNOWLEDGEMENTS

We acknowledge the contribution of Mr. George Aspromallis in evaluating some of the correlation effects.

REFERENCES

1. a) K Siegbahn, C. Nordling, A. Fahlman, R. Nordberg, K. Hamrin, J. Hedman, G. Johansson, T. Bergmark, S.-E. Karlsson, I. Lindgren, B. Lindberg, ESCA-Atomic, Molecular and Solid State Structure Studied by Means of Electron Spectroscopy, Nova Acta Reg. Soc. Sci. Upsaliensis, Ser. IV, 20 Uppsala (1967).
b) D.W. Turner, C. Baker, A.D. Baker, C.R. Brundle, Molecular Photoelectron Spectroscopy, Wiley-Interscience, 1970.
c) D.A. Shirley, editor, Electron Spectroscopy, North-Holland, Amsterdam (1972).
d) T.A. Carlson, Photoelectron and Auger Spectroscopy, Plenum Press (1975).
e) K.D. Sevier, Low Energy Electron Spectrometry, Wiley, 1972.
f) Proceedings of the Electron Spectroscopy Symposium, (Uppsala, May, 1972), Phys. Scr. 16, No. 5-6 (1977).
2. U. Gelius, J. Electr. Spectr. 5, 985 (1974).
3. S.P. Kowalczyk, L. Ley, R.L. Martin, F.R. McFeely, D.A. Shirley, Faraday Disc. Chem. Soc., Vancouver, Canada, July 15-17, 1975.
4. W. Mehlhorn, B. Breuckmann, and D. Hausamann, Phys. Scr. 16, 177 (1977).
5. M.S. Banna, D.C. Frost, C.A. McDowell, and B. Wallbank, J. Chem. Phys. 68, 696 (1978).
6. R.L. Martin, E.R. Davidson, M.S. Banna, B. Wallbank, D.C. Frost, and C.A. McDowell, J. Chem. Phys. 68, 5006 (1978).
7. P. Bisgaard, R. Bruch, P. Dahl, B. Fastrup, M. Rødbro, Phys. Scr. 17, 49 (1978).
8. M.W.D. Mansfield, Proc. Roy. Soc. A346, 555 (1975).
9. W. Lotz, J. Opt. Soc. Am. 58, 915 (1968).
10. D.A. Shirley, R.L. Martin, S.P. Kowalczyk, F.R. McFeely, and L. Ley, Phys. Rev. B15, 544 (1977).
11. M.O. Krause and C.W. Nestor, Jr., Phys. Scr. 16, 285 (1977).
12. D.R. Beck and C.A. Nicolaides, J. El. Spect. 8, 249 (1976).
13. V. Schmidt, N. Sander, W. Mehlhorn, M.Y. Adam and F. Wuilleumier, Phys. Rev. Letts. 38, 63 (1977).
14. D. Spence, J. Chem. Phys. 68, 2980 (1978).
15. H. Hartmann and E. Clementi, Phys. Rev. 133, A 1295 (1964).
16. H.A. Bethe and E.E. Salpeter, Quantum Mechanics of One and Two-Electron Atoms, Academic Press, 1957.
17. J.P. Desclaux, Comp. Phys. Comm. 9, 31 (1975).
18. I.P. Grant, Adv. Phys. 19, 747, (1970).
19. J.C. Slater, Phys. Rev. 81, 385 (1951).
20. a) A. Rosén and I. Lindgren, Phys. Rev. 176, 114 (1968).
b) D.A. Liberman, J.T. Waber, and D.T. Cromer, Comp. Phys. Comm. 2, 107 (1971).
21. C. Froese-Fischer, Comp. Phys. Comm. 4, 107 (1972).
22. a) D.R. Hartree, The Calculation of Atomic Structures,

Wiley, 1957.
b) J.P. Desclaux, D.F. Mayers, and F.O'Brien, J. Phys. B4, 631 (1971).
c) J.B. Mann and J.T. Waber, J. Chem. Phys. 53, 2397 (1970).
d) D.F. Mayers, private communication, 1971.

23. e.g. L.S. Cederbaum and W. Domcke, J. Chem. Phys. 66, 5084 (1977).
24. e.g. L. Hedin and G. Johansson, J. Phys. B2, 1336 (1969).
25. a) J. Linderberg and Y. Öhrn, Propagators in Quantum Chemistry, Academic, 1973 and this volume.
b) B.T. Pickup and O. Goscinski, Mol. Phys. 26, 1013 (1973).
26. L.S. Cederbaum, G. Hohlneicher, and W. von Niessen, Chem. Phys. Lett. 18, 503 (1973) and this volume.
27. F.P. Larkins in Atomic Inner Shell Processes, Vol. I, edited by B. Crasemann, Academic Press, 1975.
28. J.B. Mann and W.R. Johnson, Phys. Rev. A4, 41 (1971).
29. Y. Nambu, Prog. Theor. Phys. 5, 321 (1950).
30. M.H. Mittleman, Phys. Rev. 5A, 2395 (1972).
31. M. Blume and R.E. Watson, Proc. Roy. Soc. A271, 565 (1963).
32. J.P. Desclaux and Yong-Ki Kim, J. Phys. B8, 1177 (1975).
33. S.J. Rose, I.P. Grant and N.C. Pyper, J. Phys. B 11,1171 (1978).
34. N.C. Pyper, I.P. Grant and R. Gerber, Chem. Phys. Letts. 49, 479 (1977).
35. G. Das and A.C. Wahl, J. Chem. Phys. 64, 4672 (1976).
36. D.P. Chong and Y. Takahata, Int. J. Quant. Chem. 12, 549 (1977).
37. H.P. Kelly, in P.G.H. Sandars, Ed. Atomic Physics 2, Plenum Press, 1973.
38. T. Lee, N.C. Dutta, and T.P. Das, Phys. Rev. A4, 1410 (1971).
39. S. Garpman, I. Lindgren, J. Lindgren, and J. Morrisen, Phys. Rev. 11A, 758 (1975).
40. R.K.Nesbet, Phys. Rev. 175, 2 (1968).
41. T. Shibuya and V. McKoy, Phys. Rev. A2, 1108 (1970).
42. M. Ya. Amusia, N.A. Cherepkov, and L.V. Chernysheva, Soviet Physics-JETP 33, 90 (1971).
43. G. Wendin and M. Ohno, Phys. Scr. 14, 148 (1976).
44. C. Froese-Fischer, Int. J. Quant. Chem. S8, 5 (1973).
45. O.R. Platas and H.F. Schaefer III, Phys. Rev. A4, 33 (1971).
46. H.J. Silverstone and O. Sinanoğlu, J. Chem. Phys. 44, 3608 (1966).
47. D.R. Beck and C.A. Nicolaides, Int. J. Quant. Chem. S8, 17 (1974).
48. D.R. Beck and C.A. Nicolaides, Int. J. Quant. Chem. S10, 119 (1976).
49. A.W. Weiss, Phys. Rev. A9, 1524 (1974).
50. D.R. Beck and C.A. Nicolaides, the first paper by these authors appearing in this volume. Hereafter, Paper I.
51. E.J. McGuire, Phys. Rev. A9, 1840 (1974).
52. U. Fano, Phys. Rev. 124, 1866 (1961).

53. P.L. Altick and E. Neal Moore, Phys. Rev. 147, 59 (1966) and this volume.
54. Y. Komninos, D.R. Beck, and C.A. Nicolaides, unpublished (1978).
55. R.F. Bacher, Phys. Rev. 43, 264 (1933).
56. R.N. Zare, J. Chem. Phys. 45, 1966 (1966).
57. P.S. Bagus, A.J. Freeman, and F. Sasaki, Phys. Rev. Lett. 30, 850 (1973).
58. O. Sinanoğlu, Proc. Roy. Soc. A260, 379 (1961).
59. I. Öksüz and O. Sinanoğlu, Phys. Rev. 181, 54 (1969).
60. W.E. Lamb, Jr. and R.C. Retherford, Phys. Rev. 72, 241 (1947).
61. H.A. Bethe, Phys. Rev. 72, 339 (1947).
62. P.J. Mohr in Beam Foil Spectroscopy Volume I, edited by I. Sellin and D.J. Pegg, Plenum Press, 1976.
63. G.W. Erickson, Phys. Rev. Lett. 27, 780 (1971).
64. B.E. Lautrup, A. Petermann, and E. de Rafael, Phys. Repts. 3, 193 (1972).
65. G.W. Erickson and O.R. Yennie, Ann. Phys. (N.Y.) 35, 271 (1965).
66. A.M. Desiderio and W.R. Johnson, Phys. Rev. A3, 1267 (1971).
67. G.E. Brown, J.S. Langer, and G.W. Schaefer, Proc. Roy. Soc. A251, 92 (1959); G.E. Brown and D.F. Mayers, Proc. Roy. Soc. A251, 105 (1959).
68. C.A. Nicolaides, Chem. Phys. Letts. 19, 69 (1973).
69. D.R. Beck and C.A. Nicolaides, Phys. Fenn. 9S, 244 (1974).
70. D.R. Beck and C.A. Nicolaides, J. Elect. Spect. 8, 249 (1976).
71. S. Bashkin and J. O. Stoner, Atomic Energy Levels and Grotrian Diagrams Volume 1, North-Holland, 1975.
72. T.D. Thomas and R.W. Shaw, Jr., J. Electr. Spectrs. 8, 45 (1976).
73. H. Hillig, B. Cleff, W. Mehlhorn, and W. Schmitz, Z. Physik 268, 225 (1974).
74. C.E. Moore, Atomic Energy Levels, Volume 1, National Bureau of Standards, U.S. GPO (1949).
75. N.G. Krishnan, W.N. Delgass, and W.D. Robertson, J. Phys. F 7, 2623 (1977).
76. H. Peterson, K. Radler, B. Sonntag and R. Haensel, J. Phys. B8, 31 (1975).
77. D.R. Beck, J. Chem. Phys. 51, 2171 (1969).
78. D.R. Beck and H. Odabasi, Ann. Phys. (N.Y.) 67, 274 (1971).
79. L.N. Ivanov, E.P. Ivanova and U.I. Safronova, J. Quant. Spectrsc. Rad. Trans. 15, 553 (1975).
80. W.R. Johnson and C.D. Lin, Phys. Rev. 14A, 565 (1976).
81. I.P. Grant, D.F. Mayers, and N.C. Pyper, J. Phys. B 9, 2777 (1976).
82. B. Swirles, Proc. Roy. Soc. A152, 625 (1935).
83. An algorithm to do this has been completed.
84. Yong-Ki Kim, Phys. Rev. 154, 17 (1967).
85. e.g. American Institute of Physics Handbook, 3rd edition (1972).

86. K. Siegbahn, J. Elect. Spectr. 5, 3 (1974).
87. J.A. Bearden and A.F. Burr, Rev. Mod. Phys. 39, 125 (1967).
88. J.W. Gadzuk, J. Vac. Sci. Tech. 12,289 (1975).
89. P.H. Citrin and D.R. Hamann, Chem. Phys. Lett. 22, 301 (1973).
90. L. Ley, S.P. Kowalczyk, F.R. McFeely, R.A. Pollak, and D.A. Shirley, Phys. Rev. B8, 2392 (1973).
91. R.E. Watson, M.L. Perlman and J.F. Herbst, Phys. Rev. B13, 2358 (1976).
92. U. Gelius, Phys. Scr. 9, 133 (1974).
93. D.A. Shirley, Chem. Phys. Letts. 17, 312 (1972).
94. J. Friedel, Phil. Mag. 43, 153 (1952); Adv. Phys. 3, 446 (1954).
95. C.A. Nicolaides and D.R. Beck, Chem. Phys. Letts. 27, 269 (1974).
96. R.M. Wilson, J. Chem. Phys. 60, 1692 (1974).
97. D. Firsht and R. McWeeny, Mol. Phys. 32, 1637 (1976).
98. A.B. Kunz, this volume.

ATOMIC PHOTOIONIZATION CROSS SECTIONS

P. L. Altick

Physics Dept., Univ. of Nevada, Reno, Nevada 89557, USA

In these lectures I will discuss various methods of calculating atomic photoionization cross sections with emphasis on resonant structures. I will attempt to order the methods according to the complexity of the problem that they can be used on. Thus the Feshbach projection operator formalism, which is most applicable to the calculation of resonance parameters in two electron systems will be first and the R-matrix and quantum defect methods which can handle many channel photoionization will be last. Because of time limitation, the discussion will be descriptive rather than detailed. Not all useful approaches can be presented, but the most important ones left out (RPA, MBPT) will be covered by other lecturers.

The appearance of resonances or structure in various kinds of cross sections in atomic, molecular, or nuclear physics is by no means rare. The first low energy neutron-nuclei scattering experiments produced typically nothing but sequences of resonances. They are also ubiquitous in atomic physics, and already more than 40 years ago Beutler investigated resonant phenomena in various photoionization cross sections[1]. At the atomic level our basic theory is, of course, a wave theory, and so it is not surprising that various kinds of interference effects manifest themselves when a beam of particles interacts with a complex target.

Resonances come in all shapes and sizes and show up in different kinds of experiments, e.g. photoionization, elastic and inelastic electron scattering, inelastic proton scattering, etc. When attempting to understand such diverse phenomena, a helpful unifying concept is that the resonance occurs because of the structure of an excited state of the system. Since excited states

Cleanthes A. Nicolaides and Donald R. Beck (eds.), Excited States in Quantum Chemistry, 361–382.

are the main topic of this school, this point of view is quite an appropriate one to pursue.

Thus, the theoretical problem of describing resonant phenomena boils down to the problem of calculating properties of excited states of various systems. In these lectures we will look at several approaches to this problem that have been successful.

Before becoming immersed in the theory, however, we will orient ourselves by looking at some diverse experimental results. A system that has received an immense amount of theoretical attention is helium. Here, Madden and Codling[2] have found resonances in photoabsorption in the neighborhood of 200 Å. Several series were found converging to thresholds of the He^+ ion. The diagram, Fig. 1, shows the location of some of those resonances below the n = 2 He^+ level. These are examples of Feshbach Resonances, i.e. they can be associated with definite configurations. Thus the lowest ^1P resonance is denoted 2s2p and others are evidently formed from 2snp, 2pns and 2pnd configurations. Such labeling does not mean that these configurations form stationary states. Rather, one can form a physical picture of the two electrons occupying bound orbitals for, say 10 to 100 orbits and then an energy transfer takes place with the result 2s 2p → 1s kp where kp represents a free electron. Looked at in this way, the atom is said to autoionize. The lifetime of the pseudo bound state manifests itself in a resonant width, Γ, and for isolated resonances, i.e. Γ << separation between neighboring pairs, the intrinsic parameters of the resonance are its location and its width. In addition, there is a line profile index, q,

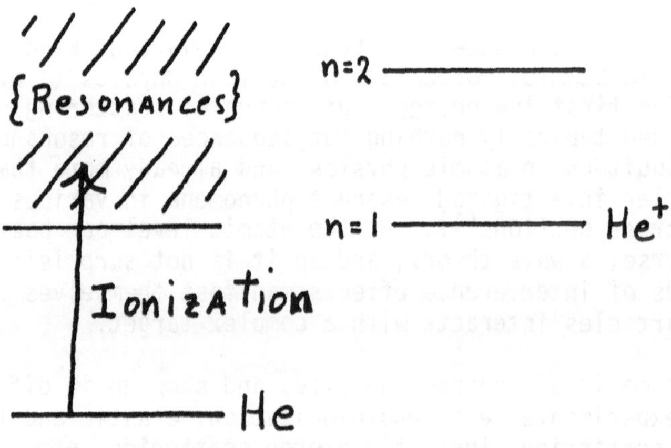

Fig. 1. A level diagram for He showing the location of the lowest Feshbach resonances.

introduced by Fano(3) whose value depends on the method of excitation, but, when calculated, allows detailed comparison between theory and experiment. Fig. 2 shows a sequence of resonances in He. The asymmetric profile is a characteristic line shape and can be reproduced by a suitable value of q. In He the near degeneracy of the levels with the same n gives rise to special configuration-interaction effects which are an interesting story, but one we will not pursue. For purposes of orientation, the width of the broadest resonance is $\Gamma \sim .04$ eV giving a lifetime of $\sim h/\Gamma \sim 2 \times 10^{-14}$ s. The narrowest resonance (2p3d) has $\Gamma \sim 10^{-6}$ eV.

If we go from $He(1s^2)$ to $Be(2s^2)$ we find drastic differences in the appearance of the photoionization cross section in the vicinity of the resonances. In Fig. 3 a broad series and a narrow series are evident. These are also Feshbach resonances arising from 2pns, 2pnd configurations. The large width of the 2pns resonances indicates a strong interaction between the bound and continuum states and so the reaction 2pns → 2s kp goes fast. The ultimate reason for this is that the 2s and 2p orbitals are from the same shell and thus overlap considerably. The widths of the 2pns resonances are so large that they can no longer be treated as isolated, and it becomes a problem to find suitable parameters to characterize the appearance of the cross section.

In both of the above cases, the resonant configuration was

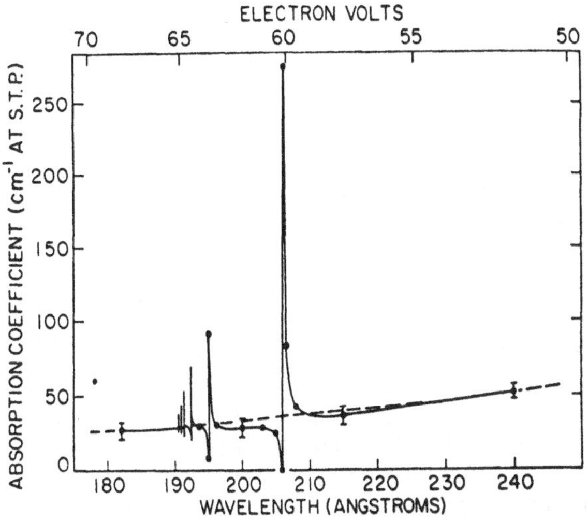

Fig. 2. The absorption coefficient of helium in the 175-245-Å.

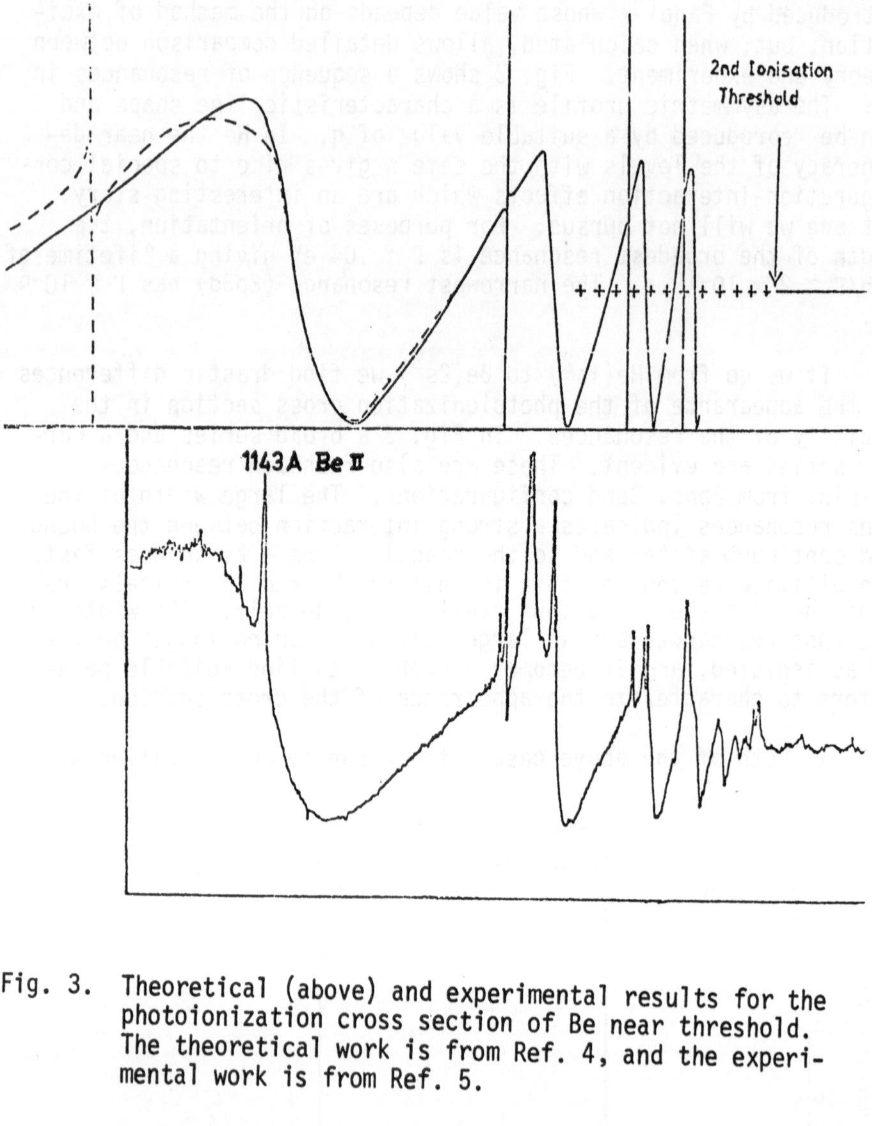

Fig. 3. Theoretical (above) and experimental results for the photoionization cross section of Be near threshold. The theoretical work is from Ref. 4, and the experimental work is from Ref. 5.

doubly excited from the ground state, but all that is necessary for autoionization is that the configuration lie above the first ionization threshold and so single excitations from inner shells are also candidates. This type of resonance is found in the photoionization cross section of Ar(6) shown in Fig. 4. Here the resonances are due to configurations of the type $3s3p^6np\ ^1P$. Note that the position of the resonances is noted by a sharp dip rather than an increase in the cross section. This feature is readily explainable in terms of theory we will present. At higher energies double excitation and inner shell excitations overlap resulting in a spectrum of considerable complexity.

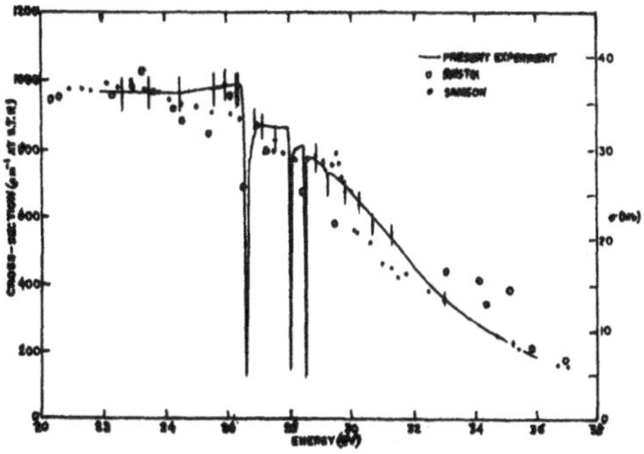

Fig. 4. The photoionization cross section of Ar from Ref. 6.

As a final example, a somewhat different kind of feature is found in the photoionization cross section of Xe sketched in Fig. 5. This large feature begins with the 4d threshold for ionization and is due to 4d → εf photon excitation. The effective potential for the f electron contains both attractive and repulsive regions separated by a barrier perhaps a few eV high. It is possible for the f electron to get "delayed" in the attractive well at certain energies and this occurrence manifests itself by a broad peak in the cross section. Such a peak is called a "shape resonance" as it is not identified with a high lying configuration but rather is a result of the dynamics of an electron in a central field.

The earliest calculations on the resonances in He and H^- used the Feshbach projection operator theory,[7,8] and papers are still appearing with very accurate results based on it. The achievement of the formalism is to reduce the calculation of the position of a resonance to a bound state problem. Thus the arsenal of weapons built up over the years for attacking such bound state problems is available. On the other hand, to complete the description of the resonance, a width and energy shift are needed and to get them, one must employ a continuum function.

The theory will be developed with an eye to applying it to He. There are difficulties in applications to heavier atoms which will be mentioned as we go along. We begin by looking for a partial wave scattering solution to the $e + He^+ \rightarrow e + He^+$ problem, i.e. we seek a function Ψ_γ where

$$(H - E)\Psi_\gamma = 0. \qquad (1)$$

The subscript γ represents an appropriate set of quantum numbers, e.g. L^2, L_z, S^2, S_z. Since these do not change during the

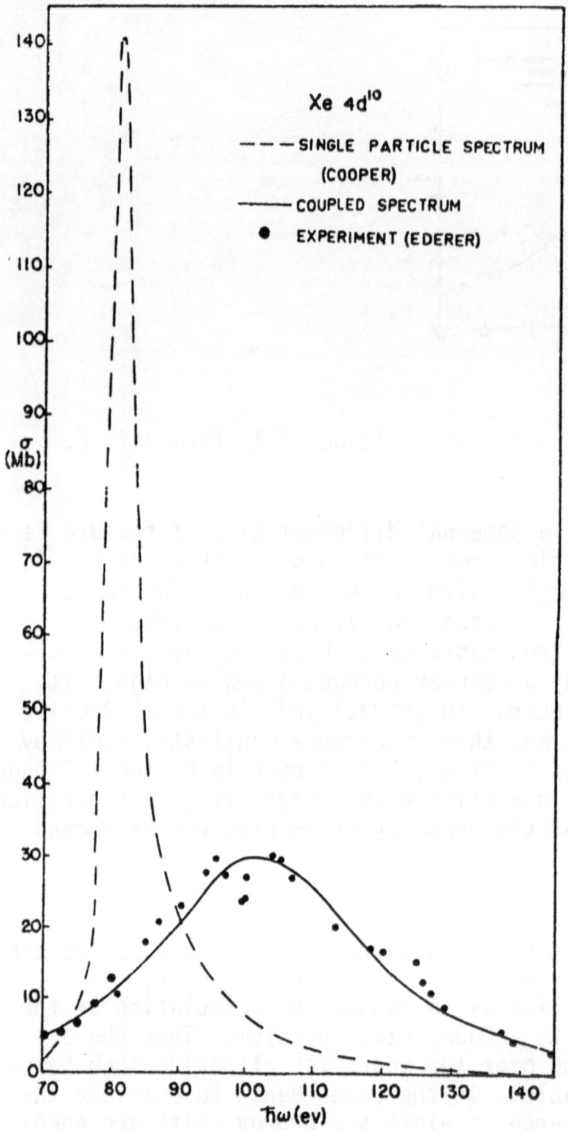

Fig. 5. The 4d photoionization cross section of Xe. (From W. Brandt, L. Eder, S. Lundqvist, JQSRT, 7, 185 (1967)).

development, we will no longer write them. If we are interested in resonances below the n = 2 level of He^+, E should be chosen so that Ψ describes elastic scattering below that threshold. Thus

$$\Psi(\underset{\sim}{r}_1,\underset{\sim}{r}_2)\underset{r_2\to\infty}{\sim} u_0(\underset{\sim}{r}_1)\sqrt{\frac{2}{\pi k}}\,\frac{\sin\,(kr_2+\phi)}{r_2} \tag{2}$$

where $E = k^2/2$. The energy normalization is chosen, and the complete phase shift ϕ includes effects of angular momentum, the Coulomb field, plus effects of the short range interaction with the atomic electron. The function $u_0(\underset{\sim}{r}_1)$ is the 1s function of He^+. In general it is an eigenstate of the target atom Hamiltonian and so is known exactly only for one electron atoms.

We now seek to portion the quantum mechanical vector space into two parts by introducing projection operators P, Q, where

$$P + Q = 1. \tag{3}$$

Thus, they span the entire space. P is then chosen so that

$$P\Psi(\underset{\sim}{r}_1,\underset{\sim}{r}_2)|_{r_2\to\infty} = \Psi(\underset{\sim}{r}_1,\underset{\sim}{r}_2)|_{r_2\to\infty} \tag{4}$$

and $Q = 1 - P$. Since $\Psi(\underset{\sim}{r}_1,\underset{\sim}{r}_2)$ is symmetrized, it doesn't matter which variable becomes large. The fact that P is specified only in an asymptotic region results in a lack of uniqueness for the choice and, in fact, there are an infinite number of choices of P which will satisfy Eq. 4. This freedom caused some confusion earlier, and to better understand it, consider the vector space with a basis of symmetrized products of He^+ eigenfunctions $u_n(\underset{\sim}{r})$. Here n represents both bound and continuum states. Now let

$$\Psi(\underset{\sim}{r}_1,\underset{\sim}{r}_2) = \mathcal{A} \sum\!\!\!\!\!\!\int_{m'} a_{nm} u_n(\underset{\sim}{r}_1) u_m(\underset{\sim}{r}_2) \tag{5}$$

where $\mathcal{A}$ performs the necessary symmetry operation. In the asymptotic region

$$\Psi(\underset{\sim}{r}_1,\underset{\sim}{r}_2)_{r_2\to\infty} \sim u_0(\underset{\sim}{r}_1) \sum\!\!\!\!\!\!\int_{m'} a_{om'}\, u_{m'}(\underset{\sim}{r}_2) \tag{6}$$

where now the sum m' includes only continuum states as the others vanish in this limit. Thus P only need project onto this subset of basic vectors and whether or not it includes any others is arbitrary. A common and workable choice of Q is

$$Q = (1 - |u_0(\underset{\sim}{r}_1)\rangle\langle u_0(\underset{\sim}{r}_1)|)\cdot(1 - |u_0(\underset{\sim}{r}_2)\rangle\langle u_0(\underset{\sim}{r}_2)|) \tag{7}$$

i.e. all sets of products in Eq. (5) with either n or m = 0 are in the P subspace.

To see the utility of introducing P and Q, we write

$$\begin{aligned} P(H-E)(P+Q)|\Psi\rangle &= 0 \\ Q(H-E)(P+Q)|\Psi\rangle &= 0 \end{aligned} \tag{8}$$

(From here on the Dirac notation will be convenient.) By some formal manipulation we get an equation for $|P\Psi\rangle$ which is

$$(H_{pp} + V_{opt} - E)|P\Psi\rangle = 0, \tag{9}$$

where $H_{pp} \equiv PHP$ and

$$V_{opt} = H_{PQ} \frac{1}{Q(E-H)Q} H_{QP}. \tag{10}$$

Recalling that $|P\Psi\rangle$ has all the scattering information, all we have accomplished so far is to define an optical potential. It proves useful to consider now the set of eigenstates of H_{QQ} which are complete within the Q space. Thus

$$(H_{QQ} - \varepsilon_n)|\Phi_n(\underset{\sim}{r}_1,\underset{\sim}{r}_2)\rangle = 0. \tag{11}$$

The spectrum of H_{QQ} is discrete below the n = 2 threshold of He^+ because of the choice of Q. Thus the choice we made is appropriate for the description of resonances in this region. Singling out one of these levels, t, we can write (9) as

$$(H_{pp} + V'_{opt} - E)|P\Psi\rangle = -H_{PQ} \frac{|\Phi_t\rangle\langle\Phi_t|H_{QP}|P\Psi\rangle}{E-\varepsilon_t} \tag{12}$$

where V'_{opt} is missing the t term. We see that the rhs of Eq. 12 varies rapidly for $E \sim \varepsilon_t$. This rapid variation will cause the phase shift to move through π radians as E passes through the region of ε_t.

To solve Eq. 12 consider the complete set of solutions to the homogeneous equation $|\pi_\epsilon\rangle$, thus

$$|\pi_\epsilon\rangle \underset{r_2\to\infty}{\to} u_0(\underset{\sim}{r}_1)\sqrt{\frac{2}{\pi k}}\frac{\sin(kr_2 + \eta)}{r_2} \tag{13}$$

where $\epsilon = \frac{k^2}{2}$ and now η is the combined phase shift due to all factors, but not the level t. Writing the Green's function

$$G = \int d\epsilon \frac{|\pi_\epsilon\rangle\langle\pi_\epsilon|}{\epsilon-E}, \tag{14}$$

a formal solution to Eq. 12 is

$$|P\Psi\rangle = |\pi_E\rangle - GH_{PQ}\frac{|\Phi_t\rangle\langle\Phi_t|}{E-\varepsilon_t}H_{QP}|P\Psi\rangle. \tag{15}$$

We are thus led immediately to

$$\langle\Phi_t|H_{QP}|P\Psi\rangle = \frac{\langle\Phi_t|H_{QP}|\pi_E\rangle(E-\varepsilon_t)}{E-\varepsilon_t-\Delta} \tag{16}$$

where

$$\Delta = \langle\Phi_t|H_{QP}GH_{PQ}|\Phi_t\rangle. \tag{17}$$

Now, Eq. 15 becomes

$$|P\Psi\rangle = |\pi_E\rangle - GH_{QP}|\Phi_t\rangle\frac{\langle\Phi_t|H_{QP}|\pi_E\rangle}{E-\varepsilon_t-\Delta} \tag{18}$$

To find the phase shift due to the level t, we look at Eq. (18) in the asymptotic region. To get the correct asymptotic behavior for $|P\Psi\rangle$, the principal value of G is taken, thus

$$G|_{r_2\to\infty} \sim P\int kdk u_o(\underset{\sim}{r}_1)\sqrt{\frac{2}{\pi k}}\frac{\sin(kr_2+\eta)\langle\pi_\varepsilon|}{r_2\left(\frac{k^2}{2}-\frac{K^2}{2}\right)}$$

$$\sim \pi u_o(\underset{\sim}{r}_1)\sqrt{\frac{2}{\pi K}}\cos(Kr_2+\eta)\langle\pi_E| \tag{19}$$

Using Eq. (19) in Eq. (18) we find

$$P\Psi|_{r_2\to\infty} \sim \frac{u_o(r_1)}{\cos\eta_r}\sqrt{\frac{2}{\pi K}}\frac{\sin(Kr_2+\eta+\eta_r)}{r_2} \quad \text{where}$$

$$\tan\eta_r = \frac{-\frac{1}{2}\Gamma}{E-\varepsilon_t-\Delta} \tag{20}$$

The η_r is a resonant phase shift, and the parameters of the resonance are its width

$$\Gamma = 2\pi|\langle\Phi_t|H_{QP}|\pi_E\rangle|^2 \tag{21}$$

and its location

$$\varepsilon_r = \varepsilon_t + \Delta. \tag{22}$$

Thus the position of the resonance is shifted from the eigenvalue ε_t by the background continuum. Many calculations on e + H and e + He^+ have been carried out using the Feshbach approach. Some early ones[9,10] just used hydrogenic functions as basis sets. If the 1s function is omitted, the resulting set forms vectors in the Q space mentioned above, and the calculation becomes a standard bound state CI problem. In Ref. 10 an alternate choice of Q was made in that some configurations contained the 1s orbital. To achieve higher accuracy, one could imagine using a variation trial function including r_{12} explicitly. This has in fact been done in a series of papers by Bhatia and co-workers[11] who solved the sticky problem of projecting Q onto a function of the form

$$\Psi_{Tr} = e^{-\alpha r_1 - \beta r_2} \Sigma\, C_{\ell mn}\, r_1{}^{\ell} r_2{}^{n} r_{12}{}^{m}. \tag{23}$$

Of course, one is not through when the eigenvalues of H_{QQ} have been found. To compare with scattering experiments, the width Γ, and shift, Δ, must also be known. For e + He^+, $\Delta \sim 10^{-3}$ eV or smaller so it is not terribly important. The widths are of the same magnitude, but can be measured in high resolution electron scattering experiments and especially in photoionization work. Bhatia et al have calculated these parameters by using a polarized orbital continuum function for $|\pi_E\rangle$.

As a sample of the kind of results obtained, we will look at some ^1P resonances in e + He^+ below the n = 2 threshold in Table I.

We are comparing eigenvalues of H_{QQ} here with experimental values so Δ is being ignored. The Bhatia-Tempkin calculation uses ~80 terms in a function of the form given in Eq. 23, and is

Table I. Properties of Low Lying ^1P Resonances in He. All figures in eV. Location is in eV above the ground state of He^+.

	$^1P(1)$		$^1P(2)$		$^1P(3)$	
	E	Γ	E	Γ	E	Γ
Ref. 11	60.152	.0363	62.759	1.17×10^{-4}	63.663	.010
Ref. 10	60.27		62.76		63.67	
Ref. 9	60.35		62.79		63.71	
Expt. (Ref. 2)	60.13	.038	62.76		63.65	.008

the most elaborate to date. The agreement with experiment, including the shifts, is quite satisfactory. In Ref. 10 ~40 hydrogenic configurations are used and in Ref. 9, ten such configurations are used. Recently theoretical results for S, P, D, F resonances below n = 2, and n = 3 thresholds have been published for isoelectronic He atoms with Z = 1-5.(12) These are CI calculations using hydrogenic basis functions.

These results are impressive; however, they are obtained on two electron systems for which exact wave functions are known for the positive ion. To treat heavier atoms one immediately confronts the problem of defining Q. If it is defined to project out an approximate ground state, e.g. a Hartree Fock state, then it is not clear what the spectrum of H_{QQ} is as the number of configurations in the Q space is increased indefinitely. Also, in some cases the widths of the resonances become sizable fractions of an eV, and the shifts too, so that one must worry about the energy dependence of Γ and Δ about ε_r, which is to say that the parametrization of the resonance breaks down. For photoionization calculations an excited state wave function is needed, and the Feshbach approach does not provide one. For all these reasons, and because it isn't too much more work, one might calculate Ψ directly instead of breaking it up, by the method of CI in the continuum. The basic formalism for this kind of calculation is due to Fano(3) and will be described in the context of a calculation carried out for magnesium by Bates and myself.(13) The resonances in Mg have a very similar appearance to those shown in Fig. 3 for Be.

The ground state configuration of Mg (Z = 12) is $1s^2 2s^2 2p^6 3s^2$. The fundamental initial approximation is that the core of 10 electrons is inert so that the calculation becomes one of the dynamics of the two valence electrons. The photoionization cross section is calculated, so the excited state to be found has ^1P symmetry. The resonances arise from 3p-ns and 3p-nd configurations interacting with a 3s-kp continuum background. The wave function Ψ is composed of symmetrized products of Hartree Fock orbitals found as follows. Orbitals to be used in the ground and autoionizing configurations were computed in the Mg III core potential, including exchange. The p orbitals which represent the continuum were found in a core +3s potential appropriate for singlet states. Using two potentials in this way results in two kinds of discrete p orbitals. Thus the basis set is not orthogonal in the one electron sense, and when computing one electron operator matrix elements, overlap integrals must be evaluated. This disadvantage is more than compensated for by the fact that the Hamiltonian is diagonal within the set of these continuum states. Let $|\epsilon\rangle$ represent a configuration $3s\epsilon p$, then

$$\langle\epsilon|H|\epsilon'\rangle = \epsilon\delta(\epsilon-\epsilon'), \tag{24}$$

a result which simplifies the computation a great deal.

Using this basis then if $|n\rangle$ represents a configuration 3s-np and if $|\nu\rangle$ represents an autoionizing configuration, i.e. either 3p-ns, 3p-nd or 3dnf, the various matrix elements are

$$\langle n|H|n'\rangle = e_n\delta_{nn'}$$

$$\langle n|H|\epsilon\rangle = 0$$

$$\langle n|H|\nu\rangle = V_{n\nu}$$

$$\langle\epsilon|H|\epsilon'\rangle = \epsilon\delta(\epsilon-\epsilon')$$

$$\langle\epsilon|H|\nu\rangle = V_{\epsilon\nu}$$

$$\langle\nu|H|\nu'\rangle = u_{\nu\nu'} \tag{25}$$

and we wish to find the excited state

$$|\Psi_E\rangle = \sum_n a_n|n\rangle + \sum_\nu f_\nu|\nu\rangle + \int d\epsilon' a_{\epsilon'}|\epsilon'\rangle. \tag{26}$$

Using the matrix elements above and $H\Psi_E = E\Psi_E$, we find

$$a_\epsilon = \sum_\nu V_{\epsilon\nu}f_\nu/(E-\epsilon), \tag{27}$$

i.e. a_ϵ is singular at $E = \epsilon$. Following Fano, we remove the singularity by the substitution

$$a_\epsilon = P\left[\frac{b_\epsilon}{E-\epsilon}\right] + \beta(E)\, b_E\, \delta(E-\epsilon). \tag{28}$$

The P indicates principal value, and the b_ϵ are now non-singular. This type of substitution generates the correct asymptotic behavior for $|\Psi_E\rangle$ as will be shown below. The coefficient $\beta(E)$ will also be interpreted. By a little manipulation, the equations to be solved are

$$(E-u'_{\nu\nu})f_\nu - \sum_{\nu'\neq\nu} u'_{\nu\nu'}f_{\nu'} - P\int d\epsilon' \frac{V_{\epsilon'\nu}b_{\epsilon'}}{E-\epsilon'} - \beta(E)b_E V_{E\nu} = 0$$

$$b_\epsilon - \sum_\nu V_{\epsilon\nu}\, f_\nu = 0 \tag{29}$$

$$u'_{\nu\nu'} = u_{\nu\nu'} + \sum_n \frac{V_{\nu n}V_{n\nu'}}{(E-e_n)}\,.$$

The a_n have been formally eliminated through the relation.

$$(E-\epsilon_n)a_n = \sum_\nu V_{n\nu}f_\nu. \tag{30}$$

To understand the form of the solution of Eqns. (29) near a resonance, suppose that there is just one resonant configuration ν and neglect also the discrete part of the continuum. If $u_{\nu\nu} \equiv e_\nu$, we have

$$(E-e_\nu)f_\nu - P\int \frac{d\epsilon' V_{\epsilon'\nu} b_{\epsilon'}}{(E-\epsilon')} - \beta(E)b_E V_{E\nu} = 0$$

$$b_\epsilon = V_{\epsilon\nu}f_\nu \tag{31}$$

or, eliminating b_ϵ in the upper equation

$$(E-e_\nu) - P\int \frac{d\epsilon' |V_{\epsilon'\nu}|^2}{E-\epsilon'} - \beta(E)(V_{E\nu})^2 = 0 \tag{32}$$

so

$$\beta(E) = \frac{E-e_\nu-\Delta}{|V_{E\nu}|^2}, \text{ where } \Delta = P\int d\epsilon' \frac{|V_{\epsilon'\nu}|^2}{(E-\epsilon')}. \tag{33}$$

This resembles Eq. 20 quite strongly, and, in fact, they are really the same thing. The asymptotic form of $|\epsilon\rangle$ is

$$|\epsilon\rangle_{r\to\infty} \propto \frac{\sin(kr+\eta)}{r} \tag{34}$$

in analogy to Eq. 13. But

$$|\Psi_E\rangle_{r\to\infty} = P\int \frac{d\epsilon' b_{\epsilon'}|\epsilon\rangle}{E-\epsilon'} + \beta(E)b_E|E\rangle; \tag{35}$$

a relation that leads immediately to

$$\tan \eta_r = -\left(\frac{\pi}{\beta(E)}\right) \tag{36}$$

by steps similar to those in Eqs. (19) and (20). Combining Eq. (33) and (36) we arrive back at (20). The explicit values of f_ν and b_ϵ can also be found for this case.[3]

Returning to Eqs. (29), they may be solved numerically by discretizing the integral in a manner described in Ref. 14. The result is a set of linear, homogeneous, algebraic equations with eigenvalue $\beta(E)$. The normalized solution yields the coefficients a_n, f , b_e. In the neighborhood of a narrow isolated resonance, we can fit the computed phase shift, which now is $\eta + \eta_r$, to the form

$$\eta_o + \eta_1 E + \tan^{-1} \frac{\Gamma/2}{(\varepsilon_r - E)} \tag{37}$$

where the first two terms are background terms. Thus Γ and ε_r can be determined immediately from the solution. The actual phase shift obtained for Mg is shown in Fig. 6 where broad (3p-ns) and narrow (3p-nd) resonances can be seen.

To see the effect of these resonances on the photoionization cross section, which is given by

$$\sigma(E) = 8.067 \frac{1}{\Delta E} |\langle \Psi_b | P_z | \Psi_E \rangle|^2, \tag{38}$$

a ground state must be determined and the matrix element evaluated.

Fano[3] has shown that the profile of an isolated, narrow, resonance is given by

$$\frac{\sigma(E)}{\sigma_B(E)} = \frac{(q+\varepsilon)^2}{(1+\varepsilon^2)} \tag{39}$$

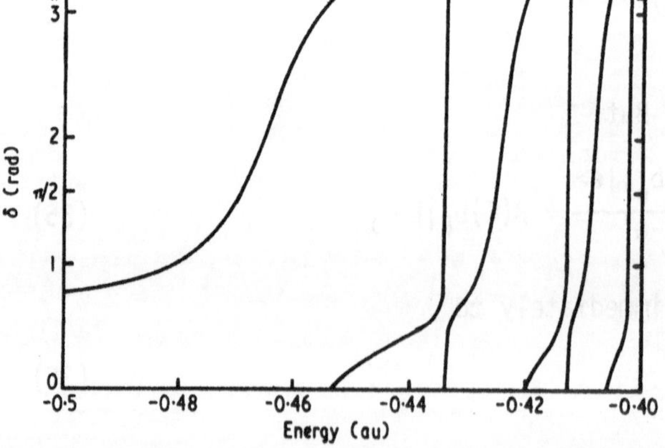

Fig. 6. Phase shift for Mg. The energy is measured relative to the MgIII ground state.

where $\sigma_B(E)$ is the background, i.e. non resonant cross section, $\varepsilon \equiv E-\varepsilon_r/(\frac{1}{2}\Gamma)$, and q, the line profile index, is a constant. Digressing for a bit, the profile in Eq. 39 has been much used in analysis of autoionizing lines. The value of q may be plus or minus, and the magnitude >1 or <1. For $q < 1$, the resonance appears as a window. An example of this was shown in Fig. 4. In this case, the oscillator strength to the resonant level is weaker than to the continuum so in that region where Ψ contains a strong admixture of the resonant level, the oscillator strength is depressed below the background. For the more common case of $q > 1$, the reasoning is reversed and an asymmetric peak, like that in Fig. 2, is seen.

For the alkaline earths, however, the broad resonances cannot be parametrized by q because q is energy dependent over the resonance. In fact, in Be, q goes from $+\infty$ to $-\infty$ over the 2p-3s resonance. A 47 configuration ground state was used in Eq. (38) to give a cross section for Mg, which is shown in Fig. 7. It agrees in general shape with experiment.[5] Analogous calculations have been carried out for He[14] and Be.[15]

By this approach we have avoided difficulties in the partition of configuration space in P and Q space and presumably the results would increase in accuracy as more configurations are added. Thus more systems are susceptible to treatment using this method than Feshbach's. However, the applications to date have been in energy regimes where there is only one open channel and

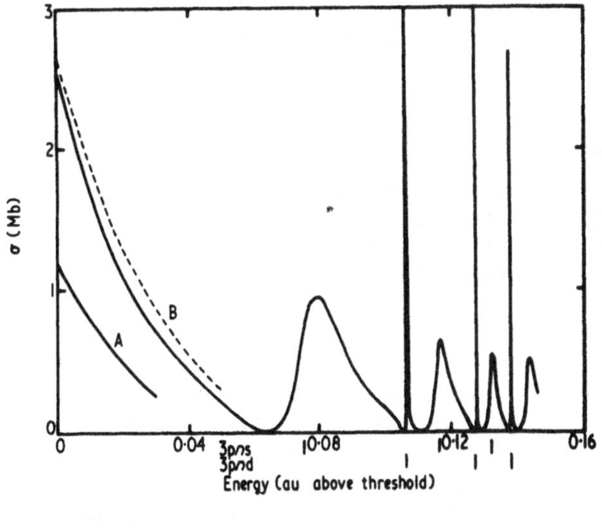

Fig. 7. Photoionization cross section of Mg (dipole-velocity formulation).

several closed ones. To treat several open channels using CI in the continuum, one encounters severe difficulties which have not yet been overcome. These will be outlined by considering an extension of the Mg problem to higher energies.

Suppose an excited state wave function of Mg is desired above the 3p threshold, i.e. both the 3s and 3p channels are open. Configurations like 3s-αp, 3p-αs and 3p-αd would be used to compose the wave function where α indicates a discrete quantum number below the threshold and energy above the threshold. To avoid the cumbersome formalisms of non orthogonal basis sets, one would like to define, once and for all, a zero order Hamiltonian, Ho, that is the same for all the members of the basis.

One possible choice for the problem being considered here would be a Hartree Fock Hamiltonian for Mg^{++}, i.e. the orbitals would be computed in a potential which has a Coulomb tail with charge 2. This is an appropriate choice for the 3s and 3p orbital in configurations where the other electron is in a high Rydberg state or in the continuum, but a poor choice for the continuum orbitals because the asymptotic Coulomb potential has the wrong charge, i.e. 2 instead of 1. Because of this matrix elements like

$$\langle 3s\ kp|H|3s\ k'p\rangle$$

develop logarithmic singularities as $k \to k'$ and the prescription given in Eq. 28 breaks down as neither term is defined in this case. Thus the analytic structure of the amplitudes a_ϵ defined earlier becomes more complicated and it appears difficult to fold this analytic structure into a feasible numerical approach. See Ref. 16 for a more detailed treatment of this problem.

So CI in the continuum has not been generally applied as yet to problems with more than one open channel, but one specific application has been made.(17) Theoretical methods have been developed, however, which are capable of handling this situation. An important one is the R matrix method adapted to atomic physics mainly by Burke and collaborators.(18)

The R matrix approach involves partitioning physical space into two regions $r < a$ and $r > a$. The internal region is where the electron interactions are large and exchange is important and so in this region essentially CI is used to construct a N+1 electron wave function. Because the space is bound, there are no divergences in matrix elements to worry about. On the spherical boundary certain boundary conditions on the free electron wave function are imposed and these lead to an R matrix of the form

$$R_{ij}(E) = \sum_{\lambda=1}^{\infty} \frac{\gamma\lambda_i \gamma\lambda_j}{E_\lambda - E} \tag{40}$$

where the i,j are channel indices, E_λ are the eigenvalues of the states found inside the sphere, and the $\gamma\lambda_i$ are amplitudes also found from the solution inside the sphere. The point in writing Eq. 40 down is to see that its energy dependence is explicit so it need not be recalculated for each E.

Now standard close coupling equations are solved outside the sphere using the long range residual potentials, and the matching of solutions (but not slopes) at r = a provides the scattering information or the excited state wave function for photoionization.

The excited states of a number of neutral atoms and ions have been calculated in this way and the method seems fully capable of yielding reliable results for light atoms with not too many open channels. As one illustration in Fig. 8 the experimental and theoretical photoionization cross section for Al is shown. The ground state is $3s^2 3p\,^2P$ and three channels ($3s^2\ ^1S$, $3s3p\ ^1P$, $3s3p\ ^3P$) were included in the R matrix calculation. Previous theoretical efforts had yielded a cross section only about 1/2 the experimental value. The value of the cross section is important

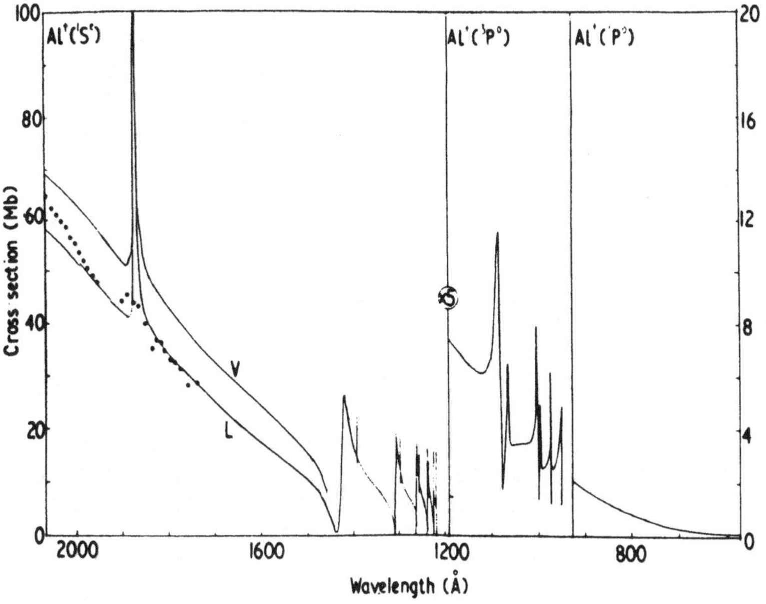

Fig. 8. Photoionization Cross Section of Al.

in studies of the solar atmosphere, so the present agreement between theory and experiment is pleasing.

Another method of treating several open channels which is closely allied in spirit to the R matrix approach is quantum defect theory or multi channel quantum defect theory (MQDT). Here the goal is to describe excited states, including series of resonances, by more fundamental parameters which have weak energy dependence.

Let us pursue this line of thought further by considering a fictitious atom with two ionization thresholds. We will call the two series with their associated continua channels a and b with threshold energies $I_a > I_b$. A complete description of the excited states of the atom would include the energies of the discrete levels below I_b, the widths and energies of the autoionizing levels between I_a and I_b and the scattering matrix for energies above I_b. Applying MQDT, one can find parameters which are either constant around the thresholds or vary slowly with energy from which all of the above data can be extracted. Thus this theory has the power to correlate seemingly distinct properties of the atom and to find many theoretical values from just a few parameters. The theory has been developed mainly by two groups headed by Seaton(19) and Fano(20) whose approaches are somewhat different. It has been applied to both atoms and molecules. In what follows we give a descriptive account and hope in so doing to suggest some of the power of the approach.

The main ideas are most clearly seen by studying a one channel system. The alkalies below the 2nd ionization threshold are a good example. It has long been known that the levels labeled by a particular L have energies given by a relation

$$E_n = -\frac{1}{2(n-\mu)^2} \tag{41}$$

where n is an integer and μ, the quantum defect, is almost constant for large n. Seaton was able to show that

$$\delta(0) = \pi\mu \tag{42}$$

where $\delta(0)$ is the phase shift at zero energy. Thus, within a few eV of threshold, <u>one</u> parameter, μ, gives the location of the discrete states and also the scattering information. When more channels are present, the same general approach can be developed, but before that is discussed, we should ask why it is that μ is constant over such an energy range of a few eV on both sides of threshold.

The answer is found by breaking the radial wave function for the channel electron into two parts as follows

$$\Psi_E(r) = \phi(E,r) \text{ for } 0 \leq r \leq r_0 \text{ and} \quad (43)$$

$$= N[f(E,r)\cos\pi\mu - g(E,r)\sin\pi\mu] \quad r \geq r_0. \quad (44)$$

The radius r_0 is chosen to be sufficiently large so that the potential energy of the electron beyond r_0 is $-1/r$, i.e. just the Coulomb potential. But the solutions of the Schrodinger equation for such a potential are well known and so in Eq. (43) $f(E,r)$ is the regular Coulomb function and $g(E,r)$ the irregular. The value of μ is found by matching the logarithmic derivative of $\phi(E,r)$ at r_0, so the near constancy of μ derives from the near constancy of this derivative. Evidently $\phi(E,r)$, being the wave function in the core, is rather difficult to calculate; in fact it doesn't really exist because the many electron wave function is not separable in the core region. Nevertheless for our purposes we assume it is to be found in some sort of effective potential $U(r)$. The curvature of the function is determined by $(U(r)-E)$, but around the first threshold $E\sim0$ and is thus very small compared to $U(r)$ for small values of r. As a consequence of this, the curvature of $\phi(E,r)$ and hence its logarithmic derivative is practically independent of E. A nice numerical illustration of this independence can be found in a recent lecture by Starace.[21]

To find the proper extension of this idea to a multi channel system, consider a two channel atom. A standard scattering calculation, above I_a, would produce two wave functions with the following asymptotic forms

$$\Psi_1 \sim \phi_a(f(\epsilon_a,r) + K_{aa}\, g(\epsilon_a,r)) + \phi_b\, K_{ab}\, g(\epsilon_b,r) \quad (45a)$$

$$\Psi_2 \sim \phi_b(f(\epsilon_b,r) + K_{bb}\, g(\epsilon_b,r)) + \phi_a\, K_{ba}\, g(\epsilon_a,r) \quad (45b)$$

The functions ϕ_a and ϕ_b are the positive ion wave functions along with angular and spin factors for the free electron. The energies ϵ_a and ϵ_b are the asymptotic kinetic energies of the free electron. The coefficients form the K matrix. Fano and coworkers have emphasized that the useful parameters for MQDT come from the characterization of the "short range eigenstates" of the system which are different from the asymptotic states in Eq. 45. The short range eigenstates are those for which the K matrix is diagonal with the form

$$K_{\alpha\beta} = \delta_{\alpha\beta}\tan\pi\mu_\alpha \quad (46)$$

where μ_α are the eigenphase shifts. These states have asymptotic form

$$\Psi_\alpha \sim [U_{\alpha a}\phi_a f(\epsilon_a,r) + U_{\alpha b}\phi_b f(\epsilon_b,r)] - \tan\pi\mu_\alpha [U_{\alpha a}\phi_a g(\epsilon_a,r) + U_{\alpha b}\phi_b g(\epsilon_b,r)] \quad (47)$$

The transformation from the asymptotic states to the short range ranges is called a frame transformation and is accomplished by the matrix $U_{\alpha i}$. Arguing by analogy to the single channel case, the μ_α and $U_{\alpha i}$, coming from the short range solution should have weak energy dependence, an assumption which is borne out by the applications.

Altogether then, the relevant parameters in the MQDT are the μ_α, the frame transformation matrix $U_{\alpha i}$, and the threshold energies I_a and I_b. If photoionization is being considered, a further vector of oscillator strengths, D_α, is necessary. It is possible to take a complex system like Ar[22] with 5 channels and by analyzing experimental data determine most of these MQDT parameters and thus be able to predict energy levels, autoionization profiles, scattering matrices, etc. For the same system, Ar, recently an ab-initio calculation was performed which produced μ_α and $U_{i\alpha}$ directly[23] and found reasonable agreement with the values fit from experiment.

In the approach used by Seaton and collaborators, solutions to close coupling equations above all relevant thresholds are found numerically. The K matrix is then related to matrix Y which has simple analytic properties and thus can be continued below the thresholds. Once below, then, one goes from Y back to K and thus effectively can continue the K matrix in this way avoiding the complexities of thresholds and series of resonances. As one example of a number of applications now made, Moores[24] found locations and widths of resonances in Be using MQDT and was able to achieve good agreement with direct close coupling results.

Granted that several-open-channel systems can be treated successfully by the above methods, let us look into what happens as the excitation energy is raised still higher. More channels will open until eventually the possibility of two or more free electrons in the field of the ionic core exists. There is at present no satisfactory theoretical way to describe such a state. The R matrix method, in its present formulation, as well as the MQDT assume one free electron for which an asymptotic wave function is known analytically. This function can be used for matching to an inner function on a spherical boundary. When there are two free electrons in an ionic field, the asymptotic wave function is not known. Further, it is not clear what the boundary between inner and outer regions should be now nor what boundary conditions to impose on such a boundary. This problem is addressed in Ref. 25 where a hyperspherical coordinate set is proposed ($R = r_1^2 + r_2^2$) with a boundary R = constant, but no applications have been forthcoming so far.

Under these circumstances, it is somewhat remarkable that sensible theoretical results can be obtained by more or less ignoring the difficulties in the excited state. Byron and Joachain,[26] ten years ago, computed the double photoionization cross section for He using an uncorrelated final state composed of Z = 2, i.e. unscreened, Coulomb functions and achieved results in good agreement with experiment. More recently Carter and Kelly[27] using many-body-perturbation-theory computed double photoionization cross sections for Ar and Ne and found some sensitivity to the final state in Ne, but still obtained good results. Their basis set for the free electrons was computed in an ionic core of one charge. The message of these calculations seems to be that ground state correlations are the dominant ones for this process.

Despite these successes, the effect of mutual screening of the free electrons is not understood, and further investigation to clarify when such effects are or are not important would be valuable.

REFERENCES

1. E.g. H. Beutler, Z. Physik 93, 177 (1935).
2. R. P. Madden and K. Codling, Ap. J. 141, 364 (1965).
3. U. Fano, Phys. Rev. 124, 1866 (1961)
4. J. Dubau and J. Wells, J. Phys. B6, 1452 (1973).
5. G. Mehlman-Balloffet and J. M. Esteva, Ap. J. 157, 945 (1969).
6. R. P. Madden, D. L. Ederer, K. Codling, Phys. Rev. 177, 136, (1969).
7. H. Feshbach, Ann. Phys. (NY) 5, 337 (1958); 19, 287 (1962).
8. T. F. O'Malley and S. Geltman, Phys. Rev. 137, A1344 (1965).
9. P. L. Altick and E. N. Moore, Phys. Rev. Ltrs. 15, 100 (1965).
10. L. Lipsky and A. Russek, Phys. Rev. 142, 59 (1966).
11. A. K. Bhatia and A. Tempkin, Phys. Rev. A11, 2018 (1975).
12. L. Lipsky, R. Anania, M. J. Conneely, Atomic Data and Nuclear Data Tables, 20, 127 (1977).
13. G. N. Bates and P. L. Altick, J. Phys. B6, 653 (1973).
14. P. L. Altick and E. N. Moore, Phys. Rev. 147, 59 (1966).
15. G. N. Bates, private communication.
16. P. L. Altick, Phys. Rev. 179, 71 (1969).
17. D. E. Ramaker and D. M. Schrader, Phys. Rev. A9, 1980 (1974).

18. For a review, see P. G. Burke, "R Matrix Theory in Atomic and Molecular Processes," in Atomic Physics 5 (Plenum Press, New York, 1977).
19. M. J. Seaton, Comments on Atomic and Molecular Physics, II, 37, (1969). Reference to the basic series of papers are given here.
20. U. Fano, J. Opt. Soc. Am. 65, 979 (1975), and references given therein.
21. A. F. Starace, "The Quantum Defect Theory Approach," in NATO Advanced Study Series, Series B, Physics V.18 (Plenum Press, New York, 1976).
22. C. M. Lee and K. T. Lu, Phys. Rev. A8, 1241 (1973).
23. C. M. Lee, Phys. Rev. A10, 584 (1974).
24. D. L. Moores, Proc. Phys. Soc. 91, 830 (1967).
25. U. Fano and M. Inokuti, Argonne National Lab. AWL-7680 (unpublished).
26. F. W. Byron, Jr. and C. J. Joachain, Phys. Rev. 164, 1 (1967).
27. R. L. Carter and H. P. Kelly, Phys. Rev. A, 16, 1525 (1978).

THEORY OF ATOMIC AND MOLECULAR NON-STATIONARY STATES WITHIN THE COORDINATE ROTATION METHOD

CLEANTHES A. NICOLAIDES and DONALD R. BECK

Theoretical Chemistry Institute
National Hellenic Research Foundation
48 Vas. Constantinou Ave., Athens 501/1, Greece

ABSTRACT.

The recently developed dilation (coordinate rotation) theory of the Coulomb Hamiltonian allows the calculation of energies and widths of nonstationary atomic and molecular states using square-integrable basis sets only. In the pioneering applications of this theory to two electron atomic autoionizing states, it was found necessary to employ large basis sets in brute force CI calculations, an expensive approach which has significant limitations when it comes to larger systems. In this paper we present a many-body theory of autoionizing and autodissociating states which implements the dilatation theory in an efficient and consistent way. In the case of autodissociating states it is not required to invoke the Born-Oppenheimer approximation. The present approach first isolates in the ϑ-plane (ϑ is the rotation angle) the "localized" correlation effects from the "asymptotic" ones by rotating the coordinates of the localized function, Ψ_o, which, in the time dependent theory, represents the initially localized state before it decays. The coordinate rotation leaves the real energy of Ψ_o invariant and allows the inclusion of "asymptotic" correlation vectors, in terms of "Gamow orbitals", which perturb Ψ_o and E_o and yield the decay energy shift and width. Our theory is supported by numerical examples on H and He.

I. INTRODUCTION

Resonances which can be associated with nonstationary states, are important phenomena in photon-atom, molecule, electron-atom, molecule and atom-atom collisions. These states are highly excited. They are above the first ionization threshold, I, of the

Cleanthes A. Nicolaides and Donald R. Beck (eds.), Excited States in Quantum Chemistry, 383–402.

same symmetry and thus decay by electronic autoionization or by molecular predissociation or both. They play a fundamental - but yet little understood quantitatively - role in collision and chemical reaction kinetics and in deexcitation mechanisms of multi-electron systems.

The standard theories for treating these states involve the continuum of the scattering states [see 1-3 and refs. therein]. For example, the width Γ is given, to lowest order, by the golden rule formula [1-3] (in a.u.):

$$\Gamma = 2\pi \left| \langle \Psi_o \,|\, (H - E_o) \,|\, U(E_o) \rangle \right|^2 \tag{1}$$

where Ψ_o is the initially localized N-electron, M-nuclei wave-function, E_o its energy and $U(E_o)$ is the δ-function normalized scattering state at energy ($E_o \stackrel{o}{-} I$). $(H - E_o)$ is the perturbation, V, which causes the decay. This can be the Coulomb operator, (Coulomb autoionization) the Breit operator, (relativistic autoionization) the nuclear motion coupling operator (predissociation) etc.

The solution of the correct $U(E_o)$ is not easy, especially for molecules, where the loss of spherical symmetry and the additional degrees of freedom present, (vibrations-rotations), render the scattering problem a formidable one. Thus, currently there is great interest in applying methods which, although they employ $\mathcal{L}^2$ basis sets as ordinary bound state calculations do, supply all the necessary information about the property under examination. One such method applicable to the calculation of resonances is the Complex Coordinate Rotation (CCR) method [4-7] which is based on the dilatation transformation theory of Aguilar-Balslev-Combes (ABC) and Simon [8-10].

The ABC theory has shown that if the transformation $r \rightarrow re^{i\vartheta}$, where $0 \leqslant \vartheta \leqslant \pi/2$, is carried out on the atomic Coulomb Hamiltonian (r stands for the electronic coordinates), the resolvent, $R(z) = [z - H(\vartheta)]^{-1}$, of the rotated Hamiltonian $H(\vartheta)$, can be analytically continued onto the second Riemann sheet of the lower half plane. This continuation reveals the pole, z_o, corresponding to a resonance and defined as:

$$z_o = E - (i/2)\Gamma \tag{2}$$

where E is the position and Γ is the width.

The pole z_o is the solution of the complex eigenvalue equation;

$$\left(z_o - H(\vartheta)\right)\Psi(z) = 0 \tag{3}$$

where $\Psi(r)$ is square integrable and $H(\vartheta)$ is non Hermitian. For an atom, the form of $H(\vartheta)$ is simple:

$$H(\vartheta) = e^{-2i\vartheta} T + e^{-i\vartheta} V = T(\vartheta) + V(\vartheta) \qquad (4)$$

where T and V are the kinetic and Coulomb potential energy operators. The spectrum of $H(\vartheta)$ has the following characteristics as a function of the rotation angle ϑ [10]:

a) The bound states remain on the real axis, at the positions of the discrete eigenvalues of H.

b) The continuum, which extends from each ionization threshold to infinity, rotates by an angle -2ϑ.

c) The resonances are exposed by the rotation at their exact positions and, once exposed, they remain fixed and are not functions of the rotation angle.

A pictorial representation of the above statements is shown in figure 1.

II. APPLICATION OF THE ABC THEORY- PREVIOUS METHODS

The aforementioned mathematical results do not, of course, lift any of the difficulties of the many-body problem. However, they do lay the foundations for treating autoionizing - and predissociating states as we propose here - without the explicit use of scattering functions since we are solving a rather simple, in principle, non Hermitian eigenvalue problem in Hilbert space. Nevertheless, there remains the problem of how to make the formal results of the ABC theory computationally attractive. Quantum

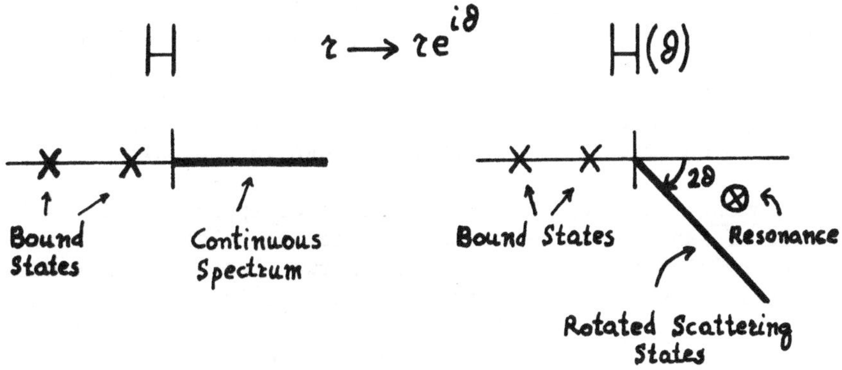

Fig. 1. Results of the ABC theory: The spectrum of $H(\vartheta)$ after the coordinate transformation $r \rightarrow re^{i\vartheta}$ is made. From ref. 10.

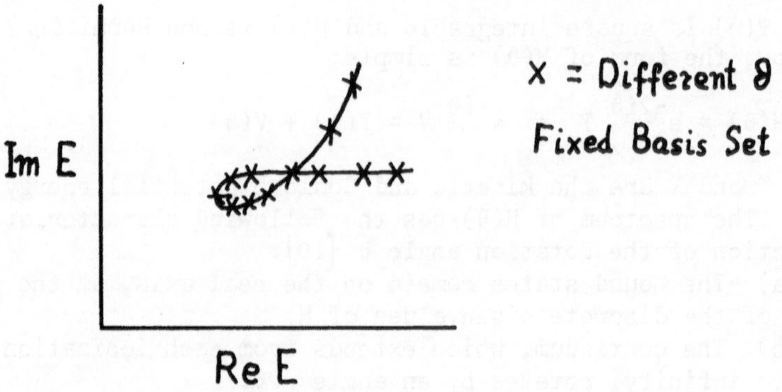

Fig. 2. Application of the ABC theory (refs. 8,9) within a fixed finite basis set: Variation of the complex eigenvalues with the angle ϑ indicates a region of stability for a particular root, with which the resonance is identified [4].

---

chemistry has as a task to produce accurate numbers. It is only when these novel theories are implemented in an optimum manner that they will prove useful to Quantum Chemistry.

Thus far, there have been applications of the CCR method to only two electron atoms and the Stark effect [4-7]. These pioneer applications have taken the following two forms - both approximate due to the fact that the basis set employed is finite and therefore the exact characteristics of the ABC theory (i.e. fig. 1) cannot be reproduced:

A) One chooses a very large square integrable basis set, usually of the Hylleraas type, and repeatedly diagonalizes $H(\vartheta)$ as a function of the angle ϑ. It turns out [4] that a particular root - which is identified with the resonance - shows a certain stability over a region of a few values of ϑ. This approach, discovered numerically by Doolen, is shown in fig. 2.

Comment: Method A is a brute force CI approach. It is incapable of handling many-electron atomic and molecular autoionizing states efficiently. In fact, even for two electron atoms there is considerable sensitivity of the final results to the choice of the basis set [7]. Besides, if one has to work so hard to obtain the position and the width, why not use formula (1) where, at least for atoms, the scattering state can be calculated reasonably accurately and without much effort? As explained in ref. 3, for He $2s2p\,^1P^0$ such a computation is orders of magnitude cheaper than a CCR computation <u>as it has been applied up to now</u>.

B) One chooses different basis sets of increasing size and, for a <u>fixed</u> angle ϑ, diagonalizes $H(\vartheta)$. A spiral-like convergence

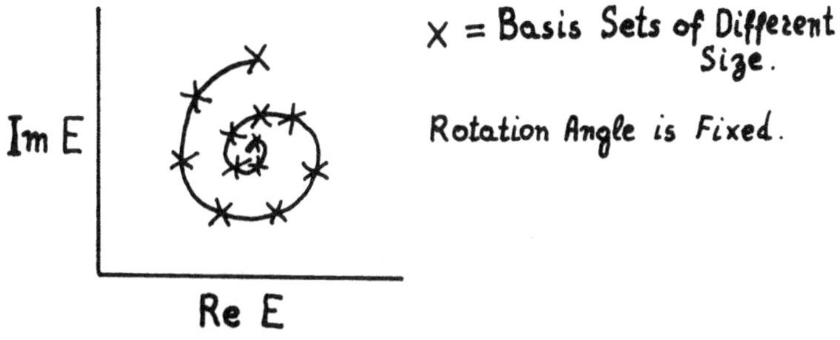

Fig. 3. Application of the ABC theory within basis sets of increasing size and a fixed rotation angle: convergence of a root with a spiral-like behavior [5,6].

---

to a complex number (identified with the resonance) is observed as a function of the size of the set [5,6]. This is shown in fig. 3.

Comment: For obvious reasons, method B has the same drawbacks when it comes to applications in arbitrary many-electron systems. Thus, both of the above pioneer, brute force computational methods seem to obey the law of diminishing returns due to the number of diagonalizations in the complex plane, (diagonalization of complex matrices is much more time consuming than ordinary diagonalizations), to the magnitude of the basis sets required or to the extreme difficulties in utilizing Hylleraas type functions for multi-electron systems. Perturbative approaches which attempt to circumvent partly these computational difficulties have already been introduced [7] with encouraging results. Nevertheless, as Ho et al [11] have recently noted, "the full advantages of the complex rotation method for more than two-electron systems have yet to be realized".

The difficulties with the size and accuracy of the basis sets required for an accurate CCR calculation indicate the absolute necessity for an alternative, breakthrough approach, capable of singling out and computing efficiently the important correlation and continuum contribution in the ϑ-plane thus reducing the magnitude of the computations.

The theory we describe below [3, 12] is in this spirit.

III. THE MANY-BODY CALCULATION OF RESONANCES IN MANY-ELECTRON-NUCLEI SYSTEMS BY THE COORDINATE ROTATION METHOD

There are two main ideas present in our proposal:
1) The isolation in the ϑ-plane, of the important correlation effects contributing to the initially localized state Ψ_o [2,3],

from the contribution of the scattering functions which interact with Ψ_o and cause the decay shift and width. This isolation and efficient computation is achieved by rotating the coordinates of $\Psi_o(r)$ so that $\Psi_o(r) \rightarrow \Psi_o(re^{i\vartheta}) \equiv \Psi_o(\vartheta)$. The previous methods have to employ large $\mathcal{L}^2$ basis sets because they must represent both the "localized electron correlation" as well as the asymptotic behavior of the resonance. This will be demonstrated by our example on the bound states of Hydrogen where only one rotated function is equivalent to an infinite, fixed, unrotated basis set.
2) The abolition of the traditional Born-Oppenheimer (B-O) approximation for highly excited, predissociating states. For such states, the B-O approximation is not necessarily a good one. The ABC theory allows in principle, a treatment of the problem of non-stationary states in N-electron, M-nuclei molecules interacting via Coulomb forces, based on purely quantum mechanical notions, i.e. by treating the nuclear positions as coordinates, i.e. dynamical variables, and not as parameters [13].

A. Molecular Excited States Without the B-O approximation (BOA) and the CCR Method

The BOA constitutes the backbone of most theoretical analysis and almost all calculations in molecular physics. Its usefulness in explaining physical and chemical phenomena need not be re-emphasized here. Yet, just like the shell (orbital) model which has also proved of fundamental importance in explaining physical chemical properties, the BOA does not represent the quantum mechanical truth rigorously. The breakdown of the BOA is considerably enhanced in excited states where, degeneracies and/or near-degeneracies may often distort the physical picture conveyed by it or may render it an inappropriate zeroth order approximation. Autodissociating (predissociating) states are usually conceptualized within a model whereby a BO excited state surface "crosses" a repulsive, dissociating surface and a transition to the molecular break-up continuum occurs caused by the corresponding nuclear motion coupling operator. The phenomenon of predissociation can also be viewed as caused by the decay of an initially localized molecular state without reference to the BOA. This picture suits the ABC theory rigorously and we would like to suggest that these two unexplored fields, i.e. quantum chemistry without the BOA and the CCR method, could be combined in the treatment of molecular autodissociating states.

We consider a molecule of N-electrons, (coordinates r), and M-nuclei, (coordinates R), in a convenient coordinate system. The usual Coulomb Hamiltonian, H(r,R), is assumed to have a complete set of stationary states, $\Psi_n(r,R)$, satisfying the time independent Schrödinger equation:

$$\left(E_n - H(r,R)\right)\Psi_n(r,R) = 0 \tag{5}$$

H(r,R) is self-adjoint and has a point as well as a continuous spectrum. This spectrum is different from the standard B-O spectrum. In particular, no energy surfaces appear. Just eigenvalues which may be distinct, degenerate or nearly degenerate. In the continuous spectrum there can be regions of "spectral concentration" (i.e. large amplitude of the localized function Ψ_o) which can be associated with resonances.

We now rotate the coordinates R and r according to the ABC theory [8-10] so that equation (3) is obtained but in this case with the rotated nuclear coordinates included. Note that no Born-Oppenheimer (BO) approximation is made. The total wavefunction simply has to satisfy fermion and boson permutation symmetry and parity invariance. The problems of symmetry and basis sets which ensue (which are outside the scope of this presentation) are assumed solvable by successive approximations to the energy problem (e.g. starting with BO or diabatic states, then using appropriate localized nuclear basis sets etc.). We expect the same characteristic ϑ-spectrum described in refs. [8-10] where now the complex poles in general need not be simple but can be overlapping (competition between autoionization and autodissociation). A two parameter (φ,ϑ) dilatation operator (one for each type of coordinate) would presumably produce the same results while it would be suitable for calculations of electronic resonances in the B-O approximation.

Once this formal operation is defined, the important question remains: Since by necessity the practical Hilbert space is finite, how should the ABC theory be applied, in a physically meaningful and computationally tractable and consistent manner, to an arbitrary nonstaionary state?

Before we present our many-body approach, we outline certain aspects of the properties of the wavefunctions for nonstationary states:

B. Remarks on Certain Characteristics of Resonance Wave-functions

Autoionizing as well as (auto)predissociating states decay to the adjacent electronic or nuclear continuum because of interactions found in their own Hamiltonian. According to time dependent theory [2,3], at t=0 of their creation they are described by a localized (square integrable) function Ψ_o, which as time evolves acquires a nonlocalizable part representing the free particle(s). At $t=\infty$, the system has decayed into its constituent free fragments. An equivalent time-independent picture is that of an energy dependent function $\Psi(r,R;z_o)$, satisfying Schrödinger's equation [14-16]:

$$\left(z_o - H(r,R)\right)\Psi(r,R;z_o) = 0 \tag{6}$$

with the boundary conditions (for notational simplicity we consider only electronic autoionization)

$$\Psi \xrightarrow[r_i \to 0]{} 0 \tag{7a}$$

$$\Psi_{r_i \to \infty} \sim c(z_o)\mathcal{A}\left[\Psi_{ion} \times e^{i\sqrt{2z_o}\, r_i}\right] \tag{7b}$$

where $\mathcal{A}$ is the N-electron antisymmetrizer and the coefficient $c(z_o)$ is determined by how the asymptotic form is matched onto Ψ_o, the localized part describing the region where the potential, $V^o \neq 0$.

The functions $\Psi(r,R;z_o)$ can be called Gamow [14] or resonance functions and have been the subject of many investigations, especially in nuclear physics [e.g. 3, 17-20]. They are characterized by the outgoing wave boundary condition through which they depend on the eigenvalue z_o. They are not square integrable and therefore matrix elements in terms of such functions must be defined in a different, but consistent way [20]. One such method is the extension of the coordinates r to the complex plane [20-22] which makes them square integrable.

We note that this operation can be thought of as an exercise in the theory of perturbations of the boundary conditions of differential equations [3] in analogy with problems related to deformed nuclei [23]. Thus, when the coordinate transformation $f(r) = re^{i\vartheta}$ is applied to the differential equation (6) we obtain eq. 3.

A comparison of the above - i.e. the wave-function characteristics of a nonstationary state - to the two methods employed in applying the ABC theory, shows that in these methods:
a) Square integrable basis sets must at the same time account for electron correlation in the localized function Ψ_o which does not contribute to the decay [2,3] as well as for the asymptotic characteristics of eq. 7b. Since there are no systematics for either of them, it is not surprising that it has been necessary to employ very large basis sets.
b) The same $\mathcal{L}^2$ set must account for electron correlation and asymptotic effects for all values of the rotation angle ϑ. Since all previous experience of choices of basis sets has been for the $\vartheta = 0$ cases, i.e. for ordinary ground or excited state calculations, the choice of a good basis set in the ϑ-plane is difficult as there is no guiding principle behind it. For example, a correlation vector which is important for $\vartheta = 0$ may lose its importance as the spectrum is rotated and vice versa.

The above observations, and comparison with concepts related to the properties of resonances as derived from time dependent decay theory [2,3,24] have led us to the following theory:

C. Many-Body Approach for N-electron, M-nuclei Nonstationary States

The position of resonances can be defined as the complex eigenvalue of the non Hermitian matrix equation [2,3, 24] in Hilbert space:

$$(z_o - H_o - A(z_o))\ \Psi(r) = 0 \tag{8}$$

where H_o is the projection of H onto the subspace defined by $\Psi_o(r)$, the localized function at t=0, and A(z) is the non-hermitian "self-energy" perturbation operator causing the shift, Δ, and the width, Γ, of the nonstationary state. A(z) contains the continuum information. By analogy, we can write eq. 3 as:

$$(z_o - H(0) - K(\vartheta))\Psi(r) = 0 \tag{9}$$

where H(0) is the projection of $H(\vartheta)$ onto the localized, unrotated (ϑ = 0) Hilbert space, with expectation value E_o, and the perturbation

$$K(\vartheta) = (e^{-2i\vartheta} - 1)\ T + (e^{-i\vartheta} - 1)\ V \tag{10}$$

contains the asymptotic information causing the shift Δ and the width Γ.

The operators in eqs. 9 and 10 should be thought of as matrices defined in terms of the function space on which they operate. The analytic form of $K(\vartheta)$ has the correct limit as $\vartheta \to 0$, i.e. K(0) = 0. This means that within the space of square integrable functions, the real Hamiltonian yields a real expectation value, $\langle\Psi_o|H|\Psi_o\rangle = \langle\Psi_o|H(0)|\Psi_o\rangle = E_o$. As ϑ is turned on, $K(\vartheta)$ is added to H(0), i.e. the total Hamiltonian becomes non Hermitian with a complex eigenvalue at z_o, (eq. 2), whose real part, $E = E_o + \Delta$. This explicit transformation of the form of the Hamiltonian from H(0) to H(0) + $K(\vartheta)$ is equivalent to stating that, in a standard CI computation using localized and scattering functions [1], H(0) represents the square integrable functions and $K(\vartheta)$ the scattering states and their interaction with the $\mathcal{L}^2$ space. (See the Q,P formalism of refs. 1 and 2.) If the function space corresponding to the scattering states is excluded, there is no decay i.e. A(z) of eq. 8 is zero. As stated earlier, this corresponds to $K(\vartheta)$=0, a condition which occurs at ϑ = 0.

The preservation of the smallness of (Δ - iΓ/2) as a function of the angle and the consistent and optimum calculation of E_o and

$(\Delta - i\Gamma/2)$ on a Hilbert space of square integrable functions constitutes the foundation of our approach.

We write for the localized function:

$$\Psi_o(r,R) = a_{HF}\,\Phi_{HF}(r,R) + \Sigma_k\, a_k\, \Phi_k(r,R) \qquad (11)$$

$\Phi_{HF}(r,R)$ is an independent particle model, (IPM), Hartree-Fock (or MCHF) function assumed calculable [25]. The corresponding Hamiltonian represents the electronic and nuclear kinetic energy as well as an average of interactions among all electrons and nuclei.

IPM Hamiltonians are separable and therefore their point spectrum can overlap the continuous one of the ((N,M) -1) Hamiltonian, which essentially coincides with the exact continuous spectrum. $\Phi_k(r,R)$ are correlation vectors which, to first order, contain single and pair correlation functions of electrons, $\Sigma(r_1)$. $\Pi(r_1,r_2)$, electron-nuclei, $\Pi(r_1,R_1)$, or nuclei $\Sigma(R_1)$, $\Pi(R_1,R_2)$. Methods of calculating $\Sigma(r_1)$ and $\Pi(r_1,r_2)$ are discussed in this volume [26]. We note that, just as with bound excited states, only a few of Φ_k carry large coefficients and are therefore important for off-diagonal matrix element calculations such as radiative and radiationless widths [2,3,27,28]. The remaining (infinite) are important only for energy positions. This observation holds for autoionized states. The calculation of $\Sigma(R_1)$, $\Pi(r_1,R_1)$ and $\Pi(R_1,R_2)$ necessary for autodissociating states, is terra incognita.

Within the B-O approximation, the separation (11) is, by now, common practice in ground state calculations. However, it has not been so common for excited states [26] and even more so for autoionizing states where, the systematic and accurate study of electron correlation in N-electron systems is in its infancy [e.g. 2,3,29]. Within this scheme, the first step is the accurate calculation of an SCF HF function. For atoms, this can be done numerically or analytically. Convergence in neutrals and negative ions is often difficult and tricks such as Z extrapolation must be used. Since there is no minimum principle, the final solution is chosen based on criteria such as orbital occupancy, nodal structure and the satisfaction of the virial theorem. We note that the third criterion is a very important computational tool at the HF level. This is because, in our opinion, any bound state approach to the calculation of resonances which uses the true Hermitian Hamiltonian can only be based on the criterion of <u>localization</u>. The virial theorem is a crucial test of localization in physical systems. Since H-F functions for ordinary bound states are known to satisfy the virial theorem rigorously, the degree of satisfaction of this condition of HF calculations or resonances has been used by us as a measure of the localizability of the corresponding Ψ_o.

The second step is the calculation of the square integrable correlation vectors Φ_k. Of the infinity of Φ_k contributing to E_o, our understanding of excited states [2,26] suggests that those important for the wave-function characteristics - i.e. for width calculations within formula 1 - are mainly those which arise from Fermi-Sea and hole-filling correlations.

The above brief discussion on the goal of our approach and on the analysis of the electronic structure of autoionizing states brings us to the related question: Since E_o is very near the exact (complex) resonance energy, and since as ϑ is turned on the correlation vectors in Ψ_o acquire in general a different and unpredictable importance, is there a transformation of coordinates in Ψ_o that preserves this closeness, preserves the electron correlation characteristics of Ψ_o and thus allows a rigorous and consistent treatment of the perturbation $K(\vartheta)$?

The answer is yes and is derived from an analogy with decay theory [2,3, 24]:

In the time dependent case, it is the unitary operator $U(t) = e^{-(i/\hbar)Ht}$, $t \geqslant 0$, (whose Laplace transform, $R(z)$, has the same properties as the resolvent $R(z,\vartheta)$) that preserves this property:

$$\langle \Psi_o | H_o | \Psi_o \rangle = \langle \Psi_o | H | \Psi_o \rangle = \langle \Psi_o | U^+ H U | \Psi_o \rangle = \langle \Psi(t) | H | \Psi(t) \rangle = E_o$$

$$0 < t < \infty \tag{12}$$

where $\Psi(t) = U(t)\,\Psi(0)$ is the rotated function in the <u>time plane</u>. The resulting perturbation, $A(z)$ is then very small (level shift and width).

By analogy, the equivalent transformation in the ϑ-plane of the wave-function Ψ_o should be given in terms of the dilatation operator, $D(\vartheta)$, which in the ABC theory was used to obtain the spectral properties of the Hamiltonian $H(\vartheta)$. For the molecular Hamiltonian used here, this property is $D(\vartheta)\ H(r,R) \rightarrow H(r\,e^{i\vartheta}, R\,e^{i\vartheta})$. Thus, the following two theorems can be stated:

<u>Theorem 1:</u> The N-electron, M-nuclei function $\Psi_o(\vartheta) \equiv \Psi_o(re^{i\vartheta}, Re^{i\vartheta})$ will leave the expectation value $E_o = \langle \Psi_o(0) | {}^oH(0) | \Psi_o^o(0) \rangle$ invariant under rotations in the ϑ-plane:

$$\langle \Psi_o | H(0) | \Psi_o \rangle = \langle \Psi_o\, D^{-1} | H(\vartheta) | D\Psi_o \rangle = \langle \Psi_o(\vartheta) | H(\vartheta) | \Psi_o(\vartheta) \rangle = E_o$$

$$0 \leqslant \vartheta \leqslant \pi/2 \tag{13}$$

<u>Theorem 2:</u> For a Coulomb Hamiltonian, in an actual Configura-

tion-Interaction (CI) type calculation with one particle square-integrable, analytic functions as basis sets, all overlap, one and two particle integrals remain invariant under transformations of the type $u(r) \rightarrow e^{i(3/2)\vartheta} u(re^{i\vartheta})$; $U(R) \rightarrow e^{i(3/2)\vartheta} U(Re^{i\vartheta})$. In the case of atoms this can be shown explicitly.

Corollary 1: Localized Hamiltonian matrices created at $\vartheta=0$ yield the same real eigenvalues and the same expansion coefficients for the resulting roots for all ϑ.

Corollary 2: In a many-body treatment of $\Psi_o(\vartheta)$, all Hartree-Fock, single and pair correlation functions which are orthogonal to core orbitals of lower configurations for $\vartheta=0$ [2] when rotated must remain orthogonalized to the rotated same core orbitals for $\vartheta > 0$.

Corollary 3: In general, the norm of $\Psi(\vartheta)$, the exact square integrable solution of $H(\vartheta)$, will be very close to the norm of $\Psi_o(\vartheta)$. The difference of their expectation values yields the level shift, Δ, and the width, Γ.

Given the above theorems and corollaries, we now turn to the calculation of the perturbation $K(\vartheta)$ and finally to examples which support the above considerations.

$K(\vartheta)$ is a one and two particle operator. Therefore, to a good approximation it is pair excitations from Ψ_o that will represent the important processes involving the continuum. This observation allows the treatment of multichannel problems: The "asymptotic pair functions" $\Pi_{as}(r_1,r_2)$, $\Pi_{as}(R_1,R_2)$ are seen to give rise to autoionization, autodissociation or both. These correlation functions should be such that they describe the asymptotic region of the resonance function yet they preserve the overall square-integrability property as required by the ABC theory. I.e. we write for the total function $\Psi(\vartheta)$:

$$\Psi(\vartheta) = a(\vartheta)\, \Psi_o(\vartheta) + b(\vartheta)\, \chi_{as}(\vartheta) \tag{14}$$

where $\Psi_o(\vartheta)$ is the rotated Ψ_o and $\chi_{as}(\vartheta)$ is the "asymptotic correlation function" i.e. the correction to $\Psi_o(\vartheta)$ which introduces explicit information about the asymptotic behavior of the resonance. Since $\chi_{as}(\vartheta)$ is the square integrable counterpart of eq. 7b, the rotation angle ϑ should be greater than the angle characterizing the complex eigenvalue.

D. Optimization of $\chi(\vartheta)$

$\chi(\vartheta)$ can in principle be calculated via perturbative or variational procedures which, however, must be different than the standard ones applied to bound states because 1) The theory contains the parameter ϑ - i.e. these procedures must be carried out directly in the ϑ-plane 2) There is no minimum principle of the Rayleigh-Ritz type on the total (complex) energy of the resonance.

Our approach is based on the use of the Virial Theorem (VT) in the ϑ-plane. In analogy with the HF calculations, we search for a constraint which can be related to the localizability of the autoionizing states and to the related stationarity as a function of ϑ of the complex pole as predicted by the ABC theory (see fig. 1) and the numerical results of Doolen (fig. 2). This stationarity can be expressed as

$$\left.\frac{\partial z(\vartheta)}{\partial \vartheta}\right|_{\vartheta=\vartheta_o} = 0 \tag{15}$$

where $z(\vartheta)$ is the complex eigenvalue

$$z(\vartheta) = \frac{\langle \Psi(\vartheta) | H(\vartheta) | \Psi(\vartheta) \rangle}{\langle \Psi(\vartheta) | \Psi(\vartheta) \rangle} \tag{16}$$

and ϑ_o is the optimum angle at which $z(\vartheta_o) = z_o$, the true resonance position.

The combination of eqs. 15 and 16 produces the condition

$$\left.\frac{\langle \Psi(\vartheta) | V(\vartheta) | \Psi(\vartheta) \rangle}{\langle \Psi(\vartheta) | T(\vartheta) | \Psi(\vartheta) \rangle}\right|_{\vartheta=\vartheta_o} = -2 \tag{17}$$

The VT of eq. 17 has also been derived and employed independently by Brändas and Froelich [30] and Yaris and Winkler [31]. In the context of our theory, since E_o is already very near the resonance position, the VT essentially fixes, in principle, the width and the shift by fixing the parameters in $\tilde{\chi}(\vartheta)$. To what extent this constraint yields a unique solution within finite basis set calculations is still unknown. However, we point out that an additional constraint may be employed for width calculations: According to the arguments of refs. [20,32] and the equivalence between eqs. 3 and 6, the equation

$$| \, \mathrm{Im} \langle \Psi(\vartheta) | H^2(\vartheta) | \Psi(\vartheta) \rangle - (-\varepsilon\gamma) \, | = \min \tag{18}$$

constitutes a constraint on the parameters of $\tilde{\chi}(\vartheta)$ ($\langle \tilde{\Psi}(\vartheta) | H(\vartheta) | \tilde{\Psi}(\vartheta) \rangle = \varepsilon - i\gamma/2$) which can be employed for the calculation of widths.

IV. APPLICATIONS

A. Hydrogen Atom

The Hydrogen spectrum has no resonances. Its discrete states are poles of $(z - H(\vartheta))^{-1}$ on the real axis. According to the ABC theory, they are independent of ϑ. In order to check the ideas described above, we did the following: First we showed (the proof is trivial) that the rotated hydrogenic functions are solutions of the rotated Hamiltonian with real eigenvalues. I.e. $(H(\vartheta) - E_n)\Phi_n(\vartheta)=0$. In this case Ψ_o is a true bound state (say the 1s state). Thus, we showed that theorem 1 holds in a case where the wave-function corresponding to the localized Ψ_o is known exactly. Secondly, we diagonalized $H(\vartheta)$ in a basis set of unrotated, real hydrogenic functions $|ns\rangle$. For the ground state we obtained complex eigenvalues from a fixed basis set:

$$1s(\vartheta) = \Sigma_{n=1}^{10} a_n(\vartheta)|ns\rangle \quad (19)$$

As ϑ increased, the coefficients $a_n(\vartheta)$, $n>1$, increased rapidly in magnitude and the width of the ground state was larger than typical resonance widths after $\vartheta>4^o$. This shows the origin of the slow convergence of the coordinate rotation method where up to now only fixed, large basis sets have been used. I.e., an arbitrarily chosen set of square integrable functions may be a good representation on the real axis but in the ϑ-plane they need not. Thus, the choice of such basis sets has up to now been arbitrary, lacking a theoretical and consistent background. Rotation of the coordinate of the 1s function makes the pseudo complex pole disappear, something that could be accomplished with real unrotated functions (e.g. Sturmians) only in the limit of completeness.

B. Gaussian Nuclear Functions and Molecular Nonstationary States

Slater type orbitals are approximations to the hydrogenic solutions and constitute suitable basis sets for calculations of the the electronic structure of atoms and molecules. Since the hydrogenic Schrödinger equation remains invariant under rotation, rotated STO's remain a basis set with fast convergence properties in the ϑ-plane. In our proposed many-body theory, we eliminate the Born-Oppenheimer approximation since a) the coordinate rotation method is perfectly suited for facing such as challenge, and b) in highly excited molecular states which can undergo autoionization or autodissociation, the B-O approximation may often be useless physically as well as computationally. In analogy with the hydrogenic Hamiltonian and wave-functions, we have rotated the harmonic oscillator and its solutions and have obtained the same eigenvalues. Thus, we suggest that the localized rotated Gaussian functions are well suited to describe the nuclear functions of molecular nonstationary states in the ϑ-plane [13].

C. The He $2s2p\,^1P^0$ Resonance

In order to explore our proposed theory within our presently limited computer limitations, we chose to study the well-known He $2s2p\,^1P^0$ resonance for which accurate results exist [e.g 33]. First we computed Ψ_o and E_o variationally by minimizing

$$\tilde{E}_o = \frac{\langle \tilde{\Psi}_o | H | \tilde{\Psi}_o \rangle}{\langle \tilde{\Psi}_o | \tilde{\Psi}_o \rangle} \tag{20}$$

where

$$\tilde{\Psi}_o = \Phi_{HF} + \tilde{\chi}\ , \langle \tilde{\chi} | \Phi_{HF} \rangle = 0,\ \langle \tilde{\chi} \,|\, He^+ 1s \rangle = 0 \tag{21}$$

Φ_{HF} is the analytic HF function of He $2s2p\ ^1P^0$ and $\tilde{\chi}$ contained 16 correlation vectors with 7 non-linear parameters [26]. The optimized E_o = - 0.692075 a.u. is only 0.023 eV higher than the energy of the large Hylleraas type function of ref. 33.

According to our theory, the complete rotated wave-function was written as

$$\Psi(\vartheta) = a(\vartheta)\ \Psi_o(\vartheta) + b(\vartheta)\ \Pi_{as}(\vartheta) \tag{22}$$

where the asymptotic pair correlation function $\Pi_{as}(\vartheta)$ represents the "hole-filling" autoionizing process $(2s2p) \leftrightarrow (1s\ \varepsilon p)$ and is taken to be:

$$\Pi_{as}(\vartheta) = \Sigma_{i=1}^{m} \mathcal{A}\left[1s(\vartheta) g_i(\vartheta, \tilde{k}_i, \tilde{a}_i)\right] \tag{23}$$

$1s(\vartheta)$ is the rotated He^+ 1s function. $g_i(\vartheta, \tilde{k}_i, \tilde{a}_i)$ represents the asymptotic behavior of the Gamow resonance (7b). It has the form:

$$g_i(\vartheta, \tilde{k}_i, \tilde{a}_i) = r^{n_i}\ e^{-\tilde{k}_i\ r}\ e^{i(\vartheta\ -\ \tilde{a}_i\ +\ 3\pi/2)} \tag{24}$$

where $(\tilde{k}, \tilde{a})$ are parameters to be determined subject to the virial theorem constraint.

Expansion of asymptotic pair correlation functions in terms of orbitals of the form (24) may prove useful for resonance studies just like Slater type orbitals have become the standard ingredient of bound state calculations. We shall call these orbitals "Gamow Orbitals" (GO).

The rotated Hamiltonian matrix was constructed after an analysis of the correlation effects of Ψ_o showed that the HF function is a good approximation to it. In fact, a straight-forward, simple calculation of the width using eq. 1 with HF functions for Ψ_o and $U(E_o)$ yielded Γ = 0.040 eV compared with the accurate [33] Γ = 0.0363 eV. Therefore, in the spirit of our theory, we took for the diagonal matrix element

$$\langle \Psi_o(\vartheta) | H(\vartheta) | \Psi_o(\vartheta) \rangle = E_o = -0.692075 \text{ a.u.} \quad (25)$$

while for the off-diagonal one, i.e. the one describing the localized-asymptotic interaction:

$$\langle \Psi_o(\vartheta) | H(\vartheta) | \Pi_{as}(\vartheta) \rangle \approx \langle \Phi_{HF}(\vartheta) | H(\vartheta) | \Pi_{as}(\vartheta) \rangle \quad (26)$$

The approximation (26), coupled with the rotation of $\Psi_o(\vartheta)$, reduces the computational magnitude of the problem considerably. We note that such approximations become valid and transparent only within our many-body scheme which is based on the explicit consideration and isolation of the important correlation effects in autoionizing states. In this case we have even been able, as a first approximation, to consider only the HF function.

The computations were carried out as follows: The Hartree-Fock function, Φ_{HF}, had an analytic form so that rotation can be carried out. Obtaining analytic Hartree-Fock functions for configurations such as 2s2p, which have states below them with identical orbital symmetry, (i.e. 1s np, 1s εp), using methods based on the Roothaan expansion approach are slow converging although possible [2,29]. Thus, we applied the following procedure: An estimate for the 2s and 2p orbitals was provided from the Be $1s^2 2s2p\,^1P^0$ sequence. A 3-configuration CI was set up where the other two configurations were $2sv_p$ and $v_s 2p$, with v_s and v_p single STO's whose exponents and coefficients were determined variationally subject to the constraint of excluding the 1sεp energy line. The CI results were then condensed to form new radial functions, 2s' and 2p' in such a way as to eliminate single excitations, (at this stage), and the process was repeated until satisfactory (here within 0.001 a.u. of the numerical HF energy) convergence was obtained.

To provide flexibility for the small and intermediate region or r, the g_i were orthogonalized to the first three members of the rotated He np Hartree-Fock functions whose unrotated forms were taken from ref. 34. We note that the g_i can be made compact or diffuse by varying the parameters k_i accordingly.

In the calculations reported here, we chose m=2 (eq. 9), $n_1=2$, $n_2=3$ (eq. 10) and $\tilde{a}_1=\tilde{a}_2=\tilde{a}$, $\tilde{k}_1=\tilde{k}_2=\tilde{k}$. For a given set of $(\tilde{k},\tilde{a})$ the complex symmetric energy matrix, which depends on ϑ only through Π_{as}, was diagonalized for $\vartheta=16^o$. A limited search around the region $\tilde{k}=1.617$, $\tilde{a}= 0.015$ - these values were chosen using the true resonance position and geometrical arguments - yielded a value of $\Gamma = 0.0538$ eV, $\Delta = -0.0157$ eV. (The accurate values of ref. 33 are $\Gamma=0.0363$ eV, $\Delta=-0.0071$ eV. However, we note that our calculation involved a 3 x 3 CI only!)

The above results, obtained at a cost of less than 1 minute

per $(\tilde{k},\tilde{a})$ point on the IBM/125 (a slow machine), suggest a satisfactory application of our theory. We expect them to improve substantially by an increase in the flexibility of the asymptotic pair function - expecially in describing the inner region - and a search in the ϑ-plane. The great advantage in this search is that the complex matrices which must be diagonalized are small since we look only for corrections to E_o which is invariant to rotations. This should be contrasted to the standard applications of the ABC theory where large complex matrices must be diagonalized [4,7].

V. CONCLUSION

The CCR method constitutes a rigorous approach to the calculation of energies and widths of nonstationary states. In this lecture we make the suggestion that it should be applicable to autodissociating molecular states if they are treated outside the standard Born-Oppenheimer approximation. In our opinion, within the framework of our many-body theory, the computational bottleneck of this proposal seems to be the HF function of a molecular excited state where one treats electrons and nuclei on an equal footing. The correlation functions could be expressed in terms of the standard STO's and nuclear Gaussians. The above proposal is of course very speculative. The B-O approximation has become standard practice in Quantum Chemistry and one hesitates to leave it even when there is no apparent reason of why it should be necessary, as in the case of predissociation and certain avoided crossings of ordinary excited states.

The other proposal of this lecture is more specific and has yielded numerical results: It deals with the many-body treatment of resonances within the ABC theory. Our approach involves the following practical steps:

1) Produce a localized, unrotated ($\vartheta=0$), compact Ψ_o and the corresponding E_o [2,3,26]. On the average, Ψ_o contains the Hartree-Fock plus 10-20 correlation vectors with optimized orbitals. E_o is very close to E, the real part of the exact energy in the complex plane, so that all the effort to improve E_o can be done at this stage. The rest of the calculation (below), involves the energy shift $\Delta \equiv E-E_o$ and width Γ only.

2) Analyze the correlation effects in Ψ_o and eliminate those which are represented by vectors with small coefficients or otherwise expected to have small matrix elements with the continuum--according to the golden rule (eq. 1). This allows the approximation $\Psi_o \rightarrow \Psi'_o$, where Ψ'_o contains mainly Fermi-Sea and those "hole-filling" correlation vectors which contribute to the localization of Ψ_o [26,3].

3) Rotate $H \rightarrow H(\vartheta)$ and retain the exact characteristics (e.g. its small size) of Ψ_o for $\vartheta > 0$ by transforming $\Psi_o(r) \rightarrow \Psi_o(re^{i\vartheta}) \equiv \Psi_o(\vartheta)$, with $\langle \Psi_o(\vartheta) | H(\vartheta) | \Psi_o(\vartheta) \rangle = E_o$, i.e. constant. The rotated, complete

Hamiltonian matrix is then formed as: $H^{11}= E_o$, $H^{1j} = H^{j1} = \langle \Psi'_o(\vartheta) | H(\vartheta) | \chi^j_{as}(\vartheta) \rangle$, $H^{jk} = \langle \chi^j_{as}(\vartheta) | H(\vartheta) | \chi^k_{as}(\vartheta) \rangle$. There are only a few $\chi^j_{as}(\vartheta)$, (3-6) "asymptotic correlation" vectors containing rotated bound as well as "Gamow orbitals" (eq. 24). Therefore, the magnitude of the complex matrix in the ϑ-plane whose diagonalization yields the energy shift Δ and width Γ, is, on the average, (4x4-8x8) only!, with a rather simple matrix element structure (typically 45-65 configurational matrix elements). This should be compared with the previous methods of the brute force CI type where: For a two electron system and 80 Hylleraas function expansion, there are 3240 complicated unrotated matrix elements to be set up (only once), followed by a series (10-20) of 80x80 complex diagonalizations at various angles, in order to choose the $E(\vartheta)$ for which the assumed basis set is most stable.

4) The "asymptotic pair functions" Π_{as}, which give rise to autoionization and autodissociation and are expanded in terms of rotated fixed H-F core orbitals and parametrized Gamow orbitals, are optimized by imposing the constraint of the Virial Theorem in the ϑ-plane (eq. 17). The choice of the optimum ϑ at which this optimization should take place most efficiently as well as the choice of the best set of Gamow orbitals (whose behavior at small and intermediate values of r is also critical) for an arbitrary system, constitute interesting future research directions within this theory.

The essence and the importance of this approach is that even for large systems, the complex matrices which are diagonalized in the ϑ-plane are small since $E_o = \langle \Psi_o(\vartheta) | H(\vartheta) | \Psi_o(\vartheta) \rangle$ is fixed, and one must optimize only the asymptotic correlation. Thus, the drastic reduction of the magnitude of the problem and the systematic and efficient manner by which localized and asymptotic correlation effects are incorporated into the complex coordinate rotation method within our many-body theory, allow considerable optimism as regards the accurate treatment of many electron-nuclei resonances by the CCR method.

Currently, the theory is being applied to inner hole states of many-electron systems.

REFERENCES

1. P.L. Altick, in this volume.
2. C.A. Nicolaides, Phys. Rev. A6, 2078 (1972); Nucl. Inst. Methods 110, 231, (1973).
3. C.A. Nicolaides and D.R. Beck, "Time Dependence, Complex Scaling and the Calculation of Resonances in Many-Electron Systems", to be published in Int. J. Qu. Chem. 14 (1978).
4. G.D. Doolen, J. Phys. B8, 525 (1975).
5. G.D. Doolen, J. Nuttall and R.W. Stagat, Phys. Rev. A10, 1612, (1974).

6. W.P. Reinhardt, Int. J. Qu. Chem. S10, 359, (1976).
7. P. Winkler, Z. Physik A283, 149, (1977); P. Winkler and R. Yaris, J. Phys. B11, 1481, (1978).
8. J. Aquilar and J.M. Combes, Comm. Math. Phys. 22, 69, (1971).
9. E. Balslev and J.M. Combes, Comm. Math. Phys. 22, 280, (1971).
10. B. Simon, Comm. Math. Phys. 27, 1, (1972); Ann. Math. 97, 247, (1973).
11. Y.K. Ho, A.K. Bhatia and A. Temkin, Phys. Rev. A15, 1423, (1977).
12. D.R. Beck and C.A. Nicolaides, Phys. Rev. Letts. submitted May 1978; paper 13, 9th EGAS Conference, Krakow, Poland, July 1977.
13. We bring to attention the recent excellent work of D.M. Bishop and coworkers, D.M. Bishop, Phys. Rev. Letts. 37, 484, (1976); D.M. Bishop and L.M. Cheung, Phys. Rev. A16, 640, (1977), where accurate calculations of the ground state of H_2^+ are performed without invoking the B-O approximation. Of course, there is considerable difference between ground state and autodissociating state calculations, including the type of variational procedures which are applicable.
14. G. Gamow, Z. Phys. 51, 204, (1928).
15. A.F.J. Siegert, Phys. Rev. 56, 750, (1939).
16. J.N. Bardsley, A. Herzenberg and F. Mandl, Proc. Phys. Soc. 89, 305, (1966).
17. Ya. B. Zel'dovich, JETP (Sov. Phys.) 12, 542, (1961).
18. T. Berggren, Nucl. Phys. A109, 265, (1968).
19. Gy. I. Szasz, Phys. Letts. 55A, 327, (1976).
20. C.A. Nicolaides and D.R. Beck, Phys. Letts. 65A, 11, (1978).
21. J. Nuttall, Bull. Am. Phys. Soc. 17, 598, (1972).
22. J.N. Bardsley and B.R. Junker, J. Phys. B5, 2178, (1972).
23. A.B. Migdal and V.P. Krainov, "Approximation Methods in Quantum Mechanics", Chapter 2, Benjamin Press (1969).
24. C.A. Nicolaides and D.R. Beck, Phys. Rev. Letts. 38, 683; 1037, (1977).
25. Solutions of this type for molecules have not been used before. We do not know whether they exist. We see little difference between them and an in principle numerical integration of the resulting H-F integro-differential equations occuring in nuclei where both protons and neutrons are treated simultaneously. Within the Born-Oppenheimer approximation, $\Phi_{HF}(r,R)\rightarrow\Phi_{HF}(r;R)$, i.e. one obtains the diabatic states discussed by T.F. O'Malley, Adv. At. Mol. Phys. 7, 223, (1971) and W.L. Lichten, Phys. Rev. 139, 27, (1965). This approximation essentially omits the nuclear coupling term, $\partial/\partial R$, which then must be included perturbatively at this stage and not after electron and nuclear correlation is taken into account.
26. D.R. Beck and C.A. Nicolaides, this volume.
27. C.A. Nicolaides and D.R. Beck, Chem. Phys. Letts. 36, 79, (1975).

28. C.A. Nicolaides and D.R. Beck, this volume.
29. C.A. Nicolaides and D.R. Beck, J. Chem. Phys. 66, 1982,(1977).
30. E. Brändas and P. Froelich, Phys. Rev. A16, 2207, (1977).
31. R. Yaris and P. Winkler, J. Phys. B11, 1475, (1978).
32. C.A. Nicolaides and D.R. Beck, Phys. Letts. 60A, 92, (1977).
33. A.K. Bhatia and A. Temkin, Phys. Rev. A11, 2018, (1975).
34. P.K. Mukherjee, S. Sengupta and A. Mukherji, Int. J. Qu. Chem. 4, 139, (1970).

USE OF CI METHODS FOR THE STUDY OF MOLECULAR DISSOCIATION PROCESSES IN VARIOUS ELECTRONIC STATES

Sigrid D. Peyerimhoff
Lehrstuhl für Theoretische Chemie
Universität Bonn, 53 Bonn, W.Germany

Robert J. Buenker
Lehrstuhl für Theoretische Chemie
Gesamthochschule Wuppertal, 56 Wuppertal 1
W. Germany

INTRODUCTION

The adequate description of bond-breaking processes is a difficult problem since it requires very good correlated wavefunctions to account for the generally quite large difference in correlation energy between the combined system and the individual fragments. The situation is especially critical if multiple bonds are broken as in N_2, for example, for which the single-configuration Hartree-Fock treatment yields only a dissociation energy of $D_e = 5.18$ eV [1], i.e 4.72 eV below the experimental result [2]. But even in systems containing only a single bond such as F_2 the use of correlated wavefunctions is essential since it is well-known that this molecule is not even found to be bound with respect to two F atoms in the Hartree-Fock approximation. And finally it is also obvious that extremely weak bonds like van der Waals interactions can only be described by methods going beyond the single-configuration approach.

The problem seems to be even more difficult for the situation in which not only ground state but also excited state potential energy surfaces up to the dissociation limits have to be treated since in this case correlation energy differences between fragments and combined system as well as between various electronic states must be accounted

Cleanthes A. Nicolaides and Donald R. Beck (eds.), Excited States in Quantum Chemistry, 403–416.

for in order to obtain a quantitativly reliable description of the behavior of the entire system. In addition further complications might arise by the interaction of various excited states, so that treatments going beyond the Born-Oppenheimer approach or those involving spin-dependent interactions (for example LS coupling) must be introduced. Hence it is clear that quite elaborate procedures are necessary to describe photochemical processes in a quantitative manner.

Fortunately the knowledge of the approximate behavior of various excited state potential surfaces is already sufficient in a number of cases in order to obtain a better understanding of various photochemical aspects, and hence it is not necessary to always push the theoretical treatment to its limits. In what follows a number of examples will be given in this context to demonstrate the accuracy and possible applicability of treatments at various levels of sophistication.

II. ACCURACY OF GROUND STATE DISSOCIATION CURVES

The accuracy with which ground state dissociation curves can be calculated depends in the main on the AO basis set employed and to some extent on the characteristics of the special system under investigation. Small systems such as He_2^+ can of course be treated extremely well without much computational (or theoretical) effort [3]. In a study of this system by an MRD-CI the dissociation energy is in error by no more than 15 meV based on a comparison of the calculated potential energy surface with experimental vibrational and rotational quantum numbers and intensities of observed levels [3]; adjustments in the calculated potential energy curve which had to be made to obtain full consistency with the measured data were less than 1.0 meV in the important bonding region between 1.3 and 3.5 bohr. The correlation energy shows much larger variations in this area (Table 1). The theoretical treatment was thereby designed in such a way that each data point on the potential energy surface required less than five minutes of CPU time on an IBM 370/168 (all steps included, i.e. AO integral generation, SCF iterations, MO transformation and CI time).

Table 1 Accuracy test for the $He_2^+(^1\Sigma_g^+)$ potential curve. Given are the energy differences (in hartree) between the experimental curve and the calculated values. For comparison the calculated correlation energy is given in the last column.

R (bohr)	ΔE (exptl.-calc.)[a)]	calculated E_{corr}
1.30	0.00001	0.06101
1.40	0.00002	0.06175
1.50	0.00001	0.06255
1.75	0.00000	0.06489
1.90	0.00000	0.06640
1.975	0.00001	0.06712
2.0626	0.00000	0.06795
2.15	0.00002	0.06875
2.25	-0.00001	0.06962
2.50	0.00000	0.07152
2.75	-0.00001	0.07297
3.00	-0.00003	0.07395
3.50	-0.00001	0.07492
4.00	-0.00017	0.07503
5.00	-0.00022	0.07448
5.50	+0.00004	0.07417
7.50	+0.00001	0.07365
10.00	+0.00056	0.07359

a) All values are normalized, so that ΔE = 0 for the minimum. Absolute error probably 0.00535 hartree (≡ 0.146 eV).

The very weak van der Waals minimum for HeH_2 in its ground state is only $0.6 \cdot 10^{-4}$ hartree [4,5] and hence the CI treatment must be designed such as to account for at least this accuracy in order to give meaningful results for the $He+H_2$ interaction. In this case it is also quite important to consider the variation in the H_2 distance with the approach of the He atom, which turns out to be a quite important factor in this reaction [5,6].

The calculations of dissociation energies for larger systems such as O_2 and N_2 exhibit of course considerably larger errors if AO basis sets of standard size (DZ plus some polarization) are employed in the MRD-CI treatment, although such basis sets have been generally found to be quite sufficient for the description of excitation energies. In O_2 the calculated dissociation energy is 5.1 eV [7] compared to the measured quantity of 5.21 eV; in N_2 the MRD-CI treatment improves the situation considerably compared to the Hartree-Fock value (5.18 eV) and finds a D_e of 9.33 eV [8] employing a relatively large basis of 72 AO's (DZ plus bond and two nitrogen d polarization species), but this value falls still short of the required 9.90 eV measured in experiments. It must be pointed out, however, that N_2 with its triple bond dissociating into two nitrogen atoms in ^4S states is probably one of the most difficult dissociation processes to treat computationally.

A further example to be chosen in this section is the hydrogen abstraction reaction [9,10] for systems like HN_2 or HCO. In this case a significant activation barrier has to be overcome but in the low C_s symmetry (bent HCO molecule) the SCF treatment is already sufficient to describe this process (18.8 kcal/mole [9] vs 17.5 $\pm$ 2 or 15.7 $\pm$ 1.5 kcal derived from experiments), so that the CI treatment adds essentially a constant lowering to all the SCF data points along the H-CO dissociation path. This reaction is also interesting from a theoretical point of view [9] since in the linear nuclear framework the lowest HCO state is of $^2\Pi$ symmetry and hence cannot dissociate (according to the usual symmetry rules for surfaces described at the electronic level) into the dissociation products $H(^2S) + CO(^1\Sigma^+)$ with combined $^2\Sigma^+$ symmetry; in the often-used language this would be a symmetry-forbidden process. On the other hand as soon as the restriction of a linear nuclear framework is relaxed (the lowest-energy occurs for the bent molecule conformation) the $^2\Pi$ state will split into an A' and A" component whereby the A' species can very readily mix with $^2\Sigma^+$- $^2A'$ configuration of the dissociation products, and the abstraction reaction would be classified as a symmetry allowed process. It is also worth noting that this mixing in the lower C_s symmetry occurs at the orbital level and hence can already be accounted for at the SCF level of treatment.

In polyatomic systems various dissociation paths might be of importance and it is generally quite difficult to treat all at the same level of accuracy and at the same time keep the amount of computational effort within reasonable limits. Standard examples found in the literature include the surface calculations on H_3^+ or H_2F [11]. Another example is given in Table 2 which compares the performance of the SCF and CI procedures in describing the stability of the dissociation products of HOCl.

Table 2 Relative stabilities (in kcal/mole) of several HOCl fragments obtained from various treatments[a)]

	D_e SCF	D_e CI	D_e best CI	D_0 best CI	D_0 Exptl.
HOCl ($^1A'$)	0.0	0.0	0.0	0.0	0.0
$H(^2S)+ClO(^2\Pi)$	80.4	98.6	97.9	91.1	98.0
$OH(^2\Pi)+Cl(^2P)$	16.0	47.1	53.6	62.0	60.3
$O(^3P)+HCl(^1\Sigma^+)$	5.5	44.7	50.9	58.3	59.3
$H(^2S)+Cl(^2P)+O(^3P)$	82.8	142.4	150.4	153.5	165.8

a) For details see the original reference [12]

It is quite obvious that again the hydrogen abstraction process is treated at least in qualitatively reasonable fashion by the SCF method, but that the single-determinantal description for the $O(^3P)$ abstraction is especially poor. Details about the CI treatment can be found in the original reference [12].

III. EXCITED STATE POTENTIAL CURVES LEADING TO DISSOCIATION

Various potential curves calculated by the MRD-CI method for the O_2 molecules are given in Fig. 1 . Two dissociation limits $O(^3P) + O(^3P)$ and $O(^3P) + O(^1D)$ are approached by the various curves and although the last data point given is only at an OO distance of 4.08 bohr, it is clearly seen from the figure that the calculations are able to describe the potential energy curves up to the dissociation limits in a quite satisfactory manner. The form of the $^3\Sigma_g^-$ ground state curve is represented very well (the calculated stretching frequency is 1621 cm^{-1} compared to 1580 cm^{-1} given experimentally) and all five curves ($^3\Sigma_g^-$, $^3\Delta_u$, $^3\Pi_u$, $^1\Pi_u$ and $^5\Pi_u$) converge to the first dissociation limit (calculated at 5.1 eV compared to the experimental 5.21 eV) to within 0.1 eV. The relative energy difference between the ground state and the $B^3\Sigma_u^-$ state is also in excellent agreement with spectroscopic measurements since the calculated 0-0 $B^3\Sigma_u^- \leftarrow {}^3\Sigma_g^-$ transition energy is 6.07 eV compared to the corresponding experimental T_0 value of 6.12 eV. Furthermore calcula-

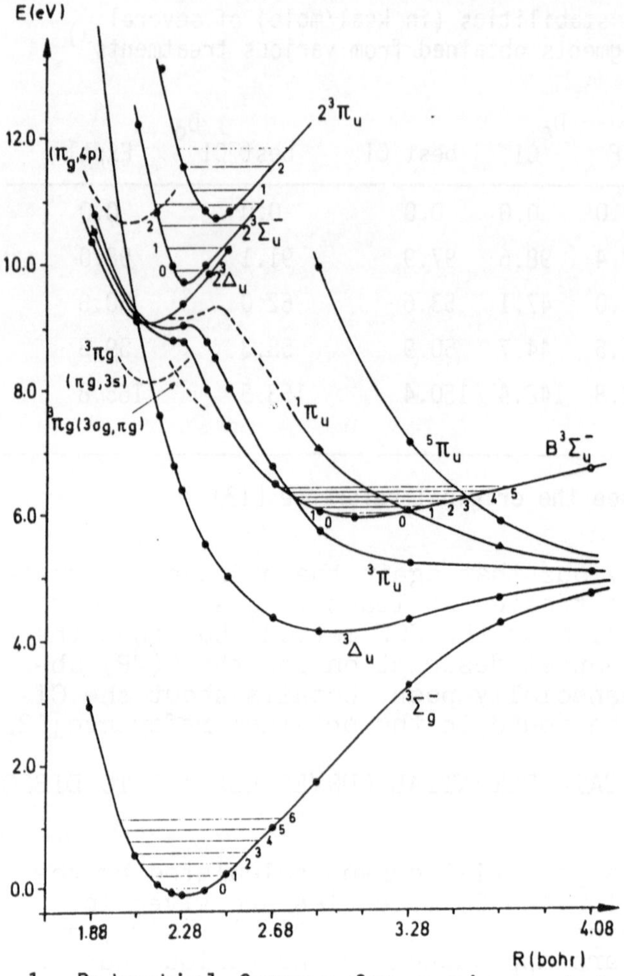

Fig. 1 Potential Curves for various states of O_2 obtained from MRD-CI calculations.

ted and experimental upper state stretching frequency compare quite satisfactorily (ν_0 = 758 cm^{-1} calc. versus 709 cm^{-1} exptl.) so that it can be assumed that the second dissociation limit $O(^3P) + O(^1D)$ is also approached to within an accuracy of 0.1 eV.

The calculated $B^3\Sigma_u^-$ potential surface is intersected by three Π_u states of singlet, triplet and quintet multiplicity, whereby the $^1\Pi_u$ is found to cross between v' = 0 and v' = 1 while the other curves cause the main perturbations around v' = 3 ($^3\Pi_u$ at the inner $B^3\Sigma_u$ limb between vibrational levels 3 and 4 and $^5\Pi_u$ at the outer branch); coupling matrix elements between the respective states would have to be evaluated in order to obtain a more quantitative theoretical picture of the predissociation phenomena which are observed for this B state experimentally.

Table 3 Dissociation limits (in eV) in N_2 (R =10 bohr) obtained from the MRD-CI treatment and comparison with experiment

State	$D_e(^4S + ^4S)$	$D_e(^4S + ^2D)$	$D_e(^4S + ^2P)$
$^7\Sigma_u^+$	9.33	-	-
$^5\Sigma_g^+$	9.37	-	-
$^1\Sigma_g^+$	9.29	-	-
2xN(^{4}S)	9.28	-	-
$^3\Pi_u$	-	12.08	
$^3\Pi_u$	-	-	13.39
Experiment	9.90	12.29	13.48

Three different dissociation limits, namely $^4S + ^4S$, $^4S + ^2D$, $^4S + ^2P$ have been studied in N_2 by the MRD-CI method [8]. Actual calculations for the energies of the lowest $^1\Sigma_g^+$, $^5\Sigma_g^+$ and $^7\Sigma_u^+$ states at NN = 10 bohr agree within 0.08 eV (Table 3) with one another and coincide for all practical purposes with the energy of two nitrogen atoms in their ^4S states once the same AO basis set is employed. In this connection it must be pointed out that such agreement is only obtained through the use of the extrapolation procedure to account for the weakly interacting species; the CI results truncated at a particular threshold ($\neq$ 0) show much larger variance. It is also interesting in this connection that the number of reference species (one for $^7\Sigma_u^+$, six for $^5\Sigma_g^+$ and 20 for $^1\Sigma_g^+$) as well as the order of the total CI space (39288 for $^7\Sigma_u^+$, 202198 for $^5\Sigma_g^+$ and 224710 for $^1\Sigma_g^+$) vary considerably from one state to another, whereby by far the smallest secular equation (order 573 at zero threshold with one reference species) is obtained for the calculation of a single nitrogen atom in its ^4S state.

The calculations for the higher dissocation limits are also quite satisfactory in the large AO basis (Table 3), but more sizeable errors [8] occur (up to 0.80 eV in the respective dissociation energies) if only a DZ basis with bond polarization functions (amounting to a total of 40 AO's instead of 72) is employed.

Various potential curves for the approach of a He atom to H_2 in a number of excited states have also been investigated and very acceptable accuracy for the MRD-CI method in the calculation of the various (predominantly van der Waals minima) has been observed [5,6]. In this study quenching of H_2 in special vibrational levels of electronically excited states by the presence of He atoms could be qualitatively explained thereby.

IV. APPLICATION TO PHOTOCHEMISTRY: NH_3 PHOTOLYSIS

A. NH_3 Excited States

The photodecomposition of NH_3 into various fragments has been studied quite extensively as a function of the incident frequency of light by various experimental procedures [13], and the CI method has recently been applied [14] for the investigation of the photochemical NH_3 fragmentation into $H(^2S)$ and NH_2 in its $^2B_1$ and $^2A_1$ states. The most important features of this study [14] will be reviewed in the present lecture.

The first excited states of ammonia, being a saturated system, can be characterized in the main as Rydberg species originating from a transition of the highest occupied MO ($3a_1$) into 3s and 3p united-atom-like AO's. Experimentally the first two, denoted by $\tilde{A}$ ($3a_1 \rightarrow 3s$) and $\tilde{B}$ ($3a_1 \rightarrow 3px, 3py$) have been well characterized while the third state, $\tilde{C}$ ($3a_1 \rightarrow 3pz$) is occasionally still considered as being part of the $B \leftarrow X$ transition, although detailed electron impact work [15] has clearly identified it as a separate electronic transition. The molecule prefers a planar geometry in all three Rydberg series and hence the appropriate spatial symmetry designation of the states is $^1A_1'(\tilde{X})$, $^1A_2''(\tilde{A})$, $^1E''(\tilde{B})$ and $^1A_1'(\tilde{C})$. Calculated vertical and 0-0 excitation energies are contained in Table 4 for comparison with experiment together with the corresponding oscillator strengths.

The calculations agree very convincingly with the well-known energy and intensity of the $\tilde{A}$ band (Table 4). There are small deviations between corresponding calculated and experimental values for the location of the 0-0 band of the $\tilde{B}$ transition, however; the origin of this $\tilde{B} \leftarrow \tilde{X}$ system was originally placed [16] at 7.455 eV and was later [17] corrected to 7.34 eV, but the CI calculations predict it to be still lower by at least one (or

Table 4 Calculated properties for the first three excited states in NH_3

	$\tilde{A}$	$\tilde{B}$	$\tilde{C}$
C_{3v} geometry symmetry	$^1A_1$	^1E	$^1A_1$
excitation	$3a_1 \rightarrow 3s$	$3a_1$ — px,y	$3a_1 \rightarrow 3pz$
ΔE (eV)	6.29	7.84	8.21
f	0.089	0.002	0.002
exptl. ΔE(eV)	6.39 (v=6)	7.91 (v=5)	≈8.14 (7.9 - 8.6)
exptl. f	0.079, 0.088 0.0696	comparable to $\tilde{C}$	30 times weaker than $\tilde{A}$
D_{3h} geometry			
symmetry	$^1A_2''$	$^1E''$	$^1A_1'$
excitation	$1a_2'' \rightarrow 3s$	$1a_2'' \rightarrow 3px,y$	$1a_2'' \rightarrow 3pz$
ΔE (eV)	5.61	6.99	7.34
exptl. T_0 (eV)	5.72	7.34	< 7.90
C_{2v} geometry symmetry	$^1B_1$	$^1A_2 + ^1B_1$	$^1A_1$

perhaps two) ν_2' quanta (Table 4) at 7.10-7.20 eV if the same accuracy is assumed as in the case of the $\tilde{A} \leftarrow \tilde{X}$ band. This renumbering (by Δv = 1) would place the intensity maximum at v' = 6 (instead of v' = 5 as assumed so far) in line with the situation for the much stronger $\tilde{A} \leftarrow \tilde{X}$ transition. The location of the 0-0 band of the $\tilde{C} \leftarrow \tilde{X}$ transition is not known experimentally while the calculations place it around 7.34 eV; i.e. 0.87 eV below the calculated vertical transition energy, in close analogy to the situation in the $\tilde{A} \leftarrow \tilde{X}$ system.

B. $NH_3 \rightarrow NH_2(^2B_1) + H(^2S)$ Decomposition

It seems to be well established [13] that production of NH_2 in its X^2B_1 ground state occurs via primary excitation into the A state of ammonia; the dissociation process $NH_3(^1A_2'') \rightarrow NH_2(^2B_1) + H(^2S)$ is a spin- and spatially-allowed process even if the relatively high C_{2v} symmetry is maintained throughout. Line broadening of the upper state indicates that the barrier toward dissociation is relatively small and the fact that no emission is seen from this Rydberg state would appear to support this assumption. Results of SCF calculations (restricted to C_{2v} symmetry) given in Fig. 2 support such an

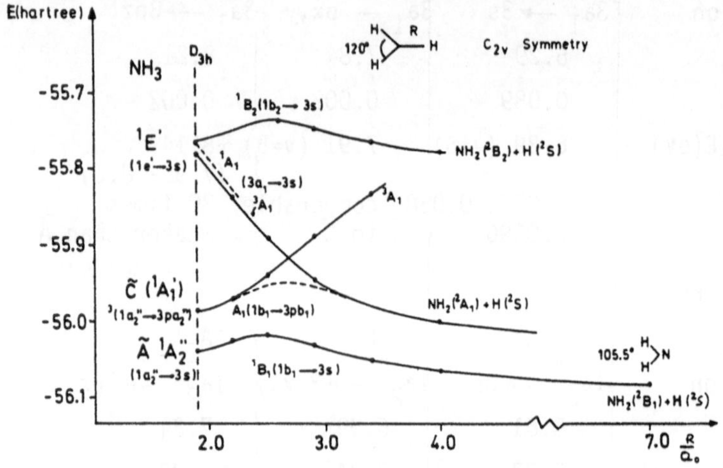

Fig. 2 Calculated SCF potential curves for fragmentation of NH_3 in various states of NH2 + H; C_{2v} symmetry assumed

assignment for the dissociation process in an almost quantitative manner: while H for this process is determined to be -1.2 eV on the basis of thermodynamic data, the calculated energy difference between NH_3 in its A($^1A_2$") state and the products NH_2 ($^2B_1$)+H^3(^2S) is -1.13 eV. From all vibrational NH_3 levels in this state tunnelling through the barrier seems to be possible.

From a more technical point of view it is interesting that the barrier to hydrogen abstraction (as in HCO, for example [9,10]) can already be obtained at the SCF level of treatment. In this case the 3s Rydberg orbital is converted along the reaction path into a 3s + σ* (NH) and finally into a pure 1s hydrogen atomic orbital at the NH2+H dissociation limit. Such a continuous transformation from a Rydberg species into one of purely valence-shell characteristics (mixing at the orbital rather than at the configuration level) is observed in various other systems such as BH [18], HNO or H_2O (Mulliken refers to this process, or rather the reverse thereof, as "Rydbergization" [18]). An important distinction in describing this type of interaction as opposed to a more commonly occurring interaction (or avoided crossing) between states is simply that there must not be a well-defined complementary electronic state which can be unambiguously identified at each value of the reaction coordinate [19]. Similar behavior is also expected for analogous states in the photolysis of water or methane.

C. $NH_3 \rightarrow NH_2(^2A_1) + H(^2S)$ Decomposition

On the basis of energy consideration it is quite often assumed that the $\tilde{B}(^1E'')$ state of ammonia induces photolysis into $NH_2^*(^1A_1) + H(^2S)$. Under the restriction of a reaction path maintaining C_{2v} symmetry, however, the $^1E''$ inducing specied correlates with either A_2 or B_1, whereas the fragments possess A_1 symmetry, hence making the overall process symmetry-forbidden in the usual sense. On the other hand on energetic grounds the $\tilde{C}(^1A_1')$ state would also be a likely candidate to induce photolysis, and indeed symmetry arguments would speak more in favor of such a reaction since $^1A_1'$ correlates directly with A_1 in C_{2v} symmetry, i.e. the symmetry of the $NH_2^*(^2A_1) \times H(^2S)$ fragments.

SCF calculations for the corresponding triplet state (Fig. 2) (which can be taken to be representative for the Rydberg singlet state of importance in the present discussion) show that after configuration mixing with a state of like symmetry, a very similar potential energy curve is obtained connecting the $\tilde{C}$ state with the $NH_2(2A_1) + H(^2S)$ fragments as has been calculated in connection with the decomposition from the $\tilde{A}$ state. A relatively small barrier for the singlet is therefore expected which again makes a tunnelling process quite probable. In this case, however, the change in character of the upper state during dissociation requires actual CI mixing in the conventional sense.In addition crossing with the $^1A_1$ ground state potential curve will occur at large NH_2-H separations in the C_{2v} framework, which will complicate somewhat the assumed reaction path. Finally it should also be mentioned that the threshold energy for $NH_2(^2A_1)$ fluorescence is 7.56 eV, i.e. a value which would also be somewaht more consistent with the location of the 0-0 band for the C state rather than that of B, which has so far been thought to be the inducing species for this process.

V. CONCLUDING REMARKS

The various examples discussed in the lecture have shown that present-day CI calculations are quite feasible for the reliable description of all portions of potential energy surfaces (including the dissociation limits) within the Born-Oppenheimer framework for ground and electronically excited states of molecules or molecular systems. They will become a very useful tool for the detailed study of photochemical processes of small systems if problems occurring in connection with the interaction of various states can also be dealt with in a quantitative manner. Such investigations will require treatments going beyond the Born-Oppenheimer approximation and/or the spin-independent hamiltonian.

Non-adiabatic matrix elements of the form $\langle\psi_i|\frac{\partial}{\partial Q}|\psi_j\rangle$ which couple two or more CI states can already be calculated using CI wavefunctions of standard size (3000 - 5000 expansion terms) [20]. Routines to evaluate matrix elements for spin-orbit and spin-spin interaction at the CI level are being written in various laboratories in order to calculate fine structure splittings as well as transitions between states of different multiplicity. And finally it will also be necessary to extend the treatment of vibrational features in order to include rotational phenomena.

Progress is being made on the solution of all of these problems so that there is a realistic hope to be able to use the theoretical treatments based on the MRD-CI procedure in the not too distant future in order to study quantitative spectroscopic and photochemical details of this nature which cannot be dealt with by purely experimental methods.

References

1. P.E. Cade, K.D. Sales and A.C. Wahl, J. Chem. Phys. 44, 1973 (1966)

2. F.R. Gilmore, J. Quant. Radiat. Transfer 5, 369 (1965)

3. J.G. Maas, N.P.F. B. van Asselt, P.J.C.M. Nowak, J. Los, S.D. Peyerimhoff and R.J. Buenker, Chem. Phys. 17, 217 (1976)

4. P. Harihan and W. Kutzelnigg, preprint communicated to the authors; A.W. Raczkowski and W. A. Lester, Jr., Chem. Phys. Letters 47, 45 (1977); J.W. Riehl, C.J. Fisher, J.D. Baloga and J.L. Kinsey, J. Chem. Phys. 58, 4571 (1973)

5. J. Römelt, S.D. Peyerimhoff and R.J. Buenker, Chem. Phys, in press

6. J. Römelt, Ph. D. thesis, Bonn (1977)

7. R.J. Buenker, S.D. Peyerimhoff and M. Perić, Chem. Phys. Letters 42, 383 (1976)

8. W. Butscher, S.K. Shih, R.J. Buenker and S.D. Peyerimhoff, Chem. Phys. Letters 52, 457 (1977)

9. S.D. Peyerimhoff and R.J. Buenker, in "The New World of Quantum Chemistry", eds. B. Pullmann and R. Parr, D. Reidel Publ. Co., Dordrecht, Holland (1976)

10. P.J. Bruna, R.J. Buenker and S.D. Peyerimhoff, J. Mol. Structure 32, 217 (1976)

11. C.F. Bender, S.V. O'Neil, P.K. Pearson and H. F. Schaefer, Science 176, 1412 (1972). See also H.F. Schaefer "The Electronic Structure of Atoms and Molecules", Addison-Wesley, Reading, Mass. (1972)

12. G. Hirsch, P.J. Bruna, S.D. Peyerimhoff and R.J. Buenker, Chem. Phys. Letters 52, 442 (1977)

13. See for example, M.B. Robin in "Higher Excited States of Polyatomic Molecules", Academic Press, New York/London (1974), Vol. 1; K.H. Becker and K.H. Welge, Z. Naturforsch. A 17, 676 (1962); 18, 600 (1963); H. Okabe and M. Lenzi, J. Chem.

Phys. 47, 5241 (1967); J. Masanet, A. Gilles and C. Vermeil, J. Photochem. 3, 417 (1974/75).

14. R. Runau, S.D. Peyerimhoff and R.J. Buenker, J. Mol. Spectry. 68, 253 (1977).

15. W.R. Harshberger, J. Chem. Phys. 54, 2504 (1971)

16. A.B.F. Duncan, Phys. Rev. 47, 822 (1935); 50, 700 (1936)

17. A.E. Douglas and J.M. Hollas, Canad. J. Phys. 39, 479 (1961)

18 R.S. Mulliken, Int. J. Quantum Chem. 5S, 83 (1971)

19. R.J. Buenker and S.D. Peyerimhoff, Chem. Phys. Letters 36, 415 (1975)

20. Results of this laboratory

THE ROLE OF THE EXCITED STATE IN ORGANIC PHOTOCHEMISTRY

Josef Michl

Department of Chemistry, University of Utah,
Salt Lake City, Utah 84112, U.S.A.

1. INTRODUCTION

Organic photochemical reactivity, along with organometallic reactivity, poses currently perhaps the most exciting challenge to those interested in theoretical organic chemistry. Because of their inherently more complicated nature, photochemical reactions are very difficult to analyze experimentally to the degree of mechanistic detail to which one has become accustomed in the case of thermal reactions. Even relatively crude theory is thus offered a good chance to make fundamental contributions to the understanding and prediction of the events occurring between initial excitation of an organic molecule to an electronically excited state (S, singlet; T, triplet) and the emergence of the first product in its thermalized ground state (S_0).

The material presented in these notes is meant to provide a brief survey of the subject for those unfamiliar with it, and can in no way be considered exhaustive. Moreover, the field of theoretical organic photochemistry is in the midst of rapid development and is full of controversies and unsettled issues. Inevitably, the examples selected for illustration will reflect the author's biases and interests, and it is important to keep in mind that many of the presently accepted concepts may well change in the future.

We shall start by outlining the current physical picture of the photochemical process for an organic molecule in solution. This is based on the Born-Oppenheimer approximation, i.e., on the concept of motion on potential energy hypersurfaces, with non-radiative jumps between surfaces added as an afterthought.

Cleanthes A. Nicolaides and Donald R. Beck (eds.), Excited States in Quantum Chemistry, 417–435.

This description will naturally introduce the distinction of two broad and partially overlapping categories of electronic excited states, which we shall refer to as "spectroscopic" and "reactive" for lack of better terms, and we shall then proceed to outline their main characteristics for selected classes of organic compounds and illustrate their role in a simple description of the photochemical process.

2. A SIMPLIFIED MODEL OF THE PHOTOCHEMICAL PROCESS

Electronic excitation, whether by photon absorption or by energy transfer, changes the electronic part of the molecular wavefunction and, to a good approximation, leaves the vibrational part of the wavefunction unchanged: $\psi_{el}.\psi_{vib} \xrightarrow{h\nu} \psi_{el}'.\psi_{vib}$. At low temperatures, ψ_{vib} is the zero-point vibrational function corresponding to the potential energy hypersurface of the ground state. In general, this will not be a vibrational eigenfunction of the excited state hypersurface, so that immediately after excitation, it will start to develop in time. Collisions with the surrounding medium will complicate this process. In a very short time, on the order of picoseconds, the molecule will end up thermally equilibrated in one or another of the accessible minima in the excited state hypersurface, with probabilities which could in principle be estimated from a trajectory calculation or an equivalent.

Three complications now need to be considered. First, even if the excited electronic state was not the lowest one of a given multiplicity, i.e. it was S_n or T_n rather than S_1 or T_1, internal conversion (radiationless jumping between surfaces of like multiplicity) will bring the molecule to the S_1 or T_1 hypersurface anyway, typically on a picosecond time-scale or even faster, and the extra electronic energy will be first converted into vibrational energy of the molecule and then lost to the surrounding thermal bath. The probabilities with which the molecule ends up in one or another minimum in S_1 (or T_1) may in principle be quite different when the initial excitation is into a higher state S_n (or T_n) (the chances of reaching remote minima usually improve as the excitation energy increases--"upper excited state reactions"), or even into a different vibrational level of the S_1 (or T_1) state ("hot excited state reactions"), but in practice such effects have been observed relatively infrequently, and organic photochemistry in dense media is usually considered to be wavelength-independent, although this appears to be a dangerous oversimplification.

A second complication arises from the fact that internal conversion from S_1 to S_0 may at times also occur extremely fast, so that the first vibrationally equilibrated species which emerges already is in the electronic ground state ("direct reactions").

It is not clear how widespread the occurrence of these reactions is in organic photochemistry. Points at which such very rapid return from S_1 to S_0 occurs have been referred to as "funnels" in S_1. In a more general sense, the expression "funnel" is used for any minimum in S_1 or avoided touching of S_1 with S_0 which efficiently returns molecules to S_0, be it after vibrational equilibration or before.

A third complication occurs in molecules containing heavy atoms or certain other structural features which accelerate intersystem crossing from the originally reached singlet manifold to the triplet manifold to such an extent that it competes with vibrational relaxation in the singlet manifold. The first thermally equilibrated species formed may then be in a minimum in the T_1 hypersurface even if the initial excitation was into S_n.

In one way or another, picoseconds after the initial excitation, the molecule will typically find itself thermally equilibrated with the bath in a local minimum in the S_1, T_1, or S_0 hypersurfaces: in S_1 if the initial excitation was by photon absorption, in T_1 if it was by sensitization, or if special structural features such as heavy atoms are present, and in S_0 if the reaction was "direct". Thus, the first two types of processes involve an intermediate and are sometimes referred to as "complex reactions", as opposed to "direct reactions" which do not. Quite commonly, but not always, the initially reached minimum in S_1 (or T_1) is located at a geometry which is close to the equilibrium geometry of the original ground-state species, so that no net chemical reaction can be said to have taken place so far, only a relaxed excited state of the starting material has been prepared.

Next, slower processes can come to play a role. The most important among these are, first, thermally activated motion from the originally reached minimum over relatively small barriers to other minima or funnels, which represent adiabatic photochemical reactions proper and can often be described by ordinary absolute reaction rate theory; second, intersystem crossing, which frequently occurs on a nanosecond time-scale and takes the molecule from the singlet to the triplet manifold and thus eventually to a new minimum in T_1; third, fluorescence ($S_1 \rightarrow S_0 + h\nu$, nanosecond time-scale) or phosphorescence ($T_1 \rightarrow S_0 + h\nu$, usually millisecond or slower time-scale) which return the molecule to the ground S_0 surface; and fourth, radiationless conversion $S_1 \rightarrow S_0$ or $T_1 \rightarrow S_0$, which is usually insignificant unless the S_1-S_0 (or T_1-S_0) energy gap in the region of the minimum is small. Other processes are possible, such as further photon absorption or either excitation or de-excitation by energy transfer (T-T annihilation, quenching), but will not be discussed here.

The description given so far is best suited for unimolecular photochemical reactions. If the process is bimolecular, both components must be considered as a "supermolecule" and the above description is then again valid except that motion along certain directions in the nuclear configuration space of the supermolecule is unusually slow since it is diffusion-limited.

Whether the eventual return to S_0 is radiative or radiationless, its most important characteristic is the location of the minimum in S_1 or T_1 from which it occurs in the nuclear configuration space, i.e. the geometry of the species at the time of the return. If the return occurs to a region of S_0 hypersurface which is sloping down back to the starting minimum, the whole process is considered photophysical since there is no net chemical change. If the return is to a region of S_0 which corresponds to a "continental divide" or which clearly slopes downhill to some other minimum in S_0, a net chemical reaction will have occurred as a result of the initial excitation, and the process is considered photochemical. In this latter case, radiative return ("adiabatic photochemistry") is uncommon. In either case, non-radiative return produces a species with considerable excess of vibrational energy, which may travel over some normally forbidding barriers to yet another minimum in S_1 before it equilibrates thermally with its surroundings ("hot ground state reaction"), but such processes appear to be very rare in dense media.

A simplified theoretical treatment of a photochemical reaction can then in principle proceed in three stages. First, the location of the minima in the S_1 (or T_1) hypersurfaces must be determined. Second, it must be determined which of the minima are accessible, i.e. not separated by insurmountable barriers, given the initial excitation conditions, and it must be estimated which of the minima will actually be populated with significant probabilities and provide points of return to the ground state. Third, it must be determined from the shape of the S_0 surface what the products of return from these important minima in S_1 (or T_1) will be. All of these steps are difficult; the second one appears to be the hardest. In practice, one is usually reduced to working backwards, i.e. to rationalizing observed experimental results rather than making a priori predictions of new types of reactions.

Already such a limited program requires considerable insight into the nature of excited state hypersurfaces, permitting estimates of the location of minima and barriers under conditions where reliable calculations are prohibitively expensive and where even crude calculations are less than straightforward, so that qualitative arguments based on devices such as correlation diagrams frequently have been the only recourse so far. It should be noted that a calculation of the shapes of the excited singlet potential energy hypersurfaces along realistic reaction paths, which usually lead through regions of biradicaloid geometries, cannot be reasonably performed by straightforward application of existing standard MO programs, since these generally begin by a closed-shell SCF calculation.

3. "SPECTROSCOPIC" AND "BIRADICALOID" MINIMA IN EXCITED STATE HYPERSURFACES

There are two basic types of geometries at which one would intuitively expect minima in S_1 and T_1 hypersurfaces to occur. First, "spectroscopic" minima near ground state geometries. After all, excitation of one electron out of dozens may well represent only a minor perturbation in the bonding, particularly in a relatively large molecule: the fluorescent S_1 and phosphorescent T_1 states of benzene have roughly the geometry of its S_0 ground state. Even in smaller molecules the difference, though noticeable, is frequently quite small: S_1 and T_1 $n\pi^*$ excited formaldehyde is pyramidal rather than planar but still quite similar to S_0 formaldehyde, etc. Radiative processes between such minima may be somewhat constrained by the Franck-Condon principle, but are generally readily observable, hence "spectroscopic" excited states. If such an excited state minimum is reached by approximately vertical excitation from a S_0 minimum, and the molecule then returns to S_0, either non-radiatively or radiatively, it will generally land in the same minimum in which it started with no net chemical change (except if a "hot ground state reaction" intervenes, but these appear to be quite rare in dense media, except perhaps for interconversion of conformers). Only if return to S_0 occurs from a "spectroscopic" minimum of one species after initial excitation of another species, after significant geometry and bonding changes while in the excited state, will a photochemical change result. These are "adiabatic photochemical reactions", quite rare for large molecules except for simple proton-transfer processes. An example of such a reaction is the rearrangement of dewarnaphthalene to naphthalene.

The other type of geometries at which one can expect minima in S_1 and T_1 hypersurfaces (or "funnels" in S_1 for rapid return to S_0) are those of biradicaloid type, i.e. those in which the simple MO picture of the molecule shows two roughly non-bonding orbitals occupied by a total of only two electrons in the ground state. Such geometries are usually highly unfavorable in the S_0 state, since the two electrons contribute nothing to bonding. As a result, one less bond is present than is effectively possible. Upon distortion to a geometry at which the two orbitals are caused to interact, one becomes bonding and holds both electrons in the ground state, while the other becomes antibonding. In the excited states S_1 and T_1 the situation is quite different. Now, the same distortion will not be stabilizing but is likely to bring about destabilization, at least initially, since only one of the two electrons can be kept in the orbital which becomes bonding, while the other is kept in the orbital which becomes antibonding and the effect of the latter usually prevails. The simplest example is the breaking of a sigma bond in its S_0, S_1 and T_1 states (Figure 1). The minimum in the S_0 state occurs at a short intermolecular distance which is clearly not biradicaloid, the σ orbital being strongly bonding, the σ^* orbital strongly antibonding. On the other hand, the minimum in the T_1 state occurs at infinite internuclear separation, a clearly biradicaloid geometry at which two non-bonding orbitals are present. The simplest example is the H_2 molecule, with a minimum in S_0 at 0.74 Å and a purely dissociative T_1 state. The S_1 minimum occurs at an intermediate but still relatively long internuclear distance (in H_2 this is 1.3 Å), at which the σ orbital is much less bonding and σ^* much less antibonding than they were at the equilibrium geometry in S_0. The difference between the best geometry of T_1 (the two non-bonding orbitals far separated in space, "loose biradicaloid geometry") and that of S_1 (the two orbitals relatively close to each other, "tight biradicaloid geometry") is easily understood when it is realized that the S_1 state at a biradicaloid geometry tends to be, and in H_2 is, of "zwitterionic" nature, and separation of the two centers requires electrostatic energy (the S_1 state is reasonably represented as $H^+H^- \leftrightarrow H^-H^+$). The location of the minimum in S_1 represents a compromise between the tendency to proceed to biradicaloid geometries, i.e. to minimize the σ_g-σ_u^* separation, and the tendency to minimize the electrostatic energy. The tendency for minima in T_1 to occur at loose biradicaloid geometries and those in S_1 to occur at tight biradicaloid geometries should result in different points of return to the S_0 surface. It is probably quite general and is responsible for a large part of the differences between singlet and triplet photochemistry.

The states of a sigma bond are more complicated if one or both atoms which it connects carry lone pair electrons, but this will not be discussed here. It will be noted that in simple molecules such as H_2 the "spectroscopic" excited states are the same as the "biradicaloid" excited states--there are no separate minima at the two types of geometries in one and the same surface. The same is true in ethylene. In general, however, this will not be so in large organic molecules. E.g., the loose biradicaloid minimum which is attained by the stretching of the benzylic C-H bond of toluene to give $C_6H_5CH_2\cdot + H\cdot$ occurs in the same T_1 surface as the phosphorescent "spectroscopic" minimum responsible for the phosphorescence of toluene, but they are quite distinct and separated by a sizeable barrier.

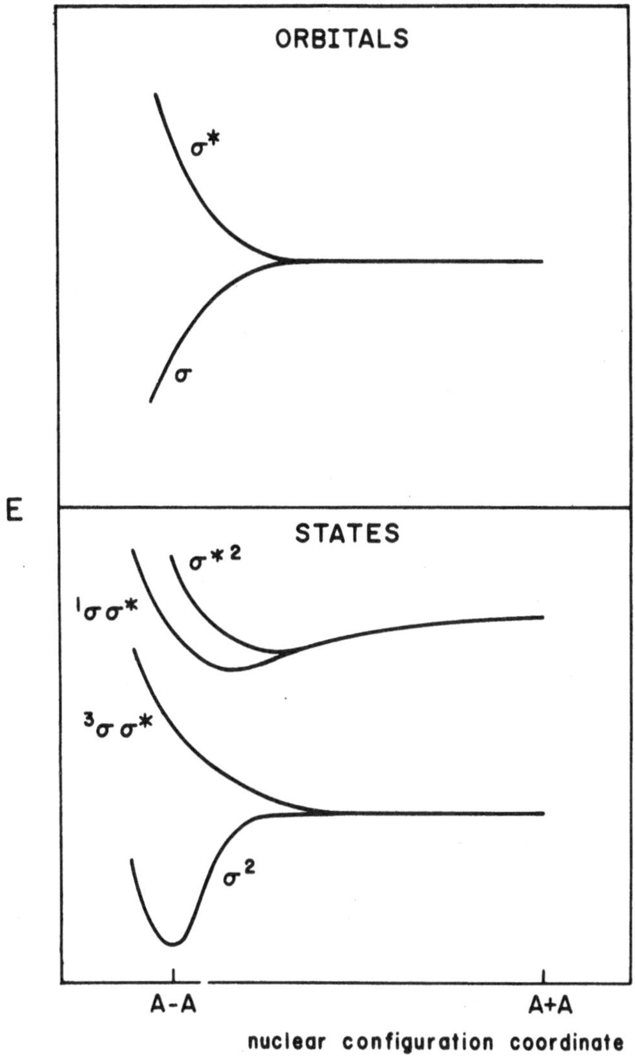

Figure 1. Electronic states of a sigma bond (schematic).

Since at biradicaloid geometries the S_0 and T_1 states generally lie close to each other and usually even the S_1 state is not much higher in energy, radiationless return from S_1 or T_1 to S_0 should be very fast and molecules in these minima should not be nearly as easy to observe directly as they are in the "spectroscopic" minima at ordinary geometries. The return to S_0 should produce highly reactive biradicaloid entities with one bond missing, and subsequent motion on the S_0 surface to one of the nearby local minima usually provides an excellent chance to produce chemical change. Thus, return to S_0 through a biradicaloid minimum is likely to lead to a net photochemical transformation for at least a fraction of the originally excited molecules.

A surface with more than one minimum can usually be thought of as resulting from an interaction of several zero-order states with one minimum in each, where the interaction results in avoidance of crossings. For instance, the T_1 surface of toluene along the path of nuclear geometries which lead to dissociation to $C_6H_5CH_2\cdot + H\cdot$ can be viewed as originating from interaction of a locally excited ππ* triplet state of the benzene chromophore and the locally excited σσ* triplet configuration of the C-H bond. The former is of lower energy at the initial geometries, the latter at the final geometries, somewhere along the way they intend to cross, but the crossing is avoided and results in a barrier separating two minima in the T_1 surface (Figure 2). Once

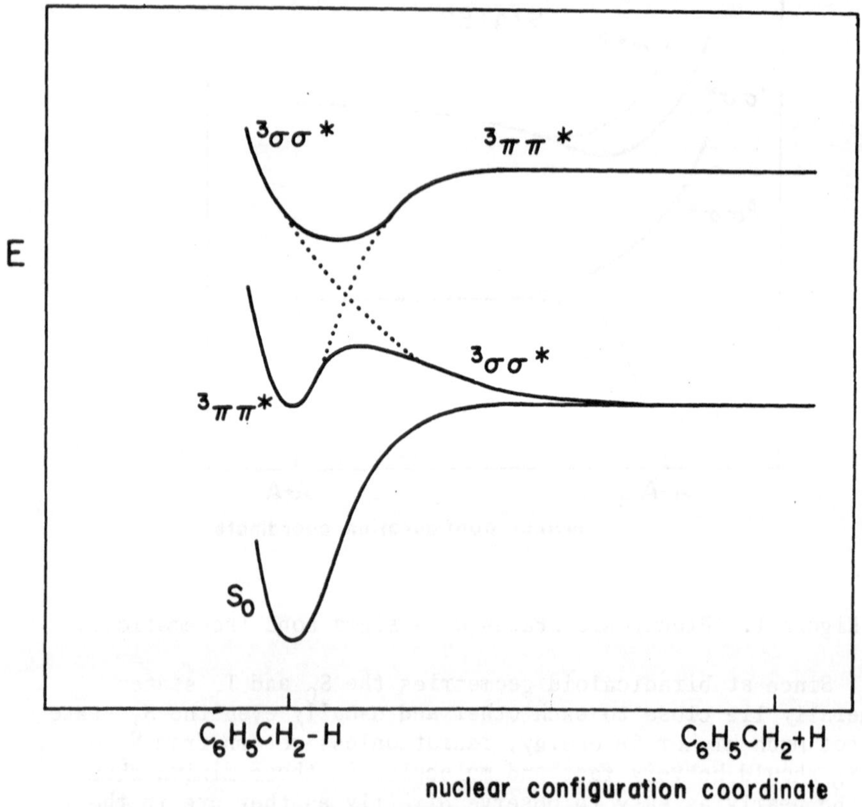

Figure 2. Triplet states of toluene (schematic).

we identify the zero order states, we can use them for nomenclature purposes. If the zero-order state responsible for the reactive minimum is described by a single configuration, it is sometimes referred to as the "characteristic configuration" for that particular photochemical path. We can now say that two excited triplet states play a role in our model reaction, the triplet photochemical conversion of toluene to $C_6H_5CH_2\cdot + H\cdot$. First, a "spectroscopic" state, into which initial excitation occurs, either by energy transfer (triplet sensitization), or indirectly by intersystem crossing from the singlet manifold; second, a "reactive" state, in which the actual dissociation occurs. In actual practice, the barrier separating the "spectroscopic" minimum from the "biradicaloid" one is too high in this particular case and the reaction proceeds only upon absorption of a second photon by the spectroscopic triplet state, but this does not detract from the illustrative value needed here.

The surfaces for both zero-order states continue to the other side of the avoided crossing, where they represent higher S_n or T_n states, and one might argue that direct vertical excitation into the "reactive" state, or more accurately, into some excited state which contains a large component of the zero-order "reactive" state, should be possible, even if unfavorable because of poor Franck-Condon overlap. In our particular case, intersystem crossing and sensitization are unlikely to populate such a high-energy state, but it is quite possible that internal conversion after the above-mentioned absorption of a second photon by a molecule in the spectroscopic T_1 state actually produces just this result.

The situation described for the dissociation of toluene appears to be widespread in photochemical reactions of large organic molecules. With exceptions such as cis-trans photoisomerization of very simple olefins, the usual means of initial excitation, i.e. photon absorption or energy transfer with production of one of the low-lying excited states, occur into a "spectroscopic" state which is different, in the zero-order sense, from the "reactive" excited state from which the return to S_0 occurs. Of course, the travel to this ultimate reactive state may proceed via a whole series of intermediate minima in S_1 or T_1, thus intermediate zero-order states. The role of the two types of states in the overall process is quite different. The spectroscopic state permits the initial excitation, and the slope of its hypersurface in the direction of the reaction coordinate determines whether a substantial barrier will stand in the way of conversion to the "reactive" state. The spectroscopic minimum preserves the excited molecule until it can escape towards the reactive minimum. Its role is particularly crucial in bimolecular processes, where this escape has to wait for diffusion to bring in a reaction partner. The initial escape probably frequently is into an excimer or exciplex state, from which the reactive minimum is reached subsequently.

A molecule in a biradicaloid minimum, i.e. in the "reactive state", has rarely been observed spectroscopically. It plays a key role in the whole photochemical process. It is by the location of its minimum (and rate of return to S_0) that the geometry reached immediately upon return to S_0 is determined. This, along with the shape of the S_0 hypersurface, will determine the nature of the products.

To complete the task of describing the role of the excited state in organic photochemistry, we turn to a brief review of the common types of spectroscopic excited states, followed by a description of reactive excited states for selected photochemical reactions.

4. "SPECTROSCOPIC" EXCITED STATES

Ground states of the vast majority of known organic molecules are closed-shell singlets. It is customary to describe excited states in terms of the dominant configuration in the CI expansion. Important occupied MOs ϕ_i are of the σ, n, and π types; important virtual MOs ϕ_j are of the σ^* and π^* types. States of $\sigma\sigma^*$, and to a considerable degree, also $\sigma\pi^*$ and $\pi\sigma^*$ types, are of limited importance for solution organic photochemistry, which is usually restricted to wavelengths longer than 200 nm. Transitions into states of $n\sigma^*$ type, important for instance in simple amines, sulfides, and halides, and particularly states of $n\pi^*$ type, important for instance in carbonyl compounds and aza heterocycles, are basically intra-atomic in nature. Although the n orbitals are not completely localized on a single atom such as O, N, S, or halogen, they generally largely extend over a different set of AOs than the σ^* and particularly π^* orbitals into which the excitation occurs. As a result, transition density may be delocalized over several atoms, but on each atom it integrates to zero, so that the overall transition dipole is a sum of small atomic dipoles of the type $\langle s|\hat{\underline{m}}|p\rangle$ and the transition density $\phi_i\phi_j$ is small. Also the self-repulsion of the overlap density $K_{ij} = \iint \phi_i(1)\phi_j(1)\,\frac{e^2}{r_{1,2}}\,\phi_i(2)\phi_j(2)d\tau_1 d\tau_2$ and therefore the singlet-triplet splitting $2K_{ij}$, are small. Transitions of this type are usually clearly present in absorption spectra only if they occur at lower energies than any of the $\pi\pi^*$ transitions, otherwise they are overlapped. At times, they can be detected by specialized techniques such as measurement of circular dichroism if the molecule is chiral. Unfortunately, in the more widely applicable magnetic circular dichroic and linear dichroic spectra, these transitions are rarely easier to detect than in ordinary absorption spectra. They are also relatively expensive to calculate, since they require explicit consideration of all valence electrons. In very small molecules, ab initio methods have been used with success. For larger molecules, it has been most common to use the semiempirical CNDO/S model, which appears to work quite well.

Transitions of the $\pi\pi^*$ type are of the interatomic type. These are truly molecular transitions, without analogy in isolated atoms. Both the π and the π^* orbitals in principle spread over the same set of AOs. If both have large coefficients on the same AOs, the transition densities have large net charges on individual atoms, frequently producing quite large transition dipoles. Also the self-repulsion of the overlap density and therefore the singlet-triplet splitting are frequently very large. If the AOs where the orbital π has large coefficients do not coincide with those where π^* has large coefficients, the transition involves interatomic charge transfer, such transitions are generally weaker and have smaller singlet-triplet splittings. Since the observation of $\pi\pi^*$ transitions is relatively easy, a very large number of them have been characterized in a great many molecules. Calculations for these transitions also are much easier, since already π-electron methods (PPP) provide quite reliable spectral predictions. Two large classes of molecules with excited states of this type will be mentioned here. One comprises linear polyenes and their derivatives, the other comprises annulenes (cyclic polyenes) and their derivatives.

Linear polyenes and their derivatives are of considerable interest in photochemistry, both because of the variety of interesting chemical transformations which they undergo and because of their relation to the process of vision. They are characterized by a relatively intense singlet $\pi\pi^*$ transition corresponding to the excitation from the highest occupied (HO) to the lowest unoccupied (LU) MO, which gradually shifts to lower energies and becomes more intense as the conjugated chain is extended. Approximately degenerate with this strongly allowed transition is an extremely weak transition into a state which is often referred to as doubly excited, since the doubly excited HOMO,HOMO $\rightarrow$ LUMO,LUMO configuration has considerable weight in it, although other configurations are also important. The existence of the state was predicted from semiempirical π-electron calculations long before it was actually observed. The direct observations have so far been limited to fairly long polyenes, where this "forbidden" state is the lowest excited singlet. The lowest triplet state of linear polyenes is well described by the HOMO $\rightarrow$ LUMO configuration.

Annulenes are the parents of a vast array of cyclic molecules, including all aromatic compounds and many others. Many of these are once again of great photochemical interest both intrinsically and because of their importance in biology (e.g., nucleic acid components). Daughter molecules can be derived not only by introduction of substituents (e.g., aniline) and heteroatoms (e.g., pyridine), but also by introducing bridges (e.g., pyrene) and cross-links (e.g., anthracene) inside the annulene ring. Each annulene is characterized by the number of atoms in the ring, n, and by the number of π electrons. Those with 4N+2 π electrons (N = 0, 1, 2, ...) and their daughter molecules are by far the most important class in practice (an example of an important class of cyclic molecules derived from 4N-electron annulene perimeters are the quinones). As shown in Figure 3, the parent (4N+2)-electron [n] annulenes are characterized by four low-energy excited configurations resulting from excitations from the degenerate HOMO to the degenerate LUMO (if N = 0, HOMO is not degenerate, and if N = n/2-1, LUMO is not degenerate, in which cases there are only two low-energy configurations). Using MOs adapted to the C_n group, which generally have non-zero angular momentum, two of these excitations can be

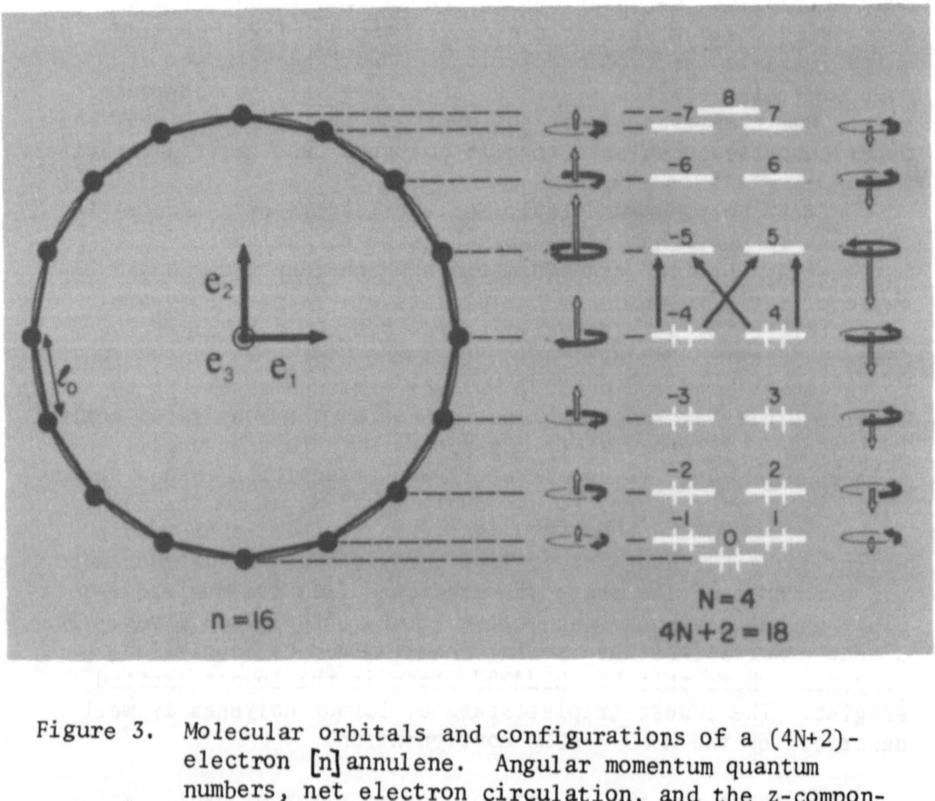

Figure 3. Molecular orbitals and configurations of a (4N+2)-electron [n] annulene. Angular momentum quantum numbers, net electron circulation, and the z-components of the magnetic moment are shown.

referred to as sense-preserving, since they preserve the sense of net circulation of the electron which is being excited. These excitations change the angular momentum quantum number of the electron which is being promoted by one and are strongly allowed. Their upper states are degenerate; this degenerate excited state is assigned the symbol B and is of E_1 symmetry in the C_n group. The other two excitations, present only if $N \neq 0$ and $N \neq n/2-1$, are referred to as sense-reversing, since they reverse the sense of net circulation of the electron. They change the angular momentum quantum number by more than one and are electric-dipole forbidden (they still appear in absorption spectra weakly, due to vibronic intensity borrowing). If the perimeter is charged, i.e. $n \neq 4N+2$, their upper state is degenerate and is assigned the symbol L (symmetry E_{2N+1} in the C_n group). If the perimeter is electroneutral, i.e. $n = 4N+2$, these two configurations mix to produce two states of B symmetry in the C_n group. In one of these, L_b, nodes pass through atoms (B_{2u} in the D_{nh} group); in the other, L_a, nodes bisect bonds (B_{1u} in the D_{nh} group). The B states lie considerably above the L states in energy. In the singlet manifold, ^1L_a lies above ^1L_b, in the triplet manifold, ^3L_b lies above ^3L_a. The ^1L_b and ^3L_b states are roughly isoenergetic.

In daughter molecules derived from (4N+2)-electron [n]-annulenes, the same four singlet and four triplet states can still be traced (if $N = 0$ or $N = n/2-1$, only the two singlet and two triplet B states are present). Except for highly symmetrical derivatives, such as coronene or metalloporphyrins, degeneracies are removed. If $n \neq 4N+2$, one observes two weak to medium-intensity closely spaced singlet transitions into L_1 and L_2 states, polarized in mutually roughly perpendicular directions, followed at higher energies by two strong transitions into B_1 and B_2 states (e.g., tropolone, indole). The transitions into the lowest singlet and triplet L_1 states correspond to predominantly HOMO → LUMO excitation. If $n = 4N+2$, two weak to medium intensity singlet transitions into L_b and L_a states appear, followed at higher energies by intense transitions to B_b and B_a states. If symmetry is relatively high, such as C_{2v} or D_{2h}, the L_b and L_a transitions are polarized at right angles to each other (B_b is parallel to L_b and B_a to L_a); if it is too low, such as C_s, they may be polarized in such a way (e.g., 1-substituted naphthalenes) or may be nearly parallel (e.g., 2-substituted naphthalenes), depending on the nature of the perturbation of the parent perimeter in a presently well understood way. Again depending on the nature of the perturbation, one or the other of the transitions may have nearly vanishing purely electronic intensity and be dominated by vibronic contributions (L_b in naphthalene, L_a in azulene). In the singlet manifold, ^1L_b usually but not always lies below ^1L_a as in the parent annulene; in the triplet manifold, ^3L_a generally is lowest. The transitions into $^{1,3}L_a$ states correspond to predominant HOMO → LUMO excitation.

Other transitions are present at higher energies, some of these calculated to contain substantial contributions from doubly excited configurations, but much less is generally known about them.

5. "REACTIVE" EXCITED STATES

Present-day knowledge of these zero-order states is based primarily on qualitative correlation diagrams and on calculations. From the way in which these states were introduced, they correspond to biradicaloid minima in S_1 and T_1. As shown on the example of a simple sigma bond (Figure 1), these do not need to occur at identical geometries even for one and the same photochemical path, but in the first approximation we shall find it useful to discuss them jointly.

Many types of biradicaloid geometries are possible and they are perhaps most conveniently listed according to one of the frequently several processes which can produce them starting from "ordinary" molecular geometries. Thus, stretching any single bond in a molecule will lead to a "stretched σ-bond" biradicaloid minimum (typically occurring for an only partially extended σ bond in S_1 and for a fully broken σ bond in T_1), with a characteristic configuration of the σσ* type. Twisting a double bond will lead to a "twisted double bond" biradicaloid minimum, with a characteristic configuration of the ππ* type. An interesting type of a biradicaloid geometry is reached by performing a pericyclic reaction of the ground-state forbidden type half-way, i.e. to the "antiaromatic" geometry where the molecule is isoelectronic with a Hückel 4N-electron annulene or a Moebius (4N+2)-electron annulene and thus biradicaloid. Along such a pericyclic path, the ground configuration of the starting material correlates with a doubly excited configuration of the product and vice versa, so that in the state correlation diagram a barrier results in the S_0 surface and a "pericyclic" minimum has a good chance of resulting in the S_1 but not T_1 surface. Various other types of biradicaloid geometries can be envisaged but will not be discussed here.

In many cases (e.g., Figure 2), the zero-order S_1 and T_1 states at the biradicaloid geometries, when followed back to the geometry of the starting material, extrapolate to states which are at very high energies (σσ*). In others, such as for the "pericyclic" minimum in S_1, they result in doubly excited states. In neither of these cases do they lead us back to one of the well known "spectroscopic" states. However, this can also happen: the $\pi\pi^*$ characteristic configuration of a "twisted double bond" biradicaloid minimum extrapolates back to one of the spectroscopic $\pi\pi^*$ states of an olefin.

Since the "reactive state" in most cases is not the state produced in the initial excitation, it is important to consider the role of the initial "spectroscopic" state in delivering the molecule to the reactive minimum. Under the usual experimental conditions, travel from the spectroscopic minimum in S_1 or T_1 has to either be downhill or over only a small barrier, with thermal activation. In terms of correlation diagrams, this requires the "spectroscopic" S_1 or T_1 state to correlate with a very low lying state at the biradicaloid geometry, and preferably with the lowest excited state of the product. Reaction paths along which the correlation is with some highly excited state of the product are said to be "orbital symmetry forbidden".

A good example is provided by pericyclic processes. In order for such a process to be photochemically feasible, one of the ground-state occupied bonding orbitals of the starting material must correlate with a ground-state unoccupied antibonding orbital of the product and vice versa. This guarantees a biradicaloid geometry half-way along the reaction path and thus a minimum in S_1 providing a productive return to S_0 at a suitable point. If it is the HOMO of the starting material which becomes LUMO of the product and LUMO of the starting material which becomes HOMO of the product (normal orbital crossing), the HOMO → → LUMO configuration of the starting material will correlate with the HOMO → LUMO configuration of the product. As noted in the preceding section, the HOMO → LUMO configuration frequently but not always predominates in the S_1 spectroscopic state of a starting or final material, and practically always predominates in its T_1 state. If this is so, no correlation-imposed barrier is expected and the conversion from the spectroscopic to the reactive minimum in S_1 or T_1 should be easy. However, in those cyclic molecules in which the ^1L_b state is below the ^1L_a state at the starting geometry, a barrier must be expected along the way. There is indeed considerable evidence for the existence of such barriers in the singlet photochemistry of benzene and naphthalene derivatives, which can at times be overcome by use of shorter excitation wavelengths.

If the orbital crossing is of the abnormal type, i.e. involves orbitals other than HOMO and LUMO, it is most unlikely that the needed configuration will predominate in the S_1 or T_1 state of the starting material and a sizeable barrier must then be expected for both the singlet and the triplet reaction along this path.

From what has been said so far about the need for conversion from a spectroscopic minimum to the reactive minimum it is hardly surprising that many photochemical reactions are suppressed at low temperatures, and conversely, that more efficient fluorescence frequently results upon cooling.

The efficiency with which typical reactive minima deliver molecules to the S_0 surface is impressive. Still, it is possible for molecules to escape on the S_1 surface and to continue towards the product geometry. This is revealed by product fluorescence (adiabatic photochemistry) and has been observed for extremely exothermic reactions for which the pericyclic minimum in S_1 probably is considerably tilted.

Little can be said at present about the electronic states of molecules at biradicaloid geometries from experiment. Few reliable calculations exist and much work is needed before a thorough understanding can be developed. The electronic states at a stretched σ-bond minimum have been already discussed (Figure 1). The states at a pericyclic minimum can be exemplified by the singlet states of hypothetical square H_4 as a model. The lowest few among these are shown in Figure 4 for a rectangular reaction path $H_2 + D_2 \rightleftarrows 2HD$. The avoided crossing of the ground (G) and the doubly excited (D) states and the resulting pericyclic minimum in S_1 and barrier in S_0 are clearly apparent. The doubly excited D state corresponds to an overall singlet coupling of two triplet H_2 molecules (triplet-triplet annihilation state). The singly excited HOMO → LUMO state (S) of the starting materials correlates with the HOMO → LUMO state of the products and contains a shallow minimum half-way through the reaction path. This state corresponds to the combination of a ground state H_2 molecule with a singlet singly excited H_2 molecule (σσ*), is referred to as the excimer state, and the minimum as the excimer minimum. In more complicated cycloaddition processes, this minimum does not occur at the same geometry as the pericyclic minimum, but earlier in the path of the approach of the two components, so that a return from it to S_0 results only in redissociation to the starting materials. To the contrary, return to S_0 from the pericyclic minimum occurs half-way along the reaction path and produces cycloaddition products as well as starting materials. The relative energies and positions of the excimer and triplet-triplet annihilation surfaces clearly play a key role in determining the photochemical process. Other states are indicated in Figure 4; the I and I' states correspond to cation-anion annihilation processes.

The comparison of Figures 1 and 4 reveals an interesting difference: at the biradicaloid geometries of H_2, the singly excited singlet state lies below the doubly excited singlet state; at those of H_4, the opposite is true. Intuitively, the state ordering in H_2 is more easily understood. Yet, the opposite ordering in H_4 is essential for the appearance of a pericyclic minimum in S_1 at the biradicaloid geometry, and for the occurrence of the photochemical process. The difference can be understood in simple terms by a combination of the MO and VB theories. It will not be discussed here in detail and it will just be noted that both excited singlets of H_2 are purely "zwitterionic", $H^+H^- \leftrightarrow H^-H^+$, while in H_4 only the singly excited state is. The doubly excited state has acquired considerable covalent character by mixing with other configurations and has been stabilized relative to the singly excited state. This is perhaps most easily understood when one remembers that in the limit of an infinitely large H_4 square this state becomes degenerate with S_0 and represents four neutral H atoms, and that in the limit of diagonal bonding this state correlates with two H_2 molecules in their ground states.

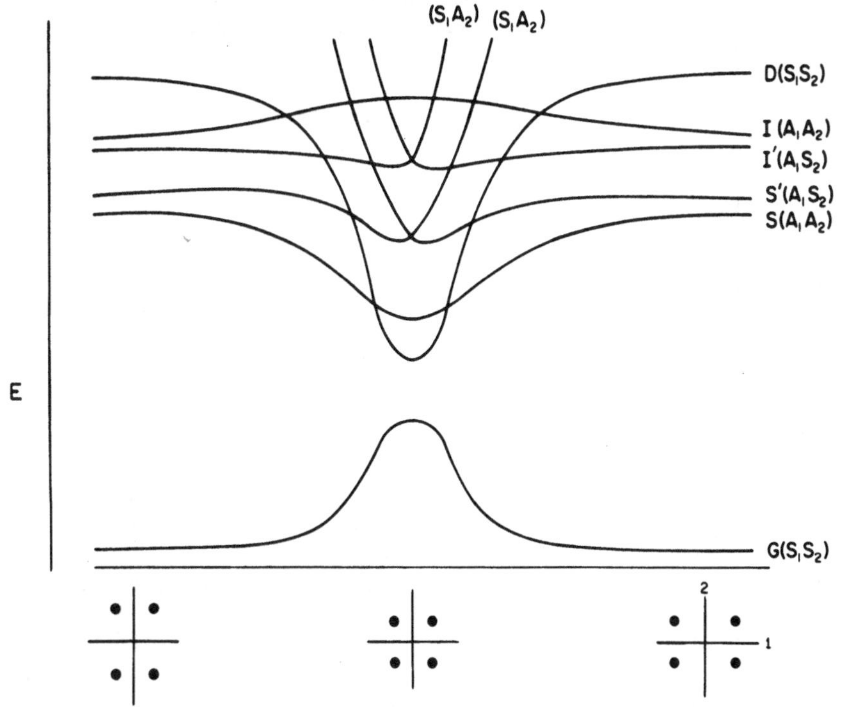

Figure 4. Singlet states of H_4 along a rectangular $H_2 + D_2 \rightleftarrows 2HD$ reaction path (schematic, symmetry labels shown refer to the two symmetry planes indicated at the bottom).

6. SUMMARY

The ideas outlined briefly above have been applied to a wide array of organic photochemical processes, and the concepts of initial excitation to a "spectroscopic" state followed by motion to a "reactive minimum", return to S_0 and thermal equilibration seem to be generally useful. The use of qualitative arguments such as correlation diagrams can, of course, only provide partial answers. To make further progress, calculations of potential energy hypersurfaces along optimized paths by reasonably reliable methods are needed, to be followed in the future by analyses of the dynamics of the processes occurring on these surfaces. Clearly organic photochemistry offers a wide-open and challenging area for theoreticians.

Since this text was written only as a study aid, no attempt was made to assign priorities to original authors and to give literature references. It appears appropriate, however, to provide a few leading references for those interested in additional detail. These are mostly to review articles: simple models of organic photochemical processes[1], the role of correlation diagrams and of barriers and minima in excited state hypersurfaces[2], the spectroscopic excited states of polyenes[3] and of cyclic π-systems[4], the excited states at biradicaloid geometries[5], and the status of quantum chemical computations for organic photochemical paths[6].

7. REFERENCES

(1) J. Michl, Topics Curr. Chem., 46, 1 (1974).
(2) J. Michl, Mol. Photochem., 4, 243, 257, 287 (1972), and in Chemical Reactivity and Reaction Paths, G. Klopman, Ed., Wiley, New York, 1974, p. 301 (Chapter 8); W. Th. A. M. van der Lugt and L. J. Oosterhoff, J. Am. Chem. Soc., 91, 6042 (1969); R. B. Woodward and R. Hoffmann, Angew. Chem. Intern. Ed. Engl., 8, 781 (1969); W. G. Dauben, L. Salem, and N. J. Turro, Acc. Chem. Res., 8, 41 (1975).
(3) B. S. Hudson and B. E. Kohler, Ann. Rev. Phys. Chem., 25, 437 (1974).
(4) J. R. Platt, J. Chem. Phys., 17, 484 (1949); W. Moffitt, J. Chem. Phys., 22, 320, 1820 (1954); M. Gouterman, J. Mol. Spectrosc., 6, 138 (1961); J. Michl, J. Am. Chem. Soc., in press.
(5) J. Michl, Photochem. Photobiol., 25, 141 (1977).
(6) J. Michl, in Semiempirical Methods of Electronic Structure Calculation. Part B: Applications (Vol. 8 of Modern Theoretical Chemistry), G. A. Segal, Ed., Plenum Press, New York, 1977, p. 99 (Chapter 3).

8. ACKNOWLEDGEMENTS

Permission to reproduce Figures 1 and 2 from Topics in Current Chemistry, 46, 1 (1974), Figure 3 from J. Am. Chem. Soc. (in press), and Figure 4 from Photochem. and Photobiol., 25, 141 (1977) is gratefully acknowledged. We are grateful to the donors of the Petroleum Research Fund, administered by the American Chemical Society, and to the National Science Foundation for support of our work in photochemistry.

EXCITONS IN SOLIDS

T. C. Collins

Directorate of Physics, Office of Scientific Research,
Bolling AFB, Washington D.C. 20332, USA

1. INTRODUCTION

The lectures here are concerned with excitations to the N-body system instead of the N+1-body system which is the normal result of an energy band calculation. Thus one has to investigate the two-body Green's function formation and not stop with the single-particle function. In the first two lectures the Green's function formalism is used to construct a set of Bethe-Salpeter equations from which the excitation energies and corresponding Bethe-Salpeter amplitudes are deduced [1]. The third part will describe one of the excited states that can be obtained using a laser beam on a semiconductor, namely the quantum mechanical fluid composed entirely of electrons and holes [2]. The exciton mechanism of superconductivity [3] and experimental evidence [4] are the subjects of the last lecture. This last part is still under study and the lecture will point out the current situation.

2. GENERAL STRUCTURE OF EXCITATIONS IN MANY-BODY SYSTEMS

There are a number of methods now being used to investigate the excited states of many-body systems as can be seen by the other lectures at the Study Institute. The method used here is to construct the particle hole (PH) polarization propagator which is similar to the quantities studied by Čižek and Paldus [5]. The poles of the PH polarization propagator correspond to excitations of the N-particle system. The propagator describes the density response of the system to an external potential and thus may be used to construct the dielectric function. One not only gets information about a system's exciton structure, but also about such quantities as collective modes and optical

Cleanthes A. Nicolaides and Donald R. Beck (eds.), Excited States in Quantum Chemistry, 437–456.

absorption structure.

In this part the Green's function formalism is used to construct a set of Bethe-Salpeter equations from which the excitation energies and corresponding Bethe-Salpeter amplitudes are deduced. From these the PH polarization propagator is constructed. The various terms of the Green's function expansion can easily be classified in such categories as hole-renormalization effects, particle-renormalization effects, and particle-hole interactions. Systematic analyses of the contributions to these categories are then possible. For the examples of atoms studied it is found that the correct description of the hole requires significant many-body corrections, but that the particle-hole interaction is unresponsive to the detailed structure of the particle or the hole and that the particle renormalization effects are small.

The polarization propagator describes the response of the system to a perturbation of the form

$$\mathcal{H}^{ex}(t) = \int \hat{\Psi}^{+}(\vec{x})\, V^{ex}(\vec{x},t)\, \hat{\Psi}(\vec{x})\, d\vec{x} \quad , \tag{2.1}$$

where $V^{ex}(\vec{x},t)$ is an external perturbing potential and $\hat{\Psi}(\vec{x})$ is a field operator of the form

$$\hat{\Psi}(\vec{x},t=0) = \sum_n u_n(\vec{x})\, \hat{c}_n(t=0) \quad . \tag{2.2}$$

Here $\{u_n(\vec{x})\}$ is a complete set of orthonormal spin orbitals, and $\{c_n(t)\}$ are Heisenberg operators obeying Fermi statistics.

The PH polarization propagator $\Pi(x,x')$ is given by

$$i\Pi(x,x') = \langle \Psi_0 | T[\hat{\Psi}^{+}(x)\hat{\Psi}(x)\hat{\Psi}^{+}(x')\hat{\Psi}(x')] | \Psi_0 \rangle \; , \qquad x = \vec{x},t \; , \tag{2.3}$$

where $|\Psi_0\rangle$ is the exact Heisenberg ground state of the system, and T is the time-ordering operator. Expanding in the complete, orthonormal set $\{u_n\}$,

$$i\Pi(x,x') = \sum_{\alpha\beta\lambda\mu} u^{*}_{\mu}(\vec{x})\, u_{\lambda}(\vec{x})\, u^{*}_{\alpha}(\vec{x}')\, u_{\beta}(\vec{x}')\; i\Pi_{\lambda\mu;\alpha\beta}(t-t') \quad . \tag{2.4}$$

The relation between the PH polarization propagator and the inverse dielectric functions is

$$\epsilon^{-1}(x,x') = \delta^{4}(x-x') + \int dx''\, V(x-x'')\, \Pi(x'',x') \quad , \tag{2.5a}$$

$$\epsilon^{-1}_{\lambda\mu;\alpha\beta}(t-t') = \delta(t-t')\,\delta_{\lambda\alpha}\,\delta_{\mu\beta} + \sum_{\gamma\delta} V_{\mu\lambda;\gamma\delta}\, \Pi_{\delta\gamma;\alpha\beta}(t-t') \quad . \tag{2.5b}$$

The frequency transform of $\Pi_{\lambda\mu;\alpha\beta}(t-t')$ is defined as:

$$\Pi_{\lambda\mu;\alpha\beta}(t-t') = \frac{1}{2\pi}\int d\omega\, \Pi_{\lambda\mu;\alpha\beta}(\omega)\, e^{-i\omega(t-t')} \quad , \tag{2.6}$$

and using the Lehmann representation $\Pi_{\lambda\mu;\alpha\beta}(\omega)$ becomes

$$\Pi_{\lambda\mu;\alpha\beta}(\omega)=\sum_n\left[\frac{\langle\Psi_0|\hat{c}_\mu^+\hat{c}_\lambda|\Psi_n\rangle\langle\Psi_n|\hat{c}_\alpha^+\hat{c}_\beta|\Psi_0\rangle}{\omega-(E_n-E_0)+i\eta}-\frac{\langle\Psi_0|\hat{c}_\alpha^+\hat{c}_\beta|\Psi_n\rangle\langle\Psi_n|\hat{c}_\mu^+\hat{c}_\lambda|\Psi_0\rangle}{\omega+(E_n-E_0)-i\eta}\right],\quad \eta=0^+, \tag{2.7}$$

where the intermediate states $|\Psi_n\rangle$ refer to excited states of the N-particle system. The poles of $\Pi_{\lambda\mu;\alpha\beta}$ give the excited state energies of the system which can be reached by a density perturbation.

Using the interaction representation the structure of $\Pi_{\lambda\mu;\alpha\beta}$ is depicted schematically as:

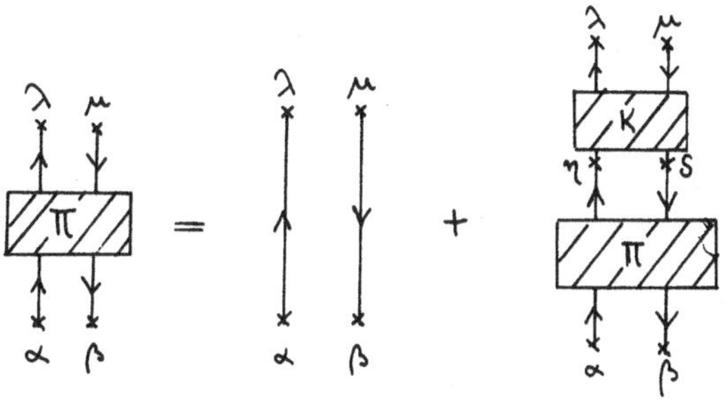

The directed lines are single-particle propagators which in general have self-energy renormalizations. $\Pi_{\lambda\mu;\alpha\beta}(\omega)$ also contains disjoint graphs, however, and these terms are independent of t-t' and contribute only at $\omega=0$. Since the main interest is for $\omega\neq 0$, these terms will be omitted from further discussion. Using the above graphs, the Bethe-Salpeter equations are:

$$\Pi_{\lambda\mu;\alpha\beta}(\omega)=\Pi^0_{\lambda\mu;\alpha\beta}(\omega)+\sum_{\varrho\nu,\eta\delta}\int\frac{d\omega_1 d\omega_2}{(2\pi)^2}\Pi^0_{\lambda\mu;\varrho\nu}(\omega;\omega_1)K_{\varrho\nu;\eta\delta}(\omega;\omega_1-\omega_2)\Pi_{\eta\delta;\alpha\beta}(\omega;\omega_2) \tag{2.8}$$

where

$$\Pi^0_{\lambda\mu;\alpha\beta}(\omega;\omega_1)\equiv -i\,G_{\lambda\alpha}(\omega_1)\,G_{\beta\mu}(\omega_1+\omega) \tag{2.9}$$

and

$$\Pi_{\lambda\mu;\alpha\beta}(\omega)=\int\frac{d\omega_1}{2\pi}\,\Pi_{\lambda\mu;\alpha\beta}(\omega;\omega_1)\ . \tag{2.10}$$

Note that $G_{\lambda\delta}$ in Eq. (2.9) refers to the full one-particle Green's function. Eq. (2.8) is an integral equation, but if the kernel $\underline{K}(\omega;\omega_1-\omega_2)$ depended only on ω, the Bethe-Salpeter equation would reduce to a Dyson's equation:

$$\Pi_{\lambda\mu;\alpha\beta}(\omega)=\Pi^0_{\lambda\mu;\alpha\beta}(\omega)+\sum_{\varrho\nu,\eta\delta}\Pi^0_{\lambda\mu;\varrho\nu}(\omega)K_{\varrho\nu;\eta\delta}(\omega)\,\Pi_{\eta\delta;\alpha\beta}(\omega)\,. \tag{2.11}$$

It is this equation which will be investigated.

In matrix notation

$$\underline{\Pi}(\omega) = \underline{\Pi}^{o}(\omega) + \underline{\Pi}^{o}(\omega)\,\underline{K}(\omega)\,\underline{\Pi}(\omega) \quad . \tag{2.12}$$

This equation is factorable, and the inverse is

$$\underline{\Pi}^{-1}(\omega) = \underline{\Pi}^{o^{-1}}(\omega) - \underline{K}(\omega) \quad . \tag{2.13}$$

The matrix elements of $\underline{\Pi}(\omega)$ have a pole at $\omega = E_n - E_o$ unless the numerators in Eq. (2.7) vanish; however, the structure of $\underline{\Pi}(\omega)$ in the vicinity of the pole is analytic, and for real ω,

$$\underline{\Pi}^{+}(\omega) = \underline{\Pi}(\omega) \quad . \tag{2.14}$$

Thus, for real ω, $\Pi(\omega)$ can be diagonalized with a unitary transformation

$$\underline{U}(\omega)\,\underline{\Pi}(\omega)\,\underline{U}^{-1}(\omega) = \underline{\Pi}^{D}(\omega) \quad . \tag{2.15}$$

From the inverse

$$\underline{U}(\omega)\,\underline{\Pi}^{-1}(\omega)\,\underline{U}^{-1}(\omega) = \underline{\Pi}^{D-1}(\omega) \quad , \tag{2.16}$$

one sees that $\underline{U}(\omega)$ also diagonalizes $\underline{\Pi}^{-1}(\omega)$. Since some of the diagonal matrix elements of $\underline{\Pi}^{D}(\omega)$ have poles at the exact excitation energies of the system, the corresponding diagonal matrix elements of $\underline{\Pi}^{D-1}(\omega)$ have zeros at the same points. Thus, the zero eigenvalues of $\underline{\Pi}^{-1}(\omega)$ correspond to the collective energy levels of the system, and one needs to solve the frequency-dependent eigenvalue problem

$$\underline{\Pi}^{-1}(\omega)\,\vec{C}(\omega) = \Lambda(\omega)\,\vec{C}(\omega) \quad , \tag{2.17}$$

with

$$\Lambda(\omega) = 0 \quad .$$

Combining Equ. (2.17) with Eq. (2.13) yields

$$\left[\underline{\Pi}^{o^{-1}}(\omega) - \underline{K}(\omega)\right]\vec{C}(\omega) = 0 \quad , \tag{2.18}$$

from which the excitation energies, ω, and the corresponding eigenvectors $\vec{C}(\omega)$ can be obtained.

To proceed further, it is necessary to study the structure of $\underline{K}(\omega)$. The first-order expression for $\underline{K}(\omega)$ is

$$K^{(1)}_{\varrho\nu;\eta\sigma} \equiv V_{\sigma\eta;\varrho\nu} - V_{\varrho\eta;\sigma\nu} \quad , \tag{2.19}$$

where

$$V_{\sigma\eta,\varrho\nu} = \int d\vec{r}_1 d\vec{r}_2\; u^{*}_{\sigma}(\vec{r}_1)\,u^{*}_{\varrho}(\vec{r}_2)\,\frac{1}{r_{12}}\,u_{\nu}(\vec{r}_2)\,u_{\eta}(\vec{r}_1) \quad . \tag{2.20}$$

Note that $K^{(1)}$ is independent of frequency so that Eq. (2.11) is correct to first-order in the electron-electron interaction. The Green's function lines may be written as

$$G_{\lambda\alpha}(\omega) = \sum_n \frac{A_\lambda^n(\omega)\, A_\alpha^{n*}(\omega)}{\omega - \omega_n(\omega)} \quad . \tag{2.21a}$$

The usual pole structure is implied with ω_n referring to the ionization energy of the N+1 body system. The $A_\lambda^n(\omega)$ is a one-particle amplitude between the N- and N+1 body systems. For free particle, Eq. (2.21a) becomes

$$G^{\circ}_{\lambda\alpha}(\omega) = \delta_{\lambda\alpha}\left[\frac{\Theta(\alpha-F)}{\omega-\omega^{\circ}_{\alpha}+i\eta} + \frac{\Theta(F-\alpha)}{\omega-\omega^{\circ}_{\alpha}-i\eta}\right] \tag{2.21b}$$

Here F refers to the Fermi level and ω°_{α} to the eigenvalue of the free-particle Hamiltonian.

To get $\underline{\Pi}^{\circ}(\omega)$, use Eqs. (2.9), (2.10) and (2.21a). After performing the frequency integration, $\Pi^{\circ}(\omega)$ becomes

$$\Pi^{\circ}_{\lambda\mu;\alpha\beta}(\omega) = \sum_{m,n}\left[\frac{g_m A_\beta^m(\omega_m) A_\mu^{m*}(\omega_m) A_\lambda^n(\omega_m-\omega) A_\alpha^{n*}(\omega_m-\omega)}{\omega-[\omega_m-\omega_n(\omega_m-\omega)]} - \frac{g_m A_\beta^n(\omega_m+\omega) A_\mu^{n*}(\omega_m+\omega) A_\lambda^m(\omega_m) A_\alpha^{m*}(\omega_m)}{\omega+[\omega_m-\omega_n(\omega_m+\omega)]}\right] \tag{2.22}$$

Here and throughout this section, the index m refers to ionizations of the N+1 particle system, while n refers to ionizations of the N-1 system; all other indices are general. The factor g_m is

$$g_m = \left[1-\frac{\partial\omega_m(\omega)}{\partial\omega}\right]_{\omega=\omega_m} \quad . \tag{2.23}$$

If the free-particle Green's function in the expression for $\underline{\Pi}^{\circ}(\omega)$ is used one obtains:

$$\Pi^{\circ\circ}_{\lambda\mu;\alpha\beta}(\omega) = \delta_{\lambda\alpha}\delta_{\beta\mu}\left[\frac{\Theta(\alpha-F)\,\Theta(F-\beta)}{\omega-(\omega^{\circ}_{\alpha}-\omega^{\circ}_{\beta})+i\eta} - \frac{\Theta(F-\alpha)\,\Theta(\beta-F)}{\omega+(\omega^{\circ}_{\beta}-\omega^{\circ}_{\alpha})-i\eta}\right] \quad . \tag{2.24}$$

Inserting $\underline{K}^{(1)}(\omega)$ into (2.12) for $\underline{K}(\omega)$ yields the structure depicted below.

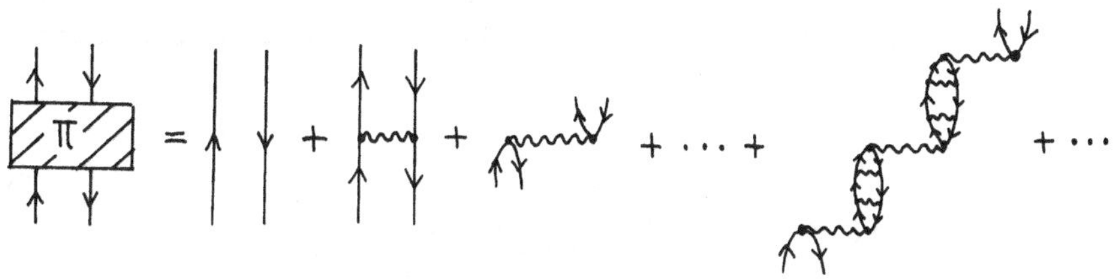

Using $\underline{\Pi}^{o,o}(\omega)$ instead of $\underline{\Pi}^{o}(\omega)$ in Eq. (2.12), and $\underline{K}^{(1)}(\omega)$ in place of $\underline{K}(\omega)$, one obtains the random-phase approximation (RPA) equations for $\Pi(\omega)$. In the RPA, Eq. (2.18) is

$$(\omega-\omega^{o}_{m\ell})C^{n}_{m\ell}-\sum_{m'\ell'}\left[(V_{m\ell,\ell'm'}-V_{mm',\ell\ell'})C^{n}_{m'\ell'}+(V_{m\ell,m'\ell'}-V_{m\ell',m'\ell})C^{n}_{\ell'm'}\right]=0, \quad (2.25a)$$

$$-(\omega+\omega^{o}_{m\ell})C^{n}_{\ell m}-\sum_{m'\ell'}\left[(V_{\ell m,\ell'm'}-V_{\ell m',\ell' m})C^{n}_{m'\ell'}+(V_{\ell m,m'\ell'}-V_{\ell\ell',m'm})C^{n}_{\ell'm'}\right]=0, \quad (2.25b)$$

$$\omega^{o}_{m\ell}\equiv\omega^{o}_{m}-\omega^{o}_{\ell} \quad .$$

In the RPA, one may make the identification

$$C^{n}_{\beta\alpha}=\langle\Psi_{o}|\hat{c}^{+}_{\alpha}\hat{c}_{\beta}|\Psi_{o}\rangle \quad . \quad (2.26)$$

Using this equation, the correct spin structure for excited state $|\Psi_{n}\rangle$ for closed-shell systems becomes:

$$C^{n}_{\beta\alpha;S}=\frac{1}{\sqrt{2}}\left(C^{n}_{\beta\uparrow\alpha\downarrow}-C^{n}_{\beta\downarrow\alpha\uparrow}\right)\ ;\ C^{n}_{\beta\alpha;T}=\begin{cases}C^{n}_{\beta\uparrow\alpha\uparrow}\\ C^{n}_{\beta\downarrow\alpha\downarrow}\\ \frac{1}{\sqrt{2}}\left(C^{n}_{\beta\uparrow\alpha\downarrow}+C^{n}_{\beta\downarrow\alpha\uparrow}\right)\end{cases} \quad (2.27)$$

The indices S and T refer to singlet and triplet, and the spin is indicated explicitly. Rearranging Eqs. (2.25) with the use of Eqs. (2.27) and assuming the basis set is real, one obtains

$$\begin{aligned}\sum_{m'\ell'}\left(\Delta E_{m\ell,m'\ell'}C^{n}_{m'\ell'}+M_{m\ell,m'\ell'}C^{n}_{\ell'm'}\right)&=\omega C^{n}_{m\ell}\\ \sum_{m'\ell'}\left(\Delta E_{m\ell,m'\ell'}C^{n}_{\ell'm'}+M_{m\ell,m'\ell'}C^{n}_{m'\ell'}\right)&=-\omega C^{n}_{\ell m}\quad ,\end{aligned} \quad (2.28)$$

where

$$\Delta E_{m\ell,m'\ell'}=\delta_{mm'}\delta_{\ell\ell'}(\omega^{o}_{m}-\omega^{o}_{\ell})+2\gamma V_{m\ell,m'\ell'}-V_{mm',\ell\ell'} \quad ,$$

$$M_{m\ell,m'\ell'}=2\gamma V_{m\ell,m'\ell'}-V_{m\ell',m'\ell} \quad ,$$

$$\gamma=1\ (\text{singlet})\ ,\quad \gamma=0\ (\text{triplet})\ .$$

In addition, using restricted Hartree-Fock orbitals, $V_{m\ell,m'\ell'}$ is the matrix element over spatial orbitals assumed to be independent of spin. The matrix $\underline{M}$ describes multipair excitations. If $\underline{M}=0$, Eqs. (2.28) are separable. Then their diagonal elements yield

$$\begin{aligned}\omega_{T}&\approx\omega^{o}_{m\ell}-V_{mm,\ell\ell}\quad ,\\ \omega_{S}&\approx\omega^{o}_{m\ell}-V_{mm,\ell\ell}+2V_{m\ell,m\ell}\quad .\end{aligned} \quad (2.29)$$

These matrix elements are equivalent to the first-order ÔÂÔ constructions [6]. In fact, the ÔÂÔ equations derived from the Â operator used in Ref. [6] are equivalent to Eqs. (2.28) if $\underline{M}$=0 and further set l=l' in ΔE. The open-shell result is more complicated than the above owing to the fact that the unperturbed ground state is multideterminantal. However, a construction can be made which gives equations similar to Eqs. (2.28) with an approximate excitation spin structure, except that in this case the different spin excitations are coupled.

The structure of Eq. (2.18) would be similar to that of Eqs. (2.28) if $\underline{\Pi}^o(\omega)$ in Eq. (2.12) is retained (instead of using $\underline{\Pi}^{oo}(\omega)$) and assume $A^n_\alpha(\omega)$ was independent of ω. This is approximately true in most cases. The major difference would be that ω^o_n would be replaced by ω_n in Eqs. (2.28). Since ω_n differs from ω^o_n by the renormalization effects coming from the particle or hole self-energies, an investigation of the self-energies would yield these effects. Also, it is clear from the structure of Eqs. (2.28) and (2.29) that K(ω) contains the particle-hole interactions. Thus one has a clear separation between structures which contribute to the relaxation and correlation of the hole or particle and those which contribute to particle-hole interactions.

To outline the above, make the replacements:

$$A^n_\lambda(\omega) = A^n_\lambda \quad , \quad g_m = 1 \quad , \quad \omega_m(\omega) = \omega_m \quad .$$

Eq. (2.22) becomes

$$\Pi^o_{\lambda\mu;\alpha\beta}(\omega) = \sum_{m,l}\left[\frac{A^m_\beta A^{m*}_\mu A^l_\lambda A^{l*}_\alpha}{\omega-(\omega_m-\omega_l)+i\eta} - \frac{A^l_\beta A^{l*}_\mu A^m_\lambda A^{m*}_\alpha}{\omega+(\omega_m-\omega_l)-i\eta}\right] \quad . \tag{2.30}$$

The poles of $\underline{\Pi}^o(\omega)$ occur at

$$\omega = \pm\,(\omega_m - \omega_l) \quad .$$

If the poles of $\underline{\Pi}(\omega)$ are not far from this value, then the above replacement is valid. Using the following definition

$$\Pi^{o'}_{ml;m'l'} \equiv \delta_{m,m'}\,\delta_{l,l'} \sum_{\lambda\mu\alpha\beta} (A^m_\beta)^{-1} (A^{m*}_\mu)^{-1}\, \Pi^o_{\lambda\mu;\alpha\beta}\, (A^{l*}_\alpha)^{-1} (A^l_\lambda)^{-1} \tag{2.31}$$

Eq. (2.12) transforms to be

$$\underline{\Pi}'(\omega) = \underline{\Pi}^{o'}(\omega) + \underline{\Pi}^{o'}(\omega)\,\underline{K}'(\omega)\,\underline{\Pi}'(\omega) \quad . \tag{2.32}$$

Here, $\underline{K}'(\omega)$ has the same structure as $\underline{K}(\omega)$ except that it contains

$$V'_{no;pq} = \sum_{\sigma\eta,\rho\nu} A^{n*}_\sigma A^o_\eta V_{\sigma\eta,\rho\nu} A^{p*}_\rho A^q_\nu \quad . \tag{2.33}$$

Thus one can solve equations for the renormalized excitations of the same form as the RPA equations

$$\left[\underline{\Pi}^{o'}(\omega)^{-1} - \underline{K}^{(1)'}\right] \vec{C}(\omega) = 0 \;,$$

except that in Eqs. (2.28), ω_n would replace ω_n^o and $V'_{no,pq}$ would replace $V_{no,pq}$.

Finally, the construction of $\Pi(\omega)$ using the Bethe-Salpeter amplitude is

$$\Pi_{\lambda\mu,\alpha\beta}(\omega) = \sum_n \left[\frac{C^n_{\lambda\mu}(\omega)\, C^{n*}_{\alpha\beta}(\omega)}{\omega - \omega^{ex}_n(\omega) + i\eta} - \frac{C^n_{\beta\alpha}(\omega)\, C^{n*}_{\mu\lambda}(\omega)}{\omega + \omega^{ex}_m(\omega) - i\eta} \right] . \tag{2.35}$$

To perform these calculations one needs the self-energy structure of the one-particle Green's function. The general structure of the one- particle Green's function presented in Eq. (2.21a) is given by the Dyson's equation

$$G_{\lambda\alpha}(\omega) = G^o_{\lambda\alpha}(\omega) + \sum_{\eta\sigma} G^o_{\lambda\eta}(\omega) \Sigma_{\eta\sigma}(\omega) G_{\sigma\alpha}(\omega) . \tag{2.36}$$

The self-energy $\Sigma_{\eta\sigma}(\omega)$ contains all many-body corrections to the one-particle Green's function. Similar to Eq. (2.18) one seeks solutions to the equation

$$\sum_\alpha \left[G^{o-1}_{\lambda\alpha}(\omega) - \Sigma_{\lambda\alpha}(\omega) \right] A^n_\alpha(\omega) = 0 \;, \tag{2.37}$$

which was used to construct Eq. (2.21a). The general structure of $G_{\lambda\alpha}(\omega)$ in the Lehmann representation is

$$G_{\lambda\alpha}(\omega) = \sum_n \left[\frac{\langle \Psi^N_o | \hat{C}_\lambda | \Psi^{N+1}_n \rangle \langle \Psi^{N+1}_n | \hat{C}^+_\alpha | \Psi^N_o \rangle}{\omega - (E^{N+1}_n - E^N_o) + i\eta} + \frac{\langle \Psi^N_o | \hat{C}^+_\alpha | \Psi^{N-1}_n \rangle \langle \Psi^{N-1}_n | \hat{C}_\lambda | \Psi^N_o \rangle}{\omega - (E^N_o - E^{N-1}_m) - i\eta} \right] . \tag{2.38}$$

Thus the $\{\omega_n\}$ refer to ionizations of the (N+1)-particle systems. One interpretation of ω_1(or ω_m) is to think of it as the energy necessary to create a hole among the occupied orbitals (or to fill a hole in the virtuals), and $\Sigma_{\lambda\alpha}(\omega)$ describes the relaxation and rearrangement of electrons around the hole.

The structure of $\Sigma_{\lambda\alpha}(\omega)$ may be determined using many-body perturbation theory. If one chooses SCF-RHF orbitals, all diagrams contributing to canonical Hartree-Fock are omitted. The second-order expression for $\Sigma_{\lambda\alpha}(\omega)$ is

$$\Sigma^{(2)}_{\lambda\alpha}(\omega) = \sum_{m\ell m'} \frac{V_{\lambda m', m\ell}\left(2V_{m\ell,\alpha m'} - V_{m'\ell,\alpha m}\right)}{\omega + \omega^o_\ell - \omega^o_{m'} + i\eta - \omega^o_m} + \sum_{m\ell\ell'} \frac{V_{\lambda\ell', m\ell}\left(2V_{m\ell,\alpha\ell'} - V_{m\ell',\alpha\ell}\right)}{\omega + \omega^o_m - \omega^o_\ell - \omega^o_{\ell'} - i\eta} . \tag{2.39}$$

A partial summation of third- and higher-order diagrams for $\Sigma_{\lambda\alpha}(\omega)$ is possible by the following procedure. The structure of these terms is given by the product of the appropriate term in Eq. (2.39) times an interaction matrix element divided by the same denominator and with appropriate sign. These elements form a geometric progression which may be summed to all orders by shifting the denominator in Eq. (2.39) to get:

$$\Sigma^{D}_{\lambda\alpha}(\omega)=\sum_{m\ell m'} \frac{V_{\lambda m',m\ell}\left(2V_{m\ell,\alpha m'}-V_{m'\ell,\alpha m}\right)}{\omega+\omega^0_\ell-\omega^0_m-\omega^0_{m'}+V_{mm,\ell\ell}+V_{m'm',\ell\ell}-V_{mm,m'm'}+i\eta} \tag{2.40}$$

$$+\sum_{m\ell\ell'} \frac{V_{\lambda\ell',m\ell}\left(2V_{m\ell,\alpha\ell'}-V_{m\ell',\alpha\ell}\right)}{\omega+\omega^0_m-\omega^0_\ell-\omega^0_{\ell'}-V_{mm,\ell\ell}-V_{mm,\ell'\ell'}+V_{\ell\ell,\ell'\ell'}-i\eta} \; .$$

The higher-order diagonal exchange terms are not included in $\Sigma^{D}_{\lambda\alpha}(\omega)$. The nondiagonal terms omitted from $\Sigma^{D}_{\lambda\alpha}(\omega)$ tend to converge rapidly, and the major contributions from diagrams of this type are included in Eq. (2.40).

To investigate further the construction of $\Sigma_{\lambda\alpha}(\omega)$, note that $\Sigma_{\lambda\alpha}$ involve ionizations of the N+2 body systems as well as excitations of the N-body system. Thus, the correct first-order structure of these excitations is

$$\text{N+2:}\quad \begin{cases} \omega^0_m+\omega^0_{m'}+V_{mm,m'm'}\pm V_{mm',mm'} \; , & m\neq m' \\ 2\omega^0_m+V_{mm,mm} \; , & m=m' \end{cases}$$

$$\text{N-2:}\quad \begin{cases} \omega_\ell+\omega^0_\ell-V_{\ell\ell,\ell'\ell'}\mp V_{\ell\ell',\ell\ell'} \; , & \ell\neq\ell' \\ 2\omega^0_\ell-V_{\ell\ell,\ell\ell} \; , & \ell=\ell' \end{cases}$$

$$\text{N:}\quad \omega^0_m-\omega^0_\ell-V_{mm,\ell\ell}+\begin{cases} 2V_{m\ell,m\ell} \\ 0 \end{cases} \; ,$$

for spin-averaged orbitals, where the upper term refers to the singlet and the lower term to the triplet. The partial summation in Eq. (2.40) may be modified to include averaged exchange interactions from the above excitations by shifting the denominators of Eq. (2.40) to reach

$$\Sigma^{DE}_{\lambda\alpha}(\omega):\; [\omega+\omega^0_\ell-\omega^0_m-\omega^0_{m'}+V_{mm,\ell\ell}-V_{m\ell,m\ell}+V_{m'm',\ell\ell}-V_{m'\ell,m'\ell'}- -V_{mm,m'm'}+i\eta] \, , \tag{2.41}$$

$$[\omega+\omega^0_m-\omega^0_\ell-\omega^0_{\ell'}-V_{mm,\ell\ell}+V_{m\ell,m\ell}-V_{mm,\ell'\ell'}+V_{m\ell',m\ell'}+V_{\ell\ell,\ell'\ell'}-i\eta] \, .$$

Acutally atomic calculations show little difference between the averaging used in Eq. (2.41) and separation into singlet and triplet excitations.

Another set of higher-order diagrams which can be thought as contributing to the correlated charge density in an HF-like model are included in Σ expressed as

$$\Sigma^{(3)}_{\lambda\alpha}(\omega) = \sum_{\ell''\neq\mu}\Big[\big(2V_{\ell\alpha,\mu\ell''} - V_{\lambda\mu,\alpha\ell''}\big)\Sigma^{(2)}_{\mu\ell''}(\omega^{o}_{\ell''}) + \big(2V_{\lambda\alpha,\ell''\mu} - V_{\lambda\ell'',\alpha\mu}\big)\Sigma^{(2)}_{\mu\ell''}(\omega^{o}_{\ell''})\Big]$$

$$-\sum_{\mu}\big(2V_{\lambda\alpha,\mu\mu} - V_{\lambda\mu,\alpha\mu}\big)\frac{\partial\Sigma^{(2)}_{\mu\mu}(\omega)}{\partial\omega}\Bigg|_{\omega=\omega^{o}_{\mu}} \quad (2.42)$$

$$+\sum_{\mu\nu m\ell\ell'}\frac{\big(2V_{\lambda\alpha,\mu\nu} - V_{\lambda\mu,\alpha\nu}\big)V_{\mu\ell',m\ell}\big(2V_{m\ell,\nu\ell'} - V_{m\ell',\nu\ell}\big)}{\big(\omega^{o}_{\ell}+\omega^{o}_{\ell'}-\omega^{o}_{m}-\omega^{o}_{\mu}\big)\big(\omega^{o}_{\ell}+\omega^{o}_{\ell'}-\omega^{o}_{m}-\omega^{o}_{\nu}\big)} .$$

In order to study the effects of screening via the RPA dielectric function in the self-energy, use Eq. (2.35) to construct $\Sigma^{SC}_{\lambda\alpha}(\omega)$.

$$\Sigma^{SC}_{\lambda\alpha}(\omega) = \Sigma^{DE}_{\lambda\alpha}(\omega) + \Sigma^{(3)}_{\lambda\alpha}(\omega) - \Sigma^{PHD}_{\lambda\alpha}(\omega)$$

$$+\sum_{n}\Bigg[\sum_{m\ell,m'\ell''}\frac{V_{\lambda\ell',m\ell}\big(C^{n}_{m\ell}+C^{n}_{\ell m}\big)\big(C^{n}_{m'\ell''}+C^{n}_{\ell''m'}\big)V_{m'\ell'',\alpha\ell'}}{\omega-\omega^{o}_{\ell'}+\omega^{ex}_{n}-i\eta} \quad (2.43)$$

$$+\sum_{m\ell,m'\ell'm''}\frac{V_{\lambda m',m\ell}\big(C^{n}_{m\ell}+C^{n}_{\ell m}\big)\big(C^{n}_{m''\ell'}+C^{n}_{\ell'm''}\big)V_{m''\ell',\alpha m'}}{\omega-\omega^{o}_{m'}-\omega^{ex}_{n}+i\eta}\Bigg] ,$$

where $\Sigma^{PHD}_{\lambda\alpha}(\omega)$ has the same structure as $\Sigma^{(2)}_{\lambda\alpha}(\omega)$ except the denominators are shifted to be

$$\Sigma^{PHD}_{\lambda\alpha}(\omega):\ \big(\omega-\omega^{o}_{m'}-\omega^{o}_{m}+\omega^{o}_{\ell}+V_{mm,\ell\ell}-V_{m\ell,m\ell}+V_{m'm',\ell\ell}-V_{m'\ell,m'\ell}+i\eta\big) ,$$

$$\big(\omega+\omega^{o}_{m}-\omega^{o}_{\ell}-\omega^{o}_{\ell'}-V_{mm,\ell\ell}+V_{m\ell,m\ell}-V_{mm,\ell'\ell'}+V_{m\ell,m\ell'}-i\eta\big) .$$

Thus $\Sigma^{SC}_{\lambda\alpha}(\omega)$ contains contributions from the screened exchange to all orders (both diagonal and off-diagonal particle-hole interaction terms), but the cross terms are between diagonal particle-particle and hole-hole ladders and diagonal particle-hole ladders only.

3. ELECTRON-HOLE LIQUIDS

It has been discovered that the carrriers of electric charge inside a crystal can exist in a state that is a liquid in a modern sense. This new liquid has many of the properties associated with an ordinary fluid such as water. The charged particles that form the new fluid can exist as a vapor; when the relative humidity ofthe vapor becomes high enough, the particles condense. Like water, the new liquid evaporates and eventually disappears when it is heated or when the density of the particles in the surrounding gas is reduced. It will not form at all at temperatures higher than a critical value. It appears as a cloud of droplets, which scatters light as a fog of water droplets does.

Although the new liquid resembles water in many ways, it is quite unusual in others. To begin with, it only exists inside a solid semiconductor and it cannot be extracted from that environment. Instead of the atoms or molecules of an ordinary liquid it consists of electrons and holes. The electrons and holes continually annihilate each other, and as a result the liquid is an inherently unstable substance. Finally, the new liquid is of particular interest because it is essentially a quantum-mechanical fluid. For these reasons the new liquid offers a unique testing ground for some of the fundamental principles of physics.

To begin the discussion of the electron-hole liquid, consider the simplest case of isotropic-nondegenerate energy bands for the electrons and holes. In the limit that the hole mass m_h equals the electron mass m_e the problem is that of a hypothetical gas of electrons and positrons in which the radiative recombination of electrons and positrons is ignored. In the limit $m_h \gg m_e$ the problem is equivalent to the metallic phase of hydrogen.

The Hamiltonian is given by

$$\mathcal{H}=-\frac{1}{2m_e}\sum_{i=1}^{N}\nabla_i^2-\frac{1}{2m_h}\sum_{j=1}^{N}\nabla_j^2+\frac{1}{2}\sum_{i\neq j}\frac{e^2}{\kappa r_{ij}^e}+\frac{1}{2}\sum_{i\neq j}\frac{e^2}{\kappa r_{ij}^h}-\sum_{i,j}\frac{e^2}{\kappa|\vec{r}_i^{\,e}-\vec{r}_j^{\,h}|}\;, \qquad (3.1)$$

where the first two terms represent the kinetic energies and the other terms represent the repulsive Coulomb interaction between like particles and the attractive Coulomb interactions between unlike particles. All of the Coulomb interactions are reduced by the static dielectric constant, κ. The unit of energy will be the binding energy of the exciton which is given by the standard hydrogenic formula $E_x=\mu/m\kappa^2$ Ry where μ is the reduced mass, $\mu^{-1}=m_e^{-1}+m_h^{-1}$. The corresponding unit of length is the Bohr radius of an exciton which is $a_x=(m\kappa/\mu)a_0$. The density of electrons n can be characterized by the dimensionless parameter r_s, the radius of a sphere whose volume is equal to n^{-1} measured in units of a_x^3.

At high densities, or small values of r_s, the electron-hole liquid will be metallic. The dominant term in the high-density limit is the kinetic energy of the degenerate electrons and holes which gives a contribution to the ground-state energy per electron,

$$E_k=\frac{3}{5}\left(\frac{k_F^2}{2m_e}+\frac{k_F^2}{2m_h}\right)=\frac{3}{5}\left(\frac{9\pi}{4}\right)^{2/3}\frac{1}{r_s^2}=\frac{2.21}{r_s^2}\;, \qquad (3.2)$$

where k_F is the Fermi momentum of the electrons. The first correction is the exchange energy which is the expectation value of the potential energy in the ground state of noninteracting electrons and holes:

$$E_{ex} = -\frac{3e^2}{2\pi K} k_F = -\frac{3}{\pi}\left(\frac{9\pi}{4}\right)^{1/3}\frac{1}{r_s} = -\frac{1.832}{r_s} \quad . \tag{3.3}$$

Note that the electrons and holes make equal contributions to the exchange energy irrespective of the mass ratio m_h/m_e. Several schemes based on the random-phase approximation (RPA) have been developed to estimate the correlation energy of the single-component electron gas in the intermediate density regime $1 \lesssim r_s \lesssim 5$. These procedures, although lacking a rigorous justification, give good agreement when compared with experiments on nearly-free-electron metals. Since the various estimate of the correlation energy of the electron gas do not differ appreciably, the Hubbard scheme for the single-component electron gas is used. The exact expression for the correlation energy is

$$E_G = -\frac{1}{2\pi n}\int_0^1 \frac{d\lambda}{\lambda}\int\frac{d\vec{k}}{(2\pi)^3}\int_0^\infty d\omega\, \mathrm{Im}[\epsilon_\lambda^{-1}(k,\omega)] - E_{ex} - 2\int\frac{d\vec{k}}{(2\pi)^3}\frac{2\pi e^2}{k^2} \quad . \tag{3.4}$$

The λ integral is over the coupling constant λe^2. In the RPA the total polarizability is given by the sum of the polarization of the electrons, Π^e, and the holes, Π^h, so that

$$\epsilon_\lambda(\vec{k},\omega) = 1 - [\Pi^e(\vec{k},\omega) + \Pi^h(\vec{k},\omega)] \quad . \tag{3.5}$$

The RPA polarizability Π_{RPA} is given by the Lindhard function

$$\Pi^e_{RPA}(\vec{k},\omega) = -\frac{4\pi e^2}{K k^2}\sum_{p,\sigma}\frac{n_{\vec{p}+\vec{k}\sigma} - n_{\vec{p}\sigma}}{\omega + [p^2 - (\vec{p}+\vec{k})^2]/2m_e + i\delta} \quad , \tag{3.6}$$

where $n_{\vec{p}\sigma}=1$, $p < p_F$, and $n_{\vec{p}\sigma}=0$ otherwise. At short wavelengths, or $k \ll k_F$ in Eq. (3.4), the RPA result for the correlation energy is seriously in error, since it treats the correlations between particles on an equal footing irrespective of their spin states. Hubbard showed that one could approximately include the diagrams that are the exchange conjugates of the RPA bubble diagrams by replacing Π_{RPA} by Π_H, where

$$\Pi_H(\vec{k},\omega) = \frac{\Pi_{RPA}(\vec{k},\omega)}{1 + f(k)\,\Pi_{RPA}(\vec{k},\omega)} \quad , \tag{3.7}$$

where $f(k)=0.5k^2/(k^2+K_F^2)$. Generalizing his arguments straightforwardly to the multicomponent system, one places Eq. (3.7) separately for each component. This complicates the integration over the coupling constant in Eq. (3.4).

In the equal-mass limit, $\Pi^e \equiv \Pi^h$ and the correlation energy can be expressed after a little algebra as

$$E_G = -\frac{1}{4\pi n}\int\frac{d\vec{k}}{(2\pi)^3}\int_0^\infty d\omega\left(\frac{2\Sigma(k,\omega)}{\Sigma'(k,\omega)}\tan^{-1}\frac{\Sigma'(k,\omega)}{1 - A'(\vec{k},\omega)} - 2\Sigma(k,\omega)\right) \quad , \tag{3.8}$$

where $\Pi = A + i\Sigma$ and $\Sigma' = [2 - k^2/2(k^2+k_F^2)]\Sigma$ with $A'/A = \Sigma'/\Sigma$. Note that the exchange contributions play a less important role than in the single-component electron gas and cancel only one-fourth of the RPA correlation energy at large k. The results are given in Fig. 1.

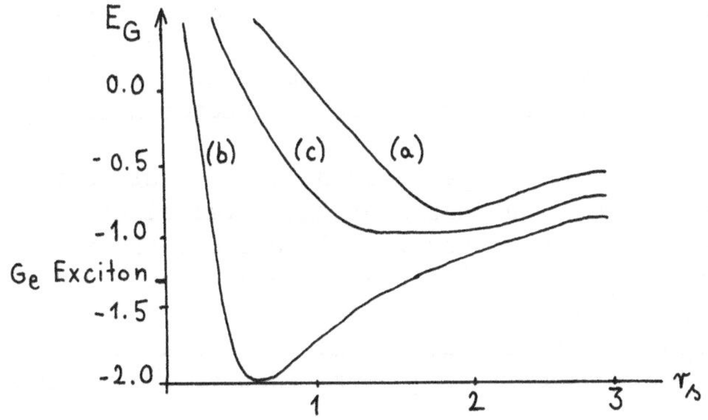

Fig. 1. Ground-state energy plotted against r_s. (a) Isotropic model with one maximum in the valence band and one minimum in the conduction band; (b) germanium and (c) germanium under a large (1,1,1) strain.

The minimum value lies substantially above the energy of free excitons so that within this approximate scheme no metallic state bound relative to free excitons (with equal masses) is found.

One major weakness of this calculation is that terms corresponding to the multiple scattering of an electron and hole have been omitted and these give rise to the exciton bound state.

4. EXCITON MECHANISM OF SUPERCONDUCTIVITY

The exciton mechanism of superconductivity is discussed with respect to a particular model, a thin metal layer on a semiconductor surface. In this model, the metal electrons at the Fermi surface tunnel into the semiconductor gap where they interact with virtual excitons, producing a net attractive interaction among the electrons in direct analogy with the phonon mechanism of superconductivity.

Two questions needing consideration are: How far are the metal electrons near the Fermi surface expected to tunnel into the semiconductor gap, and what is the effective interaction constant in the semiconductor due to exciton effects?

To investigate the first question suppose that the metal film and semiconductor are in intimate contact with no barrier separating them. This implies that there is a chemical bonding at the interface such that the tails of the wave functions of the electrons near the Fermi surface of the metal penetrate into the energy-gap region of the semiconductor. For optimum penetration, the Fermi level, E_F, of the metal should be near the center of the semiconductor gap at the interface. Band bending due to the space charge of the metal electrons in the semiconductor should be less than the order of half of the average gap. This gives some limitation on the maximum penetration of the metal electrons one can have without getting a large concentration of free carriers in the semiconductor near the interface.

To get an estimate of the penetration and calculate the band bending due to the space charge of the metal electrons in the semiconductor, a simple model will be used. Assume that the band gap E_g is small compared with the semiconductor plasma energy, so that for $\hbar\omega < E_g$, the dielectric function is large. For $\omega > \omega_g$, in the first approximation, the screening is similar to that of a metal with an electron density equal to that of the valence electrons.

For simplicity assume an isotropic energy gap in the semiconductor such that the electron energy E measured from midgap may be expressed in the form

$$E = \pm \left[\epsilon^2 + \left(\tfrac{1}{2}E_g\right)^2 \right]^{1/2} \tag{4.1}$$

in analogy with a superconductor with a gap $2\Delta = E_g$. Here ϵ is the free-electron energy, measured from the Fermi level at midgap,

$$\epsilon = \frac{\hbar^2}{2m}\left(k_z^2 + k_{\parallel}^2\right) - E_F \quad . \tag{4.2}$$

The Fermi energy may be expressed as

$$E_F = \frac{\hbar^2}{2m}\left(k_{zF}^2 + k_{\parallel}^2\right) \quad , \tag{4.3}$$

where k_{zF} is the value of k_z required to give the energy E_F for a given transverse component $k_{\parallel}$:

$$k_{zF} \equiv \left[\left(2mE_F/\hbar^2\right) - k_{\parallel}^2 \right]^{1/2} \quad . \tag{4.4}$$

For energies in the gap region, $|E| < 1/2\ E_g$, k_z becomes complex, $k_z \rightarrow k_{zF} + i\alpha$, and

$$\epsilon = i\left(\frac{\hbar^2}{m}\right) k_{zF}\,\alpha \quad , \tag{4.5}$$

where we have assumed $\alpha \ll k_{zF}$ and neglected a term of the order α^2. Solving Eq. (4.1) for α gives

$$\alpha = \frac{m\left[\left(\frac{1}{2}E_g\right)^2 - E^2\right]^{1/2}}{\hbar^2\left[k_F^2 - k_{\parallel}^2\right]^{1/2}} \quad , \tag{4.6}$$

where k_F is the magnitude of the Fermi wave vector.

The wave functions decay as $e^{-\alpha z}$ and the electron density as $e^{-2\alpha z}$. To get the average depth of penetration $D=\langle 1/2\alpha \rangle$ for electrons with energies in the gap, average over energies and over $k_{\parallel}$ values. This gives

$$D=\left\langle \frac{1}{2\alpha} \right\rangle = \left(\frac{\hbar^2}{mE_g}\right)\left(\int_{-E_g/2}^{0} [(\tfrac{1}{2}E_g)^2-E^2]^{1/2} dE\right)\left(\frac{2}{k_F^2}\int_0^{k_F} [k_F^2-k_{\parallel}^2]^{1/2} k_{\parallel}\, dk_{\parallel}\right) = \frac{\pi\hbar^2 k_F}{3mE_g} \; . \quad (4.7)$$

For $k_F=1.5\times10^8 cm^{-1}$, corresponding to an electron density of $\sim10^{23}/cm^3$ and $E_g \sim 2eV$, it is found that D is of the order of 5Å.

The voltage drop due to the space charge of the penetrating electrons in the gap is less than one might suppose because of the high dielectric constant of the semiconductor. The number of such electrons per unit area of semiconductor surface is

$$\sigma = \gamma N(0) E_g D \quad , \qquad (4.8)$$

where 2N(0) is the Fermi-level density of states for electrons of both spins in the metal, $\gamma N(0)E_g$ is the corresponding density of electrons at the interface in the energy range $-1/2\ E_g$ to 0 of the semiconductor gap, and D is the average depth of penetration.

The electric field produced by these electrons at the interface is $4\pi e\sigma/\kappa$, where κ is the dielectric constant and the voltage drop is of the order $4\pi e\sigma D/\kappa$. The change in electron energy due to this drop should be less than half the gap. Thus

$$\frac{4\pi e^2 \gamma N(0) E_g D^2}{\kappa} < \frac{1}{2} E_g \qquad (4.9)$$

or

$$8\pi e^2 \gamma N(0) D^2 < \kappa \quad .$$

For $k_F \sim 1.5\times10^8 cm^{-1}$, $N(0)\sim 3\times10^{33}\ erg^{-1} cm^{-3}$ and $D\sim5\times10^{-8}$cm, one finds $\kappa/\gamma \gtrsim 40$. For $\gamma\sim1/2$, κ_0 should be of the order of 20 or larger. Such values are typical of many narrow-gap semiconductors.

For later use, define the parameter $b\equiv \gamma D/L$, where L is the thickness of the metal film and b roughly signifies the fraction of the time the metal electrons spend in the semiconductor. Taking $L \sim 10$-15 Å, $\gamma \sim 1/2$, and $D \sim 5$Å, one gets $b \sim 1/4$-$1/6$.

Now look at the effective interaction constant in the semiconductor due to exciton effects. As it is well known from BCS theory of superconductivity, an attractive interaction between electrons near the Fermi surface in a material is necessary for a superconducting state to exist. For the metal-semiconductor system, the interaction is the combination of three contributions,

$$V(\omega) = V_{ph} + V_c + V_{ex} \quad , \tag{4.10}$$

where V_{ph} is the phonon part, V_c the Coulomb interaction, and V_{ex} the exciton part. The ω variable is an energy variable representing the energy difference between the initial and final states which occurs in the matrix elements for the scattering processes.

The Coulomb interaction is represented by the parameter μ. The Fourier transform of the screened Coulomb potential $V_c(q,\omega)$ is

$$V_c(\vec{q},\omega) = \frac{4\pi e^2}{q^2}\epsilon^{-1}(\vec{q},\omega) \approx \frac{4\pi e^2}{q^2+q_s^2} \quad \text{for } \omega \ll \omega_F \quad , \tag{4.11}$$

where $\vec{q}$, ω are the momentum and energy transfers in the electron scattering, $\epsilon^{-1}(\vec{q},\omega)$ is the inverse dielectric function for a metal of equivalent electron density, and $\vec{q}_s$ is the appropriate screening wave vector. μ is defined to be the average of $V_c(\vec{q},\omega)$ over the Fermi surface times the density of states at the Fermi surface, N(o)

$$\mu = N(0) \langle V_c \rangle \quad . \tag{4.12}$$

The average over the Fermi surface $\langle V_c \rangle$, may be expressed as an integral over q, leaving μ as a function of ω only. Choose the square well model, in which μ is a constant out to $\omega = \omega_F$:

$$\mu(\omega) = \begin{cases} \mu & \text{for } 0 \le |\omega| \le \omega_F \\ 0 & \text{otherwise} \end{cases} \tag{4.13}$$

Typical values of μ are of the order of 0.2-0.5.

The phonon interaction may be described by a parameter λ_{ph}. Adopting McMillan's definition of λ_{ph}, one has

$$\lambda_{ph} = 2\int_0^{\omega_{pm}} \frac{\alpha^2(\omega)F(\omega)}{\omega} d\omega \quad , \tag{4.14}$$

where $\alpha^2(\omega)$ is an average matrix element of the phonon interaction, $F(\omega)$ is the phonon density of states, and ω_{pm} is the maximum phonon energy. It has been assumed that the metal-semiconductor system has a single electron-phonon coupling constant which is uniform through the entire system.

Finally consider the exciton interaction. A metal electron $\vec{k}_{1\uparrow}$ is scattered to $\vec{k}_{2\uparrow}$ by exciting a semiconductor valence-band electron $\vec{k}_v$ into a state above the gap $\vec{k}_c$ and creates a virtual exciton. The paired electron $-\vec{k}_{1\downarrow}$ then absorbs the exciton and scatters into the state $-\vec{k}_{2\downarrow}$. Conservation of wave vector requires the $\vec{q}=\vec{k}_2-\vec{k}_1=\vec{k}_v-\vec{k}_c+\vec{k}$, where $\vec{k}$ is a reciprocal lattice vector. In general $\vec{q}$ may be outside of the first Brillouin zone, and thus there are several values of $\vec{k}$ and

$\vec{k}_v - \vec{k}_c$ that satisfy this condition. To estimate the exciton-electron coupling, λ_{ex}, consider the following simple model. The interaction term in the Hamiltonian for a metal electron arising from the semiconductor electrons consists of a screened Coulomb-like potential, summed over all of the semiconductor valence electrons:

$$\mathcal{H}_{int} = \sum_i \left(\frac{e^2}{|\vec{r}_i - \vec{r}|} \right)_s \quad , \tag{4.15}$$

where $\vec{r}$ is the coordinate of the metal electron and $\vec{r}_i$ is the position of the ith semiconductor-valence electron. The subscript s implies that the interaction is screened as in a metallic jellium model of equivalent electron density.

The Fourier transform for $\omega \ll \omega_F$ is

$$\mathcal{H}_{int}(\vec{r}) = \sum_i \sum_q \frac{4\pi e^2}{q^2 \, \epsilon(\vec{q}, \omega=0)} e^{i\vec{q}|\vec{r}_i - \vec{r}|} \quad . \tag{4.16}$$

Clearly, $4\pi e^2/q^2\epsilon$ is the Fourier coefficient screened interaction.

The fluctuations in potential in each cell that give rise to the band gaps will be treated in perturbation theory. The reciprocal dielectric function, $\epsilon^{-1}(\vec{k}, \vec{k}')$ is a tensor in the reciprocal lattice vectors $\vec{k}$ and $\vec{k}'$. The tensor character allows for local field effects and it has been shown that for the phonon mechanism, local fields permit an effective attractive interaction $\lambda_{ph} > \mu$ without violating stability considerations. The same applies in an analogous way to the exciton mechanism.

One can write

$$\mathcal{H}_{int} = \sum_q \left(\frac{4\pi e^2}{q^2} \right) S \, \rho_q \, e^{-i\vec{q}\vec{r}} \quad , \tag{4.17}$$

where

$$\rho_q = \sum_i e^{i\vec{q}\vec{r}_i} \quad . \tag{4.18}$$

The screening factor S in Eq. (4.17) is of order 1/2 and approximates the effects of the dielectric tensor ϵ^{-1}.

$$S = \langle \epsilon^{-1} \rangle \simeq \langle \left(\frac{q^2}{q^2 + q_s^2} \right) \rangle \sim 1/2 \quad , \tag{4.19}$$

since average q's are of the order k_F and $q_s \sim k_F$. The angular brackets denote an average over the Fermi surface.

The interacting matrix element is

$$M = \langle N, \vec{k}_2 | \mathcal{H}_{int} | 0, \vec{k}_1 \rangle \quad , \tag{4.20}$$

where $|0, \vec{k}_1\rangle$ is the initial state with no exciton and a metal electron of momentum $\vec{k}_1$ and $|N, \vec{k}_2\rangle$ is the final state with an exciton $|N\rangle$ of definite momentum $-\vec{q}$ and the metal electron scattered to $\vec{k}_2=\vec{k}_1+\vec{q}$. Assuming plane-wave states for the metal electrons, the matrix element becomes

$$M = \sum_{q'} S\left(\frac{4\pi e^2}{q'^2}\right)(\rho_{q'})_{NO}\,\delta_{qq'} \quad , \tag{4.21}$$

where $(\rho_q)_{NO} \equiv \langle N|\rho_q|0\rangle$. Actually only the transverse components of $\vec{k}$ can be defined by a wave vector and there will be only a limited number of states in the $\vec{k}_z$ direction, but this will not effect the order of magnitude of the estimates to be made. The range of interaction is less than the depth of penetration into the semiconductor.

This exciton scattering process is second order in perturbation theory, involving both the emission and absorption of a virtual exciton. Either pair of electrons may emit or absorb the exciton, implying an additional factor of 2. Thus for $\omega \ll \omega_{NO}$,

$$V_{ex} = 2\sum_N \frac{|M|^2}{\omega_{NO}} \quad . \tag{4.22}$$

However, this expression for $|M|^2$ still needs to be modified by the factor that gives the fraction of time the metal electrons are in the semiconductor b and by the decreased amplitude in the penetration region γ:

$$V_{ex} = 2S^2 \frac{4\pi e^2}{q^2} \sum_N \gamma b \frac{4\pi e^2}{q^2} \frac{(\rho_q)^2_{NO}}{\omega_{NO}} \quad . \tag{4.23}$$

Now approximate ω_{NO} by its average value $\langle\omega_{NO}\rangle = \omega_g$, the average gap width:

$$V_{ex} = 2S^2 \gamma b \frac{4\pi e^2}{q^2 \omega_g} \sum_N \left(\frac{4\pi e^2}{q^2}\right)(\rho_q)^2_{NO} \quad . \tag{4.24}$$

Consider the following two sum rules:

$$\int_0^\infty \omega \epsilon_2(\omega)\, d\omega = \frac{1}{2}\pi\omega_p^2 = \left(\frac{1}{2}\pi\right) 4\pi n_e \frac{e^2}{m} \quad , \tag{4.25}$$

$$\int_0^\infty \epsilon_2(\omega)\, d\omega = \sum_N \frac{4\pi e^2}{q^2}(\rho_q)^2_{NO} \quad , \tag{4.26}$$

where ω_p is the electronic-plasma frequency, ϵ_2 is the imaginary part of the dielectric function, and n_e is the density of semiconductor valence electrons. To approximate the first sum rule, assume that ϵ_2 is sharply peaked at $\omega = \omega_g$ and integrate over just this exciton peak:

$$\int_0^\infty \omega\, \epsilon_2(\omega)\, d\omega \longrightarrow \omega_g \int_{peak} \epsilon_2(\omega)\, d\omega \quad . \tag{4.27}$$

One may reduce n_e to some value n_{eff} to account for integrating only over the exciton peak. Then

$$\int_{peak} \frac{1}{\omega_g} \epsilon_2(\omega) d\omega = \frac{\pi}{2} \frac{\omega_p^2}{\omega_g^2} = \frac{1}{\omega_g} \sum_N \frac{4\pi^2 e^2}{q^2} (\rho_q)_{N0}^2 , \tag{4.28}$$

giving

$$V_{ex} = S^2 \gamma b \left(\frac{4\pi e^2}{q^2}\right)\left(\frac{\omega_p^2}{\omega_g^2}\right) . \tag{4.29}$$

The value of λ_{ex} is N(0) times the average of V_{ex} over the Fermi surface:

$$\lambda_{ex} = N(0) < V_{ex} > \approx [N(0) < S(\frac{4\pi e^2}{q^2}) >] < S > \gamma b \frac{\omega_p^2}{\omega_g^2} , \tag{4.30}$$

$$\lambda_{ex} = S \gamma b \mu \left(\frac{\omega_p^2}{\omega_g^2}\right) . \tag{4.31}$$

Taking favorable estimates for parameters ($b \sim 1/4$, $\gamma \sim 1/2$, $\mu \sim 1/3$, $S \sim 1/2$, $\omega_p \sim 10$eV, $\omega_g \sim 2$eV) one gets

$$\lambda_{ex} \lesssim 0.5 . \tag{4.32}$$

On the experimental side a large temperature-dependent transition in the magnitude of the ambient axial electric field inside a vertical copper tube has been observed. Above a temperature of 4.5°K, the ambient field is 3×10^{-7}V/m or greater. Below 4.5°K, the magnitude of the ambient field drops very rapidly, reaching 5×10^{-11}V/m at 4.2°K. One measures the time of flight spectra of an electron going in the center of the copper tube, giving

$$t = \left(\frac{m}{2}\right)^{1/2} \int_0^h [W - ezE_{amb}(z) - ezE_{app} - mgz]^{-1/2} dz , \tag{4.33}$$

where h is the length of the tube, E_{amb} is the effective ambient electric field assumed to consist of a constant term due to gravitationally induced distortion of the tube and a term due to the patch effect with a complicated z dependence, and E_{app} is the uniform applied field. An examination of the copper surface revealed a layer of copper oxide approximately 20 Å thick. This gives the metal-semiconductor interface which is needed for the excitonic superconductor. The screening of E_{amb} by the 2-D superconductor is a reasonable guess at this stage of the investigation.

5. CONCLUSION

The two simple descriptions for the electron-hole liquid and the exciton mechanism of superconductivity point out that one needs to have a fundamental calculation of the PH propagator in order to investigate these systems. The equations derived in part II offer a reasonable starting point. The results obtained on atomic systems were very good, and as demonstrated by Ladik [7], these approximations work quite well for complicated polymers using the ÔÂÔ approximation.

ACKNOWLEDGMENT

The author wishes to thank Profs. J. Ladik and A.B. Kunz for many very useful discussions.

REFERENCES

1. M.W. Ribarsky, Phys. Rev. A12, 1739 (1975).
2. W.F. Brinkman and T.M. Rice, Phys. Rev. B7, 1508 (1973).
3. D. Allender, J. Bray, and J. Bardeen, Phys. Rev. B7, 1020 (1973).
4. J.M. Lockhart, F.C. Witteborn and W.M. Fairbank, Phys. Rev. Letters 38, 1220 (1977).
5. J. Paldus and J. Čižek, J. Chem. Phys. 60, 163 (1974).
6. T.C. Collins, A.B. Kunz, and P.W. Deutsch, Phys. Rev. A10, 1034 (1974).
7. J. Ladik (in this volume).

ASPECTS OF THE THEORY OF DISORDERED SYSTEMS*

E. N. Economou+

Department of Physics, N. R. C. Demokritos
Aghia Paraskevi, Athens, Greece

ABSTRACT. Questions related to the structural instabilities and the disorder of non periodic solids are briefly reviewed. Disorder may cause localization of the one particle eigenstates thus affecting seriously the transport properties of the materials. The metastability of the amorphous state is revealed physically through a linear (in T) contribution to the specific heat in the limit $T \to 0$.

1. INTRODUCTION

Solid State Physics deals mainly with crystalline materials which exhibit periodicity interrupted occassionally by impurities, defects, etc. However, there are materials of great physical and technological interest which do not possess any periodicity, e.g. amorphous semiconductors, amorphous metals, non stoichiometric alloys, spin glasses, organometallic conductors such as TCNQ salts, etc. In the last ten years there has been a growing interest in these nonperiodic materials both because of the challenging problem of understanding their behavior and because of the possibilities of technological applications they offer [1]. Thus a new subfield in the Physics of Condensed Matter has been born [2-5].

Here I will attempt to present some aspects of the conceptual

* Work supported in part by NSF Grant No. DMR76-19458 at the University of Virginia.

+ Permanent Address: Department of Physics, University of Virginia, Charlottesville, VA 22901, USA.

Cleanthes A. Nicolaides and Donald R. Beck (eds.), Excited States in Quantum Chemistry, 457–469.

framework which has been developed in order to understand the properties of these materials. I will mainly concentrate on three questions: (i) the role of disorder in localizing the eigenstates; (ii) the existence in amorphous materials of many states almost degenerate to the "ground" state; (iii) the electronic structure of chalcogenide glasses. These questions were singled out because they are related with novel physical phenomena characteristic of the non-crystalline state.

2. RANDOMNESS AND LOCALIZATION

2.1 Formulation of the problem

To deal with this question we employ first the independent particle approximation (IPA). It must be stressed that such a drastic approximation has not been justified. If anything, one expects correlation effects to be more important in disordered systems (in which the IPA eigenstates may be localized) than in crystalline systems (where the IPA eigenstates are Bloch waves).

Within the IPA the problem can be formulated as follows: Given the one particle random Hamiltonian $\hat{H}$ (more correctly: given the probability distribution of the matrix elements of $\hat{H}$) find

(i) the density of states

(ii) properties of the eigenfunctions, especially whether they are localized (i.e. vanishing fast enough at infinity) or extended (i.e. non normalizable as the volume becomes infinite).

The determination of the density of states (DOS) is a very important task, since almost all physical quantities depend on the DOS. Special techniques have been developed for this purpose [6]. However, it should be noted that the DOS of a disordered system is not qualitatively different from that of the corresponding ordered system. On the other hand, the possibility of localized states forming a continuum is a novel feature not to be found in ordinary one particle quantum systems, where the continuous spectrum corresponds to extended eigenstates and the localized eigenstates form a discrete spectrum. We examine first, in Sect. 2.2, some physical consequences steming from the existence of a continuum corresponding to localized eigenstates. We then consider the question of the existence of this feature both within the framework of classical mechanics (Sect. 2.3) and the framework of quantum mechanics (Sect. 2.4).

2.2 Physical Consequences

If the Fermi energy E_F happens to be within a region of localized eigenstates, the system will exhibit insulating behavior, in the sense that

$$\sigma_{T=0} = 0, \tag{2.1}$$

where $\sigma_{T=0}$ is the zero temperature DC conductivity. Furthermore at finite temperatures the electronic transport can take place through phonon assisted hopping; Mott [7] demonstrated and others elaborated [7] that in this case the temperature dependence of the DC conductivity is given by

$$\sigma(T) = A \exp\left[-\left(\frac{T_o}{T}\right)^{1/4}\right] \tag{2.2}$$

Thus $\sigma(T) \to 0$ as $T \to 0$, which means that at T=0 the electronic mobility is zero in the region of localized states. Mott [8] proposed that the mobility jumps discontinuously to a non zero value as we enter the part of the spectrum corresponding to extended eigenstates. This non zero value gives a minimum metallic conductivity which can be estimated by assuming that the minimum mean free path, ℓ_{min}, equals to the interatomic distance, a. Starting from the expression for the conductivity

$$\sigma = \frac{n e^2 \ell}{m v_F} \tag{2.3}$$

where n is the electron density, $n = N/a^3$, N is the number of valence electrons per atom, and m is the electronic mass; v_F, the Fermi velocity, is related to n by

$$v_F = \frac{\hbar}{m} k_F \sim \frac{\hbar}{m} \quad n^{d/3} \tag{2.4}$$

d is the dimensionality of the system. Substituting Eq. (2.4) in Eq. (2.3) and taking into account that $\ell_{min} \sim a$ and that $N_{min}=1$ we obtain

$$\sigma_{min} \sim \frac{e^2}{\hbar} a^{d-2} \tag{2.5}$$

Note that for d=2 the minimum metallic conductivity depends only on universal constants. A more refined treatment [9] based on scaling considerations yields for d=2

$$\sigma_{min} = .12\, e^2/\hbar \tag{2.6}$$

For experimental evidence supporting Eq. (2.6) see ref. [10]. Eqs. (2.1,.2,.6) show the far reaching physical consequences of the assumption of the existence of continua corresponding to localized states. Having thus motivated our interest in the question of localization, we review briefly the progress which has been made in demonstrating the existence of localization as a result of randomness.

2.2 Percolation Theory

It is instructive to examine first the problem of a classical particle of energy E moving in a random potential $V(\vec{r})$. In this case the particle is localized (extended) if the surface $V(\vec{r})=E$ is closed (open). To visualize the situation consider the 2-d case; correspond to the potential $V(\vec{r})$ the height $h(\vec{r})$ of a random terrain; and to the energy E a uniform water level. The regions $h(\vec{r}) < E$ (i.e. $V(\vec{r}) < E$) are covered by water, while the region $h(\vec{r}) > E$ (i.e. $V(\vec{r}) > E$) is land. For low water levels ($E \ll E_c$) all the water areas are lakes, i.e. they are surrounded by land, a situation which corresponds to localized electrons. As the water level increases the lakes become more numerous and of larger area each. At some critical level $h=h_c$ ($E=E_c$) lakes are joined together to form the first ocean, i.e. the first extended state. This first ocean has a very complicated shape and covers a small fraction of the total area. As the water level is further raised many lakes are able to join the first ocean which becomes larger and with smoother boundaries. Finally if the level h is much higher than h_c almost all lakes join together to form a single ocean which covers almost all the total area except for some small isolated islands, which are the analogs of scattering centers in an ordinary metal. There are computer experiments (see the article by E. N. Economou et. al. in ref.[3]) which verify the picture presented above. Efforts have been made to relate the critical value E_c to a critical value, v_c, of the fraction of the water area, v, [11]; v_c depends on the dimensionality.

The mathematical treatment of this problem is facilitated by expressing it in a lattice formulation. How such a formulation arises can be seen by considering a classical electron of energy E initially placed at a particular atom. If E is larger than the saddle point of the potential between the particular atom and a specified nearest neighbor, the electron can migrate to this nearest neighbor; otherwise the direct path (bond) connecting the two atoms is blocked. In a random system a fraction $p_b = p_b(E)$ of the nearest neighbors bonds will be open and a fraction $1-p_b$ is blocked. The question is to find the probability P for an electron (initially placed at a particular atom) to escape to infinity following a path of open nearest neighbor bonds. This is the bond percolation problem. There is also the site percolation problem where we try to find P as a function of p_s, where p_s is the fraction of allowed sites in the lattice. There is a rigorous mathematical proof that a critical value of p_b, p_b^c, (or of p_s, p_s^c) exists (corresponding to the critical value E_c introduced earlier) such that for $p_b < p_b^c$ (or $p_s < p_s^c$) P=0, i.e., the electron remains localized. The critical energy E_c is termed mobility edge. The physical picture emerging from percolation theory is summarized in Fig. 1, where the mobility vs.

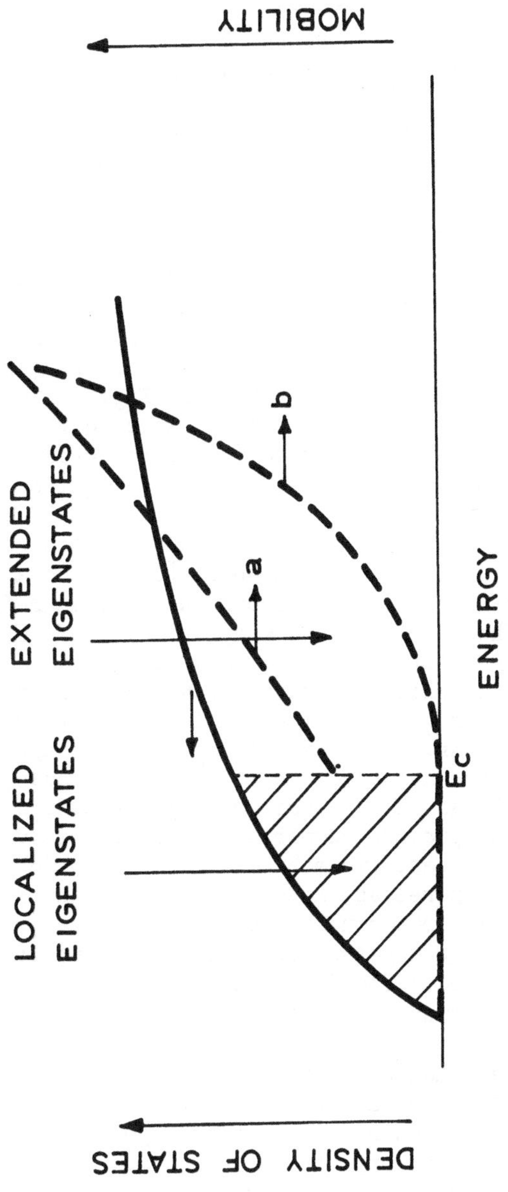

Figure 1. Density of states (solid line) and mobility according to percolation theory (curve b) and according to Mott's conjecture (curve a) vs. E for a disordered system. The mobility edge, E_c, separates a region of localized states from a region of extended states.

E is also plotted (curve b) [12].

2.3 Quantum treatment of localization.

In the last subsection the question of localization was examined within the framework of classical mechanics. What modifications quantum mechanics brings to the classical picture? One effect is that the electron can tunnel over the land area from one lake to a nearby disjoint lake creating thus the possibility of delocalization; another effect is that a wave cannot easily propagate through the very complicated shape of the first ocean because of strong surface scattering [13]. Thus the two effects oppose each other and may leave the classical picture qualitatively unchanged. Wave propagation can always be described as the transfer of the oscillation amplitude (with the help of some transfer matrix elements t_{ij}) from a region i of local oscillations of frequency ω_i to a region j of frequency ω_j. The transfer is facilitated by large values of t_{ij} and is impeded by large values of $\omega_i-\omega_j$ [14]. The simplest realization of the general problem is the coupled pendulum systems shown in Fig. 2. The localization or not of the eigen-oscillation is expected to depend on the magnitude of t_{ij} and on the spread (standard deviation) in the distribution of ω_i. The mathematical treatment of this problem is usually based on the convergence of the so called renormalized perturbation expansion for the Green's function of the system, as was originally proposed by Anderson [15]. For a review of the subject see the article by Economou et al in ref. [3], the review article by Thouless [16] and Chapter 7 in ref. [17]. For a 1-d system the eigenstates become always localized (some pathological exceptions do exist) no matter how small the degree of randomness is [18]. For higher dimensional systems there is no rigorous result. However, the available complicated approximate theories [15-17], the numerical solutions of Schrodinger equation [19-21], as well as the analysis of experimental data strongly suggest that the picture summarized in Fig. 1 is valid. In each band there are at least two mobility edges E_c which, as the degree of randomness increases, move inwards towards the center of the band. As the randomness exceeds a critical value the two mobility edges merge together and all the eigenstates in the band become localized. The disappearance of extended states as a result of randomness is called Anderson's transition.

The question of the behavior of the eigenfunctions near a mobility edge is of great importance. Mott argues that an extended eigenstate cannot follow a complicated shape like the percolation channel. On the contrary, he assumes that, if the eigenstate is extended, it is more or less uniformly spread over the whole volume. Such an assumption leads to a discontinuous drop of the mobility at E_c (curve a of Fig. 1). The value of the mobility at E_c^+ defines a minimum metallic conductivity, a concept

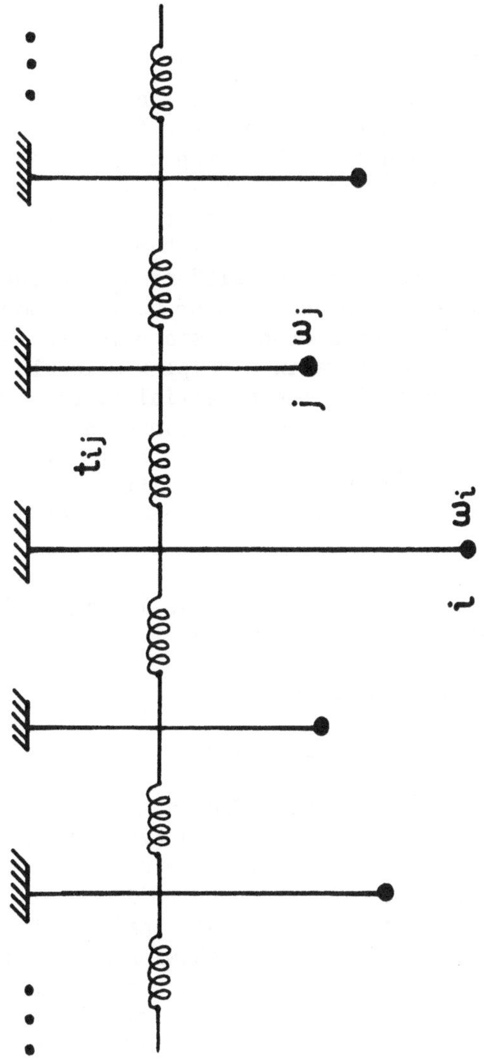

Figure 2. A system of coupled pendulums with a random distribution of the local frequencies ω_i .

which seems to be supported by experimental evidence [8,9], especially in metal-oxide-silicon field effect transistor (MOSFET) [22,23] where the position of the Fermi level can be controlled by an external voltage. It should be mentioned that in the interesting regime in MOSFETs many body effects seem to be significant; this question has not been examined yet in a satisfactory way.

3. DEGENERACY OF THE "GROUND" STATE

Many amorphous materials find themselves in a metastable "ground" state which lies higher in energy (or free energy for $T > 0$) than the true ground state which is the crystalline one; a substantial potential barrier prevents the crystallization of the material for temperature below the glass transition temperature. However, very close to the metastable "ground" state there are many other almost degenerate states differing from the "ground" state because of some local rearrangements of nuclear coordinates and bonding configurations. The amorphous state is thus somehow similar to a photoexcited state in a complex molecule [24], although for an amorphous material the potential barrier is so high that the state is stable for practical purposes. In Fig. 3 we plot schematically the potential curve vs. the totality of nuclear coordinates indicating the essential differences between a crystalline true ground state and an amorphous metastable ground state.

A direct consequence of the picture shown in Fig. 3 is that the low temperature specific heat behaves as

$$C = \alpha_3 T^3 \text{ ; crystalline insulator} \tag{3.1}$$

$$C = \alpha_1 T + \alpha_3' T^3 \text{ ; amorphous insulator} \tag{3.2}$$

where α_3 is related to speed of sound. The existence of the linear term in the specific heat was observed experimentally [25]. As a matter of fact, it was this dramatic linear term in the specific heat which led Anderson et al. [26] and Phillips [27] to postulate the existence of alternative local configurations to which the system can tunnel. Experiments showing a saturation of the ultrasonic attenuation, and measurements on thermal conductivity, temperature dependence of the sound velocity, and nuclear spin lattice relaxation provide further evidence for the existence of the local rearrangement modes (LRM). A direct experimental evidence for the existence of LRM in amorphous structures is provided by the observation of Kondo effect in non magnetic amorphous metals [28]. The Kondo effect is a logarithmic rise of the resistivity with decreasing temperature; this rise is due to the scattering of the conduction electrons by a two level system such as a magnetic impurity. Although the

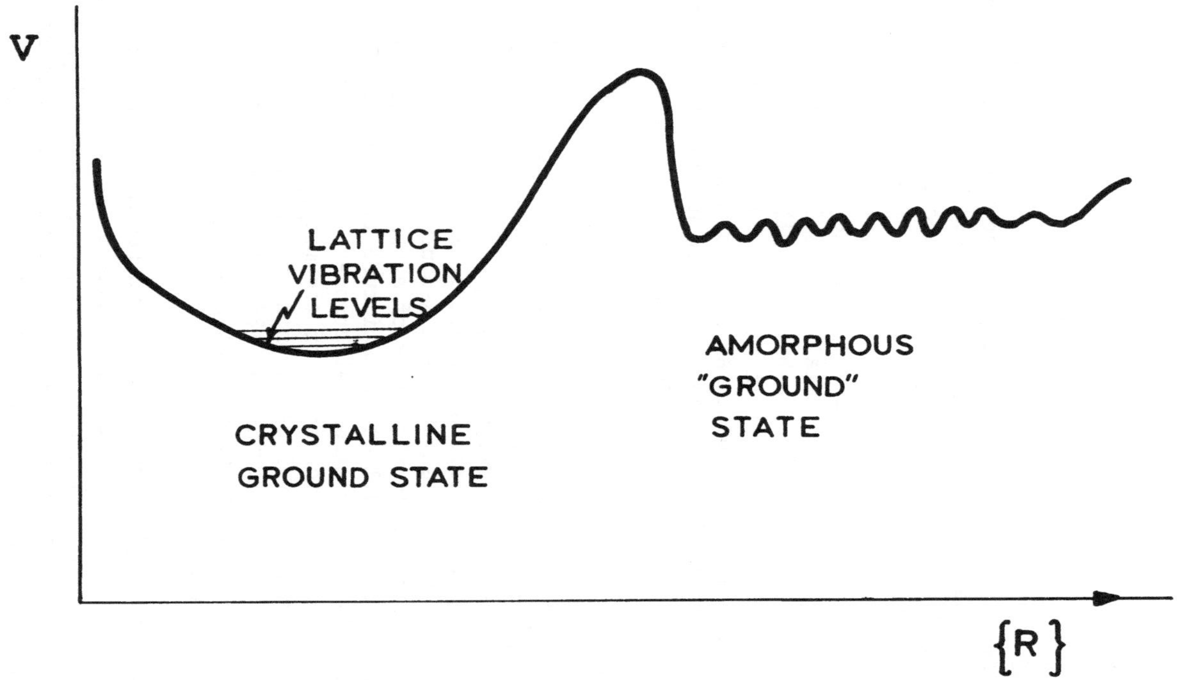

Figure 3. Schematic plot of the potential energy V vs. the totality of nuclear coordinates {R} indicating the difference between a crystalline ground state and an amorphous "ground" state.

concept of LRM has been proved very useful in analysing experimental data no one has succeeded in constructing a particular structure exhibiting explicitly LRM; such a construction is an important task for advancing our understanding of the behavior of disordered system.

4. ELECTRONIC STRUCTURE OF AMORPHOUS SEMICONDUCTORS

I am refering here to the so called chalcogenide glasses (see ref. [1] or the article by Fritzsche in ref. [3]) which can be prepared in the amorphous state in the bulk, have a low coordination number, and possess a lone pair. Examples are Se, Te, As_2Se_3, As_2S_3 etc. Understanding the electronic structure of such materials is a challenging problem. The observed pinning of the Fermi level strongly suggests the existence of a non zero density of states at the Fermi level. On the other hand, the observed diamagnetism, the non metallic behavior and the existence of an optical gap all suggest a behavior similar to the crystalline case where the Fermi level lies in the gap. Other experimental observations such as photostructural changes, midgap absorption and ESR signal after illumination, etc., are unique characteristics of the amorphous state, which are not well understood.

Anderson [29] proposed a picture which goes a long way in explaining the puzzling experimental observations. He assumed that the 1-particle density of states is very similar to that of the crystalline state. However, in the amorphous state some pairs of electrons are formed with energies around the center of the gap. Thus the electronic structure looks as shown schematically in Fig. 4. Mott et al [30] and Kastner et al [31] have proposed a similar picture on the basis of an analysis based on local molecular orbitals. In all cases a picture like that of Fig. 4 requires the existence of strong electron-electron (or hole-hole) attraction for the formation of the midgap pairs. Anderson [29] assumed that local phonons in certain special places mediate the attraction; Mott et al [30] and Kastner et al [31] talk about "defects" which provide at least partially the attractive forces required.

Economou et al [32] attempted to synthesize the unique structural characteristic of the amorphous state shown in Fig. 3 with the characteristic electronic structure shown in Fig. 4, by assuming that there is an interaction between the electrons and the LRM. Such an interaction creates both the metastable states and the pair states shown in Fig. 4; furthermore it directly relates electrons with structural features (the LRM) and consequently it accounts naturally for photostructural properties. Several of the predictions of the model are in agreement with experimental observations. The construction of a specific molecular model incorporating LRM will greatly contribute to our

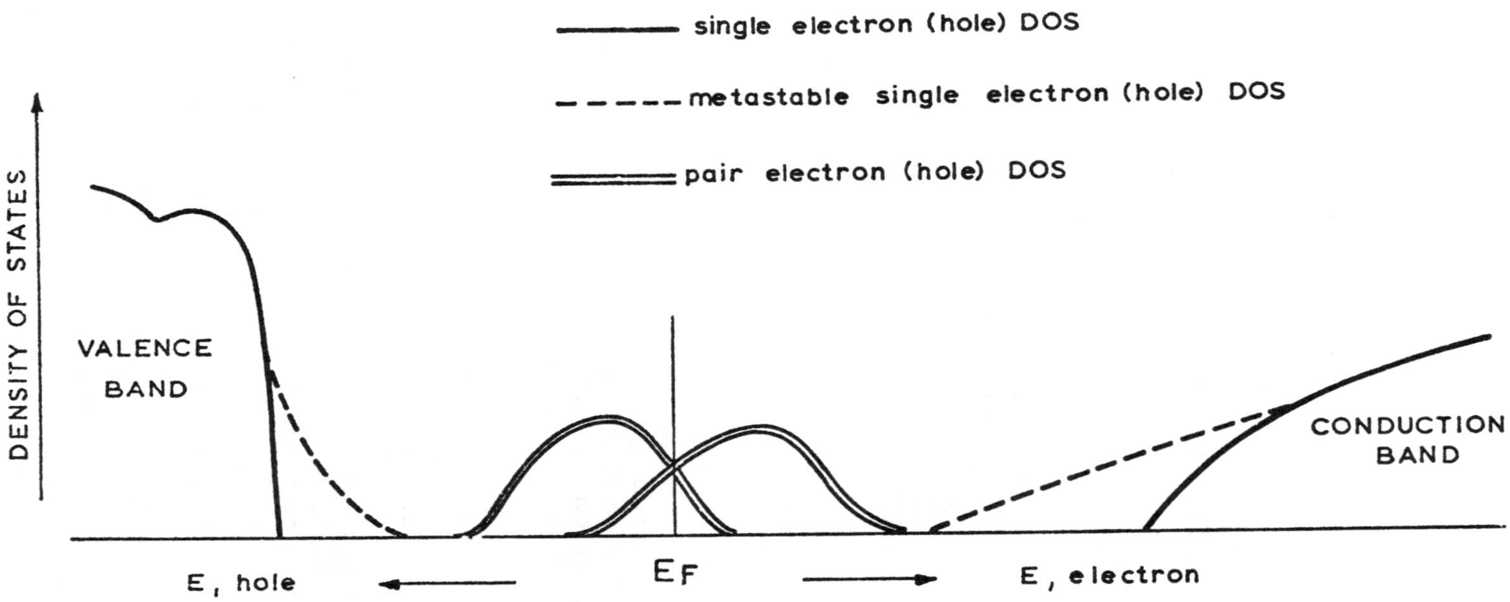

Figure 4. Schematic plot of the density of states (DOS) for a single particle or hole (solid line) and pair of particles or holes (double solid lines). The area under the dashed line represents metastable single particle or hole states, whose energy is lowered by double occupation.

understanding of the properties of amorphous materials; the role of local phonons and LRM will be revealed, and the proposed pictures will be tested explicitly.

REFERENCES

1. D. Adler, Scientific American, 236, 5, 36 (1977).
2. N.F. Mott and E.A. Davis, Electronic Processes in Non-Crystalline Materials, Clarendon Press, Oxford, England (1971)
3. J. Tauc, editor, Amorphous and Liquid Semiconductors, Plenum Press (1974).
4. P.G. LeComber and J. Mort, editors, Electronic and Structural Properties of Amorphous Semiconductors, Academic Press (1973).
5. J. Stuke and W. Brenig, editors, Amorphous and Liquid Semiconductors, Taylor and Francis, London (1974).
6. See, e.g. R.J. Elliot, J.A. Krumhansl and P.L Leath, Rev. Mod. Phys. 46, 465 (1974); J. Ladik in this volume.
7. N.F. Mott, Phil. Mag. 19, 835 (1969); V. Ambegaokar, B.I. Halperin, and J.S. Langer, Phys. Rev. B4, 2612 (1971).
8. N.F. Mott, Phil. Mag. 26, 1015 (1972).
9. D.C. Licciardello and D.J. Thouless, Phys. Rev. Lett. 35, 1475 (1975).
10. D.C. Licciardello and D.J. Thoreless, Surf. Science 58, 89 (1976); D.C. Licciardello, to be published.
11. See, e.g. V.K.S. Shante and S. Kirkpatrick, Adv. Phys. 20, 325 (1971).
12. See, e.g. S. Kirkpatrick, Rev. Mod. Phys. 45, 574 (1973).
13. Consider the ocean replaced by a membrane attached to its shores; a quantum particle propagating in the ocean will then be equivalent to an elastic wave propagating in this membrane. Tunneling can be included in this picture by allowing the boundaries (shores) to move so that the wave can be transfered (with attenuation) over the land area to nearby lakes.
14. In first order perturbation theory the transfer from i to j depends on the ration $t_{ij}/(\omega_i-\omega_j)$.
15. P.W. Anderson, Phys. Rev. 109, 1492 (1958).
16. D.J. Thouless, Phys. Rep. 13C, 94 (1974).
17. E.N. Economou, Green's Functions in Quantum Physics, to be published by Springer Verlag.
18. See, e.g. C. Papatriantafillou, E.N. Economou, T.P. Eggarder, Phys. Rev. B13, 910 (1976); C. Papatriantafillou and E.N. Economou, Phys. Rev. B13, 920 (1976) and references therein.
19. J.T. Edwards and D.J. Thouless, J. Phys. C5, 807 (1972).
20. K. Schönhammer and W. Brenig, Phys. Lett. A 42, 447 (1973).
21. D. Weaire and A.R. Williams, J. Phys. C8, 1239 (1977); D. Weaire and V. Srivastawa, J. Phys. C10, 4309 (1977).
22. N.F. Mott, Electr. Power 19, 321 (1973); F. Stern, Phys. Rev. B9, 2762 (1974).
23. D.C. Tsui and S.J. Allen Jr., Phys. Rev. Lett. 34, 1293 (1975);

N.F. Mott, M. Pepper, S. Pollitt, R.H. Wallis and C.J. Adkins, Proc. Roy. Soc. A345, 169 (1975).

24. J. Michl, this volume.
25. R.C. Zeller and R.O.Pohl, Phys. Rev. B4, 2029 (1971).
26. P.W. Anderson, B.J. Halperin and C.M. Varma, Phil. Mag. 25, 1 (1972).
27. W.A. Phillips, J. Low Temp. Phys. 7, 351 (1972).
28. R.W. Cochrance, R. Harris, J.O. Ström-Olson and M.J. Zuckerman, Phys. Rev. Lett. 35, 676 (1975).
29. P.W. Anderson, Phys. Rev. Lett. 34, 953 (1975).
30. N.F. Mott, E.A. Davis and R.A. Street, Phil. Mag. 32, 961 (1975).
31. M. Kastner, D. Adler and H. Fritzsche, Phys. Rev. Lett. 37, 1504 (1976).
32. E.N. Economou, K.L. Ngai, and T.L. Reinecke, Phys. Rev. Lett. 39, 157 (1977); K.L. Ngai, T.L. Reinecke, and E.N. Economou, to be published in Phys. Rev.

EXCITED STATES OF TRANSITION METAL OXIDES*

A. Barry Kunz

Department of Physics and Materials Research Laboratory
University of Illinois at Urbana-Champaign
Urbana, Illinois, U.S.A. 61801

ABSTRACT

A discussion of the electronic structure of the non-metallic transition metal oxides is presented. To be successful any model must quantitatively, at least, describe the following phenomena: insulating behavior of open shell systems; magnetic properties and low lying magnetic excitations; cohesion and phonon spectroscopy; "localized" excitations or excitons and; Bloch like excitations. We argue that the least sophisticated model which can attempt such a description is the Unrestricted Hartree-Fock model (UHF), and even in this limit Koopman's theorem may not be assumed but rather total energy differences of several self-consistent solutions are needed. We further show that if quantitative accuracy is needed correlation corrections beyond the UHF limit are necessary. We discuss several simple models for inclusion of correlation corrections using techniques of classical electrodynamics on one hand and of Configuration Interaction on the other hand. Detailed calculations are presented using these models and comparisons with optical spectroscopy are made. There is a reasonable comparison of theory and experiment produced by these methods, and the ground state of FeO, CoO and NiO is seen to be insulating whereas that of TiO and VO is seen to be metallic.

II. INTRODUCTION

The first transition period elements and their oxides form one of the most interesting class of materials in terms of the variety and strangeness of their physical properties. These

Cleanthes A. Nicolaides and Donald R. Beck (eds.), Excited States in Quantum Chemistry, 471–493.

systems exhibit magnetic phenomena ranging from ferromagnetism to antiferromagnetism. The oxides in particular exhibit a great variety of electrical behavior, ranging from being good conductors of electricity to being excellent insulators. Many of these properties are hard to understand on a first principal basis and a wide variety of ad hoc models are invented to account for the observed properties. An abbreviated list of some important properties of the systems TiO, VO, MnO, FeO, CoO and NiO is to be found in Table 1. A brief study of this table will enable the reader to determine quickly for himself that many of these properties behave strangely.

Consider for example the lattice parameter for the oxides. As atomic number (Z) decreases in a given row of the periodic table, the atomic radius increases, thus as the cation in oxide changes we expect the lattice constant to increase as Z decreases. This is true for TiO and VO or for MnO, FeO, CoO and NiO, but the lattice parameter of either TiO and VO is substantially smaller than for MnO. Simple crystal field arguments are able in part to justify this behavior but the detailed numerical evaluation of the radial extent of the d-orbitals negates the effect of crystal field arguments, so this property itself is yet a mystery.

A second and much more interesting anomally is associated with the occupation of the d band. All six of the oxides shown have fewer than 10 electrons in the 3d band (2-8 actually). On the basis of elementary energy band theory these systems would be metallic in nature. This is certainly true for TiO and most likely VO. One finds MnO, FeO, CoO and NiO are insulating. In addition, these four systems are antiferromagnetic and exhibit a variety of spin directions and magnetic lattice types as seen in Table 1. It is this set of properties which seems to be most interesting to investigators and it is these properties which form the basis for the investigations we discuss in this set of lectures.

The initial starting point is to recognize that these four systems, along with several others we do not consider here, fall in the category of Mott insulators (N. F. Mott, 1949, 1952, 1956, 1958, 1961, 1969). The category of system called Mott insulators may loosely be assumed to include systems which are conducting in terms of elementary energy band theory but are insulating experimentally. It is of course assumed that the experimental insulating behavior is not due to crystal imperfections, impurities or slight deviations from stoichiometrybut is an intrinsic property of these systems. It is not appropriate here to give a review of the properties of Mott insulators and the reader is referred to the works of Mott.

TABLE 1

A summary of some key experimental properties of some first period transition metal oxides. All are in a f.c.c. lattice structure.

Property \ System	TiO	VO	MnO	FeO	CoO	NiO
no. 3d electrons 10 possible	2	3	5	6	7	8
a(Å)	4.18	4.10	4.44	4.31	4.27	4.195
anti-ferromagnetic	no	?	yes	yes	yes	yes
T_N	--	?	122 K	198 K	292 K	523 K
Mag Lattice	--	?	trig.	trig.	tetr.	trig.
Spin dir.	--	?	(111)	⊥(111)	$(\bar{1}\bar{1}7)$ almost	(111)
Electronic	good metal	metal	insulator	insulator	insulator	insulator
optical gap	none	none	--	--	--	3.8 eV (2.0 perhaps)
Photo cond.	none	none	n.a.	n.a.	n.a.	3.8 eV
Surface Props.	--	--	--	--	oxidation catalyst	oxidation catalyst
Comments	Hard to get good samples Hall effects change sign with O concentration	Very hard to get good samples	--	--	Catalytic activity scales directly with excess oxygen	insulating to melting temp. less than 23% of surface sites chemisorb strongly

There have been many attempts to justify the behavior of the "Mott" insulators and it is necessary to briefly discuss several of them here. The first is due to Slater and Wilson (T. M. Wilson, 1968, 1970). This model attempts to use the antiferromagnetic lattice and ordinary energy band theory to explain the insulating behavior. In this model the cation is allowed to be spin polarized. Thus at the atomic level one has two 3d states, a majority spin state and a minority spin state, each five fold degenerate. In the crystal field of a f.c.c. system each of these levels splits into two levels, one three fold degenerate the other two fold. Normally, the threefold level is lower in energy. If one includes periodic symmetry of the lattice these levels broaden out into energy bands. Provided the bands do not overlap in this normal order case, systems of 3, 5, 6 or 8 d electrons have the possibility of remaining insulating. This evolution of energy levels is seen in Fig. 1 for the case of NiO using the band results of Wilson. There are three immediate criticisms of this result. The first is this model doesn't explain the insulating behavior of CoO. The second is that this model predicts a very small photoconductivity energy gap, which is not seen experimentally. Finally, this model is conducting above the Néel temperature whereas the systems shown in Table 1 remain insulating well above this temperature. This model is therefore not considered to be the explanation of the phenomena observed in NiO or the other systems.

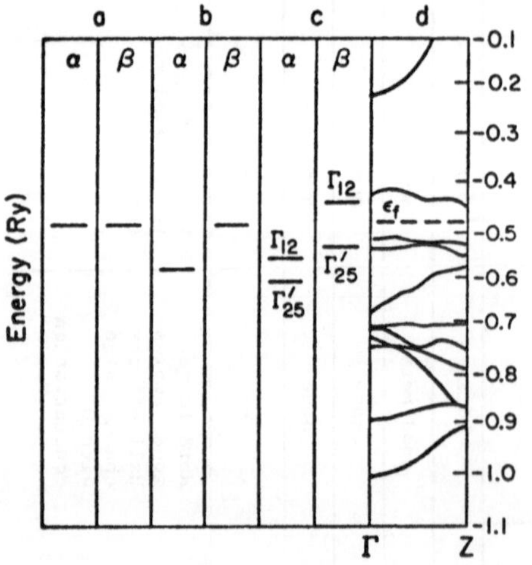

Fig. 1 The evolution of the Ni^{++} 3d level into bands is shown. In part a the 3d level of a RHF like calculation is given. In part b it is split into two levels by spin polarization. In part c the spin polarized levels are split by an octahedral field and in part d the antiferromagnetic NiO band structure of Wilson is given.

A second model is that of Hubbard. It is an ad hoc type of many body model which, provided one chooses one parameters with care is consistent with the observed conductivity data (J. Hubbard, 1963, 1964, 1965, 1966). In this model in its most simple form one assumes there is a single s orbital on each lattice site. This orbital may be either spin up or spin down. It is further assumed there is exactly one electron per lattice site. This model assumes that it takes an energy Δ to create a hole on one of the atom sites and to place a second electron on another lattice site a large distance away. This system having a hole on one site and an electron on another has periodic symmetry. This invariance gives rise to a matrix element of the Hamiltonian called, S, which is for transferring the hole and electron from their site to a neighboring one. This S parameter gives rise to an energy band of finite width. If the width is sufficiently great so that the total energy lowering due to periodic symmetry exceeds the Δ parameter one has a system in which charge hopping occurs and is thus a conductor. Clearly by judicioûs choice of Δ and S one may easily explain the properties of the Mott insulators. It is also clear that this model can use much study in order to justify its genesis and to properly define its essential parameters. This is done in this lecture series. The essential physics of a Hubbard model is simply summarized in a more general case: there are exactly n electrons per unit cell; these electrons are strongly correlated. If one moves right another exchanges with it moving left in an insulating ground state; conductivity is via a hopping mechanism.

Next we have the ad hoc model of Adler and Feinleib. This model is sufficiently broad as to cover most of the experiments and is the least justified theoretically. In this model some electrons are assumed to exist in band states, principally the oxygen 2p and the cation 4s states, and the cation 3d states are assumed to be atomically localized. This gives rise to an energy level structure as seen in Fig. 2 specifically here for NiO. In the ground state we have a filled O 2p band and a collection of Ni^{++} ions each in a $3d^8$ (^3F atomic) ground state. There are low lying excitations in which the $3d^8$ (^3F) configuration is excited to other terms within the $3d^8$ mainfold and designated as $3d^{8*}$. The next excitation is from a $3d^8$ state into the 4s conduction band. Thus here the photoconductivity is due to electrons in band states in contrast to a Hubbard model. There are also true Hubbard like excitations designated as $3d^7 + 3d^9$ etc. This model is very appealing in that it has the potential to explain much about the transition metal oxides. It also have one clear disadvantage in that Brandow has argued that for a truely periodic system no local eigenfunctions exist for a system in the independent particle limit (D. Adler 1968, D. Adler and J. Feinleib, 1970) (B. H. Brandow 1976 and references contained therein).

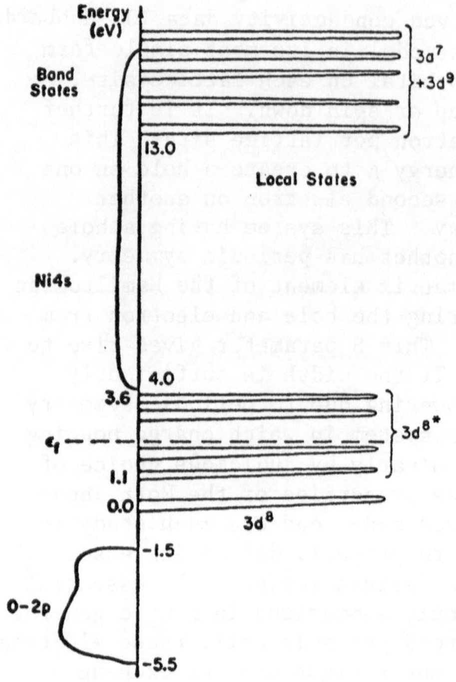

Fig. 2 The NiO energy level schematic of Adler and Feinleib is given. Both Bloch like and local states are indicated.

More recently there have been a series of attempts to develop a comprehensive theory of insulators by Brandow (B. H. Brandow 1977). This attempt is based upon detailed considerations of a Spin Polarized Hartree-Fock equation (SPHF) and uses some semi-rigorous theoretical arguments to evaluate and define the Δ parameter in a Hubbard model. This attempt furthermore refines the possibilities available to couple the ideas of Adler and Feinleib, of Hubbard and of Wilson and Slater into a coherent idea. Much of the discussion presented by Brandow is related to the magnetic properties and any attempt to actually compute the needed theoretical parameters is omitted. In this author's opinion the chief drawback to the work of Brandow is that it is incomplete, many necessary theoretical theorems are missing, thus reducing Brandow's ideas to plausability arguments. We will not review this model further.

Finally, it is necessary to mention the numerically suggested theories of Brown, Gahwiller and Kunz (1971) of Collins, Kunz and Deutsch (1974) and of Collins, Kunz and Ivey (1975). These theories were based upon developments of the generalized

Hartree-Fock theory in which the virtual orbitals were chosen from a variationally chosen operator possessing a Koopman's like theorem appropriate to excitation and including electron-hole interaction. The hole character was chosen to minimize the excitation energy. This model predicted many numbers but the calculations were in some cases crude enough to prevent very accurate quantitative answers. In addition, these works suffered in a similar way to the theories of Brandow. They lacked the necessary analytic theorems to turn the numerical studies from plausibility arguments into detailed theoretical understanding. In this manuscript we consider in detail the theoretical question of the ground state and the elementary excitations of a narrow band system and provide a more detailed set of calculations for NiO than has been available. We begin with a one electron model.

II. THE UNRESTRICTED HARTREE-FOCK METHOD (UHF)

In this study, we use the normal n-electron Hamiltonian, H, in which we neglect relativistic effects and the finite mass of the nuclei. H is found to be:

$$H = \sum_{i=1}^{N} f_i + \frac{1}{2}\sum_{i,j=1}^{N}{}' g_{ij} + \frac{1}{2}\sum_{I,J=1}^{N}{}' \frac{e^2 Z_I Z_J}{|\vec{R}_I - \vec{R}_J|} , \tag{1}$$

$$f_i = -\frac{h^2}{2m}\nabla_i^2 - \sum_{I=1}^{N} \frac{e^2 Z_I}{|\vec{r}_i - \vec{R}_I|} ,$$

$$g_{ij} = \frac{e^2}{|\vec{r}_i - \vec{r}_j|} .$$

Here there are N nuclei of charge Z_I at site $\vec{R}_I$. The electron charge is e and its mass m. Lower case letters refer to electron properties and upper case letters refer to nuclear properties. In our wavefunction we include spin degrees of freedom and thus use coordinate $\vec{x}_i$ to include space degrees of freedom $\vec{r}_i$ and spin. Ideally, we would like to solve:

$$H\Psi(\vec{x}_1 \ldots \vec{x}_n / \vec{R}_1 \ldots \vec{R}_N) = E\Psi(\vec{x}_1 \ldots \vec{x}_n / \vec{R}_1 \ldots \vec{R}_N) , \tag{2}$$

but this is too difficult. Therefore, we approximate our solution by:

$$\Psi(\vec{x}_1 \ldots \vec{x}_n / \vec{R}_1 \ldots \vec{R}_N) \cong \Psi_{HF}(\vec{x}_1 \ldots \vec{x}_N) = (N!)^{-\frac{1}{2}} \det||\phi_i(\vec{x}_j)|| , \tag{3}$$

subject to the constraints,

$$\int \phi_i^*(\vec{x})\ \phi_j(\vec{x})d\vec{x} = \delta_{ij} ,$$

and

$$\phi_i(\vec{x}) = \phi_i(\vec{r})\alpha ,$$

or

$$\phi_i(\vec{x}) = \phi_i(\vec{r})\beta .$$

The quantities α, and β are the usual Pauli eigenfunctions for spin and spin down respectively. The functional form of $\phi_i(\vec{r})$ is chosen variationally without requiring any symmetry adaption on the part of the φ's or requiring spin up spin down orbitals to occur in pairs. Thus, this model has the potential to go beyond the SPHF methods of Brandow.

If we choose the φ's by requiring the expectation value of H be minimized with respect to Ψ_{HF} and our given constraints, one obtains in canonical form the normal UFH equation, defining $\phi_i(r_j)$:

$$F\ \phi_i = \varepsilon_i\ \phi_i , \tag{3}$$

$$F_1 = f_1 + \int g_{12}\rho(\vec{x}_2\vec{x}_2)d\vec{x}_2 - \rho(\vec{x}_1\vec{x}_2)g_{12}\ \hat{P}(2,1),$$

$$\rho(\vec{x}_1\vec{x}_2) = \sum_{i=1}^{N} \phi_i(\vec{x}_1)\ \phi_i^\dagger(\vec{x}_2) .$$

In defining ρ the φ's act as operators as needed and $\hat{P}(2,1)$ is the operator which interchanges coordinate $\vec{x}_1$ with $\vec{x}_2$. If one were to take this function and constrain the φ's to be symmetry adapted to the nuclear geometry, the SPHF solution occurs and if one additionally constrains the solutions to exist in symmetry adapted spin up spin down pairs the Restricted Hartree-Fock (RHF) solution occurs. The reader is cautioned that most other authors simply refer to the RHF solution as being the Hartree-Fock solution.

Let us briefly consider some possible problems of using a UHF solution. In general there are operators called symmetry operators, S_i, which commute with H. If we ignore the question of degeneracy one finds that if,

$$[H, S_i] = 0 ,$$

then one has

$$H \Psi_i = E_i \Psi_i , \tag{4}$$

and

$$S_j \Psi_i = \Delta_i \Psi_i .$$

Therefore in the case of atoms the H given in Eq. (1) commutes with orbital angular momentum, L, and spin, S, and the z component of each. Thus, the eigenfunction of H are also eigenfactors of L, L_z, S and S_z. In the RHF model the one-electron orbitals are also required to be symmetry adapted to the same symmetry operators as H thus guaranteeing that Ψ have proper symmetry properties. In the UHF case the one electron orbitals are only eigenfunctions of S_z and thus the total wavefunction doesn't necessarily have the proper symmetry characteristics. We note this defect can be easily eliminated by forming a projected linear combination of the possible ground state UHF solutions if the ground state is degenerate. If it is not degenerate the problem vanishes. However, it is also found by many calculations that the UHF solution often has almost correct symmetry properties. In any event as Löwdin (1966) has shown, since one doesn't demand an exact eigenstate of H it is somewhat inconsistent to demand an exact eigenstate of the item commuting with H, and to do so may compromise the quality of the energy solution.

The next general consideration concerns the meaning of the eigenvalue in the UHF Eq. (3). The meaning is given by Koopman's theorem here. That is, if we assume the orbitals of the system are unchanged if either an electron is removed or added to the system then the eigenvalue, ε_i, for an occupied orbital is the negative of the energy needed to remove that electron, while the eigenvalue, ε_a, of a virtual orbital is the energy lost in adding that electron to the system. If the change in remaining orbitals is not negligible one can calculate the energy charge by performing separate self-consistent solutions for the normal and ionized system. The difference in ionization energy computed in this self-consistent way and in the Koopman's limit is termed the relaxation energy, (D. J. Mickish et al, 1974). This quantity is of considerable importance in this study.

There is one important final point concerning the eigenvalue in a UHF scheme as compared to either a SPHF or an RHF scheme. This is in a given subshell (i.e. the Ni 3d or B 2p) if the subshell is not full, each one electron orbital is allowed to have a different eigenvalue whereas in SPHF or RHF this is not so. This says that in general there is a large energy gap between the occupied orbitals of a subshell and the virtual ones. This fact and its importance to Mott insulators was partly recognized by Brandow (1977). Ignoring other considerations for a moment, the important point can be seen by considering a pair of well separated Ni^{++} ions ($3d^8$ ground state). In the RHF Koopman's limit it requires $-\varepsilon_{3d}$ energy to remove an electron from one of them and one gains back ε_{3d} energy in adding it to the other Ni for a net energy charge of 0. Thus one can easily find an assembly of such Ni^{++} ions are metallic. This is not so in the UHF limit and here it takes $-\varepsilon_{3d\,occ}$ to remove an electron from one and only again of $\varepsilon_{3d\,virt}$ is recovered upon adding the electron to the second for a finite charge. Thus unless this energy charge is overcome by energy recovery due to forming bands as one gets an assembly of overlapping Ni^{++} ions the ground state for the ensemble is insulating.

This final consideration consists of how to include electron correlations. Basically, these are due to the fact that for an insulator if a carrier is instantaneously resident on a lattice ion, the surrounding ions polarize dynamically in response to this charge. It is possible to accurately compute this correlation energy for insulators using the electron polaron model (A. B. Kunz, 1972). This model is based upon doing a two particle excitation configuration interaction calculation (CI) by means of ordinary Rayleigh-Schrodinger perturbation theory or by means of Brillouin-Wigner Perturbation theory. We include this correlation in the calculations presented here. It is possible similarly to correlate metallic systems using a field of virtual plasmons to describe the system's excitations (A. W. Overhauser, 1971).

III. DEVELOPMENT OF THE UHF METHOD FOR NARROW BAND MATERIALS

A narrow band material is one in which the overlap of the atomic orbitals from site to site is small ($\sim$ 0.1 or so) such that the band widths are narrow (0-8 eV roughly). In the case where bands are narrow the features of the UHF solution discussed in Section II, makes it possible for a system of partly filled atomic subshells to have all filled energy bands and thus be insulating. Let us consider this possibility in detail. We note that Seitz (1940) has shown that for systems of filled bands one may rotate the eigenfunctions of the Fock operator into local orbitals (a Heitler-London representation) and there is a one to one relationship between the band states and the local orbitals. This fact characterizes an insulator. If the bands are only partly filled there is only a many to one relationship between the localized orbitals (Wannier functions usually) and the band states. This fact characterizes a metal. We note it is possible for the insulator to have a zero actual gap and hence be really a semimetal or have a small gap characteristic of a semiconductor as well as a large gap. The size of this gap is a subject for numerical solution and is called the photo conducting gap. There may be other gaps of smaller size in which one forms excitons or local excitations.

Consider a translationally invariant system containing one or more atom with a partly filled subshell per unit cell. We assume each unit cell has the same atoms in the same location as any other one. We use the variational theorem. Let the functions $x_{\alpha m}(r)$ refer to a set of local orthonormal basis functions. The label α refers to site and the m to which function at that site. Then if the eigensolutions of Eq. (3) obey Bloch's theorem and if the resultant bands are partly filled, the eigenfunctions, $\phi_{\vec{k}j}$ are;

$$\phi_{\vec{k}j}(\vec{x}) = \frac{1}{\sqrt{N}} \sum_{\alpha,m} e^{i\vec{k}\cdot\vec{R}_\alpha} A_{jm\alpha}(\vec{k})\, x_{\alpha m}(\vec{r})\begin{Bmatrix}\alpha \\ \beta\end{Bmatrix} \tag{5}$$

The label $\vec{k}$ refers to symmetry labels and m is a band index. The coefficients, $A_{jm_\alpha}(\vec{k})$ are chosen to minimize total energy. If we define:

$$V_{NN} = \frac{1}{2}\sum_{IJ}{}' \frac{e^2 Z_I Z_J}{|\vec{R}_I - \vec{R}_J|} ,$$

$$f_{\alpha i\beta j} = \langle x_{\alpha i}|f_1|x_{\beta j}\rangle ,$$

$$g_{\alpha i\beta j\gamma k\delta \ell} = \langle x_{\alpha i}\, x_{\beta j}|g_{12}|x_{\gamma k} x_{\delta \ell}\rangle \tag{6}$$

and

$$S_{\alpha i\beta j} = \sum_{\substack{\ell \\ \ell,\vec{k}\ \text{occupied}}} \sum_{\vec{k}} \frac{1}{N} A_{\ell i\alpha}(\vec{k})\, A_{\ell j\beta}(\vec{k})\, e^{i\vec{k}\cdot(\vec{R}_\beta - \vec{R}_\alpha)} .$$

Then the total energy in the symmetry adapted case, E_T^S, is:

$$E_T^S = \sum_{\substack{\alpha m \\ \beta n}} S_{\alpha m\beta n}\, f_{\alpha m\beta n} + \frac{1}{2}\sum_{\substack{\alpha p \\ \beta q \\ \gamma r \\ \delta s}} [S_{\alpha p\gamma r}\, S_{\beta q\delta s} - S_{\alpha p\delta s}\, S_{\beta q\gamma r}] \times g_{\alpha p\beta q\gamma r\delta s} + V_{NN} \tag{7}$$

This is the exact energy for a single determinant trial function using one electron orbital of the form of Eq. (5). In Eq. (7)

there are terms independent of $(\vec{R}_I-\vec{R}_J)$, terms which go as $|\vec{R}_I-\vec{R}_J|^{-1}$ and terms which fall of as $\exp(-|\vec{R}_I-\vec{R}_J|)$. For sufficiently large lattice constant only the constant terms remain here and these yield a total energy (keeping in mind all unit cells are identical).

$$E_T^S \doteq N \sum_{mn} S_{\alpha m\alpha n} f_{\alpha m\alpha n} + \frac{1}{2} \sum_{\substack{pq\\rs}} [S_{\alpha p\alpha r} S_{\alpha q\alpha s} - S_{\alpha p\alpha s} S_{\alpha q\alpha r}] \times g_{\alpha p\alpha q\alpha r\alpha s} = N \tilde{E}_{ps} . \quad (8)$$

We may use an alternate starting trial eigenfunction. In this one there are assumed to be m electrons per unit cell and these can be considered to occupy m of the x's. Thus there is a one to one correspondence between occupied x's, electrons and eigensolutions of Eq. (3). Assuming here a finite gap this is an insulating state. Let us for convenience define a coefficient $\sigma_{i\alpha}$ to be unity if $x_{\alpha i}$ is occupied and 0 otherwise. Therefore, the total energy, E_T^A, in the non-symmetry adapted case, given exactly is:

$$E_T^A = \sum_{i\alpha} \sigma_{i\alpha} f_{i\alpha i\alpha} + \frac{1}{2} \sum_{\substack{i\alpha\\j\beta}} \sigma_{i\alpha} \sigma_{j\beta} [g_{i\alpha j\beta i\alpha j\beta} - g_{i\alpha j\beta j\beta i\alpha}] . \quad (9)$$

Again, in the limit of large lattice constant for a system of N identical unit cells one has

$$E_T^A = N \sum_i \sigma_{i\alpha} f_{i\alpha i\alpha} + \frac{1}{2} \sum_{i,j} \sigma_{i\alpha} \sigma_{j\alpha} [g_{i\alpha j\alpha i\alpha j\alpha} - g_{i\alpha j\alpha j\alpha i\alpha}] . \quad (10)$$

$$= N \tilde{E}_{At}$$

We note here for neutral unit cell systems this expression is correct up to and including terms falling off as $|\vec{R}_I-\vec{R}_J|^{-1}$ but not exponentially fast.

The essential question is which total energy is lower that of Eq. (10) or Eq. (8). If we choose the x's to produce the least energy solution for (10), the occupied x's just become the ground state UHF orbitals for the atoms in the unit cell and the total energy is just N times the atomic energy. In the case of Eq. (8) the total energy may be shown to be of the form of an atom in the UHF limit in which the ground state orbitals are only partly occupied unless the a's of Eq. (5) are such that a one-to-one relationship between electrons and occupied local orbitals occurs. This being the case it can be shown that

$$E_{At} \leq E_{ps} \, , \tag{11}$$

and thus the ground state for large lattice constant is one in which there is an insulating UHF trial function. This doesn't imply that the eigenfunctions of (3) are local; they have been shown for a filled band case such as here to be Bloch like (Brandow 1977 and references contained therein), only that all bands are filled or empty. Thus the ground state is inherently insulating here. If one retains terms of the form $|\vec{R}_I-\vec{R}_J|^{-1}$ in reducing Eq. (7) one finds the insulating state persists if we have only a single non-filled band and describe it by a single Wannier function per site per possible electron in the subshell until,

$$e^2/|\vec{R}_I-\vec{R}_J>_{\substack{\text{Nearest}\\ \text{Neighbor}}} \langle x_{\alpha i} s_{\alpha i}| g_{12} | x_{\alpha i} x_{\alpha i}\rangle \, . \tag{12}$$

Here $x_{\alpha i}$ is the Wannier function for the band in question. Since there is always some lattice constant for which this occurs, the UHF model has an insulator to metal transition. Typically for the transition metals the value of the matrix element in Eq. (12) for a 3d shell is about 1 Ry and thus the nearest neighbor distance for the transition is about 2 au. For such a distance overlaps are far from negligible so that Eq. (12) cannot predict where the transition occurs only that a transition is possible. We see that including overlap terms causes the transition to occur at larger separation. In particular we add twice the band width to $e^2/|\vec{R}_I-\vec{R}_J|$ for the partly filled subshell and then see if it exceeds the matrix element of (12). This is seen in Table 2 for several systems.

Before we can use this model, solving for Eq. (3), self-consistently there is one last detail to consider. The eigenvalues and hence energy bands are defined by ionization processes. Therefore, we must consider whether the new hole (electron) is local or Bloch like. If it is Bloch like the Koopman's eigenvalue corrected for correlation yields the correct definition, if a local calculation using relaxation corrections yields the eigenvalue and may include correlation corrections also. Note this hole (electron) localization is a artifact of the UHF solution. In a more correct description by projecting on the UHF trial wave function one has a solution in which the hole (electron) hops from site to site. We don't consider this here. None the less, there is a great distinction here. If localization occurs, the hole (electron) conduction is via hopping but

TABLE 2

Width, relaxation, and other relevant parameters for severed transition metal oxide systems are given as well as some compiled values of other relevant parameters discussed in the text. Results in eV.

System / Property	TiO	FeO	CoO	NiO
3d width	6.8 eV*	1.3 eV*	2.5 eV*	3.0 eV*
3d relaxation	1.4 eV*	2.9 eV*	3.7 eV*	5.0 eV*
predicted photo-produced hole conduction type	conduction	hopping	hopping	hopping
e^2/n.n. dist.	6.9 eV	6.7 eV	6.7 eV	6.9 eV
$\langle 3d3d \vert g \vert 3d3d \rangle$	17.5 eV	23.7 eV	25.1 eV	26.5 eV
predicted conductivity type	metal	insulator	insulator	insulator
experimental conductivity type	metal	insulator	insulator	insulator

*From symmetry adapted calculations. This is the appropriate limit if true Bloch like behavior occurs.

if not, is via the normal Bloch electron conductivity mechanism. Techniques similar to those used in studying the ground state nature show that the hole (electron) is local provided the hole (electron) lies below the vacuum level, and that the relaxation energy for that subshell on an ion is greater in magnitude than the width for that band. We now study the energy bands in detail for NiO to test these ideas and to study the NiO spectra. Table 2 shows results for the values of the 3d hole on several systems.

IV. CALCULATIONS FOR NiO IN SEVERAL LIMITS

In order to test the quantitative utility of the preceeding considerations we have made detailed calculations for TiO, VO, FeO, CoO and NiO. We will discuss the NiO results here briefly. Early attempts were made to look at this system in a RHF like limit in which the non-local exchange was replaced by a local one using the Slater prescription. The first calculation was a non-self-consistent one and is due to Mattheiss (1972). It is seen in Fig. 3. The second attempt of this type is a self-consistent result due to Collins, Kunz and Ivey (1975). It is

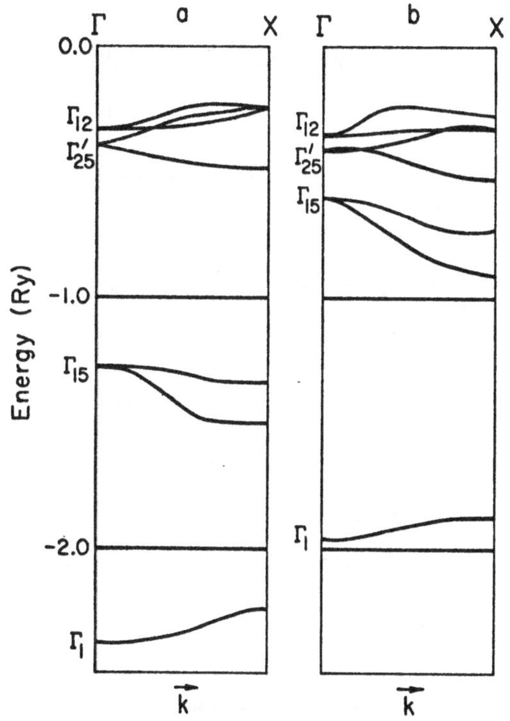

Fig. 3 Non-magnetically ordered NiO band using the Slater exchange are shown. In part a the non-self-consistent results of Mattheis are given and in part b the self-consistent results of Collins, Kunz and Ivey are shown.

seen also in Fig. 3. The essential features are: self-consistency is vital; there is a gap between the O 2p and the Ni 3d band; both results predict conducting behavior; there is an energy gap between the occupied 3d level and the virtual 4s level of about 5 eV.

The next level of sophistication is the SPHF type results of Wilson (1968, 1970) discussed earlier and shown in Fig. 1. Here again the local approximation is made for the exchange.

The self-consistent RHF bands have been evaluated for NiO (Kunz and Surratt 1978) and are seen in Fig. 4. These bands have also been corrected for relaxation and correlation effects. These results also appear in Fig. 4. We discuss these latter results here. The essential features are: correlation greatly modifies the band structure; the system here is predicted to be conducting; the width of the Ni 3d band is similar to that in the local exchange calculations; the width of the O 2p band is much

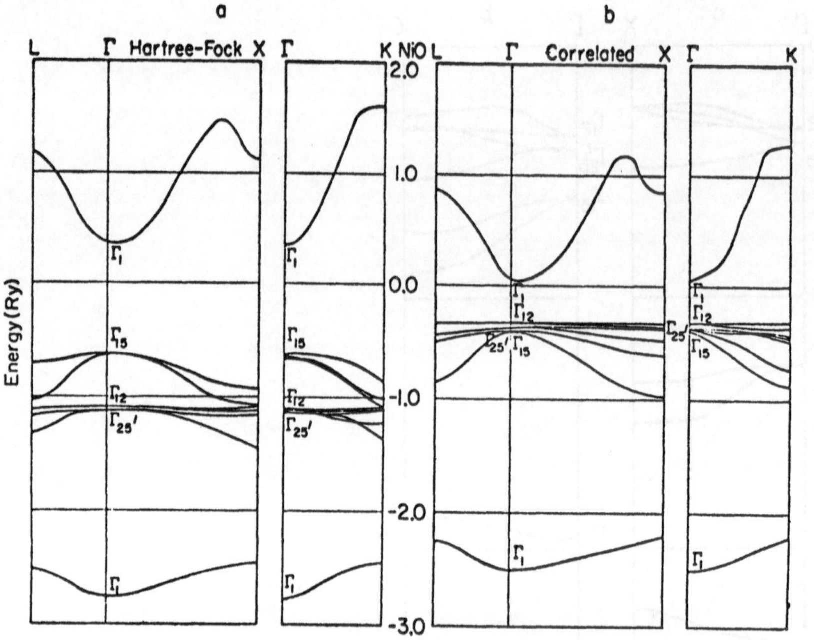

Fig. 4 The RHF results for NiO are in part a and the correlated RHF results are in part b. The results are self-consistent.

greater than in the local exchange calculation. The Ni-3d to Ni4s gap is about 5 eV; the Ni 3d and O 2p band overlap in energy.

The next result is for NiO in an antiferromagnetic state in which we symmetry adapt the solution producing a SPHF result. This calculation is self-consistent. This is a non local exchange equivalent to the Wilson result. The SPHF results appear in Fig. 5 as do the SPHF results corrected for correlation and relaxation. We discuss here these later results. We find: correlation strongly modifies the bands; the system is conducting due to Ni 3d band Ni 3d band overlaps as well as Ni 3d band O 2p band overlap. This is due to slightly greater 3d band widths here than in the Wilson result. Otherwise, the comments to our RHF solution apply here also.

The final result considered here is the true UHF result. In order to consider this result it is first necessary to find if the hole (electron) states will be Bloch like or local in the one electron picture (using one Slater determinant). We compute the band widths and relaxation energies first. This is seen for the 3d level in Table 2. From this study we deduce that: core holes in general localize; the O 2s and 2p levels

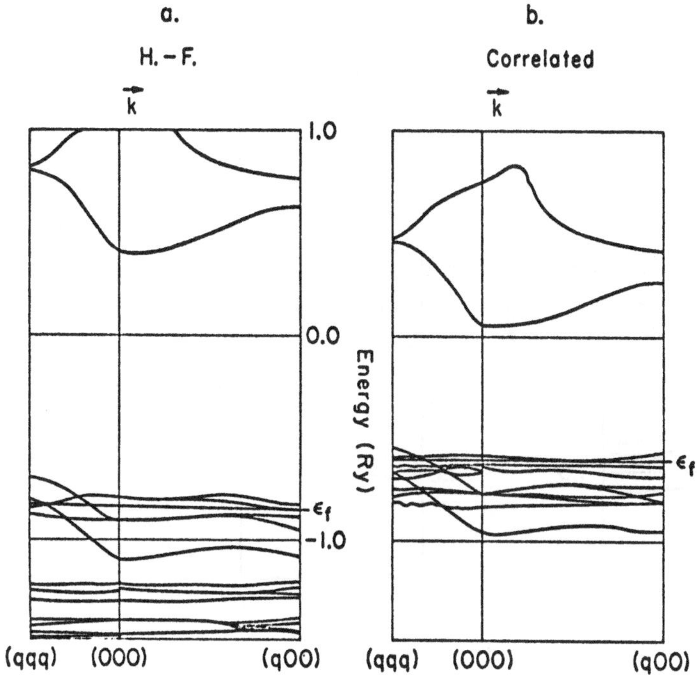

Fig. 5 The SPHF results for NiO areshown in part a and the correlated SPHF results are in part b. The results are for antiferromagnetic ordering and are self-consistent.

have Bloch holes; the Ni 3d holes or localize; the virtual levels form Bloch states. This behavior gives rise to some computational difficulties which are surmounted by using local orbital techniques, which are valid here since for our insulating NiO we have a "filled subshell" case, to construct a first order density matrix for the Fock operator self-consistently. We then diagonalize the resulting Fock matrix using a basis of local orbitals plus virtual local states subject to the constraint that the Bloch states be orthogonal to the occupied 3d manifold. This produces the O 2p and the conduction bands (Ni 4s like mostly) shown in Fig. 6 The 3d levels are then obtained by diagonalizing the 3d Fock matrix in a cluster model such that the 3d eigenvalues using Koopman's theorem here correspond to the formation of "local unrelaxed" hole states. These levels are seen in Fig. 6 for the filled and the virtual 3d manifold. Finally, relaxation effects (actually needed to have a local 3d level) are added along with correlation corrections to form the Bloch and local level scheme shown in Fig. 7. The important observations to be reached from Figs. 6 and 7 are that: there is substantial spin polarization of the occupied 3d manifold; due to the absence of a central potential about a Ni ion there is

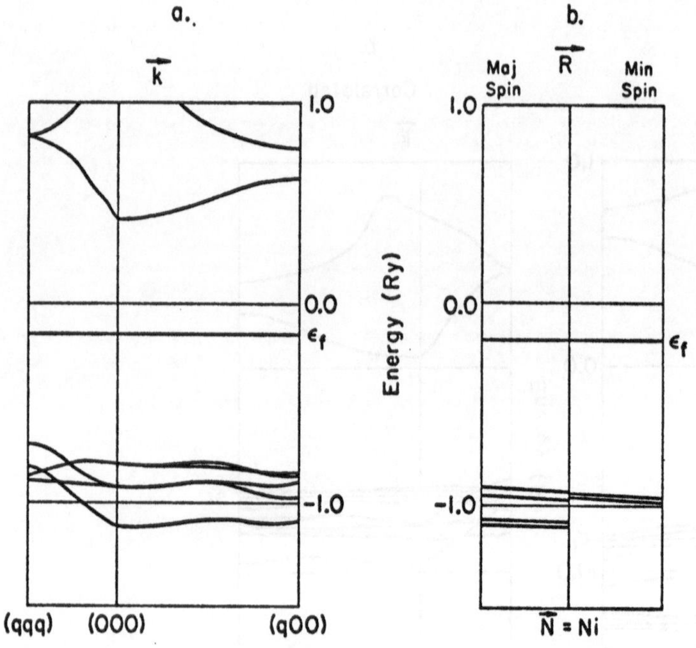

Fig. 6 The self-consistent UHF results are shown for NiO. In part a the band states (hole or electron) are shown and in part b the local states (hole or electron) as defined by ionization processes are seen.

splitting in the 3d local states due to this as well as to the crystal field effect; there is a substantial gap between the occupied and the virtual 3d manifold; this calculation finds NiO to be an insulator; relaxation and correlation effects are important; hole conduction in the Ni 3d level is by hopping; electrons in the 4s level conduct by normal band processes.

In carrying out this calculation we find each Ni ion has five electrons in the 3d level of one spin and 3 in the opposite spin. Thus due to the small hybridization computed for the 3d level on an ion with other occupied ions we find site by Ni site that Hund 's rule is obeyed. Thus each Ni ion has a magnetic moment. Furthermore, by calculation of total energies including nearest O neighbor super exchange we find each Ni likes to have as many of its neighboring Ni's of opposite spin as possible. Thus, each Ni has six nearest neighbor of like spin and six of opposite. This is consistent with (111) sheets of Ni having opposite spin. Furthermore, at this level of calculation a Néel temperature of 100 K is predicted. Due to diffuseness of the O 2p orbitals, the inclusion of superexchange to nearest neighbor O is not converged. We are currently extending this superexchange calculation to second and third neighbors. This

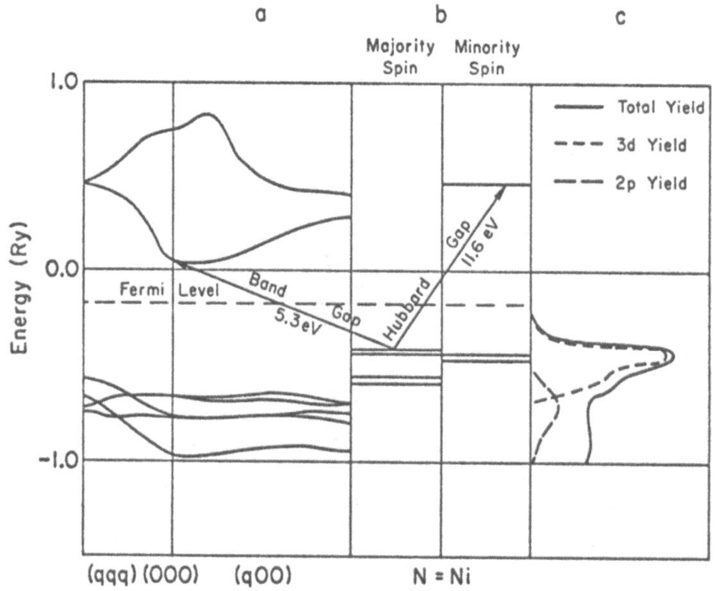

Fig. 7 The correlated UHF results for NiO are shown. The Bloch states are in part a, the local states as defined by ionization are in part b, and the photoemission data of Eastman and Freeouf are shown in part c to be compared to the occupied energy bands.

also means we label the 3d levels by a majority and minority spin label where which spin this is varies from site to site. This also implies above the Néel temperature spin ordering vanishes due to randomization of local moments but the dominant character remains insulating. Since the O 2p levels band, we don't distinguish site to site for majority, minority spins here.

We can make several predictions here. The photo gap is as shown from the Ni 3d level to the Ni 4s conduction band and has a predicted gap of 5.3 eV (experiment is 3.8-4.0 eV). If we did a symmetry projection on the hole state, the 3d levels would broaden and would lower the predicted value, so the error is of the right sign. The Hubbard gap is as shown here and is about 15 eV. This is also the predicted value for Δ since the hopping widths are not included in this figure. The photo emission data of Eastman and Freeouf (1975) are also shown in Fig. 7. The deconvolution of experiment into a 3d and 2p manifold is due to Eastman and Freeouf. Since the absolute energy of the experiment is ill defined we align the 3d peaks in theory and experiment. Agreement is excellent. Finally, we show the optical absorption data of Powell and Spicer (1977) for NiO in

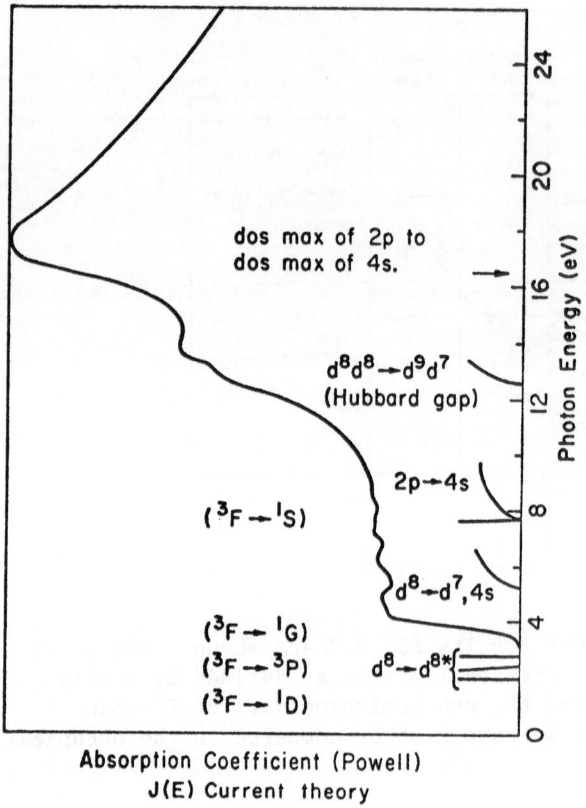

Fig. 8 A summary of the theoretical and experimental results for optical absorption spectrum of NiO is given.

Fig. 8. The theoretical spectral features of importance are included here as we compute them including the $3d^8 \rightarrow 3d^{8*}$ structure. This $3d^8 \rightarrow 3d^{8*}$ transitions lie below the observed threshold, but are either spin or dipole forbidden so we don't expect them to be easily seen experimentally. The figure is largely self explanatory. In conclusion we find a UHF model adequate for qualitative features of the Mott insulators but to achieve quantitative accuracy inclusion of correlation is essential.

*Research supported in part by the U.S. Air Force Office of Scientific Research, Grant AFOSR-76-2989.

V. A MOLECULAR TEST OF THESE IDEAS

It is possible to study the phenomena discussed here for molecular systems and there is one advantage in doing so. In Section III we discuss the criteria to be applied in determining if a system in a UHF limit is insulating or metallic and if a hole (electron) will be Bloch like or localized. The criteria in all cases is based upon total energy considerations and for solids exact evaluation is not practical, however, a set of approximate criteria are developed based upon some well defined but approximate calculations such as atomic relaxation energies, band widths, Coulomb self-repulsions, etc. have been given. In the case of a molecular system, it should be possible to perform, within the UHF limit, exact evaluations of total energy for all types of state and also evaluations of the quantities needed for the approximate criteria presented and thus test in a concrete way the approximate criteria discussed in Section III.

Several systems were studied here, He_2, He_4, He_8, H_2, H_4, Ni_2 and Ni_{10}. In all cases similar qualitative results were obtained. The most complete study performed was for He_8 and it will be discussed here as the prototype for all the other studies. For each He atom a basis of 6 s-like gaussian orbitals was used and the gaussian exponents were varied to minimize the energy of the He atom. The orbitals were not contracted. The eight He atoms were placed on the corners of a cube so that all He atoms were equivalent. The cube edge was varied in length. Since in a finite molecular system the UHF solution always has a finite energy gap between the lowest virtual state and the highest occupied state a study of the insulator metal transition is not possible since one cannot tell whether the gap is due to finiteness of system or other causes. However, one can ionize the system singly and see if the resulting hole in the system ground state is local or Bloch like. Therefore, as a function of cube edge length, I calculate the total energy of He_8, and He_8^+ using both Bloch and local hole starting input for He_8^+. The hole character is reexamined at the end of the run to see if the He_8^+ local input retains its local hole character and also to see which input produces the lowest energy state. In keeping with the approximate criteria, compute the He atomic relaxation energy and the width of the 1s He band. These factors and the predictions are seen in Table 3. In this table the 8th He atom is the one chosen at input to have the local hole.

The lesson of Table 3 is clear. The simple criteria for hole type work rather well, predicting the actual crossover to a cube edge length charge of 0.05 au. We also see that the relevant parameters are a very sensitive function of distance so the theory seems to work well. Finally, we note that in the exact solution the local hole is just that, very local. Even at the edge of

Table 3

He_8 results are given as a function of cube edge length, a, in a.u. Energies are in eV here.

quantity \ a	2.0	3.5	3.65	3.70	3.75	4.0	6.0
a.o. overlap	.3267	-	-	-	-	.0497	.0094
occ. band width	32.12	2.39	1.68	1.49	1.32	0.64	0.00
at. rel. energy	1.42	1.42	1.42	1.42	1.42	1.42	1.42
Approximate theory hole type	Bloch	Bloch	Bloch	Bloch	local	local	local
exact UHF hole type	Bloch	Bloch	Bloch	local	local	local	local
no. electrons on He (8)	1.88	1.88	1.88	1.04	1.01	1.00	1.00

going delocal, the hole is still almost perfectly localized on the eighth He atom. Similar results to this were obtained for other molecular systems studied.

REFERENCES

Adler, D. (1968) Solid St. Phys. 21, 1

Adler, D. and Feinleib J. (1970) Phys. Rev. B 2, 3112.

Brandow, B. (1976) Int. J. Quant. Chem. Symp. 10, 417.

Brandow, B. (1977) Adv. in Phys. 26, 651.

Brown, F. C., Gahwiller, C. and Kunz, A. B. (1971) Sol. St. Comm. 9, 487.

Collins, T. C., Kunz, A. B. and Deutsch, P. W. (1974) Phys. Rev. A 10, 1034.

Collins, T. C., Kunz, A. B. and Ivey, J. (1975) Int. J. Quant. Chem. Symp. 9, 519.

Eastman, D. E. and Freeouf, J. L. (1975) Phys Rev. Lett. 34, 395.

Hubbard, J. (1963) Proc. R. Soc. A 276, 238.

Hubbard, J. (1964) Proc. R. Soc. A 277, 237.

Hubbard, J. (1965) Proc. R. Soc. A 285, 542.

Hubbard, J. (1966) Proc. R. Soc. A 296, 82.

Kunz, A. B. (1972) Phys. Rev. B 6, 606.

Kunz, A. B. and Surratt, G. T. (1978) Sol. St. Comm. 25, 9.

Löwdin, P. O. (1966) Quantum Theory of Atom Molecules and the Solid State, P. O. Löwdin, Editor, 601.
Mattheiss, L. F. (1972) Phys. Rev. B 5, 290.
Mickish, D. J., Kunz, A. B. and Collins, T. C. (1975) Phys. Rev. B 9, 4461.
Mott, N. F. (1949) Proc. Phys. Soc. A 62, 416.
Mott, N. F. (1952) Prog. Metal Phys. 3, 76.
Mott, N. F. (1956) Can. J. Phys. 34, 1356.
Mott, N. F. (1958) Supplto. Nuovi. Ani. 7, 312.
Mott, N. F. (1961) Phil. Mag. 6, 281.
Mott, N. F. (1969) Phil. Mag. 20, 1.
Overhauser, A. W. (1971) Phys. Rev. B 3, 1888.
Powell, R. J. and Spicer, W. E. (1970) Phys. Rev. B 2, 2182.
Seitz, F. (1940), Modern Theory of Solids.
Wilson, T. M. (1968) Int. J. Quant. Chem. Symp. 2, 269.
Wilson, T. M. (1970) Int. J. Quant. Chem. Symp. 3, 757.

ELECTRONIC STRUCTURE AND EXCITED STATES OF POLYMERS

J. Ladik

Lehrstuhl für Theoretische Chemie der Friedrich-Alexander-Universität Erlangen-Nürnberg, 852 Erlangen, FRG and Laboratory of the National Foundation for Cancer Research at the University Erlangen-Nürnberg

ABSTRACT. The SCF LCAO Crystal Orbital (Hartree-Fock) method is reviewed in the cases of polymers with simple translational symmetry and with a combined symmetry operation. As example the band structure of the periodic DNA model polycytosine is discussed.

For the treatment of aperiodic (multicomponent) polymers the coherent potential approximation with a k- and energy-dependent self energy is described. As illustrative example calculations for the $(SN)_x$ - $(^{SN}_{H})_x$ two-component mixed polymer are presented. To investigate the extra levels of a cluster of impurities embedded in a periodic crystal the SCF resolvent method has been developed.

For the calculations of excited states of polymers applications of the $\hat{O}\hat{A}\hat{O}$ method are shown. Further the applicability of the more general Green's function formalism for the investigation of localized excitations in polymers is discussed. Finally to treat partially delocalized excitations in polymers the formalism of the intermediate (charge transfer) exciton theory and its application to polymers is reviewed.

1. INTRODUCTION

Polymers are of great interest both for practical purposes and from the point of view of physical considerations. Polymers are forming plastics, biopolymers like nucleic acids (DNA and RNA) and proteins play a fundamental role in life processes and recently highly conducting polymers like $(SN)_x$ or the TCNQ-TTF, TCNQ-NMP systems with pseudo one-dimensional chains embedded in a three-dimensional molecular crystal seem to be the candidates for the discovery of such new physical phenomena which do not occur im simpler solids. In the last years a vast amount of

Cleanthes A. Nicolaides and Donald R. Beck (eds.), Excited States in Quantum Chemistry, 495–529.

experimental information has been gathered especially about the different properties of the highly conducting polymers. To interpret these different physical and chemical properties (which underlie in the case of biopolymers also their biological functions) one has to obtain a fair knowledge of the electronic structure of polymers in their ground- and excited states.

If a polymer is completely aperiodic one can treat it only approximately. If the polymer is built up from 2 or more components in a random sequence, one can apply to it the coherent potential approximation (CPA) with a k- and energy dependent self energy [1]. On the other hand if we have a cluster of impurities or a distortion embedded in a periodic system the resolvent method can be used in its SCF form with an AO basis (ab initio SCF LCAO resolvent method [2]). Since in both cases gaps may occur in the density of the parent periodic system the effects of disorder can be considered as certain types of excited states.

If we are able to define in a polymer or in a molecular crystal an elementary cell, by the repetition of which (the symmetry operation is not necessarily a simple translation, it may be a combined operation of a simultaneous translation and rotation) we can build up the whole system, we can treat this system with the aid of a proper combination of quantum chemical and solid state physical methods.

In this paper we shall review the ab initio SCF LCAO crystal orbital (CO) method for the treatment of the ground state of periodic polymers. It will be shown that the derivation presented is valid also in the case of a combined symmetry operation. As next point the application of the CPA method for aperiodic polymers will be discussed and the SCF resolvent method will be formulated. In the following section the treatment of excited states of polymers assuming local excitations in the subunits ($\hat{O}\hat{A}\hat{O}$ method, Green's function treatment) will be briefly reviewed (for a more detailed discussion of these methods see the papers of T.C. Collins and W. von Niessen in this book). Finally the formalism of the intermediate exciton theory and its application to polymers will be presented.

2. TREATMENT OF THE GROUND STATE OF POLYMERS

2.1 Periodic Polymers

2.1.1 Ab initio SCF LCAO Crystal Orbital (CO) Method in the Case of Simple Translation.

If we have m orbitals in the unit cell of a three-dimensional polymer or molecular crystal and the number of unit cells in the direction of each crystal axis is 2N+1, we can write the delocalized crystal orbitals (CO-s) in the LCAO approximation in the form

$$\Phi_k^{\vec{p}} = \sum_{\vec{q}} \sum_{g=1}^{m} C(\vec{p})_{k;\vec{q},g} \chi_g^{\vec{q}} \quad , \tag{1}$$

where $\vec{p}=(p_1,p_2,p_3)$, $\vec{q}=(q_1,q_2,q_3)$, the integers p_j and q_j $(j=1,2,3)$ run from $-N,\ldots 0,\ldots,N$ and $\sum_{\vec{q}}$ means the summation over all cells, $\sum_{q_1=-N}^{N} \sum_{q_2=-N}^{N} \sum_{q_3=-N}^{N}$. Further $\chi_g^{\vec{q}} = \chi_g(\vec{r}-\vec{R}_{\vec{q}}-\vec{r}_{g_A})$ is the gth AO (which belongs to the atom with position vector $\vec{r}_{g_A}$) of the cell characterized by the vector $\vec{R}_{\vec{q}}=q_1\vec{a}_1+q_2\vec{a}_2+q_3\vec{a}_3$.

Writing down the expectation value

$$\frac{\langle \Phi_k^{\vec{p}} | \hat{F} | \Phi_k^{\vec{p}} \rangle}{\langle \Phi_k^{\vec{p}} | \Phi_k^{\vec{p}} \rangle} = \varepsilon(\vec{p})_k \tag{2}$$

of the Fock operator $\hat{F}$ and performing a Ritz variation procedure, we obtain in the standard way for the whole polymer the matrix equation

$$\underline{\underline{F}}\, \underline{C}(\vec{p})_k = \varepsilon(\vec{p})_k\, \underline{\underline{S}}\, \underline{C}(\vec{p})_k \ . \tag{3}$$

The hypermatrices $\underline{\underline{F}}$ and $\underline{\underline{S}}$ have the dimensions MxM with $M=m\times(2N+1)^3$ and their elements are defined as

$$F_{\vec{p},f;\vec{q},g} = \langle \chi_f^{\vec{p}} | \hat{F} | \chi_g^{\vec{q}} \rangle \ , \tag{4a}$$

$$S_{\vec{p},f;\vec{q},g} = \langle \chi_f^{\vec{p}} | \chi_g^{\vec{q}} \rangle \ , \tag{4b}$$

respectively. (The mxm blocks of $\underline{\underline{F}}$ and $\underline{\underline{S}}$ give the interactions between the orbitals belonging to the different unit cells.)

Taking into account the translational symmetry and introducing periodic boundary conditions it is easy to show that $\underline{\underline{F}}$ and $\underline{\underline{S}}$ are cyclic hypermatrices, i.e. they are cyclic in their blocks (for a detailed derivation see [3, 4]). It is possible to show [5] that with the aid of the unitary matrix $\underline{\underline{U}}$ which has the mxm blocks

$$\underline{\underline{U}}_{\vec{p},\vec{q}} = (2N+1)^{-3/2} \exp\left(\frac{i 2\pi \vec{p}\vec{q}}{2N+1}\right) \underline{\underline{1}} \tag{5}$$

we can blockdiagonalize $\underline{\underline{F}}$ and $\underline{\underline{S}}$. To perform this blockdiagonalization we can write

$$\underline{\underline{U}}^{\dagger}\underline{\underline{F}}\,\underline{\underline{U}}\,\underline{\underline{U}}^{\dagger}\underline{C}(\vec{p})_k = \varepsilon(\vec{p})_k\, \underline{\underline{U}}^{\dagger}\underline{\underline{S}}\,\underline{\underline{U}}\,\underline{\underline{U}}^{\dagger}\underline{C}(\vec{p})_k \ , \tag{6}$$

or

$$\underline{\underline{F}}'\, \underline{D}(\vec{p})_k = \varepsilon(\vec{p})_k\, \underline{\underline{S}}'\, \underline{D}(\vec{p})_k \ , \tag{7}$$

where

$$\underline{\underline{F}}' = \underline{\underline{U}}^{\dagger}\underline{\underline{F}}\,\underline{\underline{U}} \quad , \quad \underline{\underline{S}}' = \underline{\underline{U}}^{\dagger}\underline{\underline{S}}\,\underline{\underline{U}} \quad , \quad \underline{D}(\vec{p})_h = \underline{\underline{U}}^{\dagger}\underline{C}(\vec{p})_h \ . \tag{8}$$

The $\vec{p}$th diagonal block of the blockdiagonal matrices $\underline{\underline{F}}'$ and $\underline{\underline{S}}'$ is given by the expression [3,4]

$$\underline{\underline{F}}'(\vec{p}) = \sum_{\vec{q}} \exp\left[i2\pi\vec{p}\vec{q}/(2N+1)\right] \underline{\underline{F}}(\vec{q}) \ , \tag{9a}$$

$$\underline{\underline{S}}'(\vec{p}) = \sum_{\vec{q}} \exp\left[i2\pi\vec{p}\vec{q}/(2N+1)\right] \underline{\underline{S}}(\vec{q}) \ , \tag{9b}$$

respectively, where $\underline{\underline{F}}(\vec{q})$ and $\underline{\underline{S}}(\vec{q})$ stand for the submatrices of the original cyclic hypermatrices.

Using the fact that $\underline{\underline{F}}'$ and $\underline{\underline{S}}'$ are blockdiagonal, equ. (7) can be decomposed to the equations

$$\underline{\underline{F}}'(\vec{p})\,\underline{d}(\vec{p})_h = \varepsilon(\vec{p})_h \underline{\underline{S}}'(\vec{p})\,\underline{d}(\vec{p})_h \tag{10}$$

corresponding to the different blocks $\underline{\underline{F}}'(\vec{p})$ and $\underline{\underline{S}}'(\vec{p})$.

If $N \to \infty$ we can introduce the continous variables

$$k_j = \frac{2\pi p_j}{a_j(2N+1)} \qquad (j=1,2,3) \ . \tag{11}$$

Since the p_j-s had the values $-N,\dots,N$, the k_j will have values between $-\pi/a_j$ and π/a_j. Defining the vector $\vec{k}=k_1\vec{b}_1+k_2\vec{b}_2+k_3\vec{b}_3$, where the $\vec{b}_j$-s are the basis vectors of the reciprocal space (by definition $\vec{a}_i\vec{b}_j=2\pi\,\delta_{ij}$)$^{+}$ we can now write instead of (10)

$$\underline{\underline{F}}(\vec{k})\,\underline{d}(\vec{k})_h = \varepsilon(\vec{k})_h \underline{\underline{S}}(\vec{k})\,\underline{d}(\vec{k})_h \qquad (h=1,2,\dots,m) \tag{12}$$

with

$$\underline{\underline{F}}(\vec{k}) = \sum_{\vec{q}} \exp(i\vec{k}\vec{R}_{\vec{q}})\,\underline{\underline{F}}(\vec{q}) \ , \tag{13a}$$

$$\underline{\underline{S}}(\vec{k}) = \sum_{\vec{q}} \exp(i\vec{k}\vec{R}_{\vec{q}})\,\underline{\underline{S}}(\vec{q}) \ . \tag{13b}$$

To solve equ.-s (12) we can eliminate the overlap matrix $\underline{\underline{S}}(\vec{k})$ with the aid of Löwdin's symmetric orthogonalization procedure in analogy to the molecular case. The only difference is that both $\underline{\underline{F}}(\vec{k})$ and $\underline{\underline{S}}(\vec{k})$ are now not real, but Hermitian complex matrices (for the details see again [4]). We end up in this way with the eigenvalue equation

$$\underline{\underline{\tilde{F}}}(\vec{k})\,\underline{b}(\vec{k})_h = \varepsilon(\vec{k})_h\,\underline{b}(\vec{k})_h \tag{14}$$

where

$^{+}$ By identifying the vectors $\vec{k}$ with the crystal momentum we demand that they have to belong to the first Brillouin zone of the crystal.

$$\tilde{\underline{\underline{F}}}(\vec{k}) = \underline{\underline{S}}(\vec{k})^{-1/2}\,\underline{\underline{F}}(\vec{k})\,\underline{\underline{S}}(\vec{k})^{-1/2}\ ,\quad \underline{b}(\vec{k})_h = \underline{\underline{S}}(\vec{k})^{1/2}\,\underline{d}(\vec{k})_h\ . \tag{15}$$

Substituting into (14), respectively (4a) the expression

$$\begin{aligned}\hat{F} &= -\tfrac{1}{2}\Delta - \sum_{\vec{q}}\sum_{\alpha=1}^{M_\alpha}\frac{Z_\alpha}{|\vec{r}-\vec{R}_\alpha^{\vec{q}}|} + \sum_{\vec{p}}\sum_{h=1}^{n^*}\left[2\hat{J}(\vec{p},h) - \hat{K}(\vec{p},h)\right] \\ &= \hat{H}^N + \sum_{\vec{p}}\sum_{h=1}^{n^*}\left[2\hat{J}(\vec{p},h) - \hat{K}(\vec{p},h)\right]\end{aligned} \tag{16}$$

of the Fock operator (where M_A is the number of atoms in the unit cell, n^* is the number of filled bands,

$$\hat{J}(\vec{p},h;\vec{r}_1)\,\Phi(\vec{r}_1) = \langle \Phi_h^{\vec{p}}(\vec{r}_2)|\frac{1}{r_{12}}|\Phi_h^{\vec{p}}(\vec{r}_2)\rangle\,\Phi(\vec{r}_1)\ , \tag{17a}$$

$$\hat{K}(\vec{p},h;\vec{r}_1)\,\Phi(\vec{r}_1) = \langle \Phi_h^{\vec{p}}(\vec{r}_1)|\frac{1}{r_{12}}|\Phi(\vec{r}_2)\rangle\,\Phi_h^{\vec{p}}(\vec{r}_1) \tag{17b}$$

are the Coulomb and exchange operators, respectively), the LCAO form (1) of the crystal orbitals and introducing the charge bond order matrix of the polymer as

$$\underline{\underline{P}} = 2\sum_{\vec{p}}\sum_{h=1}^{n^*}\underline{C}(\vec{p})_h\,\underline{C}(\vec{p})_h^{\dagger}\ , \tag{18}$$

one can derive [3,4] for the $[\underline{\underline{F}}(\vec{q})]_{r,s}$ matrix elements the expression

$$\begin{aligned}[F(\vec{q})]_{r,s} &= \langle\chi_r^{\vec{0}}|\hat{H}^N|\chi_s^{\vec{q}}\rangle + \sum_{\vec{q}_1}\sum_{\vec{q}_2}\sum_{u,v=1}^{m} p(\vec{q}_1-\vec{q}_2)_{u,v}\ \cdot \\ &\cdot\left(\langle\chi_r^{\vec{0}}\chi_u^{\vec{q}_1}|\chi_s^{\vec{q}}\chi_v^{\vec{q}_2}\rangle - \tfrac{1}{2}\langle\chi_r^{\vec{0}}\chi_u^{\vec{q}_1}|\chi_v^{\vec{q}_2}\chi_s^{\vec{q}}\rangle\right)\ .\end{aligned} \tag{19}$$

Here $p(\vec{q}_1-\vec{q}_2)_{u,v}$ is the u,vth element of the submatrix $\underline{p}(\vec{q}_1-\vec{q}_2)=\underline{p}(\vec{q}_1,\vec{q}_2)$ for which (if we take into account the $\underline{C}(\vec{p})_h=\underline{\underline{U}}\ \underline{D}(\vec{p})_h$ transformation, the definition (5) of $\underline{\underline{U}}$ and introduce again instead of the vector $\vec{p}$ with discrete components the vector $\vec{k}$ with continously varying components (see equ. (11)) we can write

$$\underline{\underline{p}}(\vec{q}_1-\vec{q}_2) = \frac{2}{\omega}\int_\omega \sum_{h=1}^{n^*}\underline{d}(\vec{k})_h\,\underline{d}(\vec{k})_h^{\dagger}\exp[i\vec{k}(\vec{R}_{\vec{q}_1}-\vec{R}_{\vec{q}_2})]\,d\vec{k}\ , \tag{20}$$

where ω is the volume of the first Brillouin zone.

Equations (12),(13),(19) and (20) define then the ab initio SCF LCAO crystal orbital (Hartree-Fock) method with a non-local exchange for crystals or polymers with a simple translation symmetry. Using this method a number of calculations have been performed for linear chains [6] and also for simple three-dimensional crystals (simple metals and ionic crystals [7]). Different semiempirical forms (PPP CO, CNDO/2 CO, MINDO/2- and /3 CO etc.) of the described SCF LCAO CO method have been developed and applied to a large number of polymers also with more bulky unit cells [6,8,9].

2.1.2. The SCF LCAO CO Method in the Case of a Combined Symmetry Operation

We can apply the formalism developed in the proceeding point also in the case of a combined symmetry operation. To show this let us assume that we have a helix in which we get from one unit to the next by a translation $\vec{\tau}$ and a simultaneous rotation α . We can introduce then the helix operator

$$\hat{S}(\alpha,\vec{\tau}) = \hat{D}(\alpha) + \vec{\tau} \quad , \tag{21}$$

where $\hat{D}(\alpha)$ stands for the operator of the rotation around the main axis of the helix by an angle of α [10]. For the sake of simplicity let us assume further that after n repetitions[+] of the helix operation we obtain the "large" translation $\hat{T}$,

$$\hat{S}^n(\alpha,\vec{\tau}) = \hat{T} \quad . \tag{22}$$

We can further introduce again the Born-von Kármán periodic boundary conditions in the form

$$\hat{S}^{2N+1} = \hat{1} \quad , \tag{23}$$

where N is a large integer and measures the number of unit cells.

If $\hat{F}$ is the Fock operator of the helix it holds further that

$$[\hat{S},\hat{F}] = \hat{0} \tag{24}$$

and so we can classify the eigenfunctions of $\hat{F}$ according to the one-dimensional representations of the finite Abelian group $G=\{\hat{S}^m;\ m=1,\ldots,2N+1\}$. The kth representation of this group is thus $\xi_{1k}=\exp(\frac{i2\pi k}{2N+1})$. This means that the eigenvalue equation

$$\hat{S}^m \psi_k = exp[i2\pi mk/(2N+1)]\,\psi_k = \xi_{mk}\,\psi_k \tag{25}$$

has to be fulfilled, where ψ_k may have again an LCAO form (see equ. (1)), but k is now defined on the combined symmetry

---

[+] It should be mentioned that the following considerations hold also in the case, when (22) is not fulfilled, i.e. $\alpha/2\pi$ is not integer.

operation[+]. We can easily write down also the three-dimensional analogue of equ. (25) as

$$\hat{S}_1^{m_1}\hat{S}_2^{m_2}\hat{S}_3^{m_3}\Psi_{\vec{k}}=\exp[i2\pi(m_1k_1+m_2k_2+m_3k_3)/(2N+1)]\Psi_{\vec{k}}=\xi_{\vec{m}\vec{k}}\Psi_{\vec{k}}\quad,\tag{26}$$

where $S_j^{m_j}$ (j=1,2,3) means m_j repeated applications of the helix operator in the j-th direction.

To generate the eigenfunctions Ψ_k of S^m (in the simpler one-dimensional case) we can introduce the projection operator O_k [10]

$$\hat{O}_k=(2N+1)^{-1}\sum_{m=1}^{2N+1}\xi_{mk}\hat{S}^{-m}\tag{27}$$

which fulfills the relation $S^m\hat{O}_k=\xi_{mk}O_k$, $O_kO_l=\delta_{kl}$ and $\sum_k O_k=1$. If $\chi_g^q=\chi_g(\vec{r}-\vec{R}_q-\vec{r}_{g_A})$ stands again for the gth AO of the qth cell we can generate the generalized LCAO Bloch orbitals of the helix with the aid of the expression

$$\begin{aligned}\Psi_{k,g}(\vec{r})&=\hat{O}_k\chi_g(\vec{r}-\vec{R}_q-\vec{r}_{g_A})=\hat{O}_k\chi_g(\vec{r}-\vec{R}_q^{g_A})=\\&=(2N+1)^{-1}\sum_{m=1}^{2N+1}\xi_{mk}\hat{S}^{-m}\chi_g(\vec{r}-\vec{R}_q^{g_A})\qquad(\vec{R}_q^{g_A}=\vec{R}_q+\vec{r}_{g_A})\;.\end{aligned}\tag{28}$$

The same procedure can be applied also if we have in our reference cell not only a single AO but a linear combination of them (LCAO MO).

To be able to apply equ. (28) we have to express

$$\hat{S}^{-m}\chi_g(\vec{r}-\vec{R}_q^{g_A})=\chi_g[\hat{S}^m(\vec{r})-\vec{R}_q^{g_A}]\quad,\tag{29}$$

where the right hand side of (29) follows from the well known relation that S^{-m} applied to a function is identical with the transformation of the coordinate system under the inverse operation [11]. Taking into account that

---

[+]In other words the hypermatrices $\underline{\underline{F}}$ and $\underline{\underline{S}}$ will be again cyclic hypermatrices, if we construct their blocks $\underline{F}(\vec{q})$ and $\underline{S}(\vec{q})$ with the aid of the combined symmetry operation $\hat{S}$ (i.e. in the matrix elements (4a) and (4b) we can get from the cell characterized by $\vec{p}$ to the cell $\vec{q}$ by the repeated application of $\hat{S}$) and introduce periodic boundary conditions. These cyclic hypermatrices can be blockdiagonalized again with the aid of the unitary matrix (5) and we can introduce again instead of the vectors $\vec{p}$ the vectors $\vec{k}$ which have according to (11) continously varying components.

$$\hat{S}^m \vec{r} = \hat{D}(m\alpha)\vec{r} + m\vec{\tau}$$

we can further write

$$\chi_g[\hat{S}^m(\vec{r}) - \vec{R}_q^{g_A}] = \chi_g[\hat{D}(m\alpha)\vec{r} + m\vec{\tau} - \vec{R}_q^{g_A}] \quad . \tag{30}$$

Using the identity [10]

$$\hat{D}(m\alpha)\vec{r} - \hat{D}(m\alpha)\hat{S}^{-m}(\vec{R}_q^{g_A}) = \hat{D}(m\alpha)\vec{r} - \hat{D}(m\alpha)[\hat{D}(-m\alpha)\vec{R}_q^{g_A} - m\vec{\tau}] =$$
$$= \hat{D}(m\alpha)\vec{r} - \vec{R}_q^{g_A} + m\vec{\tau}$$

we can write down our final result

$$\hat{S}^{-m}\chi_g(\vec{r} - \vec{R}_q^{g_A}) = \chi_g[\hat{D}(m\alpha)[\vec{r} - \hat{S}^{-m}(\vec{R}_q^{g_A})]] \quad . \tag{31}$$

This means that by applying the helix operator S^{-m} to an AO we have to (1) perform m times the helix operation on the position of the nucleus and (2) we have to rotate the argument of the AO with the angle $m\alpha$ around the axis of the helix.

Taking into account the result (31) the ab initio SCF LCAO CO program for linear chains has been modified to be applicable also for the case of a combined symmetry operation. This modified program has been applied for a single polycytosine helix [12]. The basis used in this pilot calculation was a linear combination of Gaussian lobes of the form $\chi(r)=\left(\frac{2\alpha}{\pi}\right)^{3/4} e^{-\alpha r^2}$. The exponents α of the uncontracted Gaussian lobes and the contraction coefficients were taken for the heavy atoms from [13], for the hydrogen atoms from [14]. Though the applied basis was a minimal one (5 contracted Gaussians on the C,N and O atoms and one on the H atom) comparative calculations on molecules (for which calculations with better basis sets are also available) have shown that the overall description of the valence-shell one-electron properties (valence-shell contributions to the dipole moments, ionization potentials etc.) is with this basis rather good [12].

In the band structure calculation of polyC the nearest neighbors' interactions approximation (which in the case of this stacked chain is rather probably not very poor) has been applied. It should be emphasized that in this computation the non-local Hartree-Fock exchange term has been used without any approximation to it and all integrals in an absolute value larger than 10^{-8} a.u. have been retained. To obtain consistent results 9 different k values in the first Brillouin zone were needed. The geometry applied forthe polyC is the same as that of a Watson-Crick-type helix for which the structural data of DNA B [15] were used.

The energy band structure obtained consists of forty-five bands (corresponding to the 45 basis functions per unit cell) of which twenty-nine are doubly filled. The correspondence between the individual MO levels (ε^{MO}) and between the bands is always unambigous as it can be seen from an inspection of Table I which shows the physically most interesting four bands around the Fermi level (for further details see [12]).

Table I

The physically most important four energy bands around the Fermi level of periodic polycytosine. The original molecular energy levels (ε^{MO}), the band minima and maxima (ε^{CO}_{min} and ε^{CO}_{max}, respectively) with the corresponding ka values and the band widths($\Delta\varepsilon$) are given in eV

ε^{MO}	ε^{CO}_{min}(ka)	ε^{CO}_{max}(ka)	$\Delta\varepsilon$
4.585	4.813(0)	5.129(π)	0.316
1.929	1.5346(0)	2.775(π)	1.241
-9.766 [a)]	-9.665(π)	-9.113(0)	0.552
-11.488	-11.115(0)	-10.898(π)	0.218

a) Highest filled level.

Inspection of the wave functions shows that though the symmetry of the original MO-s is broken in polyC because of the stacked arrangement of the units, one can still define quasi-π - and σ-type bands. In this sense the valence band, the conduction band and the second lowest unfilled band are of quasi-π-type and the second highest filled band is of quasi-σ -type. Though the eight quasi-π -type bands are located mainly around the Fermi level, as the more detailed results [12] show, the σ-π separation is not fulfilled in polyC.

It is important to point out that the valence- and conduction bands resulting from the present ab initio calculation are much broader ($\sim$0.5 and $\sim$1.2 eV, respectively) than those obtained previously with different semiempirical CO methods [16]. If this trend will be valid also for other periodic DNA models (for which calculations are in progress) and also for calculations with larger basis sets, it will be inevitable to recalculate the transport properties of DNA [17] with these new band structures.

Concerning the positions of the bands it should be mentioned that though the description of the filled bands seems to be satisfactory (the theoretical first ionization potential of 9.11 eV obtained applying Koopman's theorem agrees quite well with the experimental value of 8.90 eV [18]), the gap ($\Delta E_g \sim$ 10.65eV) is clearly too large. This very large gap is a direct consequence of the failure of the

Hartree-Fock method in describing the virtual levels with the aid of an N particle potential instead of the appropriate N-1 particle potential [19]. To the correction of this gap we shall return at the treatment of the excited states of polymers.

2.2 Aperiodic Polymers

2.2.1 Application of the Coherent Potential Approximation (CPA) to Polymers

As it is well known, most polymers are aperiodic consisting of two or more components (for instance in DNA we have four different units, in proteins twenty). For the treatment of these substitutionally disordered polymers in which the overall geometry is the same (the double helix of DNA, the α-helix or β pleated sheet structure of proteins), but the units are different, the Coherent Potential Approximation (CPA) seems to be the most suitable method.+

The basis assumption of the CPA method is [1] that one substitutes the average of the multicomponent system by an effective medium determined so that the average fluctuation through the medium is zero. This can be achieved by replacing each site in the crystal but one++ with an unknown coherent potential. One embeds then at the reference site an A or a B component (in a simple case of an A, B two-component compositionally disordered system) with the probability f and 1-f, respectively. We solve then the problem of this single impurity embedded in the effective medium described by the coherent potential. The coherent potential on the other hand is determined by the self-consistency requirement that the average scattering (or fluctuation) from the chosen reference site is also zero.

\+ Until the different units are rather similar the simpler virtual crystal approximation (which essentially averages the Fock matrices according to the composition and the nearest neighbors' frequencies; see [2]) gives tolerable results, but in most practical cases this method does not provide an acceptable description of the aperiodic polymers [20]. On the other hand if the units of the aperiodic polymer are not very small, direct cluster calculations [21] cannot be performed, especially not in an *ab initio* form.

++ In this single site approximation of CPA it is assumed that after a scattering form this site no repeated scattering occurs before a scattering from another site has taken place.

To formulate mathematically the method[+] we can start from the Dyson equation of the single particle Green's function G of the disordered system,

$$G = G^0 + G^0 \Delta G \quad , \tag{32}$$

where G^0 is the Green's function of the unperturbed perfectly periodic system which is in its Fourier transformed form

$$G^0(\vec{k},E) = [E - \varepsilon^A(\vec{k})]^{-1} \; , \quad G^0(E) = \Omega^{-1} \sum_{\vec{k}} [E - \varepsilon^A(\vec{k})]^{-1} \tag{33}$$

(Ω is the volume of the crystal). Further the deviation from the perfectly periodic system A $\Delta(\vec{k})$ can be written[++] in the case of a single site diagonal perturbation [1] as

$$\Delta(\vec{k}) = \varepsilon^B(\vec{k}) - \varepsilon^A(\vec{k}) \; . \tag{34}$$

Equ. (32) (which can be derived [1] from the perturbation expansion

$$G = G^0 + G^0 \Delta G^0 + G^0 \Delta G^0 \Delta G^0 + \cdots \tag{35}$$

of G) can be written also in the form

$$G = G^0 + G^0 T G^0 \tag{36}$$

where the one-dimensional scattering matrix T, is defined as

$$T = \Delta (1 - \Delta G^0)^{-1} \; . \tag{37}$$

In the case of CPA we can define a Green's function G_e for the effective medium again through the Dyson equation

$$G_e = G^0 + G^0 \Sigma G_e \tag{38}$$

(which defines also the self-energy Σ). Solving (38) for G^0 we obtain

$$G^0 = G_e (1 + \Sigma G_e)^{-1} \; . \tag{39}$$

[+] We write here down only the basic equations, for their derivation see [1].

[++] By writing (32) in a scalar form we consider a single band, or in other words our single reference site has only one orbital. In the general case with more than one orbital per site or with defects extending over more sites (many band case) (32) has to be written in a matrix form.

Substituting this expression of G^o into the Dyson equation (35) of the exact Green's function of the system one can write

$$G = G^o(1+\Delta G) = G_e(1+\Sigma G_e)^{-1}(1+\Delta G) \tag{40}$$

which leads after rearrangement to the equation

$$G = G_e + G_e \Delta_e G \quad , \tag{41}$$

where $\Delta_e = \Delta - \Sigma$ [1]. Thus at the unperturbed sites (Δ=0) of the original system $\Delta_e = -\Sigma$ and of the perturbed sites $\Delta_e = \Delta - \Sigma$. Using the definition (37) of T we can now write for its average (by substituting instead of Δ Δ_e and instead of G^o G_e)

$$\langle T \rangle = (1-f)(-\Sigma)[1+\Sigma G_e]^{-1} + f(\Delta-\Sigma)[1-(\Delta-\Sigma)G_e]^{-1} \quad , \tag{42}$$

where f is the probability of the perturbed sites.

According to the fundamental approximation of CPA the average fluctuation from the effective medium is zero,

$$\langle T \rangle = 0 \quad . \tag{43}$$

Putting equ. (42) equal to zero one obtains after some manipulations the CPA equation [1, 22]

$$\Sigma(\vec{k},E) = f\Delta(\vec{k}) / \{1 + [\Sigma(\vec{k},E) - \Delta(\vec{k})]\, G_e(E)\} \quad . \tag{44}$$

With (33) we obtain from (38) for G_e

$$G_e(\vec{k},E) = [E - \varepsilon^A(\vec{k}) - \Sigma(\vec{k},E)]^{-1} \tag{45a}$$

and finally

$$G_e(E) = \Omega^{-1} \sum_{\vec{k}} G_e(\vec{k},E) = \Omega^{-1} \sum_{\vec{k}} [E - \varepsilon^A(\vec{k}) - \Sigma(\vec{k},E)]^{-1} \quad . \tag{45b}$$

In the case when the $\vec{k}$ dependence of Σ can be neglected $G_e(E)$ is simply [23]

$$G_e(E) = G^o(E-\Sigma) \quad . \tag{46}$$

Equ. (46) can be applied, however, only in cases when the density of states curves of the periodic systems A and B have the same shape, only one curve is shifted with respect to the other by a constant Σ value. In the case of substitutionally disordered polymers due to the complexity of their subunits we cannot expect, however, that this condition will be fulfilled.

Returning to the general formalism with $\vec{k}$-dependent self-energy we can write for the spectral density [23] and for the average density of states per molecule, respectively,

$$A(\vec{k},E) = -\pi^{-1}\,\mathrm{Im}\,G_e(\vec{k},E+i0) \quad , \tag{47}$$

$$\varrho(E) = \Omega^{-1}\sum_{\vec{k}} A(\vec{k},E) = -\pi^{-1}\,\mathrm{Im}\,G_e(E+i0) \quad . \tag{48}$$

Using further the Kramers-Kronig [24] relation one obtains

$$G_e(E) = \int_{-\infty}^{\infty} \varrho(E')/(E-E')\,dE' \quad . \tag{49}$$

Finally using the rule

$$\frac{\alpha_1 + i\beta_1}{\alpha_2 + i\beta_2} = \frac{\alpha_1\alpha_2 + \beta_1\beta_2}{\alpha_2^2 + \beta_2^2} + \frac{\alpha_2\beta_1 - \alpha_1\beta_2}{\alpha_2^2 + \beta_2^2}$$

for the ratio of two complex numbers we can write

$$\mathrm{Im}\,G_e(\vec{k},E) = \mathrm{Im}\,\Sigma(\vec{k},E)\,\{[E-\varepsilon^A(\vec{k})-\mathrm{Re}\,\Sigma(\vec{k},E)]^2 + \\ + [\mathrm{Im}\,\Sigma(\vec{k},E)]^2\}^{-1} \quad . \tag{50}$$

Equations (44) - (50) are the expressions which can be used for an actual calculation with a k-dependent self-energy. (In the case of a linear aperiodic polymer we can write instead of the vector $\vec{k}$ everywhere in these equations the scalar k). One starts the iterative procedure with a guess for $G_e(E)$⁺, one obtains the first values for $\Sigma(k,E)$ from (44) for different k and E values. Substituting this into (50) and the resulting Im $G_e(k,E)$ values into (47) and (48), respectively, one obtains the first approximation for $\varrho(e)$. Putting finally this into (49) one gets the next approximation for $G_e(E)$. This numerical procedure (which only recently was worked out [25] for the treatment of an aperiodic polymer with k-dependent self-energy) has to be repeated until self consistency is reached in $\varrho(E)$.

The CPA method with k-dependent Σ has been applied for the treatment of hydrogen impurities in the highly conducting polymer $(SN)_X$ (polysulphurnitride) which becomes superconductive at 0.26°K [26]. Several authors performed band structure calculations for this quasi one-dimensional system (both semiempirical [27] and minimal basis ab initio [28] ones). There are also in the literature non-self-consistent OPW and LCAO calculations for the 3-dimensional system [29]. Recently we have

---

⁺ Taking either the Green's function of pure A($\Sigma=0$, $G_e(E)=G^o(E)$) or the virtual crystal Green's function ($\Sigma(k)=f\Delta(k)$ which one obtains in a good approximation from (44) if $\Delta(k)$ is small) computing it by (49) with the virtual crystal $\varrho_{VC}(E')$ density of states which belongs to the energy band

$$\varepsilon_{VC}(k) = (1-f)\,\varepsilon^A(k) + f\,\varepsilon^B(k) \quad . \tag{51}$$

executed also an ab initio double ζ band structure calculation [30] for the linear chain which has given rather good agreement with experiment for the effective electronic mass and density of states at the Fermi level.

Recently at IBM (San Jose) five to ten mol per cent hydrogen was found in $(SN)_X$ [31]. The position of the hydrogen impurities is unknown (as it is of the Br_2, I_2 and ICl molecules with which $(SN)_X$ was modified also [32]), but most probably the H atoms bind to the N atoms. In this way they change the hybridization state of the N atoms and the number of π-electrons in the partially filled band of $(SN)_X$ (in a \S≠N\ unit there are 3 π-electrons, while in a \S/N(H)\ unit there are 4).

To find out the shift of the Fermi level and the change in the density of states the above described form of CPA has been applied using as input the ab initio minimal basis band structures of the periodic $(SN)_X$ and $(\overline{SN}_H)_X$ chains [33]. From these the densities of states $\varrho(E)$ of the two periodic chains have been computed using the method of Delhalle [34] (see Fig. 1)

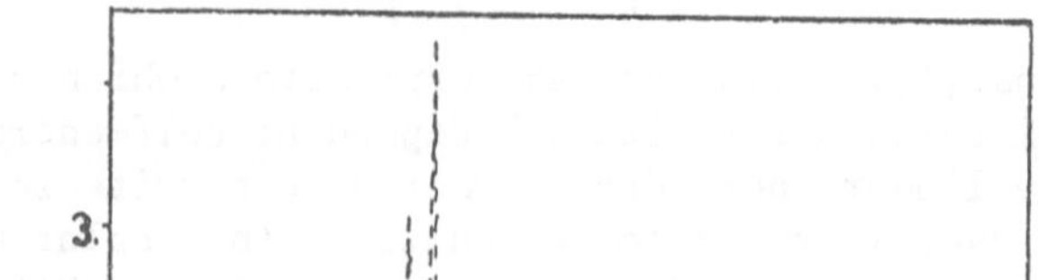

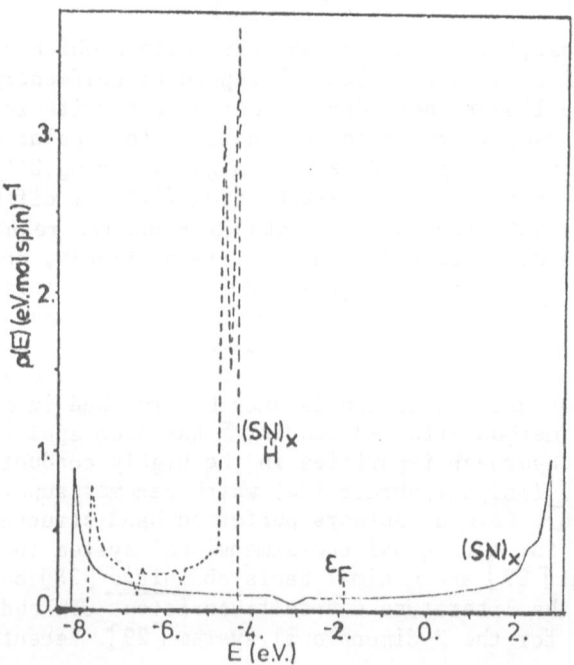

Fig. 1

The density of states curves of the pure $(SN)_X$ and $(\overset{SN}{\underset{H}{|}})_X$ systems in $(\text{eV. mol. spin})^{-1}$ units.

As we can see from Fig. 1. the shape of the two density of states curves is very different. Our first attempts to apply CPA to the mixed system with a constant (k-independent) Σ have therefore failed. As next step we applied the above described procedure with a k- and E-dependent Σ, $\Sigma(k,E)$ [25].

In the actual calculations 93 different E_i-values and k_i points ($0 \leq k_i \leq \pi/a$) were chosen for which $\Sigma(k_i,E_i)$ was computed in every iteration step. To reach self-consistency between 24 (f=0.3) and 70 (f=0.5) iteration steps were needed using the SCF criterion

$$\left| \rho_e^{(n)}(E_i) - \rho_e^{(n-1)}(E_i) \right| \leq 10^{-3}\ [\text{a.u. mol}]^{-1} \quad \forall E_i \ . \tag{52}$$

Finally after self-consistency was reached the position of the Fermi level was determined for each value of f by numerical integration using the relation

$$\int_{\varepsilon_{min}}^{\varepsilon_F} \rho_e^{(SCF)}(E)\, dE = 1 + f \ . \tag{53}$$

The calculations for a given f value have taken between 100 and 200 seconds on a CYBER 172 computer (for further details of the numerical calculations see [25]).

Fig.-s 2 and 3 present the density of states curves of the mixed system for different hydrogen concentrations f obtained with the aid of the described procedure .

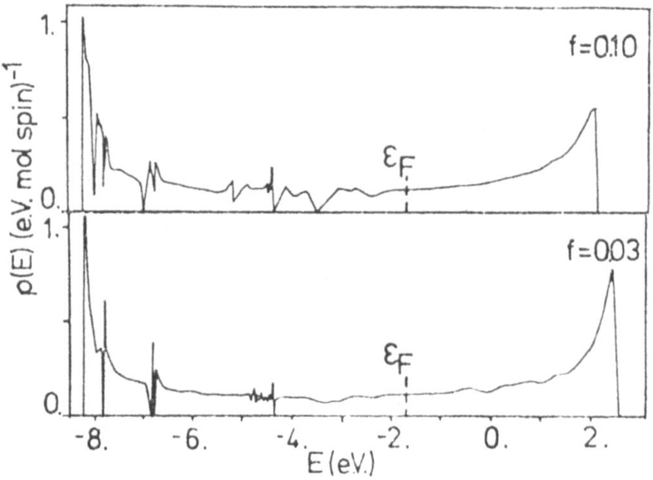

Fig. 2

The density of states curves of the $(SN)_X$ and $(\frac{SN}{H})_X$ mixed system obtained in the CPA approximation with f=0.03, and 0.10 in (eV. mol. spin)$^{-1}$ units.

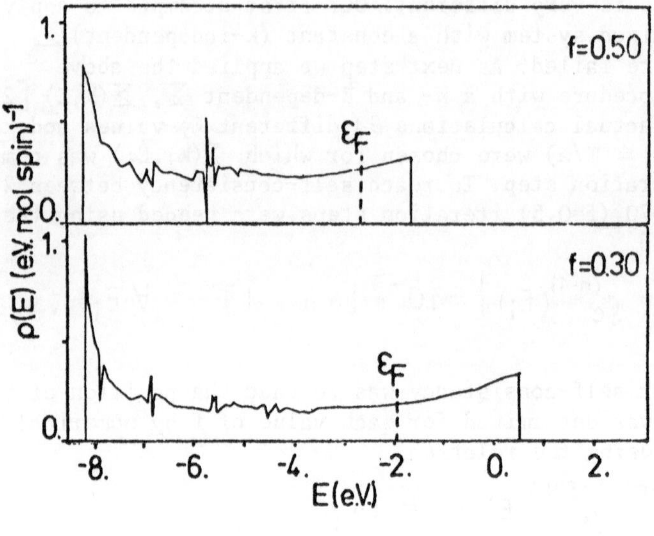

Fig. 3

The density of states curves of the $(SN)_X$ and $(\overset{SN}{\underset{H}{|}})_X$ mixed system obtained in the CPA approximation with f=0.30 and f=0.50 in (eV. mol. spin)$^{-1}$ units.

Further density of states curves for f=0.05, 0.07, 0.20 and 0.40 can be found in [25].

The most interesting feature of the $\varrho_e(E)$ curves in the Fig.-s is their complicated structure with spikes and dips which in many cases go to zero producing gaps in the band of the mixed system. In this connection it should be emphasized that there are no zero dips in the $\varrho(E)$ curves of the two periodic systems (see Fig. 1) and therefore their occurrence (already at f=0.03) is a genuine effect of the aperiodicity. Such structures are well known from computer experiments on linear chains [22] and from cluster-CPA calculations [22] (in which the impurities are extended over more sites), but could not be obtained until now with the one-site CPA with a k-independent Σ.

Looking at the positions of these spikes and dips it is easy to recognize (that they always occur at such E values at which the $\varrho(E)$ curves of the periodic systems have a large curvature (compare Fig. 2 and 3 with Fig. 1). Therefore probably the small gaps occurring in the band of the mixed system are not due to Anderson localization, but to the fact that levels of the original periodic systems which lie in a high curvature region of $\varrho(E)$ are more sensitive to perturbations due to the other component, than the other levels.

From Fig. 1 one can see that the $(\overset{SN}{\underset{H}{|}})_X$ periodic chain has a much narrower valence band than $(SN)_X$ (with corresponding large peaks in the density of states). The Fermi level of $(SN)_X$ lies at -1.09eV, while the upper limit of the completely filled valence band of $(^{SN}_{H})_X$ is at -4.38eV. One would expect on the basis of the rather large differences in the density of states curves of the two systems that in the mixed system already a small percentage of $(^{SN}_{H})_X$ would have a comparatively large influence on the density of states curve of pure $(SN)_X$.

This expectation is fulfilled as one can see from Fig. 2 where the density of states curves obtained in the CPA approximation for the mixed system with 10% (or less) hydrogen are shown. At very low hydrogen concentration (f=0.03) already new peaks in the density of states curve start to develop in the region between -4.4eV and -8.0eV. At higher f values (see Fig. 3) peaks of $(SN)_X$ and $(^{SN}_{H})_X$ in the region between -7 and -8eV fuse to one broader peak. On the other hand in consequence of the very high but extremely narrow peaks of $(^{SN}_{H})_X$ between -4.8 and -4.4eV new peaks develop but they are shifted to the region around -5.6eV (see especially the curve belonging to f=0.50). In all these cases one can see a clear demonstration for the fact that the CPA method gives essentially different results than the simple virtual crystal approximation.

The position of the Fermi level of the mixed system is not a sensitive function of f at low concentrations (ε_F=-1.9eV for f=0.00, ε_F=-1.6eV for f=0.03 and 0.10. At high concentrations of course its position shifts towards lower energies (ε_F=-2.1eV at f=0.30 and ε_F=-2.6eV at f=0.50). One should mention that the density of states at the Fermi level monotonously increases with the increase of f ($\varrho(E)$=0.10, 0.11, 0.13, 0.18 and 0.26 at f=0.00, 0.03, 0.10, 0.30 and 0.50, respectively).

Finally one should point out that by doping $(SN)_X$ with hydrogen one could expect to find the theoretically obtained spikes and dips in the $\varrho(E)$ curve of the mixed system by photoelectron spectroscopy (for the complications which can occur by the interpretation of these spectra see, however, [25]). Further since according to the BCS theory of superconductivity the transition temperature T_c depends exponentially on the density of states at the Fermi level [35],

$$T_C = 1.14\, \Theta_D \exp\left(-\frac{1}{\varrho(\varepsilon_F)V}\right) \tag{54}$$

(here Θ_D is the Debye temperature and V is the BCS electron-electron interaction parameter), one would expect that due to the increase of $\varrho(\varepsilon_F)$ with the concentration of hydrogen, the transition temperature would increase also with increased H doping.

2.2.2 SCF LCAO Resolvent Method for the Treatment of a Cluster of Impurities Embedded in a Periodic Polymer

To be able to treat a cluster of impurities, an extended distortion or another type of local perturbation of a three-dimensional periodic crystal we have to introduce a proper labeling of the different cells. For that purpose let us assume again that we have m orbitals in the unit cell of a periodic 3-dimensional crystal and the number of unit cells should be $(2N+1)^3$ (2N+1 cells in each direction). Let us introduce the row vector

$$\underline{\chi}^{\vec{p}} = \underline{\chi}^{p_1,p_2,p_3} = (\chi^{\vec{p}1}, \ldots, \chi^{\vec{p}m}) \tag{55}$$

for the m orbitals of the cell characterized by $\vec{p}$. From the row vector $\underline{\chi}^{\vec{p}}$ one can build up the hyper-row-vector containing the AO-s of the chain characterized by p_1 and p_2

$$\underline{\chi}^{p_1,p_2} = (\underline{\chi}^{p_1,p_2,(-N)}, \ldots, \underline{\chi}^{p_1,p_2,p_3}, \ldots, \underline{\chi}^{p_1,p_2,N}) \tag{56}$$

which will have m(2N+1) components. From the vectors (56) we can form such hyper-row-vectors which contain all the AO-s of the cells lying in the plane characterized by the vector component p_1:

$$\underline{\chi}^{p_1} = (\underline{\chi}^{p_1,(-N)}, \ldots, \underline{\chi}^{p_1,p_2}, \ldots, \underline{\chi}^{p_1,N}) \tag{57}$$

which have $m(2N+1)^2$ components. Finally from the vectors (57) one can construct the hyper-row vector $\underline{\chi}$ containing all the AO-s of the whole crystal [3],

$$\underline{\chi} = (\underline{\chi}^{-N}, \ldots, \underline{\chi}^{p_1}, \ldots \underline{\chi}^{N}) \quad . \tag{58}$$

With the aid of these conventions we can easily find out the labeling of any mxm block, of the matrix $\underline{\underline{F}} = \langle \underline{\chi}^+ | \hat{F} | \underline{\chi} \rangle$.$^+$ Thus $\underline{\underline{F}}_{q_1}^{p_1} = \langle \underline{\chi}^+_{q_1} | \hat{F} | \underline{\chi}^{p_1} \rangle$ is that submatrix of $\underline{\underline{F}}$ which contains the AO-s of the planes characterized by q_1 and p_1, respectively. Similarly the submatrix $\underline{\underline{F}}_{q_1,q_2}^{p_1,p_2} = \langle \underline{\chi}^+_{q_1,q_2} | \hat{F} | \underline{\chi}^{p_1,p_2} \rangle$ gives the interaction between the AO-s belonging to the chains characterized by q_1, q_2 and p_1, p_2, respectively. Finally $\underline{\underline{F}}_{\vec{q}}^{\vec{p}} = \langle \underline{\chi}_{q_1,q_2,q_3} | \hat{F} | \underline{\chi}^{p_1,p_2,p_3} \rangle$ is the submatrix between the AO-s belonging to the cells labeled by $\vec{q}$ and $\vec{p}$, respectively.

Let us assume that at the cell characterized by $\vec{p}$ we have an extended impurity (or distortion) instead of the original cell and for the sake of simplicity let us assume that the impurity cell has only m orbitals.

$^+$ Since $\underline{\chi}$ is a row vector and thus $\underline{\chi}^+$ a column vector, $\langle \underline{\chi}^+ | \hat{F} | \underline{\chi} \rangle$ forms a quadratic matrix of the dimension $\tilde{N}x\tilde{N}$, where $\tilde{N}=m(2N+1)^3$.

Then we can introduce the submatrix

$$\tilde{\underline{\underline{F}}}{}^{\vec{p}}_{\vec{p}} = \underline{\underline{F}}{}^{\vec{p}}_{\vec{p}} - \underline{\underline{F}}{}^{\vec{p}\,0}_{\vec{p}} \qquad (\vec{q} = \vec{p}) \tag{58}$$

which gives the deviation of the Fock matrix of the impurity ($\underline{\underline{F}}{}^{\vec{p}}_{\vec{p}}$) from that of the unperturbed unit cell ($\underline{\underline{F}}{}^{\vec{p}\,0}_{\vec{p}}$).
We can introduce in a similar way for the deviations of the intercell interaction matrices caused by the different potential of the impurity the matrices

$$\tilde{\underline{\underline{F}}}{}^{\vec{p}}_{\vec{q}} = \underline{\underline{F}}{}^{\vec{p}}_{\vec{q}} - \underline{\underline{F}}{}^{\vec{p}\,0}_{\vec{q}} \quad , \tag{59}$$

where $\underline{\underline{F}}{}^{\vec{p}\,0}_{\vec{q}}$ refers again to the corresponding matrix the unperturbed crystal.

In the case of first neighbors' interactions we can write $q_1=p_1\pm1$, $q_2=p_2$, $q_3=p_3$; $q_1=p_1$, $q_2=p_2\pm1$, $q_3=p_3$; $q_1=p_1$, $q_2=p_2$, $q_3=p_3\pm1$ (which defines altogether 6 different deviation matrices $\tilde{\underline{\underline{F}}}{}^{\vec{p}}_{\vec{q}}$). In a similar way we have 12 second neighbors ($q_1=p_1\pm1$ and $q_2=p_2\pm1$, $q_3=p_3$; $q_1=p_1\pm1$ and $q_3=p_3\pm1$, $q_2=p_2$; $q_1=p_1$, $q_2=p_2\pm1$ and $q_3=p_3\pm1$) and 8 third neighbors ($q_1=p_1\pm1$ and $p_2=p_2\pm1$ and $q_3=p_3\pm1$).

To construct a deviation hypermatrix $\underline{\underline{F}}'$ from the deviation submatrices (58) and (59) we have to find out the labels of the corresponding blocks. For the sake of simplicity let us restrict ourselves to first neighbors' interactions (the extension to the second and third neighbors goes similarly in a straightforward way [36]). In this case we can write (taking into account the labeling of the hypermatrix components introduced above)

$$\underline{\underline{F}}' = \left(\begin{array}{ccccccc}
 & & & \tilde{\underline{\underline{F}}}(-3) & & & \\
 & & & \vdots & & & \\
 & \underline{\underline{0}} & & \tilde{\underline{\underline{F}}}(-2) & & \underline{\underline{0}} & \\
 & & & \vdots & \xleftarrow{\quad (2N+1)^2 \quad} & & \\
 & & & \hat{\underline{\underline{F}}}(-1) & \xleftarrow{\; 2N+1 \;} & & \\
\hat{\underline{\underline{F}}}(-3)^{tr} \cdots & \hat{\underline{\underline{F}}}(-2)^{tr} \cdots & \hat{\underline{\underline{F}}}(-1)^{tr} & \tilde{\underline{\underline{F}}}{}^{\vec{p}}_{\vec{p}} & \hat{\underline{\underline{F}}}(1) \cdots & \hat{\underline{\underline{F}}}(2) \cdots & \hat{\underline{\underline{F}}}(3) \\
 & & & \hat{\underline{\underline{F}}}(1)^{tr} & & & \\
 & & & \vdots & & & \\
 & \underline{\underline{0}} & & \hat{\underline{\underline{F}}}(2)^{tr} & & \underline{\underline{0}} & \\
 & & & \vdots & & & \\
 & & & \hat{\underline{\underline{F}}}(3)^{tr} & & &
\end{array}\right) , \tag{60}$$

where $\tilde{\underline{\underline{F}}}(1) = \tilde{\underline{\underline{F}}}_{\vec{p}}^{p_1,p_2,p_3+1}$, $\tilde{\underline{\underline{F}}}(2) = \tilde{\underline{\underline{F}}}_{\vec{p}}^{p_1,p_2+1,p_3}$, $\tilde{\underline{\underline{F}}}(3) = \tilde{\underline{\underline{F}}}_{\vec{p}}^{p_1+1,p_2,p_3}$,
$\tilde{\underline{\underline{F}}}(-1) = \tilde{\underline{\underline{F}}}_{\vec{p}}^{p_1,p_2,p_3-1}$ etc. and $\tilde{\underline{\underline{F}}}(1)^{tr} = \tilde{\underline{\underline{F}}}^{\vec{p}}_{p_1,p_2,p_3+1}$ etc.
Since a change in p_2 means a change of the chain to which the cell characterized by $\vec{p}$ belongs and a change in p_1 gives a change in the plane given by p_2 and p_3, the distances between the non-zero blocks indicated in $\underline{\underline{F}}'$ are easily understandable.

After the described construction of $\underline{\underline{F}}'$ we can formulate our problem as

$$\underline{\underline{F}}\,\underline{c} = (\underline{\underline{F}}^o + \underline{\underline{F}}')\,\underline{c} = \lambda \underline{\underline{S}}\,\underline{c} = \lambda(\underline{\underline{S}}^o + \underline{\underline{S}}')\,\underline{c}\,, \tag{61}$$

where $\underline{\underline{F}}^o$ and $\underline{\underline{S}}^o$ are the Fock- and overlap hypermatrices, respectively, of the unperturbed crystal and $\underline{\underline{F}}'$ and $\underline{\underline{S}}'$ are the corresponding deviation matrices ($\underline{\underline{S}}'$ can be constructed in the same way as $\underline{\underline{F}}'$). We can blockdiagonalize again the cyclic hypermatrices $\underline{\underline{F}}^o$ and $\underline{\underline{S}}^o$,

$$\begin{aligned} &\underline{\underline{U}}^+\underline{\underline{F}}^o\underline{\underline{U}}\,\underline{\underline{U}}^+\underline{c} - \lambda\underline{\underline{U}}^+\underline{\underline{S}}^o\underline{\underline{U}}\,\underline{\underline{U}}^+\underline{c} = \\ &\quad = (\bar{\underline{\underline{F}}}^o - \bar{\underline{\underline{S}}}^o)\,\underline{\underline{U}}^+\underline{c} = -\underline{\underline{U}}^+(\underline{\underline{F}}' - \lambda\underline{\underline{S}}')\,\underline{c}\,, \end{aligned} \tag{62}$$

where

$$\bar{\underline{\underline{F}}}^o = \underline{\underline{U}}^+\underline{\underline{F}}^o\underline{\underline{U}}\,, \quad \bar{\underline{\underline{S}}}^o = \underline{\underline{U}}^+\underline{\underline{S}}^o\underline{\underline{U}}\,. \tag{63}$$

The unitary matrix was defined before (see equ. (51) for $\underline{\underline{U}}$). Multiplying (62) from the left by $\underline{\underline{U}}(\underline{\underline{F}}^o - \underline{\underline{S}}^o)^{-1}$ and introducing the resolvent [37, 2]

$$\underline{\underline{Z}} \equiv \underline{\underline{U}}\,(\bar{\underline{\underline{F}}}^o - \lambda\bar{\underline{\underline{S}}}^o)^{-1}\,\underline{\underline{U}}^+ \tag{64}$$

we obtain the equation

$$\underline{c} = -\underline{\underline{Z}}\,(\underline{\underline{F}}' - \lambda\underline{\underline{S}}')\,\underline{c}\,. \tag{65}$$

If one takes into account that $\underline{\underline{F}}'$ and $\underline{\underline{S}}'$ have with a few exceptions zero blocks, the hypermatrix equ. (65) reduces to a set of 7m homogeneous linear equations, if we write down separately the equations for those subvectors of $\underline{c}$ which correspond to the non zero blocks of $\underline{\underline{F}}'$ and $\underline{\underline{S}}'$. In this way we obtain in a straightforward way in the first neighbors' interactions approximation

$$\underline{\underline{\Lambda}}(\lambda)\,\underline{\omega} = \underline{\underline{\Lambda}}(\lambda)\begin{pmatrix} \underline{c}_{-3} \\ \underline{c}_{-2} \\ \underline{c}_{-1} \\ \underline{c}_{0} \\ \underline{c}_{1} \\ \underline{c}_{2} \\ \underline{c}_{3} \end{pmatrix} =$$

$$= \begin{pmatrix} \underline{\underline{\Lambda}}_{-3}^{-3}+\underline{\underline{1}} & \underline{\underline{\Lambda}}_{-3}^{2} & \cdots & \underline{\underline{\Lambda}}_{-3}^{0} & \cdots & \underline{\underline{\Lambda}}_{-3}^{3} \\ \vdots & & & & & \\ \underline{\underline{\Lambda}}_{0}^{-3} & \cdots & & \underline{\underline{\Lambda}}_{0}^{0}+\underline{\underline{1}} & \cdots & \underline{\underline{\Lambda}}_{0}^{3} \\ \vdots & & & & & \\ \underline{\underline{\Lambda}}_{3}^{-3} & \cdots & & \underline{\underline{\Lambda}}_{3}^{0} & \cdots & \underline{\underline{\Lambda}}_{3}^{3}+\underline{\underline{1}} \end{pmatrix} \begin{pmatrix} \underline{c}_{-3} \\ \underline{c}_{-2} \\ \underline{c}_{-1} \\ \underline{c}_{0} \\ \underline{c}_{1} \\ \underline{c}_{2} \\ \underline{c}_{3} \end{pmatrix} = \underline{0} \quad , \tag{66}$$

where according to the detailed calculation [3, 36]

$$\underline{\underline{\Lambda}}_{K}^{L} = \underline{\underline{Z}}_{K}^{0} (\underline{\underline{F}}' - \lambda \underline{\underline{S}}')_{0}^{L} \qquad \begin{matrix} K=-3,\ldots,0,\ldots,3 \\ L=-3,\ldots,3 \quad \text{but } L\neq 0 \end{matrix} \quad , \tag{67}$$

$$\underline{\underline{\Lambda}}_{K}^{0} = \sum_{i=-3}^{3} \underline{\underline{Z}}_{K}^{i} (\underline{\underline{F}}' - \lambda \underline{\underline{S}}')_{i}^{0} \quad . \tag{68}$$

In this simplified notation 0 corresponds to $\vec{p}$ and for instance $i=-3$ corresponds to the vector $\vec{q}$ with components $q_1=p_1-1$, $q_2=p_2$ and $q_3=p_3$.

The blocks of $\underline{\underline{Z}}_{\vec{q}}^{\vec{p}}$ are given by

$$Z_{\vec{q}}^{\vec{p}} = \omega^{-1} \int_{\omega} \exp[i(\vec{R}_{\vec{q}} - \vec{R}_{\vec{p}})\vec{k}]\, [\underline{\underline{F}}(\vec{k}) - \lambda \underline{\underline{S}}(\vec{k})]^{-1}\, d\vec{k} \quad , \tag{69}$$

where the mxm matrices $\underline{\underline{F}}(\vec{k})$ and $\underline{\underline{S}}(\vec{k})$ which belong to the periodic problem were defined by equ.-s (12) and (13) (to make these definitions consistent with our present notation we have to write in equ.-s (13) instead of $\underline{\underline{F}}(\vec{q})$ and $\underline{\underline{S}}(\vec{q})$ $F_{\vec{p}+\vec{q}}^{\vec{p}}$ and $S_{\vec{p}+\vec{q}}^{\vec{p}}$, respectively).

To solve the equ. (66) for λ we can write

$$\det[\underline{\underline{\Lambda}}(\lambda)] = 0 \quad . \tag{70a}$$

If the order of the matrix 7m is not small (which is the case already if we have 10 electrons in the unit cell) the solution of (70a) may be rather difficult, because the determinant in certain regions of λ may be a very quickly varying function of λ. To overcome this difficulty we can multiply $\underline{\underline{\Lambda}}$ (which is not a Hermitian matrix) from the left by $(\tilde{\underline{\underline{F}}}' - \lambda\tilde{\underline{\underline{S}}}')$ to obtain the Hermitian matrix $\underline{\underline{\Lambda}}^{H}(\lambda) = (\tilde{\underline{\underline{F}}}' - \lambda\tilde{\underline{\underline{S}}}')\underline{\underline{\Lambda}}(\lambda)$ [38]. (It is easy to show that the matrices $\tilde{\underline{\underline{F}}}'$ and $\tilde{\underline{\underline{S}}}'$ of dimension 7mx7m can be constructed from the non-zero blocks of (60) in the form of

$$\tilde{\underline{\underline{F}}}' = \begin{pmatrix} & & & X & & & \\ & \underline{\underline{0}} & & X & & \underline{\underline{0}} & \\ & & & X & & & \\ X & X & X & X & X & X & X \\ & & & X & & & \\ & \underline{\underline{0}} & & X & & \underline{\underline{0}} & \\ & & & X & & & \end{pmatrix} \quad ,$$

where X stands for a non-zero block

of $\underline{\underline{F}}'$.) One then can determine the roots λ of $\det[\underline{\underline{\Lambda}}^H(\lambda)]=0$ by solving the eigenvalue equ.

$$\underline{\underline{\Lambda}}^H(\lambda)\,\underline{\alpha}_j(\lambda) = E_j(\lambda)\,\underline{\alpha}_j(\lambda) \tag{70b}$$

for the zero eigenvalues, $E_j(\lambda)=0$ [38]. After the energy values λ have been computed they can be substituted back into (66) to determine the subvector $\underline{c}_i$ $(i=-3,\ldots,3)$.

One should mention that instead of using (69) for the blocks of (68) one can write down an expression for its elements in which the denominator will be simply $[\varepsilon(\vec{k})_j - \lambda]$ (see equ. (3.23) of [2])instead of $\underline{\underline{F}}(\vec{k})-\lambda\underline{\underline{S}}(\vec{k})$. To obtain this expression one has to repeat the derivation leading to (66) [2] by not simply blockdiagonalizing $\underline{\underline{F}}^o$ and $\underline{\underline{S}}^o$ but completely diagonalizing them (in other words one has to solve first the problem of the periodic crystal). One would expect that this "fully diagonalized" version of the theory will make easier the determination of the impurity levels, than the above in detail described "block diagonalized" version.

One can perform the whole described procedure also in an SCF way if one writes down for the elements of $\underline{\underline{F}}'$ the appropriate Hartree-Fock expressions:

$$\begin{aligned}(\tilde{\underline{\underline{F}}}^{\vec{p}}_{\vec{q}})_{f,g} &= \langle \bar{\chi}_{\vec{q},f}|\hat{F}|\tilde{\chi}^{\vec{p},g}\rangle - (\underline{\underline{F}}^{\vec{p}\,o}_{\vec{q}})_{f,g} = \\ &= \langle \bar{\chi}_{\vec{q},f}|\hat{H}^N|\tilde{\chi}^{\vec{p},g}\rangle + \sum_{\vec{p}_1,\vec{p}_2=-3}^{3}\ \sum_{u,v=1}^{m} p^{(imp.)}(\vec{p}_1-\vec{p}_2)_{u,v}\,\cdot \\ &\cdot\left(\langle \bar{\chi}_{\vec{q},f}\,\bar{\chi}_{\vec{p}_1,u}|\tilde{\chi}^{\vec{p},g}\,\bar{\chi}^{\vec{p}_2,v}\rangle - \tfrac{1}{2}\langle \bar{\chi}_{\vec{q},f}\,\bar{\chi}_{\vec{p}_1,u}|\bar{\chi}^{\vec{p}_2,v}\tilde{\chi}^{\vec{p},g}\rangle\right).\end{aligned} \tag{71}$$

Here the elements of the impurity charge-bond order matrix are defined as

$$p^{(imp.)}(\vec{p}_1-\vec{p}_2)_{u,v} = 2\sum_{j=1}^{n^*} c^*_{j;\vec{p}_1,u}\, c_{j;\vec{p}_2,v}\quad , \tag{72}$$

where the coefficients $c_{j;\vec{p}_1,u}$ etc. are components of the subvectors $\underline{c}_{j;\vec{p}_1}$ of the perturbed problem (in the submatrix the symbolic notation $\sum^3_{\vec{p}_1,\vec{p}_2=-3}$ means that the components of the vector $\vec{p}_1$ and $\vec{p}_2$, respectively, have to take the values ennumerated after equ. (59)). Further $\tilde{\chi}^{\vec{p},g}$ stands for the g-th impurity orbital in the impurity cell given by $\vec{p}$ and

$$\bar{\chi}_{\vec{q},f} = \begin{cases}\tilde{\chi}_{\vec{q},f} & , \text{ if } \vec{q}=\vec{p} \\ \chi_{\vec{q},f} & \text{ if } \vec{q}\neq\vec{p}\end{cases}\quad . \tag{73}$$

Equations (71) and (72) determine then in the usual way the SCF procedure.

To summarize to find the impurity levels (or surface bands+)

+One can obtain the self-consistent surface bands of a three-dimensional crystal as a special case of the described theory [39].

of a crystal one (1) starts by solving the periodic problem (which is necessary to apply the numerically easier "fully diagonalized" version). (2) Then one has to find the roots λ_j of $\det[\underline{\underline{\Lambda}}(\lambda)]=0$. (3) Substituting then back into (66) one can compute the vector $\underline{\omega}$. (4) Finally substituting the components of its subvectors $\underline{c}_j$ (i=-3,...,3) into (72) one can construct the impurity submatrices $\underline{\underline{F}}^{\underline{\underline{p}}}_q$. The steps (2)-(4) of this procedure have to be repeated until self consistency is reached. In this way the impurity levels lying outside the bulk bands can be calculated.+

A detailed investigation of the numerical aspects of the described procedure [20] and its application first to the case of an one-dimensional chain is in progress.

3. EXCITED STATES OF POLYMERS

3.1 Intracell Excitations

If our polymer contains subunits not very strongly interacting like in the stacked chain of polycytosine and we have a completely filled valence band, one can approximate the excited states of the polymer on the basis of those of its constituents.

3.1.1 The Application of the Excitation Hamiltonian ($\hat{O}\hat{A}\hat{O}$ Method) to Polymers

In the closed-shell Hartree-Fock (HF) theory the singlet excitation energy from a filled level with energy ε_i to a virtual level with energy ε_a is given by the well known expression

$$^1\Delta E_{i\to a} = \varepsilon_a - \varepsilon_i - J_{ia} + 2K_{ia} \quad , \tag{74}$$

where $J_{ia} = \langle \Phi_i(1)\,\Phi_a(2)\,|1/r_{12}|\,\Phi_i(1)\,\Phi_a(2)\rangle$

$K_{ia} = \langle \Phi_i(1)\,\Phi_a(2)\,|1/r_{12}|\,\Phi_a(1)\,\Phi_i(2)\rangle$. The expression (74) is of course only an approximation to the correct excitation energy, because (1) it does not take into account the change of the distribution of the other electrons (this so-called relaxation energy can be treated if one makes a separate open-shell calculation for the excited state and takes the difference of the total energies of the excited state and of the ground state; ΔSCF method) and (2) does not take into account the change of correlation energies in the excited and ground state, respectively. To take into account both effects one has to make a separate

---

+ In the case if impurity levels lie within the bulk bands, i.e. the eigenvalues $E_j(\lambda)$ of $\underline{\underline{\Lambda}}^H$ have poles at $\lambda = \varepsilon(\vec{k})$, the λ values still can be determined in a somewhat more complicated way [20] which we do not discuss here.

calculation together with correlation (and possibly, if it is known, taking also the different geometry of the excited state) for the excited and ground states, respectively, and then one has to compute the difference of the total energies of both states. Such rather extensive calculations (together with geometry optimization) have been already performed for some simple molecules [40] but cannot be executed yet for larger molecules and polymers.

For polymers and solids we have the additional problem that as it is easy to show [41] the integrals J_{ia} and K_{ia} vanish, if the number of electrons n goes to infinity,

$$\lim_{n\to\infty} J_{ia} = \lim_{n\to\infty} K_{ia} = \lim_{n\to\infty} \frac{\ln n}{n} \to 0 \ . \qquad (75)$$

This means that we are left with the singlet excitation energy

$$^{1}\Delta E_{i\to a}(\vec{k}) = \varepsilon_a(\vec{k}) - \varepsilon_i(\vec{k}) \qquad (76)$$

of the solid or polymer+ which of course is too large.

To overcome this difficulty and to take into account (at least the larger part) of the relaxation, one can calculate more correct virtual energy levels of the constituent molecules of a polymer than the Hartree-Fock method provides taking instead of the incorrect n-particle V^n potential of the HF method the correct n-1 particle V^{n-1} potential [19] with the aid of the modified Fock-operator

$$\hat{F}' = \hat{F} + \hat{O}\hat{A}\hat{O} \qquad (77)$$

[42]. In this expression the projection operator $\hat{O}$ defined by

$$\hat{O} = \hat{1} - \hat{\rho} = \hat{1} - \sum_{i=1}^{n^*} |\Phi_i^{HF}(1)\rangle\langle\Phi_i^{HF}(1)| = \sum_{a=n^*+1}^{m} |\Phi_a^{HF}(1)\rangle\langle\Phi_a^{HF}(1)| =$$
$$= \sum_{a=n^*+1}^{m} |\Psi_a(1)\rangle\langle\Psi_a(1)| \qquad (78)$$

projects into the subspace of the virtual orbitals. The HF orbitals $|\Phi_i^{HF}(1)\rangle$ are defined by the equation

$$\hat{F}\,|\Phi_i^{HF}(1)\rangle = \varepsilon_i^{HF}\,|\Phi_i^{HF}(1)\rangle \qquad (79)$$

while the orbitals $|\Psi_a(1)\rangle$ are the eigenfunctions of the modified operator $\hat{F}'$,

$$(\hat{F} + \hat{O}\hat{A}\hat{O})\,|\Psi_a(1)\rangle = \varepsilon'_a\,|\Psi_a(1)\rangle \ . \qquad (80)$$

+ As it is well known in a solid at an excitation from band i to band a the crystal momentum $\vec{k}$ has to be conserved.

In the case of a singlet-singlet $i \rightarrow a$ excitation we can choose our $\hat{A}$ operator as

$$\hat{A}_i = - \langle \Phi_i^{HF}(2) | 1/r_{12} (1 - 2\hat{P}_{12}) | \Phi_i^{HF}(2) \rangle \quad .$$

Expanding the orbitals $|\Psi_a(1)\rangle$ in terms of the HF orbitals

$$|\Psi_a(1)\rangle = \sum_{\ell=1}^{m} c_{a,\ell} | \Phi_\ell^{HF}(1)\rangle \quad , \tag{81}$$

we obtain the matrix equation

$$\underline{\underline{F}}'_i \, \underline{c}_a = \varepsilon'_a \, \underline{c}_a \quad , \tag{82}$$

where the elements of the matrix $\underline{\underline{F}}'_i$ are defined as

$$(\underline{\underline{F}}'_i)_{k,\ell} = \langle \Phi_k^{HF} | \hat{F} + \hat{O}\hat{A}_i\hat{O} | \Phi_\ell^{HF} \rangle \quad . \tag{83}$$

From equ. (83) it is easy to see that for the filled orbitals $\underline{\underline{F}}'_i$ has the same eigenvalues ε_i as the Fock-matrix of the unmodified Fock-operator $\hat{F}$, the virtual levels, however, will be changed. The corrected excitation energy (the difference of the total energies in the excited and in the ground state) can be obtained in this way directly as the difference of the corrected one-electron energies [42]

$${}^1\Delta\tilde{E}_{i\rightarrow a} = E_{i\rightarrow a} - E_G = \varepsilon'_a - \varepsilon_i \quad . \tag{84}$$

Since we can assume in a rather good approximation that in a stacked polymer like polyC an excitation occurs locally on a single cytosine molecule, after performing an $\hat{O}\hat{A}\hat{O}$ calculation for the single molecule one can shift the centers of the empty bands to the corresponding corrected positions of the virtual levels. Applying the $\hat{O}\hat{A}\hat{O}$ method for the excitation from the HOMO level, one obtains in this way a corrected gap between the valence band and the approximated exciton band (assuming it has the same width as the conduction band). In Table II we give the HOMO and LEMO levels of the four nucleotide bases and their with the $\hat{O}\hat{A}\hat{O}$ method corrected LEMO levels.

Table II

The minimal basis HOMO and LEMO levels and with the ÔÂÔ method corrected LEMO levels (in eV-s) of the four nucleotide bases.

Level	Cytosin(C)	Guanin(G)	Thymin(T)	Adenin(A)
LEMO	1.931	3.612	3.203	2.332
LEMO(corrected)	-2.816	-3.495	-4.488	-3.582
HOMO	-9.768	-8.209	-9.602	-9.288
Corrected gap	6.952	4.714	5.154	5.706
Experimental HOMO→LEMO excitation energy [43]	4.3	4.2	4.7	4.9

Comparing the corrected $\varepsilon_{LEMO} - \varepsilon_{HOMO}$ energy differences in Table II with the experimental values one can see that though they are still too large (especially in the case of C), the ÔÂÔ procedure has corrected the major part of the error, if we compare the band gap with the experimental excitation energy (as we have to do in the case of a polymer).

If we shift now the center ot the conduction band of polyC (see Table I) to its corrected LEMO level and further if we assume that the valence and exciton bands of the other three nucleotide bases have the same widths as the valence and conduction bands of polyC and we place them around (as centers) the HOMO and (with ÔÂÔ corrected) LEMO levels of the single molecules, we obtain the estimated valence and exciton bands of the four homopoly-nucleotides which are given in Table III.

Table III

The valence and exciton bands of polyC and those estimated in the case of the other three homopolynucleotides (in eV-s)

		polyC	polyG	polyT	polyA
Exciton band	center	-2.816	-3.495	-4.448	-3.582
	width	1.241	1.241	1.241	1.241
Valence band	center	-9.768	-8.209	-9.602	-9.288
	width	0.552	0.552	0.552	0.552
Gap		6.056	3.818	4.257	4.810

In connection with Table III it should be mentioned that the experimental UV absorption spectrum of DNA corresponds in a good approximation to the superposition of the absorption spectra of its constituents (with correspondingly smaller intensity). This fact seems to be in contradiction to the existence of a band structure of DNA. On the other hand if the valence and exciton bands of the homopolynucleotides would have rather similar dispersions (i.e. the k-values belonging to the lower and upper limits of the valence and exciton band of a given homopolynucleotide would be the same (which is not the case for the valence and conduction band of polyC; see Table I) and the widths of the two bands would be the same) this could be understood also in the case of broad valence and exciton bands. To be able to answer finally this question one has to perform a detailed exciton band calculation of the four homopolynucleotides (see section 3.3.3) starting from their ab initio HF band structures (which has not been done yet). Such a calculation would yield not only the position of the exciton bands, but also their width and dispersion.

3.1.2 Possibility of the Application of the Green's Function and Coupled Cluster Expansion Formalism

In this point we want to mention only briefly that in a polymer of weakly interacting units one can apply instead of the $\hat{O}\hat{A}\hat{O}$ method the more general Green's function formalism to calculate the excited states of the units. After that one could try to estimate the excited states of the whole polymer starting from the knowledge of the excited states of the monomers in a similar way as it was outlined in the previous point at the discussion of the $\hat{O}\hat{A}\hat{O}$ method.

The Green's function formalism is discussed in several papers in this volume [44]. As it is well known the excitation energy of a system of electrons having the Hamiltonian $\hat{H}$, are given by the poles of the corresponding hole-particle Green's function [45]

$$G(\alpha,\beta,\gamma,\delta;t)=\langle\Psi_0|\hat{T}[a^+_\delta(t)\,a_\gamma(t)\,a^+_\alpha(0)\,a_\beta(0)]|\Psi_0\rangle \quad . \tag{85}$$

Here $|\Psi_0\rangle$ stands for the exact ground state in the Heisenberg representation, $a^+_i(t)$ and $a_i(t)$, respectively, are the creation and annihilation operators defined on the states of an approximate separable Hamilton-operator H_0 and $\hat{T}$ is the time-ordering operator. Substituting the exact time-dependence of the operators in the Heisenberg representation into (85), introducing the step function $\Theta(t)=1$ if $t>0$, $\Theta(t)=0$ if $t<0$ and performing a Fourier transformation with respect to the time on the resulting expression one obtains for the connected part of the Green's function [45, 46]

$$G_c(\alpha,\beta,\gamma,\delta;E)=\lim_{\varepsilon\to 0}\int_{-\infty}^{\infty}dt\,\exp(iEt-\varepsilon|t|)\,G_c(\alpha,\beta,\gamma,\delta;t)=$$

$$=-i\lim_{\varepsilon\to 0}\sum_{n\neq 0}\left[\frac{\rho_n^+}{E_n-E_0-E-i\varepsilon}+\frac{\rho_n^-}{E_n-E_0+E+i\varepsilon}\right] \quad . \tag{86}$$

Here

$$G_c(\alpha,\beta,\gamma,\delta;t) = G(\alpha,\beta,\gamma,\delta;t) - \rho_0 \quad , \tag{87}$$

where ρ_0 is the disconnected part of the Green's function

$$\rho_0 = G(\alpha,\beta;0)\, G(\delta,\gamma;0) \quad . \tag{88}$$

Further

$$\rho_n^+ = \langle \Psi_0 | a_\delta^+ a_\gamma | \Psi_n \rangle \langle \Psi_n | a_\alpha^+ a_\beta | \Psi_0 \rangle \quad , \tag{89a}$$

$$\rho_n^- = \langle \Psi_0 | a_\alpha^+ a_\beta | \Psi_n \rangle \langle \Psi_n | a_\delta^+ a_\gamma | \Psi_0 \rangle \quad .^+ \tag{89b}$$

From (86) we can observe that the exact excitation energies of the system are given by the poles of the connected part of the particle-hole Green's function (in the energy representation).

To find these poles, at least approximately, Paldus and Čižek [46] have expressed the Green's function in a perturbation form. For that purpose they have expressed the Hamiltonian $\hat{H}$ as

$$\hat{H} = \hat{H}_0 + g\hat{V} , \tag{90}$$

where g is the coupling constant and they choose for $\hat{H}_0$ the Fock operator, $\hat{H}_0 = \hat{F}$. After this they have rewritten the Green's function (85) in the interaction representation. Using Wick's theorem and diagrammatical techniques they could express the individual terms of the resulting perturbation series through the set of the pertinent Feynman-diagrams. Without going into more details (for them see [46]) it should be mentioned only that they have derived the analogue of the Dyson equ. for the hole-particle Green's function. They have given in this way explicit expressions to the third order of perturbation theory for the excitation energy [46]. The method was applied until now only to a system of 6 π electrons using the PPP(Pariser-Parr-Pople) model Hamiltonian [47]. There would be no larger difficulties, however, to apply this method to the units of a polymer using as one-electron wave functions their ab initio (Hartree-Fock) MO-s.

It should be mentioned only that another promising possibility for the direct determination (taking into account also the change in the correlation energy) of the excited states of larger monomers (like the nucleotide bases in DNA) in a polymer is the recently worked out open-shell version of the coupled cluster ($|\Psi\rangle = e^T |\Phi_0\rangle$) method [48] (for the coupled cluster method in its closed-shell form see [49]).

$^+$From the last expressions one can see that in deriving (86) the identity $\sum_n |\Psi_n\rangle\langle\Psi_n| = 1$ has been applied.

This extension of the method starts with the Ansatz

$$|\Psi^{O.S.}\rangle = \hat{W}|\Psi\rangle = \hat{W}e^{\hat{T}}|\Phi_0\rangle \ , \tag{91}$$

where in

$$\hat{W} = \sum_{i=1}^{m} \hat{W}_i \tag{92}$$

the first operator W_1 provides an appropriate zero order wavefunction of the open shell system by acting on $|\Phi_0\rangle$ and the further operators $W_i (i=2,3,\ldots,m)$ contain i more creation-annihilation operator pairs than W_1 (in other words they describe the changes in the correlation with respect to the reference closed-shell state $|\Psi_0\rangle$). The Ansatz (91) can provide good results (also if we take into account only the first few terms in the series (92)) if the excitations involve directly only one or two valence electrons in which case one can assume that the majority of the pair (or higher cluster) correlations will not be drastically changed. Without going into further details (for them see [48]) it should be mentioned that by expressing the excitation energy as the change of the expectation values of $\hat{H}$ formed by $|\Psi\rangle$(ground state) and $|\Psi^{O.S.}\rangle$(excited state), respectively, it was possible to show [48] by cancellation of diagrams that one can obtain a direct expression of the energy difference (excitation). Since the programming of the closed shell version of this method in its ab initio form and its application to polymers (using instead of HF MO-s HF Wannier functions obtained from the HF Bloch orbitals of the band structure calculation) are in progress in cooperation with J. Čižek [50], it will not be very difficult to extend the programs also to the case of excitations using the new open-shell version of the coupled cluster expansion method.

Determining with the help of these methods in the future the excited states of the monomers of a polymer with weakly interacting units, one could place in a rough approximation around these excited states as centers the conduction band of the polymers pertaining its width. In this way one would obtain at least for the position (but of course again not for its width and dispersion) of the exciton band a better approximation than with the aid of the simple $\hat{O}\hat{A}\hat{O}$ method.

3.1.3 The Application of the Intermediate (Charge Transfer) Exciton Theory to Polymers

If we have performed an approximate Hartree-Fock (ab initio SCF LCAO CO) calculation we can transform the delocalized Bloch functions obtained into around the site $\vec{R}_i$ localized Wannier functions with the aid of the expression

$$a_k(\vec{r}-\vec{R}_i) = \frac{1}{\omega}\int \exp(-i\vec{k}\vec{R}_i)\, u_{k,\vec{k}}(\vec{r})\, d\vec{k} \tag{93}$$

[51], where the LCAO Bloch function

$$u_{h,\vec{k}}(\vec{r}) = (2N+1)^{-3/2} \sum_{\vec{q}} e^{i\vec{k}\vec{R}_{\vec{q}}} \sum_{g=1}^{m} d(\vec{k})_{h,g} \chi_g^{\vec{q}} \tag{94}$$

and h stands for the band index.

With the aid of these Wannier functions we can write down the many electron wavefunction for the excited configuration in which an electron with spin σ has been excited from the state n at site $\vec{R}_i$ to the state m with spin σ' at site $\vec{R}_j$

$$\Phi_{nm}(\vec{R}_i\sigma \rightarrow \vec{R}_j\sigma') = \hat{A}[a_{n\vec{R}_1}\alpha \ \ldots \ a_{m\vec{R}_j}\sigma' \ldots] , \tag{95}$$

where the Wannier function $a_{m\vec{R}_j}\sigma'$ takes the place in the antisymmetrized product of $a_{n\vec{R}_i}\sigma$ $(\sigma,\sigma'=\alpha \text{ or } \beta)$. Introducing the notation $\vec{R}_j=\vec{R}_i+\vec{\beta}$ we can form from the configurations (95) the translationally invariant exciton wavefunction.

$$\begin{aligned}\Phi_{nm}(\vec{K},\vec{\beta}) &= (2N+1)^{-3/2} \sum_{\vec{k}} e^{-i\vec{\beta}\vec{k}} \Phi_{nm}(\vec{k}-\vec{K} \rightarrow \vec{k}) = \\ &= (2N+1)^{-3/2} \sum_{\vec{R}_i} e^{i\vec{K}\vec{R}_i} \Phi_{nm}(\vec{R}_i \rightarrow \vec{R}_i+\vec{\beta}) \ .\end{aligned} \tag{96}$$

Here $\Phi_{nm}(\vec{k}-\vec{K}\rightarrow\vec{k})$ stands for the configuration (antisymmetrized product) of Bloch functions in which $u_{n,\vec{k}-\vec{K}}$ has been substituted by $u_{m,\vec{k}}$. This means that our electron-hole pair (exciton) has a momentum of $\vec{K}$ and a separation of $\vec{\beta}$ (for the sake of simplicity we have supressed the spin indices in (96)).

Finally we can take also a linear combination of the different separations of the electron-hole pair (in the simple localized or Frenkel exciton theory $\vec{\beta}=\vec{0}$). So we can write for the wavefunction of our intermediate (charge transfer) exciton

$${}^{M}\Psi_{nm}(\vec{K}) = \sum_{\vec{\beta}} {}^{M}U_{nm}(\vec{K},\vec{\beta})\, {}^{M}\Phi_{nm}(\vec{K},\vec{\beta}) \ , \tag{97}$$

where M=1 or 3 stands for the singlet or triplet exciton state with momentum $\vec{K}$, respectively.

Following the work of Takeuti [52] and its generalization to off-diagonal elements $(\vec{\beta}'\neq\vec{\beta}'')$ of the matrix $V_{nm}(\vec{K})$ by Kertész [53] one can derive using perturbation theory the expression

$$\begin{aligned}{}^{M}U_{nm}(\vec{K},\vec{\beta}) = \sum_{\vec{\beta}'} \sum_{\vec{\beta}''} G_{nm}(E,\vec{K},\vec{\beta}-\vec{\beta}')\, {}^{M}V_{nm}(\vec{K},\vec{\beta}',\vec{\beta}'') \cdot \\ \cdot\, {}^{M}U_{nm}(\vec{K},\vec{\beta}'') \quad (M=1,3)\end{aligned} \tag{98}$$

for the determination of the coefficients ${}^{M}U_{nm}(\vec{K},\vec{\beta})$. Here the electron-hole Green's function G_{nm} is given by

$$\begin{aligned}G_{nm}(E,\vec{K},\vec{\beta}-\vec{\beta}') = (2N+1)^{-3/2} \sum_{\vec{k}} \exp[i\vec{k}(\vec{\beta}-\vec{\beta}')] \cdot \\ \cdot [E(\vec{K}) - (\varepsilon_m(\vec{k}) - \varepsilon_n(\vec{k}-\vec{K}))]^{-1}\end{aligned} \tag{99}$$

[52, 53] and the matrix elements ${}^{M}V_{nm}(\vec{K}, \vec{\beta}', \vec{\beta}'')$ are defined as

$$ {}^{M}V_{nm}(\vec{K},\vec{\beta}',\vec{\beta}'') = -\sum_{\vec{R}_i} e^{i\vec{K}\vec{R}_i} \Big(\langle a_n(\vec{r}_1) a_m(\vec{r}_2-\vec{R}_i) | \frac{1}{r_{12}} | a_n(\vec{r}_1-\vec{\beta}') a_m(\vec{r}_2-\vec{R}_i-\vec{\beta}'') \rangle - 2\delta_M \langle \quad \rangle_{exch} \Big) . \quad (100) $$

Here the integral $\langle \; \rangle_{exch}$ is the exchange integral corresponding to the first term and δ_M=1 or 0 for a singlet or triplet excited state, respectively. To solve the system of equations (98) one has to find the $E(\vec{K})$ value for which a solution exists.

The described theory has been applied by Kertész [54] for polyethylene (the -HC=CH-CH=CH chain) and for polyC and polyU (U stands for uracil) using semiempirical SCF LCAO π electron (PPP) band structures [55]. For polyethylene alternating bond distances and for polyC and polyU the geometrical data of DNA B [56] have been applied.

In Table IV we give following Kertész [54] for polyethylene the singlet and triplet intermediate exciton bands for the valence band→conduction band transition including first neighbors' charge transfer (β=0,$\pm$1)

Table IV

The first singlet and triplet intermediate exciton bands of polyethylene in the first neighbors' charge transfer approximation after Kertész [54] (in eV-s)

κ	${}^{1}E(\kappa)$	${}^{3}E(\kappa)$	$\varepsilon_c(\kappa) - \varepsilon_v(\kappa)$
0	2.17	0.03	15.02
$2\pi/5$	3.70	1.27	13.10
π	6.06	3.22	3.62

Looking at the Table we can see first of all that the excitation energies belonging to different values of κ are very different from the corresponding one-electron level differences. The physically most significant theoretical ${}^{1}E(\kappa=0)$ values are 3.50 eV in the case of a Frenkel exciton (β=0), 2.17 eV in the first neighbors' and 2.24 eV in the second neighbors' (β=0,$\pm$1,$\pm$2) intermediate exciton case [54]. The intermediate exciton values agree quite well with the corresponding experimental value of 2.32 eV [57].

An analysis of the coefficients ${}^{1}U_{vc}(\kappa,\beta)$ and ${}^{3}U_{vc}(\kappa,\beta)$ obtained for polyethylene from the same calculation [54] shows that the singlet exciton is rather delocalized (${}^{1}U_{vc}(\pm1)$ and ${}^{1}U_{vc}(\pm2)$ are -0.52 and 0.26, respectively) while the triplet exciton is more strongly localized.

In contrary to polyethylene in the cases of the stacked

polyC and polyU chains the $^1E(\kappa=0)$ values calculated with the aid of the intermediate exciton theory do not differ strongly from the excitation energies of the single C and U molecules calculated with the aid of the closed shell Hartree-Fock formalism (see equ. (74)). The difference $^1E(\kappa=0)-{}^1\Delta E^{HF}_{HOMO\rightarrow LEMO}$ is for polyC 0.07eV in the case of the intermediate exciton ($\beta=0,\pm1$) and -0.03eV in the case of the Frenkel exciton ($\beta=0$), while for polyU the shift given by the intermediate exciton theory is 0.16eV [54]. The smallness of these energy shifts (if the shifts are similar also in the case of ab initio band structures and also for the other homopolynucleotides) could explain that the UV absorption spectrum of DNA corresponds in a good approximation to the superposition of the spectra of the constituent bases (see 3.1.1).

Finally it should be mentioned that though for the calculation of the excitation energies the application of the intermediate exciton theory does not seem to be necessary in the case of the weakly interacting homopolynucleotides, the situation is completely different if one computes the more sensitive intensities. According to the calculations of Kertész [54] the oscillator strength (f) values differ strongly (in the case of polyU by a factor greater than 2) if someone applies the intermediate exciton theory instead of assuming localized (Frenkel) excitons. The hypochromicity (H) values calculated on the basis of these f values differ by a factor of 5 for U. Further for polyC and for polyU the H values calculated with the aid of the intermediate exciton theory agree much better with experiment than the Frenkel exciton H-values [54].

We can conclude that the inclusion of charge transfer type excitations is necessary also in the case of covalently not-bonded (stacked or hydrogen bonded) polymers if one desires to describe also the finer details of their electronic spectra with the help of an exciton theory. Such calculations starting from our ab initio (HF) band structure are in progress.

ACKNOWLEDGMENT

The author wishes to express his gratitude to Professors T.C. Collins F. Martino and J. Čižek and to Drs. S. Suhai and M. Seel for very useful discussions. He is further indebted to the Fonds of Chemical Industry (Fond der Chemischen Industrie, BRD) for the financial support of this research.

REFERENCES

1. For a review of this method see: F. Martino in Quantum Theory of Polymers, J.-M. André, J. Delhalle and J. Ladik, Eds., D. Reidel Publ. Co., Dordrecht-Boston, p. 169 (1978).
2. J. Ladik and M. Seel, Phys. Rev. B 13, 5338 (1976).

3. G. Del Re, J. Ladik and G. Biczó, Phys. Rev. 155, 967 (1967); J.-M. André, L. Gouverneur and G. Leroy, Int. J. Quant. Chem. 1, 427 and 451 (1967); R. N. Euwema, D.L. Wilhite and G.T. Surrat, Phys. Rev. B7, 818 (1973).
4. J. Ladik in "Electronic Structure of Polymers and Molecular Crystals", J.-M. André and J. Ladik, Eds., Plenum Press, New York-London, p. 23 (1975).
5. G. Biczó, unpublished result .
6. For a review of these calculations see: J.-M. André in "Electronic Structure of Polymers and Molecular Crystals", J.-M. André and J. Ladik, Eds., Plenum Press, New York-London p. 1 (1975); J.-M. André and J. Delhalle in "Quantum Theory of Polymers", J.-M. André, J. Delhalle and J. Ladik, Eds., D. Reidel Publ. Co., Dordrecht-Boston, p. 1 (1978).
7. For a review see: T.C. Collins in "Electronic Structure of Polymers and Molecular Crystals", p. 389.
8. For a review see: J. Ladik in "Electronic Structure of Polymers and Molecular Crystals", p. 663.
9. J. Ladik in "Quantum Theory of Polymers" p. 279.
10. Ch. Merkel in "Elektronische Eigenschaften vom Molekülkristallen" (Electronic Properties of Molecular Crystals) Thesis, Technical University Munich, 1977.
11. See for instance: M. Hamermesh "Group Theory and its Application to Physical Problems", Adison Wesley Publ., Reading, Mass. p. 80 (1964).
12. S. Suhai, Ch. Merkel and J. Ladik, Phys. Lett. 61A, 487 (1977).
13. B. Mely and A. Pullman, Theor. Chim. Acta (Berl.) 13, 278 (1969).
14. S. Huzinaga, J. Chem.Phys. 42, 1293 (1965).
15. S. Arnott, S.D. Dover and A.J. Wonacott, Acta Cryst. B25, 2192 (1969).
16. J. Ladik in "Adv. Quantum Chem.", P.-O. Löwdin, Ed., Academic Press, New York-London 7, p. 397 (1973). S. Suhai and J. Ladik, Int. J. Quant. Chem. 7, 547 (1973).
17. For a review see: S. Suhai in "Quantum Theory of Polymers", p. 335.
18. C. Lifschitz, E.D. Bergmann and B. Pullman, Tetrahedron Lett. 46, 4583 (1967).
19. W.J. Hunt and W.A. Goddard III, Chem. Phys. Lett. 3, 414 (1969); H.P. Kelly, Adv. Chem. Phys. 14, 129 (1969).
20. M. Seel, "Theoretische Untersuchungen zur Aperiodizität und zum Korrelationsproblem in Polymeren und Molekülkristallen" (Theoretical Investigations of Aperiodicity and Correlation in Polymers and Molecular Crystals) Thesis, University Erlangen-Nürnberg 1978.
21. See for instance: W.L. McCubbin in "Quantum Theory of Polymers" p. 185 (1978).
22. A.J. Elliot, J.A. Krumhansl, P.L. Leath, Rev. Mod. Phys. 46, 465 (1974).
23. P. Soven, Phys. Rev. 156, 809 (1967); S. Kirkpatrick, B. Velický, H. Ehrenreich, Phys. Rev. B1, 3250 (1970).

24. See for instance: J. Callaway, Quantum Theory of the Solid State, Academic Press, New York-London, p. 517 (1974).
25. M. Seel, T.C. Collins, F. Martino, D.K. Rai and J. Ladik, Phys. Rev. B (submitted).
26. For recent review papers see: H.P. Geserich and L. Pintschovius, Adv. Solid State Phys. 16, 65 (1976); G.B. Street and R.L. Greene, IBM J. Res. Develop. 21, 99 (1977).
27. H. Kamimura, A.J. Grant, F. Levy, A.D. Yoffe, C.D. Pitt, Solid State Comm. 17, 49 (1975); D.E. Parry, J.M. Thomas, J. Phys. C3, L45 (1975); A. Zunger, J. Chem. Phys. 63, 4854 (1975); S. Suhai, M. Kertész, J. Phys. C9, L347 (1976); N.T. Rajan and L.M. Falicov, Phys. Rev. B12, 1240 (1975).
28. C. Merkel and J. Ladik, Phys. Lett. 56A, 395 (1976); M. Kertész, J. Koller, A. Ažman and S. Suhai, Phys. Lett. 55A, 107 (1975).
29. W.E. Rudge and P.M. Grant, Phys. Rev. Lett. 35, 1799 (1975); W.Y. Ching, J.G. Harrison and C.C. Lin, Phys. Rev. B15, 5975 (1977); I.P. Batra, S. Ciraci and W.E. Rudge, Phys. Rev. B15, 5858 (1977).
30. S. Suhai and J. Ladik, Solid State Comm. 22, 227 (1977).
31. B. Györffy and S. Faulkner (personal communication).
32. W.D. Gill, W. Bludan, R.H. Geiss, P.M. Grant, R.L. Greene, J.J. Mayerle and G.B. Street, Phys. Rev. Lett. 38, 1305 (1977).
33. S. Suhai (unpublished results).
34. J. Delhalle, Bull. Soc. Chim. Belg. 84, 135 (1975).
35. See for instance: G. Rickayzen, "Theory of Superconductivity", Interscience-Wiley, New York-London (1965).
36. J. Ladik, unpublished result.
37. G.F. Koster and J.C. Slater, Phys. Rev. 96, 1208 (1954); M. Kertész and G. Biczó, in "Proceedings on Computers in Chemical Research and Education", J. Hadži, Ed., Ljubljana, p. 4195 (1973).
38. A. van der Avoird, S.P. Liebmann and J.M. Fassaert, Phys. Rev. B10, 1230 (1974).
39. J. Ladik, Phys. Rev. B17, 1663 (1978).
40. See the paper of R.J. Buenker and S.D. Peyerimhoff in this volume.
41. J. Avery, J. Packer, J. Ladik and G. Biczó, J. Mol. Spectr. 29, 194 (1969); A. Bierman and J. Ladik (unpublished result).
42. For a detailed description of this formalism and for its connection with the general Green's function theory see [7], T.C. Collins in "Quantum Theory of Polymers", p. 75 and the paper of T.C. Collins in this volume.
43. L.B. Clark, G.G. Peschel and I. Tinoco Jr., J. Phys. Chem. 69, 3615 (1965).
44. See the papers of T.C. Collins, of J. Linderberg and Y. Öhrn and of W. v. Niessen in this volume.
45. See for instance: J.D. Thouless, "The Quantum Mechanics of Many Body Systems", Academic Press, New York-London (1961); A.L. Fetter and J.D. Walecka, "Quantum Theory of Many Particle

Systems", McGraw-Hill, New York (1971).
46. J. Paldus and J. Čižek, J. Chem. Phys. 60, 149 (1974).
47. R.G. Parr, "The Quantum Theory of Molecular Electronic Systems", Benjamin, New York (1963).
48. J. Paldus, J. Čižek, M. Saute and A. Laforque, Phys. Rev. A17, 805 (1978).
49. J. Čižek, J. Chem. Phys. 45, 4256 (1966); ibid 14, 35 (1969); J. Čižek and J. Paldus, Intern. J. Quant. Chem. 5, 359 (1971); J. Paldus, J. Čižek and I. Shavitt, Phys. Rev. A5, 50 (1972); J. Paldus and J. Čižek, Adv. Quant. Chem. 9, 105 (1975).
50. J. Čižek and S. Suhai (to be published).
51. See for instance: E.I. Blount, Solid State Phys. 13, 305 (1963); J. Des Cloizeaux, Phys. Rev. 135, A698 (1964).
52. Y. Takeuti, Progress of Theor. Phys. 18, 421 (1957)
53. M, Kertész, phys. stat. soli. (b) 62, K75 (1974).
54. M. Kertész, Kémiai Közlemények 46, 393 (1976) (in Hungarian).
55. J. Ladik and G. Biczó, Acta Chim. Acad. Sci. Hung. 67, 297 (1971).
56. S. Arnott, S.D. Dover and A.J. Wonacott, Acta Cryst. B25, 2192 (1969).
57. J.N. Murrell "The Theory of Electronic Spectra of Organic Molecules", Methuen Co., London (1963).

THEORY OF SURFACE STATES AND CHEMISORPTION

Jaroslav Koutecky

Institut fuer Physikalische Chemie und Quantenchemie der Freien Universitaet Berlin

INTRODUCTION

Let us start with an obvious and consequently trivial statement. The theoretical methods useful for an investigation of solid surfaces must be closely related to the methods used for a description of both molecules and crystals, due to the fact that the same natural laws determine electronic structures of a solid surface as well as of molecules and crystals. For the same reason, important observable properties of a crystal surface can be considered as a synthesis or compromise between the properties of a solid bulk and of a molecule. Besides the properties which are customarily named physical, the existence of characteristic chemical properties offers a foundation for the large extremely important field of technical chemistry known as heterogeneous catalysis.

It was evident a long time ago that, in principle, a quantum mechanical treatment of the electronic structure of surfaces is a challenging problem. But twenty years ago a detailed theoretical investigation was very difficult and not very rewarding because the direct experimental verification of tediously obtained predictions was absolutely out of the question. The technically superb catalysts available a long time ago have such a complicated geometry and electronic structure so that they are practically useless for a com-

Cleanthes A. Nicolaides and Donald R. Beck (eds.), Excited States in Quantum Chemistry, 531–556.

parison of calculated and experimentally determined properties. The flood of theoretical surface science work in the very recent literature is not only due to the demanding and complex nature of the theoretical problem, but is due to the only recently manageable preparation and experimental investigation of clean surfaces. Along with the recent possibilities for a preparation of clean surfaces, appropriate experimental methods for measurements of spectroscopical and structural properties of surfaces are currently available (e.g., ultra violet and x-ray photoelectron spectroscopy, Auger electron spectroscopy, appearance-potential spectroscopy, vacuum-tunneling spectroscopy, field-ion spectroscopy and low-energy electron spectroscopy).

Interesting and very specific properties of a solid surface resulting from the fact that a surface represents a compromise between a molecule and a solid, have attracted the attention of many scientists as an interesting intellectual exercise in the early days of quantum mechanics. This can be illustrated by the example of the so called electronic surface states. The surface states are one-electron states in which the probability of finding an electron in the surface region is finite and the probability of finding an electron far away from the surface is negligible. Different types of surface states were predicted and investigated in sporadic single papers (e.g., by Tamm in 1932, Shockley and Goodwin in 1939, Coulson and Hoffmann in 1950).

Already in these pioneering contributions, a variety of methodical approaches were employed such as the Kronig-Penney potential, the nearly free-electron model, and the tight-binding approximation which is analogous to the LCAO-MO method of quantum chemistry. This manifold of approaches which are customarily used in solid state theory and in quantum chemistry correspond to a variety of aspects which emphasize either the physical or chemical nature of surface phenomena.

It seems natural that the chemical aspects of the electronic structure of surfaces can be better described using quantum chemical approaches. Different kinds of MO-LCAO approaches play a dominant role because they preserve the chemical individuality of the crystal constituents to a certain degree. The generalization of the MO-LCAO approach as the expansion of one-electron functions in terms of more or less localized basis functions offers a fundament for nearly all ab-initio quantum chemical techniques. It is clear that the specificity of the surface problem requires some methodical modification of the current methods used in molecular physics. The number of atoms and therefore of atomic orbitals (basis functions) is de facto infinite. Also the translation symmetry of the system which simplifies the solution of the problem in the solid state theory is lost at least in the direction perpendicular to the surface.

In these lectures, the emphasis will be put on the MO-LCAO approach to the problem of electronic structure of solid surfaces. We will pay special attention to the one-electron surface states and localized chemisorption states because they are good examples of the specific properties of the solid surface. It is of course useful to "embed" the surface state theory in a more general theory of the electronic structure of solid surfaces, and for this reason we will start with some general considerations.

GREEN'S OPERATOR

The Green's function technique and the so called resolvent method are quite extensively used to describe the electronic states of very large systems. Several methodical approaches to the problem of solid surfaces are based on the Green's operator theory. For this reason, some fundamental definitions and properties of the Green's operator will be reviewed here.

The time independent Schroedinger equation

$$(\hat{H}-E_j)|E_j,a\rangle = \hat{\Delta}(E_j)|E_j,a\rangle = 0 \qquad (1)$$

can be also written as

$$\hat{H}\hat{P}_j = E_j\hat{P}_j \qquad (2)$$

In equation (2) the operator

$$\hat{P}_j = \sum_a |E_j,a\rangle\langle E_j,a| \qquad (3)$$

is the projector on the space spanned by all eigenstates

$|E_j,a\rangle$ with the eigenvalue E_j.

The Green's operator is defined as

$$\hat{G}(z)=(z\hat{I} - \hat{H})^{-1}\equiv\hat{\Delta}^{-1}(z) \quad (4)$$

at any point z in the complex plane except at $z=E_j$ for which Eq.(1) is fullfilled. From equations (1) and (4) it follows that:

$$\hat{G}(z)\hat{P}_j = (z - E_j)^{-1}\hat{P}_j \quad (5)$$

Because the basis formed by the eigenvectors of an observable is by definition complete, the so called spectral expansion of the Green's operator exists:

$$\hat{G}(z) = \sum_j (z - E_j)^{-1}\hat{P}_j \quad (6)$$

On the other hand, the projection operator $\hat{P}_j$ can be expressed in the form of an integral over the contour $\Gamma(E_j)$ in the complex plane which encloses the single eingenvalue E_j lying of course on the real axis:

$$\hat{P}_j = \frac{1}{2\pi i}\oint_{\Gamma(E_j)}\hat{G}(z)dz \quad (7)$$

Evidently, the projector on the space of all states with energies lying in the interval on the real axis enclosed by the contour Γ is:

$$\hat{P}_\Gamma = \frac{1}{2\pi i}\oint_\Gamma \hat{G}(z)dz \quad (8)$$

As a consequence of definitions (4) and (8), important for the energy calculations of large systems, it follows:

$$\frac{1}{2\pi i}\oint_\Gamma (z\hat{I}-\hat{H})\hat{G}(z)dz = \frac{1}{2\pi i}\oint_\Gamma z\,\hat{G}(z)dz-\hat{H}\hat{P}_\Gamma=0 \quad (9)$$

GREEN OPERATORS FOR TWO SYSTEMS AND DYSON EQUATION

Let us compare two systems S and S_o with Hamiltonians $\hat{H}$ and $\hat{H}_o$, respectively. The system S can be for example the crystal C with the adsorbed atom A.

The system S_O consists of the same crystal C and the atom A separated by a large distance so that there is no interaction between them. If the difference between both Hamiltonians can be formulated as a perturbation potential $\hat{V}$

$$\hat{V} = \hat{H} - \hat{H}_o \tag{10}$$

then the relation between the Green's operator $\hat{G}$ assigned to $\hat{H}$ and the operator $\hat{G}_o$ assigned to $\hat{H}_o$ is given by the Dyson equation

$$\hat{G}(z) = \hat{G}_o(z)(\hat{I}+\hat{V}\hat{G}(z)) \tag{11}$$

The formal solution of Dyson's equation can be written as:

$$\hat{G}(z) = \hat{L}(z)^{-1}\hat{G}_o(z) \tag{12}$$

where

$$\hat{L}(z)=\hat{\Delta}(z)\hat{G}_o(z)=\hat{\Delta}(z)\hat{\Delta}_o^{-1}(z)=\hat{I}-\hat{G}_o(z)\hat{V} \tag{13}$$

The difference between the Green's operator $\hat{G}(z)$ and $\hat{G}_o(z)$ is:

$$\hat{G}(z)-\hat{G}_o(z)=(\hat{L}^{-1}-\hat{I})\hat{G}_o(z)=\hat{L}^{-1}\hat{G}_o\hat{V}\hat{G}_o \tag{14}$$

If the Schroedinger equation (1) is rewritten with the help of the perturbation potential $\hat{V}$ (10) as:

$$(\hat{E}_j\hat{I}-\hat{H}_o)|E_j,a\rangle = \hat{V}|E_j,a\rangle \tag{1a}$$

the multiplication of Eq. (1a) by $\hat{G}_o(E_j)$ yields

$$\hat{L}(E_j)|E_j,a\rangle=0 \tag{15}$$

if the operator $\hat{L}(E_j)$ defined by Eq.(13) exists for $z=E_j$.

REPRESENTATION OF GREEN'S OPERATORS

As an illustration we will consider a simple complete orthonormal basis set $\underline{B}$ formed by the kets $|\mu\rangle \in \underline{B}$:

$$\hat{I} = \sum_{\mu} |\mu\rangle\langle\mu| \tag{16}$$

The representation of the Green's operator $\hat{G}(z)$ in the basis $\underline{B}$ is the matrix:

$$\underset{\sim}{G}(z)=(\langle\mu|G(z)|\nu\rangle)=(G_{\mu\nu}) \tag{17}$$

with

$$G_{\mu\nu}=(-1)^{\mu+\nu}|\Delta|_{\nu\mu} \Big/ |\Delta| \tag{18}$$

where the symbol $|\Delta|$ denotes the determinant of the matrix representation of the operator defined by Eq. (4). The subdeterminant of the determinant $|\Delta|$, in which the ν-th row and μ-th column are omitted is labelled by $|\Delta(z)|_{\nu\mu}$.

It is evident that the points on the real axis for which it holds $z=E_j$ (so called poles of $\hat{G}$) are given by the determinantal equation

$$|\Delta(E_j)|=0 \tag{19}$$

The representation of the Eq. (15) in the basis $\underline{B}$ gives a linear homogenous system of equations of the form:

$$\sum_{\nu} \langle\mu|\hat{L}(E_j)|\nu\rangle\langle\nu|E_j,a\rangle=0 \tag{20}$$

with the following condition for obtaining a non-trivial solution:

$$|\langle\mu|\hat{L}(E_j)|\nu\rangle|=|(\delta_{\mu\nu}-\sum_{\chi} G^{o}_{\mu\chi}(E)V_{\chi\nu})|=0 \tag{21}$$

In the above equation $G^{o}_{\mu\chi}$ and $V_{\chi\nu}$ are matrix elements of the operators $\hat{G}_o$ and $\hat{V}$, respectively in the basis $\underline{B}$. Evidently the relation (21) is mainly useful in the shown form if the operator $\hat{G}^o(E_j)$ assigned to the Hamiltonian $\hat{H}_o$ exists (i.e., if the eigenvalue E_j of the Hamiltonian $\hat{H}$ describing the investigated system is not degenerate with any eigenvalue E^o_k of the Hamiltonian $\hat{H}_o$ assigned to the unperturbed system S_o).

THE BASIS FORMED BY LOCALIZED KETS

If the basis kets $|\mu\rangle \in \underline{B}$ are sufficiently localized and simultaneously the perturbation potential is sufficiently local so that $V_{\chi\nu} = \langle\chi|\hat{V}|\nu\rangle = 0$ holds when $|\chi\rangle \in \bar{Q}$ and $|\nu\rangle \in Q \subset \underline{B}$ where $Q \cup \bar{Q} = \underline{B}$, then the partition of the basis $\underline{B}$ into parts Q and $\bar{Q}$ is advantageous for a treatment of the determinantal equation (21).

Since $V_{\chi\nu} = 0$ if $|\nu\rangle \in \bar{Q}$, it follows that

$$\langle\mu|\hat{L}(E_j)\;|\nu\rangle = \delta_{\mu\nu} \qquad (22)$$

In other words the columns of the determinant $|\langle\mu|\hat{L}(E_j)|\nu\rangle, |\nu\rangle \in \bar{Q}$ contain zeros as the off-diagonal matrix elements and the value 1 as the diagonal elements. The dimension of the determinant is reduced to the dimension of the subspace spanned by the kets $|\mu\rangle \in Q$. The dimension of Q can be finite even when the dimension of the basis $\underline{B}$ is infinite. If the projector on the system Q is named $\hat{Q}$, the determinantal equation (21) can be written in the form:

$$|\underset{\sim}{L(E)}| = |\langle\mu|\hat{Q}(\hat{I}-\hat{G}_o(E)\hat{V})\hat{Q}|\nu\rangle| = 0 \qquad (23)$$

Solutions of the Eq. (23) yield eigenvalues of the Hamiltonian $\hat{H}$ which are distinct from the eigenvalues of the Hamiltonian $\hat{H}_o$ describing the unperturbed system S_o.

Two examples should illustrate this procedure. The system S_o in the first example is composed of the crystal C and the chemisorbate molecule A which do not mutually interact. In the system S, the crystal C and the molecule A <u>do</u> <u>interact</u>. The formulation of the problem is especially simple if the model Hamiltonian is a Hueckel type effective one-electron Hamiltonian and the basis is formed by the orthogonal (or orthogonalized) well localized orbitals. The number of orbitals localized at any single center can be greater than one. Hence, the perturbation potential $\hat{V}$ has the form

$$\hat{V} = \hat{V}_A + \hat{V}_{AQ} + \hat{V}_{QA} + \hat{V}_Q \qquad (24)$$

In Eq. (24), the notation has been introduced that the product of three operators $\hat{K}\hat{Z}\hat{L}$ is denoted by $\hat{Z}_{KL}$ if the operators $\hat{K}$ and $\hat{L}$ are projectors on the vector spaces K and L respectively ($\hat{Z}_{KK} \equiv \hat{Z}_K$). A denotes in Eq. (24) the vector space spanned by atomic orbitals associated with the molecule, and $Q \in C$ is the subspace of atomic orbitals associated with the crystal for which holds

$$\hat{Q}\hat{V} \neq \hat{0} \quad \text{and} \quad \hat{V}\hat{Q} \neq \hat{0} \tag{25}$$

The first and fourth term in Eq. (24) describes the perturbation inside the chemisorbate molecule A and in the neighborhood of the chemisorption site inside the crystal C, respectively. The second and third term on the right-hand side of Eq. (24) describe the chemical interaction between chemisorbate and chemisorbent. The Green's operator $\hat{G}_o \equiv \hat{G}^o$ can be written as:

$$\hat{G}_o \equiv \hat{G}^o = \hat{G}^o_A + \hat{G}^o_B \tag{26}$$

The matrix $\underset{\sim}{L}(E)$ in the determinantal equation (23) takes the block form:

$$\underset{\sim}{L}(E) = \begin{pmatrix} \delta_{\mu\nu} - \langle\mu|\hat{G}^o_A \hat{V}_A|\nu\rangle & -\langle\mu|\hat{G}^o_A \hat{V}_{AQ}|\nu\rangle \\ -\langle\mu|\hat{G}^o_Q \hat{V}_{QA}|\nu\rangle & \delta_{\mu\nu} - \langle\mu|\hat{G}^o_Q \hat{V}_Q|\nu\rangle \end{pmatrix} \begin{matrix} A \\ Q \end{matrix} \tag{27}$$

$$\qquad A \qquad\qquad\qquad Q$$

For illustration let us consider an oversimplified model with one atomic orbital $|\sigma\rangle$ in the chemisorbate molecule A forming a chemical bond with one AO $|\chi\rangle$ of the chemisorption site in the crystal surface. The Coulomb perturbation is also acting only on the atomic orbitals $|\sigma\rangle$ and $|\chi\rangle$. The perturbation potential in the Hueckel model takes the simple form:

$$\hat{V} = \Delta\alpha_\sigma |\sigma\rangle\langle\sigma| + \beta_{\sigma\chi} (|\sigma\rangle\langle\chi| + |\chi\rangle\langle\sigma|) + \Delta\alpha_\chi |\chi\rangle\langle\chi| \tag{28}$$

Hence, Eq. (27), for the determination of the energies distinct from energies of the unperturbed crystal C and the unperturbed adsorbate molecule A, reduces to

$$(1-\Delta\alpha_\sigma\, G^o_{\sigma\sigma})(1-\Delta\alpha_\chi\, G^o_{\chi\chi})-|\beta_{\sigma\chi}|^2 \cdot G^o_{\sigma\sigma}G^o_{\chi\chi}=0 \tag{29}$$

Dividing Eq. (29) by $G^o_{\sigma\sigma}G^o_{\chi\chi}$, one obtains the relation

$$\left[(G^o_{\sigma\sigma})^{-1}-\Delta\alpha_\sigma\right]\left[(G^o_{\chi\chi})^{-1}-\Delta\alpha_\chi\right]-|\beta_{\sigma\chi}|^2=0 \tag{29a}$$

which has a form very similar to the Hueckel equation for a diatomic molecule:

$$(E-\alpha_\sigma)(E-\alpha_\chi) - \beta^2_{\sigma\chi} = 0 \tag{30}$$

This similarity indicates that in general the chemisorbate molecule together with the chemisorption site can be considered in the one-electron approximation as a so called chemisorption molecule embedded in its neighborhood. This is particularly true if we are interested in a localized one-electron state with an energy located quite far from the energy band (or from the energy bands in more complicated models) of the unperturbed crystal. If in addition the chemisorbate consists of a single atom described by the orbital $|\sigma\rangle$, then

$$(G^o_{\sigma\sigma})^{-1}= E-\alpha^o_\sigma;\ (G^o_{\chi\chi})=\sum_j \frac{|\langle\chi|j\rangle|^2}{E-\varepsilon_j} \doteq \frac{1}{E-\varepsilon_c} \tag{31}$$

where α^o_σ is the Coulomb integral of unperturbed AO and ε_c is the mean energy of the energy band associated with the unperturbed crystal C. Substitution of the relations (31) into Eq. (29a) yields Eq. (30) with

$$\alpha_\sigma = \alpha^o_\sigma + \Delta\alpha_\sigma \text{ and } \alpha_\chi = \alpha_c + \Delta\alpha_\chi \tag{32}$$

In the second example we will start with the infinite crystal described by the one-electron Hamiltonian $\hat{H}^o$. Let us cut the crystal in two halves C and R. The purpose is to obtain the surface states starting from the information which we assume to have for the infinite crystal with complete translation symmetry in three dimensions. The perturbation potential forbids the transition of an electron from the parts C to R and vice versa:

$$\hat{V} = -\hat{H}^{o}_{CR} - \hat{H}^{o}_{RC} + \hat{V}_{R} + \hat{V}_{C} \tag{33}$$

The Green's operator $\hat{G}$ has the property

$$\hat{G} = \hat{G}_{C} + \hat{G}_{R} \tag{34}$$

By multiplication of the Dyson equation (11) by the projector $\hat{C}$ from the right and left side, we obtain the following equation:

$$\hat{G}_{C} = \hat{G}^{o}_{C}(1 + \hat{V}_{C})\hat{G}_{C} - \hat{G}^{o}_{CR}\hat{H}^{o}_{RC}\hat{G}_{C} \tag{35}$$

which has the formal solution

$$\hat{G}_{C} = (\hat{L}_{C})^{-1}\,\hat{G}^{o}_{C} \tag{36}$$

with

$$\hat{L}_{C} = (\hat{C} - \hat{G}^{o}_{C}\hat{V}_{C} + \hat{G}^{o}_{CR}\hat{H}^{o}_{RC}) \tag{37}$$

Hence, Eq. (23) takes the form:

$$|\hat{L}_{C}(E)| = <\mu|\ \hat{Q} - \hat{G}^{o}_{C}\hat{V}_{C} + \hat{G}^{o}_{CR}\hat{H}^{o}_{RC}|\nu> = 0 \tag{38}$$

with the dimension of the region $Q \subset C$ which is perturbed by cutting the infinite crystal. Eq. (38) was used in the literature for a determination of surface state energies. It is worth mentioning that if we multiply Eq. (11) from the right by the projector $\hat{C}$ and from the left by the projector R, the following relation is obtained:

$$\hat{G}^{o}_{RC}(E)\left[\hat{I} + \hat{V}_{C}\hat{G}_{C}(E)\right] = \hat{G}^{o}_{R}(E)\hat{H}^{o}_{RC}\hat{G}_{C}(E) \tag{39}$$

It is evident that Eq. (38) yields only a necessary but not sufficient condition for the energies of surface states and consequently the solutions of the Eq. (38) can also yield spurious states. Therefore, it is necessary to carry out a careful examination of the wave functions of the surface states obtained by the Eq. (38).

PARTITION OF THE INVESTIGATED SYSTEM

The Dyson equation, in general, and especially the determinantal Eq. (28) are useful only if we are able to determine the matrix element $G^{o}_{\mu\nu}$. The determination of the Green's operator matrix elements for a large system is not a simple task if we do not know explicitely the energies of the one-electron functions. For this reason several techniques for a direct calculation of the Green's operator matrix elements will be demonstrated in this section.

Let us divide the vector space spanned by the basis $\underline{B}$ in two subspaces A and R. The Hamiltonian and the corresponding Green's operator can be decomposed according to the above mentioned partition scheme:

$$\hat{H}=\sum_{C,D=A}^{R} \hat{H}_{CD},\quad \hat{G}=\sum_{C,D=A}^{R} \hat{G}_{CD} \tag{40}$$

The multiplication of the Green's operator definition

$$(z\hat{I}-\hat{H})\,\hat{G}(z) = \hat{I} \tag{4a}$$

from the left and right by the projector $\hat{A}$ yields the following relation:

$$(z\hat{I}-\hat{H}_A)\hat{G}_A - \hat{H}_{AR}\hat{G}_{RA} = \hat{A} \tag{41}$$

The Green's operator $\hat{G}_{RA}$ can be calculated from the relation obtained by multiplication of Eq. (4a) from the left by $\hat{R}$ and from the right by $\hat{A}$:

$$(z\hat{I}-\hat{H}_R)\hat{G}_{RA}-\hat{H}_{RA}\hat{G}_A=0 \tag{42}$$

For the Green's operator $\hat{G}_A$ it follows that

$$\hat{G}_A= \left[(z\hat{I}-\hat{H}_A)-\hat{H}_{AR}\hat{G}_R^{(1)}\ \hat{H}_{RA}\right]^{-1} \tag{43}$$

where

$$\hat{G}_R^{(1)} = (z\hat{I} - \hat{H}_R)^{-1} \tag{44}$$

The quantity $\hat{G}_R^{(1)}$ is the Green's operator for the subsystem R. Notice, that the $\hat{G}_R^{(1)} \neq \hat{G}_R$ because $\hat{G}_R$ is the projection of the Green's operator, assigned to the whole system, on the subsystem R.

As a very simple example let us formulate the Green's operator matrix element of a linear chain C composed of a single AO in each elementary cell. All Coulomb integrals α_μ and resonance integrals $\beta_{\mu\nu}$ are equal to α and β, respectively with the exception that $\alpha_1 \neq \alpha$ and $\beta_{12} \neq \beta$. According to Eq. (43) the first matrix element of $\hat{G}$ is:

$$<1|\hat{G}|1> = G_1 = \left[(z-\alpha_1) - \beta_{12}^2 G_2^{(1)}\right]^{-1} \tag{45}$$

If the system C is a semiinfinite chain with $\alpha_1 = \alpha$ and $\beta_{12} = \beta$, then

$$G_1 = G_2^{(1)} \tag{46}$$

and G_1 can be calculated directly from Eq. (45) as:

$$G_1 = \frac{1}{2\beta^2}\left[(z-\alpha) \pm \sqrt{(z-\alpha)^2 - 4\beta^2}\right] \tag{47}$$

Since in our model the subsystem R is a linear chain with $\alpha_1 = \alpha$ and $\beta_{12} = \beta$, Eq. (47) determines $G_2^{(1)}$ as well, and Eq. (45) for the semiinfinite chain C reads as:

$$G_1 = \left[z - \alpha_1 - \frac{\beta_{12}^2}{2\beta^2}\left((z-\alpha) \pm \sqrt{(z-\alpha)^2 - 4\beta^2}\right)\right]^{-1} \tag{48}$$

The values of z can be real or complex. The sign in front of the square root in Eqs. (45), (47) and (48) must be chosen in such a way that

$$|G_1| < 1 \tag{49}$$

Evidently if z is real the Green's operator matrix element can be complex when $|z-\alpha| < 2|\beta|$. This is the case when z lies inside the energy band of the nonlocalized states for which this simple model yields the energy expression:

$$E = \alpha + 2\beta \cos\xi \tag{50}$$

with $0 \le \xi < 2\pi$.

The other application of Eq. (23) is the so called continuous fraction expression for the diagonal matrix elements of the Green's operator. Let us partition the system S into the parts U_j where $j=1,\ldots.L$ with the corresponding projectors $\hat{U}_j$. The system composed from all parts U_k for $k=l+1\ldots.L$ is denoted by $\bar{U}_l$. The Green's operator $\hat{U}_1\hat{G}\hat{U}_1 = \hat{G}_{(1)}$ is

$$\hat{G}_{(1)} = \left[\hat{\Delta}_{(1)} - \hat{H}_{(1,\bar{1})}\hat{G}_{(\bar{1})}\hat{H}_{(\bar{1},1)}\right]^{-1} \tag{51}$$

with

$$\hat{\Delta}(j) = \hat{U}_j\hat{\Delta}\hat{U}_j, \hat{G}(j) = \hat{U}_j\hat{G}\hat{U}_j, \hat{H}(j,\bar{j}) = \hat{U}_j\hat{H}\hat{U}_{\bar{j}}, \hat{H}_{(\bar{j},j)} = \hat{U}_{\bar{j}}\hat{H}\hat{U}_j \tag{52}$$

Repeating the partition procedure for the vector space $U_{\bar{1}}$ one obtains

$$\hat{G}_{(1)} = \left[\hat{\Delta}_{(1)} - \hat{H}_{(1,\bar{1})}\left[\hat{\Delta}_{(2)} - \hat{H}_{(2,\bar{2})}\hat{G}_{(2)}\hat{H}_{(\bar{2};2)}\right]^{-1}\hat{H}_{(\bar{1},1)}\right]^{-1} \tag{53}$$

If the vector spaces U_j can be chosen in such a way that $\hat{H}(j,\bar{j}) = \hat{H}(j,j+1)$, then Eq. (53) takes the form

$$\hat{G}_{(1)} = \left[\hat{\Delta}_{(1)} - \hat{H}_{(1,2)}\left[\hat{\Delta}_{(2)} - \hat{H}_{(2,3)}\hat{G}_{(3)}\hat{H}_{(3,2)}\right]^{-1}\hat{H}_{(2,1)}\right]^{-1} \tag{54}$$

If this procedure is continued, an operator expression which reminds us of the continuous fraction expressions is obtained:

$$\hat{G}_{(1)} = \left[\Delta_{(1)} - \hat{H}_{(1,2)}\left[\hat{\Delta}_{(2)} - \hat{H}_{(2,3)}\left[\hat{\Delta}_{(3)} - \hat{H}_{(3,4)}\left[\ddots \hat{G}_{(\bar{n})}\right]^{-1}\hat{H}_{(4,3)}\right]^{-1}\hat{H}_{(3,2)}\right]^{-1}\hat{H}_{(2,1)}\right]^{-1} \tag{55}$$

In the case of a linear chain with a single AO in each elementary cell, the matrix element $\langle 1|\hat{G}_1(z)|1\rangle$ can be written directly in the form of a continuous fraction expression

$$\langle 1|\hat{G}(z)|1\rangle = \left[z-\alpha_1-\beta_{12}^2\left[z-\alpha_2-\beta_{23}^2\left[z-\alpha_3-\beta_{34}^2\cdot\left[\ddots\ \beta_{n,n+1}^2 G_{\bar{n}}\right]^{-1}\right]^{-1}\right]^{-1}\right]^{-1} \qquad (56)$$

where $G_{\bar{n}} = \langle n+1|\hat{G}_{\bar{n}}(z)|n+1\rangle$

For example, if the following relation holds:

$$\langle n+1|\hat{G}_{\bar{n}}(z)|n+1\rangle = \langle n+2|\hat{G}_{\overline{n+1}}(z)|n+2\rangle = \frac{1}{2\beta^2}\left[(z-\alpha)\pm\sqrt{(z-\alpha)^2-4\beta^2}\right] \qquad (57)$$

the sequence can be determined and the explicit expression for $\langle 1|\hat{G}_1(z)|1\rangle$ is obtained. The continuous fraction procedure can be started at an arbitrary appropriate AO of the chain and proceed towards the proximate chain end as well as parallely in the oposite direction (going away from the nearest end of the chain).

The representation of the operator relation (55) can be, in principle, determined if for a given n, the matrix

$$(\langle\mu|\hat{G}_{\bar{n}}(z)|\nu\rangle) \quad \text{with } |\mu\rangle,|\nu\rangle\in\bar{U}_n$$

is known. Then, the matrix representation of the inverse operator $\hat{M}^{-1}$ can be successively evaluated according to the following relation

$$(\underset{\sim}{M}^{-1})_{\mu\nu} = (-1)^{\mu+\nu}|\underset{\sim}{M}|_{\nu\mu}^{-1} \Big/ |\underset{\sim}{M}| \qquad (58)$$

HARTREE-FOCK PROCEDURE

Similarly as in the quantum theory of molecules

an explicit consideration of the electron-electron interaction, at least in an approximate way, is of importance for the treatment of electronic surface phenomena. For this reason the application of the Hartree-Fock approach will be discussed in this section.

In the restricted Hartree-Fock procedure (RHF), two electrons with opposite spins always occupy the lowest possible one-electron states obtained as eigenfunctions of the RHF one-electron operator $\hat{F}$. In the UHF approach the Hartree-Fock operators for electrons with spins α and β are different. Because of its simplicity, we will discuss the RHF approximation only. As it is well known, the UHF approach has the advantage of respecting the correlation effects to some degree, but it treats the spin properties in an inappropriate way.

In the Green's operator theory briefly described in these lectures, no specific properties are attributed to the observable $\hat{H}$ to which the Green's operator $\hat{G}(z)$ has been assigned. Therefore, all relations obtained until now can be also used for the Hartree-Fock operator $\hat{F}$. The H-F operator in the LCAO approach with zero differential overlap approximation depends, of course, upon the matrix elements of the one-electron density, i.e., upon the atomic charges and the bond orders in the quantum chemistry language. The analogs for a very large system can be formulated according to Eq. (8) with the help of the projector on the Fermi sea (i.e. the manifold of doubly occupied energy levels) as:

$$P_{\mu\nu} = \frac{1}{\pi i} \oint_{\Gamma(HF)} \langle \mu | \hat{G}(z) | \nu \rangle dz \qquad (59)$$

where the operator $\hat{G}(z)$ is assigned to the Hartree-Fock operator $\hat{F}$ and the contour $\Gamma(HF)$ in the complex plane encloses the Fermi sea. According to Eq. (9), the sum of doubly occupied eigenvalues of $\hat{F}$ is equal to:

$$\mathcal{E} = \frac{1}{\pi i} \oint_{\Gamma(HF)} z \mathrm{Tr} \hat{G}(z) dz \qquad (60)$$

The electronic energy in the H-F theory must be, of course, corrected in such a way that the electron-electron interaction is not counted twice. Therefore, the electronic energy of the system is:

$$E_{el} = \frac{1}{2\pi i} \oint_{\Gamma(HF)} \left[z\mathrm{Tr}\hat{G}(z)dz + \mathrm{Tr}(\hat{h}_1 \hat{G}(z)) \right] dz \qquad (61)$$

where h_1 is the one-electron part of the Hamiltonian H. In the ZDO approximation, the operator $\hat{G}(z)$ depends upon the orbital charges $P_{\mu\mu}$ and upon the bond orders $P_{\mu\nu}(\mu \neq \nu)$.

In the theory of chemisorption, the differences between properties of the system S and S_o are important (S_o and S are models of the chemisorbate and the chemisorbent without as well as with mutual interaction, respectively). The chemisorption energy can be written in the form:

$$\Delta E = \frac{1}{2\pi i} \oint_{\Gamma(HF)} \left[z\mathrm{Tr}(\hat{G}(z) - \hat{G}_o(z)) + \mathrm{Tr}(\hat{h}_1 \hat{G}(z) - \hat{h}_1^o \hat{G}_o(z)) \right] dz + CT(\Delta\Gamma) \qquad (62)$$

where $CT(\Delta\Gamma)$ is the correction term due to the fact that the Fermi energies of the systems S and S_o are not the same. This correction term includes mainly the contributions from the changed occupation of the localized levels. The operator $\hat{h}_1$ and $\hat{h}_1^o$ is the one-electron part of the Hamiltonian $\hat{H}$ and $\hat{H}_o$ describing the system S and S_o, respectively.

$$\hat{H} = \hat{h}_1 + \hat{h}_2 \quad \text{and} \quad \hat{H}_o = \hat{h}_1^o + \hat{h}_2^o \qquad (63)$$

The difference between the Green's operators which figures in the first term on the right-hand side of the Eq. (62), can be written as:

$$\mathrm{Tr}(\hat{G}(z) - \hat{G}_o(z)) = \mathrm{Tr}\left[\hat{L}^{-1} \hat{G}_o \hat{V} \hat{G}_o \right] = \mathrm{Tr}(\hat{L}^{-1} \hat{G}_o^2 \hat{V}) =$$

$$\mathrm{Tr}\left(\hat{L}^{-1} \frac{\partial \hat{L}}{\partial z} \right) = \frac{1}{|L|} \sum_{\mu,\nu} |L|_{\mu\nu} \langle \mu | \frac{\partial L}{\partial z} | \nu \rangle =$$

$$\frac{\partial |L|}{\partial z} \Big/ |L| \qquad (64)$$

In Eq. (64) the following simplification was possible to introduce:

$$\frac{\partial \hat{L}}{\partial z} = - \frac{\partial \hat{G}_o(z)}{\partial z} \hat{V} = \hat{G}_o^2(z) \hat{V} \qquad (65)$$

because of the relation

$$\frac{\partial \hat{G}_o(z)}{\partial z} = -\sum_j \frac{\hat{P}_j}{(z-E_{jo})^2} = \sum_{j,k} \frac{\hat{P}_j}{(z-E_{jo})} \cdot \frac{\hat{P}_k}{(E_{ko}-z)} = - \hat{G}_o^2(z) \qquad (66)$$

The chemisorption energy can be consequently written in the form:

$$\Delta E = \frac{1}{2\pi i} \oint_{\Gamma(HF)} \left[z \frac{\partial |L|}{\partial z} \Big/ |L| + Tr(\hat{h}_1(\hat{G}(z)-\hat{G}_o(z)) + Tr(\Delta\hat{h}_1 \hat{G}_o(z)) \right] dz + CT\ (\Delta\Gamma) \qquad (67)$$

where $\Delta\hat{h}_1 = \hat{h}_1 - \hat{h}_1^o$. The first term in Eq. (67) gives the difference between all poles (below the common Fermi level) of the Green's operator $\hat{G}(z)$ assigned to the system S in which the chemisorption interaction is switched on and all poles (below the same Fermi level) of $\hat{G}_o(z)$ assigned to the system S_o with no chemisorption interaction. Only differences of quantities describing the systems S and S_o figure directly in all terms entering the expression for the chemisorption energy ΔE.

The calculation of the eigenvalues and eigenfunctions of the Hartree-Fock operator can be made easier when the whole model under consideration is partitioned in two parts. The region near to the surface (including chemisorbate if it is present) forms part A where the electron density differs substantially from the electron density in the crystal bulk. The region more distant from the crystal surface forms part R where the electron density is roughly equal to the electron density in the bulk and, therefore, assumed to be known. The relation similar to Eq. (43)

$$\hat{F} = \hat{F}_A - \hat{F}_{AR}\, \hat{G}_R^{(1)}\, \hat{F}_{RA} \qquad (68)$$

makes possible a reduction in the dimensionality of the problem if the Green's operator for the hypothetical crystal with the same one-electron density in the surface region as within the crystal bulk is known.

ROLE OF SYMMETRY IN SURFACE SCIENCE

Some models which are of interest in surface science exhibit two-dimensional translation symmetry. For example, models for clean surfaces without large surface reconstruction as well as models for ideal surfaces with a complete monolayer of chemisorbate have the translation symmetry given by the surface geometry.

The one-electron functions $|j,\vec{k}\rangle$ must belong to a given representation of the two dimensional translation group of the surface. A representation is characterized by the two-dimensional wave vector $\vec{k}$

$$\vec{k} = (k_1 \vec{a}_1^{\,*} + k_2 \vec{a}_2^{\,*}) \tag{69}$$

where the $\vec{a}_s^{\,*}$ are the elementary vectors of the reciprocal surface lattice. The eigen kets of the RHF operator have the form:

$$|j;\vec{k}\rangle = \sum_{\rho=1}^{nc} \sum_{m_3=1}^{N_3} \langle\rho,m_3;\vec{k}|j;\vec{k}\rangle|\rho,m_3;\vec{k}\rangle =$$

$$\sum_{\rho=1}^{nc} \sum_{m_3=1}^{N_3} c_{j;\rho,m_3}(\vec{k})|\rho,m_3;\vec{k}\rangle \tag{70}$$

The number of layers considered in the model is given by N_3. The label ρ distinguishes AO's localized in the same elementary cell and nc is the number of such AO's. Layers parallel to the surface are labelled by m_3. The layer orbital $|\rho,m_3;\vec{k}\rangle$ is obtained from a projection of the operator $\hat{P}_{\vec{k}}$ on the selected AO $|\rho,m_3;\vec{o}\rangle$ of type ρ localized in the arbitrarily chosen elementary cell $\vec{o}$ in the m_3'th layer:

$$P_{\vec{k}} = K \sum_{m_1,m_2=-\infty}^{+\infty} \exp(\vec{m}\vec{k})\ \hat{t}_{\vec{m}} \tag{71}$$

where K is a normalization constant and the operator $\hat{t}_{\vec{m}}$ traslates the orbital $|\rho,m_3;\vec{o}\rangle$ from the cell labelled $m_1=m_2=o$ parallel to the surface into the cell labelled by m_1,m_2.

The introduction of "symmetry adapted" layer orbitals in the H-F equations gives(if we drop the assumption on the orthogonality of AO's) the classical secular equation of the form:

$$\sum_{m_3=1}^{N_3} \sum_{\rho=1}^{nc} C_{j;m_3\rho}(\vec{k}) \Big[F_{m_3',\rho';m_3\rho}(\vec{k}) - E_j(\vec{k})\, S_{m_3'\rho';m_3\rho}(\vec{k}) \Big] = 0 \tag{72}$$

where $m_3'=1....N_3$ and $\rho'=1.....nc$. The Fourrier transforms of HF matrix $\hat{F}$ and of the overlap matrix S are:

$$F_{m_3'\rho';m_3\rho}(\vec{k}) = K \sum_{m_1,m_2=-\infty}^{+\infty} \exp(i\vec{k}\vec{m}) \langle \rho', m_3'; \vec{o} | \hat{F}\hat{t}_{\vec{m}} | \rho, m_3; \vec{o} \rangle \tag{73}$$

and

$$S_{m_3',\rho';m_3\rho}(\vec{k}) = K \sum_{m_1,m_2=-\infty}^{+\infty} \exp(i\vec{k}\vec{m}) \langle \rho', m_3; \vec{o} | \hat{t}_{\vec{m}} | \rho, m_3; \vec{o} \rangle \tag{74}$$

Equations (72) are very often used in theoretical surface work in order to solve the model of a crystal in which the finite numbers of layers N_3 is considered.

It is evident for the crystal bulk that the SCF problem can be analogously formulated in secular equation form:

$$\sum_{\rho=1}^{nc} C_{j;\rho}(\underline{\vec{k}}) \Big[F_{\rho';\rho}(\underline{\vec{k}}) - E_j(\underline{\vec{k}}) S_{\rho';\rho}(\underline{\vec{k}}) \Big] = 0 \tag{75}$$

where $\rho'=1,....nc$. In Eq. (75) the Fourier transforms corresponding to three-dimensional wave vector $\underline{\vec{k}}$

$$\underline{\vec{k}} = \sum_{i=1}^{3} k_i \vec{a}_i^{\,*} \tag{76}$$

have analogous meaning as in the two-dimensional case (cf. Eqs. (73) and (74)).

The MO crystal orbitals have the form:

$$|j;\vec{\underline{k}}> = \sum_{\rho=1}^{nc} <\rho;\vec{\underline{k}}|j;\vec{\underline{k}}>|\rho;\vec{\underline{k}}> =$$

$$\sum_{\rho=1}^{nc} c_{j;\rho}(\vec{\underline{k}})|\rho;\vec{k}> \qquad (77)$$

with the three-dimensional Fourier transforms $\rho,\underline{k}$ of AO's $|\rho;\vec{\underline{m}}>$ with

$$\vec{\underline{m}} = \sum_{i=1}^{3} m_i\vec{a}_i \qquad (78)$$

The solutions (77) of secular equations (75) can be used as a starting-point of the procedure based on the Dyson equation or at least for sake of comparison between surface and bulk properties.

EXPLICIT CONSIDERATION OF A NON-ORTHOGONAL BASIS.

In the LCAO-type approximation commonly used in quantum chemistry and in solid state theory the non-orthogonal one-electron basis is adjusted to the studied problem. Therefore, for instance, the two systems considered in the derivation of Dyson equation are characterized in the approximative methods (ab-initio as well as semiempirical) with different basis sets: $|\mu^o>\in\underline{B_o}$ and $|\mu>\in\underline{B}$ for the systems S_o and S, respectively. The system S_o can be composed from a crystal and a chemisorbed molecule at infinite distance and the system S can be the same molecule and the same crystal in mutual interaction. Very often it is possible to find an one-to-one correspondence between $|\mu^o>$ and $|\mu>$, so that the notation $|\mu> \doteq |\mu^o>$ can be used.

Nevertheless it holds:

$$<\mu^o|\nu^o> = S^o_{\mu\nu} \neq <\mu|\nu> = S_{\mu\nu} \qquad (79)$$

The corresponding identity operators for systems S and S_o have different meaning:

$$\hat{I} \neq \hat{I}_o \qquad (80)$$

Obviously, the following relation holds:

$$(z\hat{I}-\hat{H})-(z\hat{I}_o-\hat{H}_o) = \hat{G}^{-1}(z)-(\hat{G}^o(z))^{-1} =$$

$$- \hat{V}+z(\hat{I}-\hat{I}_o) = - \hat{V}_s(z) \tag{81}$$

where the definition (10) of the perturbation potential V is used. The multiplication of Eq. (81) by $\hat{G}^o$ from left and by $\hat{G}$ from right followed by multiplication of the resulting relation by the operator $\hat{L}_s$ defined as:

$$\hat{L}_s = \hat{I}_o-\hat{G}^o\left[\hat{V}-z(\hat{I} - \hat{I}_o)\right] \tag{82}$$

gives the resolved Dyson equation for a nonorthogonal basis:

$$\hat{I}_o\hat{G} = \hat{L}_s^{-1}\hat{G}^o\hat{I} \tag{83}$$

The dual basis to the nonorthogonal basis $\underline{B}$ and $\underline{B_o}$ is defined as:

$$|\mu_1\rangle=\sum_{\chi}(S^{-1})_{\chi\mu}|\chi\rangle \tag{84}$$

and

$$|\mu_1^o\rangle=\sum_{\chi}(S_o^{-1})_{\chi\mu}|\chi\rangle \tag{85}$$

respectively. The mentioned assignment $|\chi^o\rangle \doteq |\chi\rangle$, was utilized. The corresponding identity operators $\hat{I}$ and $\hat{I}_o$ for the systems S and S_o are:

$$\hat{I} =\sum_{\mu} |\mu\rangle\langle\mu_1|=\sum_{\mu} |\mu_1\rangle\langle\mu|$$

$$\hat{I}_o=\sum_{\mu} |\mu\rangle\langle\mu_1^o|=\sum_{\mu} |\mu_1^o\rangle\langle\mu| \tag{86}$$

According to the choice of the basis the different representations of the Green's operator $\hat{G}(z)$ exist:

$$G^{[k,l]}(z) = (\langle\mu_k|\hat{G}(z)|\nu_l\rangle)\equiv(G_{\mu\nu}^{[k,l]}) \qquad k,l=0,1. \tag{87}$$

with the convention $|\mu_o\rangle$ for the kets of the original basis $\underline{B}$. Kets $|\mu_1\rangle$ from the corresponding dual basis are defined by Eq. (84). The representation of the relation (4) defining $\hat{G}(z)$ can be written in the form:

$$\sum_{\chi} \langle \mu_1 | \hat{G} | \chi_1 \rangle \langle \chi | \hat{\Delta} | \nu \rangle = \langle \mu | \nu \rangle = \delta_{m\nu} \qquad (88)$$

or

$$\underset{\sim}{G}^{[1,1]} \underset{\sim}{\Delta}^{[0,0]} = \underset{\sim}{I} \qquad (89)$$

Therefore the matrix $G^{[1,1]}$ can be obtained directly from the "secular" matrix $\Delta^{[0,0]}$ but frequently the matrix

$$\underset{\sim}{G}^{[1,0]} = \underset{\sim}{G}^{[1,1]} \underset{\sim}{S} \qquad (90)$$

is more useful.

For example the local density of states in the one-electron approximation is proportional to the imaginary part of the $\mathrm{Tr}(G^{[1,0]})$.

The Dyson equation (83) has the following representation

$$\underset{\sim}{G}^{[1,1]} = (\underset{\sim}{L}_S^{-1})^{[1,0]} \; (\underset{\sim}{G}^{o})^{[1,1]} \qquad (91)$$

with

$$\langle \mu_1^o | \hat{L}_s | \chi \rangle = \delta_{\mu\chi} - \sum_{\lambda} (G^o_{\mu\lambda})^{[1,1]} (V_{\lambda\chi} - z(S_{\lambda\chi} - S^o_{\lambda\chi})) \qquad (92)$$

If the perturbation $\hat{V}=\hat{H}-\hat{H}_o$ as well as the difference between the basis $\underline{B}$ and B_o characterized by the quantity $\Delta S_{\mu\nu} = S_{\mu\nu} - S^o_{\mu\nu}$ are nearly vanishing outside the region Q, then analogously to Eq. (22) the following relation holds:

$$\langle \mu | \hat{L}_s(E) | \nu \rangle = \delta_{\mu\nu} \qquad , \nu \in \bar{Q} \qquad (93)$$

Therefore, the basic equation of the Koster-Slater resolvent method can be written in the form:

$$|L_s(E_j)| = |\langle \mu_1^o | \hat{Q} \; \hat{L}_s(E_j) \hat{Q} | \nu \rangle| \qquad (94)$$

If the partitioning scheme is applied for the investigated system in the nonorthogonal basis, the relation (43) can be written in the matrix form (cf.

also definition (44):

$$(\underset{\sim}{G}_A^{-1})^{[0,0]} = \underset{\sim}{\Delta}_A^{[0,0]} - \underset{\sim}{\Delta}_{AR}^{[0,0]} (\underset{\sim}{G}_R^{(1)})^{[1,1]} \underset{\sim}{\Delta}_{RA}^{[0,0]}$$

$$= \underset{\sim}{H}_A^{[0,0]} - E\underset{\sim}{S}_A - (\underset{\sim}{H}_{AR}^{[0,0]} - E\underset{\sim}{S}_{AR})(\underset{\sim}{G}_R^{(1)})^{[1,1]}$$

$$\cdot (\underset{\sim}{H}_{RA}^{[0,0]} - E\underset{\sim}{S}_{RA}) \qquad (95)$$

DISCUSSION AND CONCLUSION

In this series of lectures the fundamentals of the tight binding (LCAO-MO) theory of electronic structure of surfaces and of chemisorption have been demonstrated. The orthogonal basis spanning the space of one-electron functions was used in the major part of these lectures. As it has been also shown the theory can be easily generalized for a nonorthogonal basis by introducing the dual basis. This procedure is shown in detail in the review article [1] (compare also reference [2]). In comparison with the matrix representation the direct use of operator formalism is advantageous because the origin and the meaning of individual terms due to the approximative character of LCAO approach are more transparent.

The relations derived in the present paper can be used directly in the theory of localized one-electron states of different kinds [3] and localized chemisorption states [4]. The number of papers utilizing the LCAO-MO theory for the investigation of surface states is at present quite large [5]. Also, the basic relations of the tight binding method which have been presented here in context with the surface theory can be applied more generally for a determination of the local density of states using the Green's function formalism [6]. The concept of the surface molecule [7] as well as the continuous fraction [8] and momentum [9] methods for direct calculation of local density of states in semiconductors and transition metals are based on the discussed partition technique.

It is worthwhile to emphasize that the Green's function technique for problems of the electronic structure of surfaces with and without chemisorbed species must be used with extreme care, mainly if the model Hamiltonian does not describe the electron-electron interaction in a sufficiently efficient way. If the transition from the realistic quasicontinuous spectrum to the (nonrealistic idealized) continuous spectrum is made not carefully enough, the nonrealistic unsufficient screening of the surface charges may easily occur[10].

The consideration of models taking into account the finite number of two-dimensionally infinite layers can reproduce some surface properties in a very nice manner[11]. This type of models allows a straightforward implementation of the SCF procedure. The simulation of surfaces and surface defects can be carried out in the form of cluster calculations which can utilize almost any method known from quantum chemistry.

The Roothaan-like one-electron approaches using one-electron function expansion in terms of localized orbitals have many disadvantages when applied on large systems (e.g., over completeness problem, improper description of the nonbounded and nearly nonbounded electronic states, insufficient consideration of correlation effects). The methods using one-electron density functional formalism[12], pseudopotential approaches[13] as well as procedures with explicit introduction of electron correlation in the valence bond fashion[14] attempt to overcome the mentioned shortcomings of the one-electron LCAO-MO type methods.

Nevertheless, the complete investigation of the LCAO approach with all mentioned disadvantages is very important and fruitful because of the substantial interpretation power of the tight binding methods for description of chemical phenomena. The systematic comparison with other approaches should give more insight in the nature of physical and catalytic properties of solid surfaces.

ACKNOWLEDGEMENT

This work has been partly supported by a grant from Deutsche Forschungsgemeinschaft.The author thanks to Mr. O. Fromm and Dr. P. Schneider for helpful discussions and to Mrs. A. Polinske for careful typing of the manuscript.

LITERATURE

1. J. Koutecky, Prog. Surface Mem. Sci., 11, 1 (1976).
2. T.B. Grimley, J.Phys.C:Solid State Phys., 3, 1934 (1970).
J. Ladik, in "Electronic Structures of Polymers and Molecular Crystals", Ed. J. Andre, J. Ladik and J. Delhalle, Plenum Publishing Corporation, 1975, p. 23
J. Ladik, Phys. Rev. B 17, 1663 (1978)
3. I. Tamm, Z. Physik, 76, 849 (1932).
W. Shockley, Phys. Rev., 56, 317 (1939).
4. T.A. Hoffmann, J.Chem.Phys., 9, 85 (1950).
5. J. Koutecky, Adv.Chem.Phys., 9, 85 (1965).
S.G. Davidson and J.P. Levine, Solid State Phys., 25, 37 (1970).
J.D. Levine and S. Freeman, Phys.Rev., B 2,3255 (1970).
P. Kalkstein and D. Soven, Surf.Sci.,26, 85 (1971).
E-Ni Foo and How-Sen Wong, Phys.Rev., B 9, 1859 (1974).
6. D.M. Newns, Phys. Rev., 178, 1123 (1969).
T.B. Grimley, J.Vac.Sci.Technology, 8, 31 (1971).
M.J. Kelly, Surface Sci., 43, 587 (1974)
J.W. Gadzuk, Surface Sci., 43, 441 (1974)
M.J. Kelly and R. Haydock, Surf. Sci., 38, 139(1973).
7. T.B. Grimley in Structure and Properties of Metal Surfaces, Eds. S.Shimodaira, M. Maeda,G. Okamoto, M. Onchi and V. Tamai (Makruzen Tokyo,1973) p. 72
M.J. Kelly, Surf.Sci., 43, 587 (1974).
M.J. Kelly, J.Phys. C.:Solid State Physics, 7, 157 (1974).
8. R. Haydock, V. Heine and M.J. Kelly, J.Phys.C.: Solid State Physics, 5, 2845 (1972).
J.L. Moran-Lopez, J.Phys.F: 5, 1977 (1975).
9. F. Cyrot-Lackmann, M.C. Desjonqueres and J.P.Gaspard J.Phys.C: Solid State Physics, 7, 925 (1974).
10. O. Fromm, unpublished results
11. K. Hirabayaski, J.Phys.Soc.,(Japan)27, 1475 (1969).
K.C. Panday and J.C. Phillips, Surf.St.Com. 14, 439 (1970).
I. Alstrup, Phys.Stat.Sol.(b),45,209 (1971).
12. W.Kohn and L.J. Sham, Phys.Rev.A, 140, 1133 (1965).
M.P. Lang, Solid State Commun., 9,1015 (1965).
Phys. Rev. B,4, 4234 (1971).
S.C. Ying, J.R. Smith and W. Kohn, Phys. Rev., B, 11, 1483 (1975).
N.P. Lang and A.R. Williams, Phys.Rev.,Lett. 34, 531 (1975).
13. J.A. Appelbaum and D.R. Hamann, Rev.Mod.Phys., 48, 479 (1976).

L.M. Kahn and S.C. Ying, Solid State Commun., 16, 799 (1973).
M. Schlueter, R. Chelikowsky, G. Loute and L. Cohe Phys. Rev. B, 12, 4200 (1975).
M. Schlueter and M.L. Cohen, Phys.Rev. B, 17,716 (1978).
14. J.R. Schrieffer, J.Vac.Sci., Technol. 9, 561(1972)
J.R. Schrieffer and R. Gomer, Surf.Sci., 25, 315 (1971)
T.L. Einstein and J.R. Schrieffer, Phys. Rev. B, 7, 3629 (1973).

INDEX